Artificial Intelligence for Molecular Biology

Muhammad Nabeel Asim • Sheraz Ahmed •
Andreas Dengel

Artificial Intelligence for Molecular Biology

Fundamental Methods and Applications

Springer

Muhammad Nabeel Asim
German Research Cntr for AI
Kaiserslautern, Germany

Sheraz Ahmed
German Research Cntr for AI
Kaiserslautern, Germany

Andreas Dengel
German Research Cntr for AI
Kaiserslautern, Germany

ISBN 978-3-031-90449-3 ISBN 978-3-031-90450-9 (eBook)
https://doi.org/10.1007/978-3-031-90450-9

This Springer imprint is published by the registered company Springer Nature Switzerland AG
The registered company address is: Gewerbestrasse 11, 6330 Cham, Switzerland

Foreword

The genetic code is the fundamental distinction between living and non-living entities. This code encompasses a comprehensive repository of information about organism's traits, physiological processes, and evolutionary history. It also encompasses complete set of instructions essential for an organism's structure, function, and development. Within the landscape of genetic code a complex interplay of multiple factors including genetic mutations, environmental exposures, epigenetic alterations, and viral integrations contribute to the development of a wide range of diseases such as cancer, genetic disorders, autoimmune diseases, neurodegenerative conditions, and cardiovascular diseases. Genetic code understanding holds the promise to unlock the mysteries of life and pave the way for groundbreaking advancements in therapies, medicine, biotechnology, and beyond. Traditional experimental methods have been instrumental in advancing our understanding about genetic code but these methods are expensive, time-consuming, and error-prone. Artificial intelligence (AI) offers a promising avenue to accelerate genetic code understanding efforts by complementing wet-lab experiments. The thorough realization of AI potential within the realm of genetic code requires a robust integration of AI and molecular biology disciplines. This book embarks on a comprehensive exploration of the complex relationship between AI and molecular biology disciplines. We aim to equip readers with the necessary tools to decipher the complex patterns embedded within genetic code (DNA, RNA, and Proteins) by providing comprehensive fundamental principles of molecular biology and the cutting-edge advancements in Artificial Intelligence (AI).

The journey begins with a foundational understanding of molecular biology, crafted specifically for AI researchers. Molecular biology concepts are extensively covered in standard biology textbooks. However, AI researchers often struggle in grasping relevant biological context from these resources because these textbooks are crafted to equip biologists with in-depth knowledge. The extensive knowledge is often overwhelming for AI researchers who seek more concise and task-oriented information. The initial chapter is dedicated to provide a comprehensive biological foundation and motivation for various genetic sequence analysis tasks. It translates complex biological processes and concepts into AI researchers' understandable knowledge. A comprehensive detail about these fundamental concepts in a clear and concise manner is important to set the stage for AI researchers to develop biological knowledge informed AI applications for genomics and proteomic sequence analysis tasks. Following the imperative to integrate biological knowledge within AI frameworks, the subsequent chapter provides a detailed description of DNA, RNA, and protein structures. Since structural details of biomolecules (DNA, RNA, Protein) influence their function and behavior, consequently their understanding is important for AI researchers to develop AI applications by simultaneously taking raw sequences and structural features of biomolecules.

The pivotal third chapter serves as a bridge between molecular biology and AI disciplines. The prime objective of this chapter is to provide a comprehensive overview of diverse spectrum of DNA, RNA, and protein sequence analysis tasks. Additionally, it establishes a foundational stage for the subsequent seven chapters by presenting a high-level overview of an AI-driven genomics and proteomics sequence analysis pipeline. Specifically, it offers essential biological knowledge and motivations for the development of AI applications across 47 DNA, 45 RNA, and 62 protein sequence analysis tasks. To streamline the application of AI methodologies within genetic sequence analysis landscape, it aligns DNA, RNA, and protein sequence analysis tasks with fundamental AI paradigms such as classification, regression, and clustering. Additionally, the chapter presents a holistic architecture for an end-to-end AI-driven genomics and proteomics sequence analysis pipelines. Within this framework, it illustrates various strategies for benchmark datasets development and provides a high-level overview of contemporary sequence representation learning techniques that effectively transform raw genetic sequences into meaningful numerical representations. It delves into the intricacies of feature engineering methodologies and a comprehensive knowledge of machine learning and deep learning models along with associated evaluation metrics. The chapter concludes with a holistic overview of existing genomics and proteomics sequence analysis frameworks available to the research community.

The foundation of AI-driven genomics and proteomics sequence analysis relies on the availability of comprehensive sequence data. Fortunately, a vast amount of genetic sequence data, including DNA, RNA, and protein sequences, is readily available in public databases. Researchers can leverage these databases to create benchmark datasets for a wide array of genetic sequence analysis applications. Following the need of a comprehensive databases knowledge for development of benchmark datasets, Chap. 4 provides in-depth knowledge of diverse databases including data formats, retrieval systems, species, organisms, sequence characteristics, annotations, and sequence statistics. A thorough understanding of data formats and retrieval systems is essential for efficient data extraction and interpretation. The primary objective of this chapter is to equip AI researchers with necessary knowledge to make informed decisions regarding database selection for the creation of task-specific benchmark datasets within the extensive landscape of genetic sequence databases including 47 peptides, 35 DNA, 64 RNA, 68 protein, and 12 CRISPER system-oriented databases.

The application of AI algorithms to raw genetic sequences presents a significant computational challenge. Unlike human cognition, AI models require numerical representations. Consequently, the transformation of genetic sequences into meaningful statistical vectors is a crucial step in the development of AI-driven genetic sequence analysis applications. The quality of these vectors profoundly impacts AI algorithms performance. High-quality representations, encompassing informative patterns within the sequence data, can enhance the performance of even relatively simple algorithms. Conversely, low-quality vectors lacking in informative patterns of nucleotides or amino acids can hinder the performance of even sophisticated models. Following the need of a comprehensive literature that brings existing sequence encoding methods at a single platform, Chaps. 5 and 6 illustrate details of diverse types of sequence representation learning methods. A comprehensive information about these methods equip AI researchers with essential knowledge to select and develop the most effective sequence representation learning methods capable of transforming raw genetic sequences into statistical vectors by acquiring complex and informative patterns.

In genetic sequence analysis landscape, the CRISPR-Cas9 system holds immense potential for the development of targeted therapies for a wide range of diseases, including cancer, genetic disorder, and hereditary diseases. This system leverages a programmable guide RNA (gRNA) and Cas9 proteins to precisely cleave problematic regions within DNA sequences. This editing is a multi-step process such as designing a precise gRNA, selecting the appropriate Cas protein, and thoroughly evaluating both on-target and off-target activity of the Cas9-gRNA complex. To ensure the accuracy and effectiveness of CRISPR-Cas9 system, after the targeted DNA cleavage, the process requires careful analysis of potential errors and the resultant outcomes such as indels and deletions. Following the success of AI in various fields, researchers are now leveraging AI algorithms to catalyze and optimize the multi-step process of CRISPR-Cas9 system. To achieve this goal AI-driven applications are being integrated into each step, but existing AI applications have limited performance and many steps still rely on expensive and time-consuming wet-lab experiments. The primary reason behind low performance of AI applications is the gap between CRISPR and AI fields. Effective integration of AI into multi-step CRISPR-Cas9 system demands comprehensive knowledge of both domains. Chapter 7 of this book is dedicated to bridge the knowledge gap between AI and CRISPR-Cas9 research. Primarily, this chapter presents a unique platform for AI researchers to grasp deep understanding of the biological foundations behind each step in the CRISPR-Cas9 multi-step process. Furthermore, it provides details of 65 available datasets that can be utilized to develop AI-driven applications for each step. It also provides details of existing 49 predictive pipelines in terms of used representation learning methods and machine or deep learning predictors. Moreover, it provides performance values of 49 predictive pipelines across 65 benchmark datasets. A thorough analysis of existing datasets and details of predictive pipelines is utilized to highlight different steps in CRISPR-Cas9 multi-step process where the performance of AI-driven applications is less. Finally, it presents multiple reasons behind the low performance of predictors and suggests different directions that can be followed to develop more powerful AI applications for each step.

The second volume of this book illustrates the application of advanced AI techniques in genetic sequence analysis. It begins by introducing the foundational concepts of word embeddings language models and how these techniques can be applied to biological sequence data. It covers wide array of word embeddings and language models based applications in DNA, RNA, and protein sequence analysis tasks. Additionally, it delves into peptide classification landscape and offers a comprehensive overview of existing AI models, datasets, and methods used in peptide research.

In conclusion this book emphasizes the interplay between data, algorithms, and biological knowledge. By understanding how AI can be harnessed to analyze and interpret biological data, researchers can gain unprecedented insights into the mechanisms of life. Ultimately, this book aims to empower readers to become active participants in the ongoing revolution of bioinformatics, where the convergence of biology and computer science is unlocking the secrets of life itself.

Kaiserslautern, Germany
November 2024

Muhammad Nabeel Asim
Sheraz Ahmed
Andreas Dengel

Preface

The realm of Artificial Intelligence (AI) colliding with the world of biology forms a captivating nexus of two seemingly disparate realms—one molded by algorithms and data, the other by life forms and nature. Yet concealed behind this apparent dichotomy lies a boundless potential that can plunge us deeper into the enigma of existence than ever before. Within the pages of this book, we delve into the harmonious convergence of AI and molecular biology, merging the marvels of evolution with the boundless vistas of technology.

In recent years, the advent and progression of next-generation sequencing techniques have ushered in a substantial surge in the accessibility of genomic and proteomic data, repositories brimming with invaluable insights into biological processes and maladies. As biology researchers actively wield a synergy of wet-lab experimentation and computational tools to glean significant revelations from genomics and proteomics data, a parallel wave of artificial intelligence revolutionizes global paradigms, trading laborious human tasks for swift computational processes. Yet, the application of artificial intelligence methods in genomic and proteomic sequence analysis remains confined. The primary impediment to AI's adoption in this domain lies in the gap separating AI experts from biologists.

It is, however, the exquisite harmony of AI and biology that presents the promise of unfurling novel frontiers, embarking upon uncharted trails. From unraveling genetic codes to sculpting intricate biological processes, AI empowers us to grasp the seemingly inscrutable. This book extends an invitation to partake in this exhilarating odyssey, where we traverse the bridges interlinking these two disciplines and forge connections between AI virtuosos and biologists.

The initial chapter of this book addresses the identified void and endeavors to impart fundamental biological concepts, specifically tailored for AI researchers. In the realm of crafting AI-driven applications for sequence analysis within genomics and proteomics, a crucial distinction surface—AI methodologies hinge upon statistical values and aren't inherently designed to directly process raw DNA, RNA, and protein sequences. The pursuit of refining AI applications geared toward gleaning insightful data from the analysis of genomics and proteomics sequences has spurred the innovation of more adept sequence coding methodologies. The driving ambition behind each new approach is to effectively extract distinctive nucleic and amino acid patterns from raw sequences, subsequently generating a statistical representation for each sequence. This endeavor has yielded over a hundred coding methods, yet they remain dispersed.

This book deals in detail with traditional encoding methods, examining their use in DNA, RNA, and protein sequence analysis and highlighting their various applications. We put a special focus on word embedding techniques and language models that originated in the field of Natural Language Processing (NLP). These techniques are able to transform text data into a statistical feature space due to their ability to learn word distributions from large text datasets. While raw sequences and text data offer similar processing capabilities, the integration of word embedding methods into AI-assisted analysis of genome and proteome sequences remains limited. This limitation stems in part from the lack of all-encompassing explanations of their use in this context, as well as the lack of a unified platform to host these techniques. This book meticulously explicates the application of word embedding methods and large and language models to raw sequence data. It demystifies their operation and unveils the plethora of variations at play. With a firm grasp of these methodologies, researchers are poised to cultivate more resilient and precise AI applications for genomics and proteomics analysis, ushering in novel insights and knowledge.

While a substantial volume of genomics and proteomics sequence data exists across numerous biological databases, there appears to be a gap in awareness within the AI research community regarding the availability of such data. To bridge this gap and foster a comprehensive understanding of these databases, the book offers a high-level overview of biological databases. These databases are thoughtfully categorized based on their distinct characteristics, empowering researchers to efficiently extract the required data.

Moreover, recognizing the imperative to expedite research in genomics and proteomics sequence analysis and to bridge the gap between AI researchers and biologists, we included a high-level summary of diverse genomic and proteomics sequence analysis tasks, underscored by the rationale and necessity for developing AI applications tailored to these tasks.

In a bid to further support these tasks, the book encompasses over 100 publicly available benchmark datasets. This consolidation of datasets under a single platform offers AI researchers seamless access to a wide array of datasets. We hope that this accessibility will not only facilitate the development and evaluation of AI-based predictors across multiple dataset types but also promote the dissemination of research findings within the broader research community.

As you embark on this literary journey, may these pages serve as a gateway to a world of ideas, insights, and inspiration, seamlessly bridging the realms of AI and molecular biology. Let the contents ahead not only spark your imagination but also empower you to navigate the intricate interplay between these two domains. As the preface unfolds, consider it an invitation to venture further, to engage with the narratives that follow, and to forge your own connection with the tapestry of knowledge woven within these chapters, harmonizing the worlds of AI and molecular biology. With anticipation and gratitude, we invite you to delve into the heart of this book and uncover the treasures it holds for both AI enthusiasts and molecular biologists alike.

Kaiserslautern, Germany
November 2024

Muhammad Nabeel Asim
Sheraz Ahmed
Andreas D engel

Acknowledgments

First and foremost, we thank our families and friends, whose unwavering support and encouragement provided the strength and motivation needed throughout this journey. A special thank you goes to Dr. Faiza Mehmood for her invaluable guidance and mentorship, and to Muhammad Ali Ibrahim, Summra Saleem, Tayyaba Asif, and Ahtisham Fazeel Abbasi, whose insights and support were instrumental in bringing this book to completion. Each of their contributions has left an indelible mark on this work, and we are truly grateful for their presence along this journey.

Contents

Acronyms

DNA	Deoxyribonucleic Acid
RNA	Ribonucleic Acid
ncRNA	Non-coding Ribonucleic Acid
lncRNA	long Non-coding Ribonucleic Acid
SNcRNA	Small Non-coding Ribonucleic Acid
mRNA	Messenger Ribonucleic Acid
miRNA	Micro Ribonucleic Acid
PPI	Protein-Protein Interaction
VHPPI	Viral Host Protein-Protein Interaction
NLP	Natural Language Processing
DNN	Deep Neural Network
CNN	Convolution Neural Network
RNN	Recurrent Neural Network
LSTM	Long Short-Term Memory
SVM	Support Vector Machine
RF	Random Forest
ET	Extra Tree
SGD	Stochastic Gradient Descent
bP	Base Pair
AE	Autoencoder
ACC	Accuracy
PR	Precision
SP	Specificity
SN	Sensitivity
MCC	Mathew's Correlation Coefficient
FPR	False Positive Rate
TPR	True Positive Rate
AUC	Area Under The Curve
RMSE	Root Mean Square Error

Chapter 1
Fundamentals of Molecular Biology

Biology and computer science, two seemingly disparate fields, actually share a significant connection. Life, with all its complex processes, and computation, with its ability to simulate and analyze complex systems, have fundamental similarities that allow for interdisciplinary collaboration and exploration. I am certain that there are great discoveries waiting for those who explore the interface between biology and computer science.

The main goal of this chapter is to equip computer scientists with fundamental knowledge of biology, which will enable them to develop more sophisticated applications supported by artificial intelligence for analyzing biological sequences. Imagine any general or specific purpose software that solves a single or multiple problems ranging from finding your location, your friends food preferences, or translating your valuable documents into different languages for business purposes. What is the fundamental building block on which even the most complex, efficient, and accurate software is built? Undoubtedly, the basic building block of any kind of software is a piece of instruction called a statement. Individual statements are combined to form expressions and functions, and these expressions and functions are further combined to create larger modules that serve specific functionalities in software programs. Similarly, the fundamental building block of the biological world is the cell.

Just like formation of a software program, the amazing biological world is a huge maze of interconnections on a large diversity of scales. It is shown in Fig. 1.1 that cells interact to form tissues, tissues assemble to form organs, organs combine to form organs system and ultimately organism. The interplay continues well beyond organisms as organisms form populations, populations inhabit the ecosystems, and ecosystems come together to create the world in which we live. A common phenomena which runs across all the layers of connectivity is that, communication and transfer of information occurs from cell-to-cell, organism-to-organism, and eventually generation-to-generation.

M. Nabeel Asim et al., *Artificial Intelligence for Molecular Biology*,
https://doi.org/10.1007/978-3-031-90450-9_1

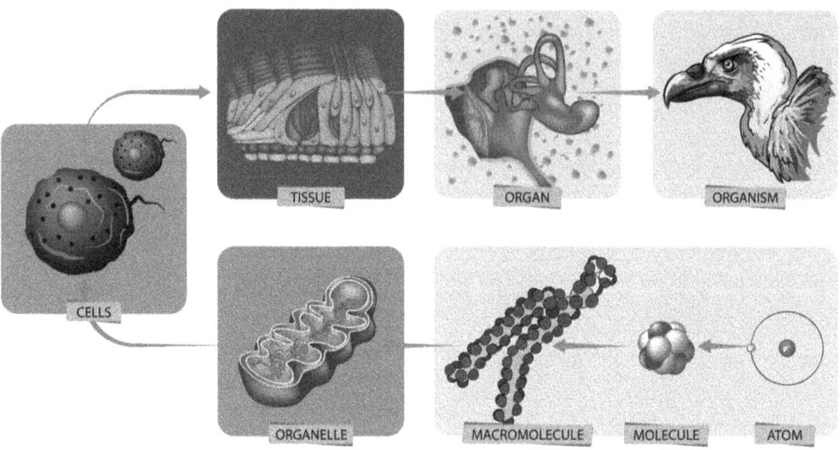

Fig. 1.1 A precise illustration of complete hierarchy of the biological organization—from atom to organism

While developing a software program, we write statements, expressions and functions. On the other hand, in living organisms, an important question is how information is communicated and transferred between cells, tissues, organs, and organ systems. The essential information about various biological processes in living organisms is contained within their genomes, which consist of deoxyribonucleic acid (DNA) molecules present in all living cells. To understand how the entire genome becomes active and alive, it is crucial to comprehend the intricate sequence of smart processes in which different molecular components collaborate. These processes are the core of molecular biology, and are the key players in the following chapter.

Before starting our journey to decode processes, this chapter discusses the basic biological themes on which the study and understanding of molecular biology entirely relies. Furthermore, it gives an overview of life, most common molecular building blocks of life, biochemical processes, central dogma of molecular biology, regulation of gene expression and its relation with cellular compartments, genome expression, sequencing of complete genome, diversity and classification of living organisms, potential and benefits of computational biology over experimental methods.

1.1 Defining Life from a Multidimensional Perspective

Scientists and researchers have been trying to unravel the mystery of how life began and to discover fundamental units of living and non-living things. After extensive research, almost all scientists concur that the atom is the fundamental building block of non-living matter, while the cell is the fundamental building block of living organisms. Recent research findings reveal that cells are the fundamental structural unit of all kinds of living organisms, which contain diversified information about life. Although cells are the tiniest entities but they contain a large world that possesses the essential qualities of life. This is the foundation of what scientists call as cell theory: everything on the earth that is alive is either a cell or a group of cells. The cell is indeed the simplest thing which can be certainly declared alive, hence in simple terms, all those entities which have a cell or a group of cells can be classified as living organisms.

Another perspective about living creatures is inheritance, which reveals the resemblance among parents and their offspring. Aristotle and other ancient philosophers realized the concept of inheritance long ago, however, the foundation of biological inheritance remained a mystery. Several explanations were presented, such as Aristotle believed that the only mother's influence during pregnancy is comparable to the impact of soil characteristics on the growth of a particular plant from a seed. In other words, just as soil plays a crucial role in the development of a plant, Aristotle believed that only mothers play a significant role in shaping the development of their unborn offspring. Some other scientists believed that children acquire an average combination of their parents' traits. A basic unit of heredity, which is passed from parents to children and carries important information that determines certain characteristics of organism (e.g. eye color), is called a gene. The identification of the gene as a segment of DNA opened a new door for understanding how inheritance occurs. The expression of genes provides a mechanism to make sense of the complex blend of striking similarities and unique traits that flow through generations. Gene expression is also the primary source of information that life utilizes to construct, maintain, and replicate cells.

Another significant characteristic of living organisms is evolution, which refers to the ability of living organisms to adapt and change over time in response to environmental pressures. Chameleons are known for their ability to change their skin color to match their surroundings. This adaptation allows them to blend into their environment and avoid detection by predators. Similarly, Arctic foxes change the color of their fur from brown or gray in summer to white in winter. This adaptation allows them to blend in with their surroundings and avoid detection by predators. Arctic foxes also have a thick layer of fur and a compact body shape, which helps them conserve body heat in harsh winter climates. Consider another example, like children of same biological parents usually have strikingly different appearance, qualities, and characteristics, many of which they have adapted to their environment eventually enabled them to behave different and bring next revolution in the world. The bigger questions in this regard are where these variations come from and also how these variations occur? Genetic variation occurs due to the random assortment of genes during sexual reproduction. When two parents produce offspring, each offspring will receive a unique combination of genes from their parents. This can result in variations in physical characteristics such as eye color, hair color, and height. In addition to genetic variation, environmental factors can influence the development of children from the same biological parents. For example, a child's diet, exposure to toxins or chemicals, and physical activity level can all impact their development. Even small differences in these factors can result in significant differences in physical and mental development.

Furthermore, epigenetic changes can also play a role in how children of the same biological parents differ from each other. Epigenetic changes refer to modifications to the DNA molecule that do not change the genetic code but can affect how genes are expressed. These modifications can occur in response to environmental factors and can result in differences in gene expression between siblings. We are aware of the fact that for the creation of a new human being, we need mother's egg and father's sperm, each of which contains a single set of chromosomes bearing our 23,000-odd genes from 46,000 genes of genome. The fusion of both kinds of cells forms a single cell called zygote which has two sets of chromosomes based on two different genomes coming from each parent. The zygote becomes an embryo when it starts to divide and differentiate into unique cell types. Embryo grows and develops into a multi-cellular organism based on different cell types, tissues, and organs. In this process, it is the random mixing as well as recombination of genetic codes of parents that results in unique characteristics in their offspring, with every child getting a distinct mix of genes from father and mother. The answer to how these variations occur at first place is mutations, which are random changes or copying errors in the genetic code due to several factors like exposure to radiations, chemicals, food patterns, etc. A gene can exist in several different variants referred as alleles. A mutation is passed to next generation only if it has occurred in sex cells, that are egg and sperm cells. If mutations occur in other cell types, then they fully get lost once the body dies.

From several scientific theories, one of the scientific theories states that evolution happens through natural selection in which those characteristics of organisms are survived as well as passed to next generations which are highly suited to their respective environments. Regardless of the mechanisms of evolution, entities that evolve over time can be considered as living. Furthermore, only living organisms continuously acquire and utilize information about both their external environment and their internal states with the help of energy produced by a set of chemical processes collectively called metabolism. Scientists considered that metabolism is a reflection of the cellular life, and without it, life can neither exist nor can be defined at all. Considering various biological aspects of living organisms, such as reproduction, metabolism, growth, adaptability, stimulus responsiveness, genetic inheritance, and evolution, many definitions of the life have been presented. According to the Noble prize winner Physicist Erwin Schrodinger, living organisms are not merely self-reproducing entities because living cells involve more than just simple replication of genetic code. Some definitions predominately centered on reproduction ability. While other definitions defined living entities as autonomous systems which can acquire and use energy, evolve over time, control the in-coming and out-going of biomolecules through membrane, regulate chemical reactions, carry and store diverse kind of information using various macro-molecules.

However, existing definitions of "life" are problematic and inaccurate due to the emergence of new categories like artificial or synthetic life. In addition, the definition of life must apply to both single-cellular and multi-cellular organisms. To better facilitate the readers, we have tried to formulate a precise yet comprehensive definition of life, any entity which store and transmit different kind of biological information using various macro-molecules, has the ability to evolve, communicate with the environment, and can be regarded as physical, chemical, and informational machine which construct its own metabolism to maintain itself, grow, as well reproduce is called a living thing. All other entities which do not operate according to these conditions can be classified as non-living.

Now, when we have cracked the definition life, lets understand various core components of living organisms and biological processes, starting from macromolecules that are common across almost all kinds of living organisms.

1.2 Common Macromolecules of Living Organisms

The entire diversity of modern life is thought to have evolved from a common hypothetical living organism known as the last universal common ancestor (LUCA). LUCA is believed to have lived around 4 to 5 billion years ago, when Earth's environment was quite different from today's environment. Scientists believe that the molecular components of LUCA have been retained from generation to generation and species to species as diverse living organisms have evolved over the years. Through comprehensive analysis, scientists have found that organic molecules of all living organisms are made up of the same collection of simple compounds, which have masses around 100 to 1000 daltons and 30 or more carbon atoms. In general, cells have four major kinds of tiny compounds: nucleotides, amino acids, fatty acids, and sugars. The repeated subunits of these tiny compounds make up large biological molecules called macromolecules, which perform various essential cellular functions. Figure 1.2 indicates four basic categories of tiny compounds and their corresponding huge macromolecules:

- Nucleotides make up Nucleic Acids
- Amino Acids make up Proteins
- Fatty Acids make up Fats and Lipids
- Sugars make up Carbohydrates

1.2.1 Nucleic Acids

Nucleic acids are molecules that play a vital role in storing and transferring genetic information within cells. There are two main types of nucleic acids: Deoxyribonucleic Acid (DNA) and Ribonucleic Acid (RNA). DNA is responsible for carrying genetic information, while RNA is a central player that uses the information stored in DNA to produce proteins. Nucleic acids have a very simple

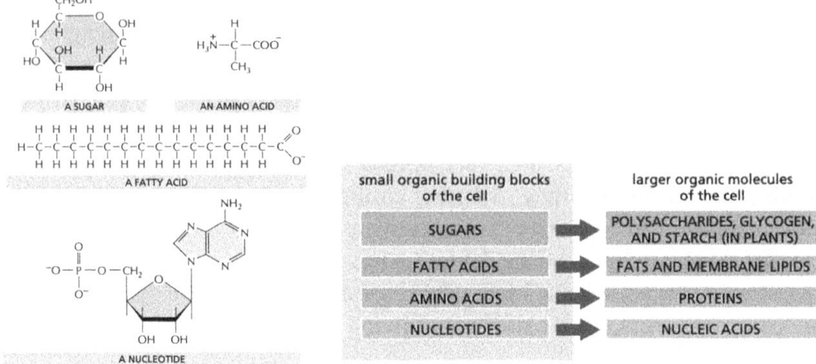

Biological Molecules

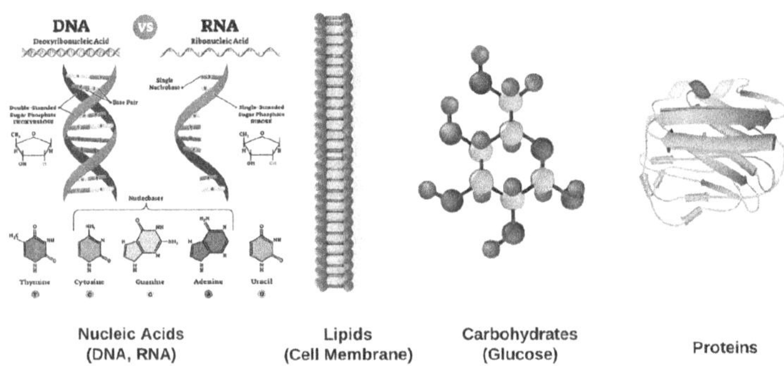

Fig. 1.2 The four primary families of tiny organic molecules within cells. These tiny molecules constitute the subunits or monomeric building blocks for almost all macromolecules including Nucleic Acids, Proteins, Lipids, and Carbohydrates

linear structure composed of monomer building blocks called nucleotides. Every nucleotide is composed of three parts: a nitrogenous base, a five-carbon sugar compound, and a phosphate group. The nitrogenous base can be one of four different types: adenine, guanine, cytosine, or thymine (or uracil in RNA). The five-carbon sugar compound in DNA is called deoxyribose, while in RNA, it is called ribose. The phosphate group contains one phosphorus atom and four oxygen atoms.

Figure 1.3 graphically represents double helix structure of DNA molecule that contains genetic code which provides the instructions for the development and function of all living organisms. It can be seen in Fig. 1.3, the DNA molecule consists of two complementary strands of nucleotides, each with a sugar-phosphate backbone and nitrogenous bases. The phosphate based backbone mainly have a negative charge because of the presence of phosphate ions. Resultantly, the entire DNA molecule has an overall negative charge. The interior region of double helix structure is made up of nitrogen bases namely adenine (A), cytosine (C), guanine

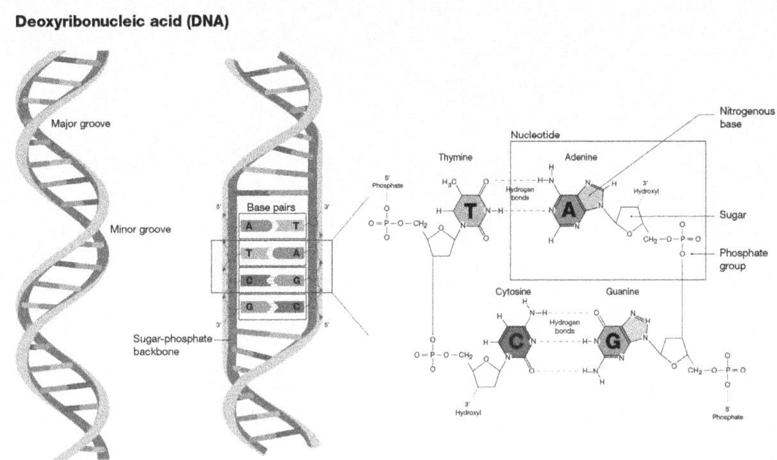

Fig. 1.3 DNA is a double stranded molecule which stores genetic information (Adapted from National Human Genome Research Institute, 2023 [876])

(G), and thymine (T), whereas, sugar as well as phosphate groups form outer region of double helix structure. In both strands, the nitrogen bases are stacked on top of one another in pairs, where A always pairs up with T, and C always pairs up with G. The complementary shapes of these bases and double helix structure makes DNA a high stable molecule and an ideal candidate for the storage of genetic information. On the other hand, RNA is a single-stranded nucleic acid. It is composed of a chain of nucleotides, each with a sugar-phosphate backbone and nitrogenous bases that are similar to those found in DNA, but with uracil replacing thymine. The less stable structure of RNA makes it an ideal functional molecule that drives the process of protein synthesis.

In living organisms that belong to eukaryotic family, DNA exists inside tiny compartment of cell called nucleus and in prokaryotic family related living organisms, DNA exist inside a different compartment of the cell called nucleoid. According to most recent studies, eukaryotic DNA is more bigger and complex as compared to prokaryotic DNA. Both strands of eukaryotic DNA are 67 billion miles long. In other words eukaryotic DNA is as long as the distance of 150,000 round trips from earth to moon. The most dominant question about eukaryotic DNA that pops up in every mind is that how such a huge molecule is stored inside tiny nucleus that is around 5 to 10 micrometers (μm) in diameter? The answer of this question is well illustrated in Fig. 1.4. Figure 1.4 shows that in order to fit the long negatively charged DNA within the tiny nucleus of cell, the entire negatively charged DNA is wrapped around small, positively charged histone proteins. This wrapping builds a structure known as nucleosome that consists of a segment of DNA wrapped around eight histone proteins, also known as histone octamer. To simply put, histone

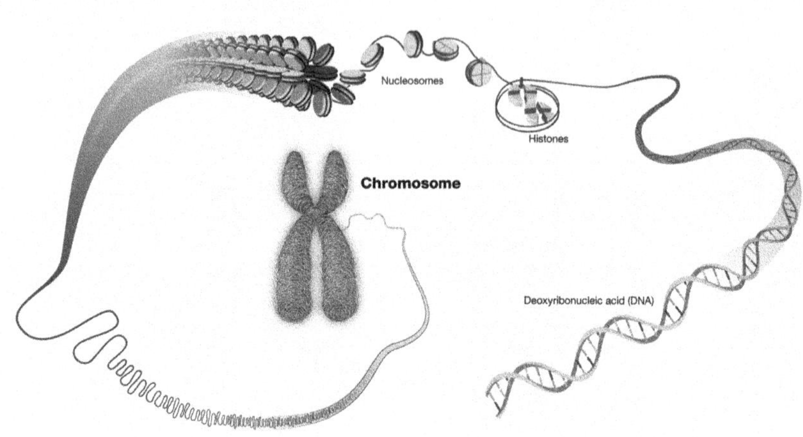

Fig. 1.4 DNA organization in form of chromosomes (Adapted from National Human Genome Research Institute, 2023 [875])

proteins octamer or nucleosome is just like a ball and DNA is like a thread, so a very long thread is wrapped around the ball.

As we know DNA sequence contains important instructions about diverse types of biological processes. These instructions can not be read from wrapped DNA. In fact to read these instructions, DNA follows the loop of unwrapping and wrapping. There are several factors that can influence the charge of DNA or histone proteins. Strong opposite charges make DNA very tightly wrapped that hinder the process of unwrapping DNA and prevents the biological processes to read any instructions. This leads to the failure of various biological processes that causes initiation and propagation of several diseases. Artificial Intelligence is being extensively utilized to detect whether DNA is tightly wrapped or loosely wrapped. Furthermore, it is also being utilized to detect the positions of nucleosomes. Similar to charges, when there are irregularities in the positioning of nucleosomes, problems arise in the process of wrapping and unwrapping DNA.

Multiple nucleosomes together make a chromosome. The number of nucleosomes that make up a chromosome can be estimated by dividing the total length of the DNA molecule by the length of a single nucleosome, which is about 147 base pairs long. Based on this calculation, each human chromosome contains millions of nucleosomes. However, it is important to note that the number of nucleosomes can vary depending on the specific chromosomal region being examined. For example, regions of the chromosome that are actively transcribed into RNA tend to have a more open chromatin structure called euchromatin with fewer nucleosomes, while in heterochromatin state, regions that are silenced may be more tightly compacted with more nucleosomes. Many living organisms have several, independent linear shaped chromosomes, while other have the solo circular shaped chromosome.

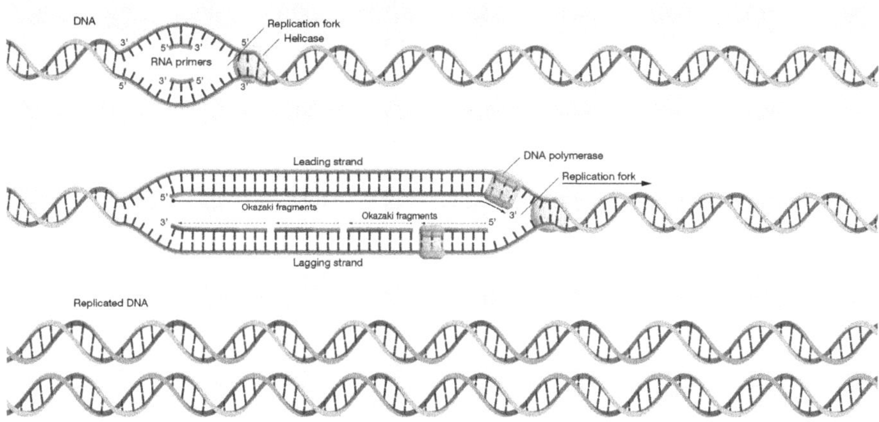

Fig. 1.5 DNA replication process is common across all organisms, the complementary nature of two DNA strands allow the DNA to be copied (Adapted from National Human Genome Research Institute, 2023 [220])

As we have seen that DNA molecule is made up of two strands, and one strand contains pairs of nucleotides that happen to be opposite pairs of nucleotides in another strand. This means that if we know the nucleotides sequence of one strand, we can easily determine the sequence of the other strand. For example, if one strand nucleotides sequence is ACGTGCAT then in the other strand, nucleotides sequence will be TGCACGTA. This example illustrates sequence information at one strand is similar to other stand but other stand has opposite pairs of nucleotides. To perform DNA sequence analysis, researcher only take nucleotides sequence of one strand. Although both strands contain similar information but DNA double helix structure is important to smooth diverse types of cellular process that are important for the maintenance of organisms lives and to produce offspring.

Figure 1.5 illustrates how the physical structure of nucleotides in DNA helix stranded structure enables the accurate replication and transmission of genetic material. Replication and transmission of DNA are made possible by the complementary base pairing and double helix shape of DNA. These features ensure that each strand of DNA can serve as a template for the synthesis of a new complementary strand during replication, leading to the faithful transmission of genetic information from one generation to the next. Replication and transmission of genetic material are essential processes for repairing tissues, organs in same organisms and transfer parent characteristics to their offspring. DNA replication is a biological process where cells create an exact copy of their inherent DNA before the cell division. Considering, many tissues in the body, such as skin, blood, and digestive tissue, require constant renewal due to wear and tear, injury, or disease. Cell divide and make an exact copy of their genetic information through DNA replication to support the constant growth, repair, and maintenance of living organisms. Cell division ensures that each new cell will have the same genetic information as the original cell to properly implement all cellular functions in living organisms.

Figure 1.5 graphically illustrates the process of DNA replication that happens in three stages namely initiation, elongation, and termination. During the initiation phase, a group of enzymes or catalysts known as helicases unwinds double-stranded DNA molecule at particular locations known as replication origins, resulting in the formation of a replication fork. During elongation, DNA polymerase enzyme adds the nucleotides to the growing DNA strand in a sequence complementary to template strand. The polymerase mainly reads the template strand from 3' to 5' direction and creates the new strand from 5' to 3' direction. This is because, two strands of DNA actually run in opposite directions, one is continuously synthesized in a process known as the leading strand, while the other is synthesized in short fragments known as Okazaki fragments in a process known as the lagging strand. During the process of termination, the replication fork encounters a specific termination site, and the newly produced DNA strands are separated.

The process through which genetic information is conveyed from generation to generation is known as DNA transmission. It is demonstrated in Fig. 1.6, DNA transmission completely relies on DNA replication, chromosomal segregation, and chromosome distribution to daughter cells during cell division. DNA replication allows the copying of double-stranded DNA molecule to produce two identical copies of the genetic material. Following replication, two copies of DNA molecule remain linked together to form a structure known as chromosome. During the process of cell division, the replicated chromosomes are segregated into daughter

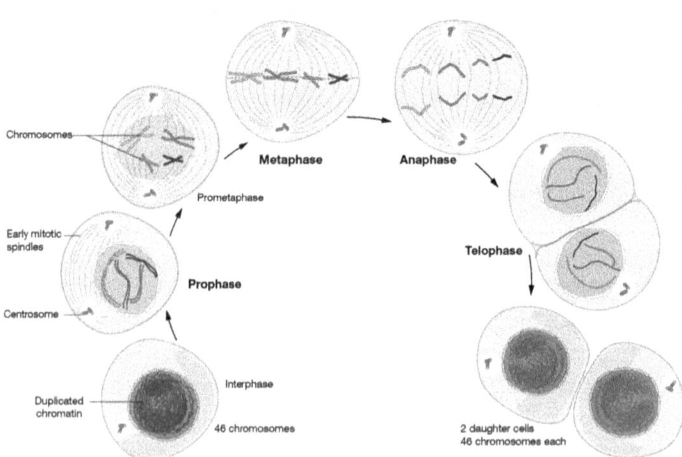

Fig. 1.6 Replicated chromosomes are separated into two progeny cell through the process of segregation (Adapted from National Human Genome Research Institute, 2023 [220])

cells, with each daughter cell obtaining one copy of each chromosome. This assures that each daughter cell is genetically identical to the parent cell.

Any errors in replication and transmission processes can lead to various types of genetic mutations that can cause a range of problems in newborn babies, depending on the nature and location of the mutation. Some possible effects of DNA replication errors in newborns such as changes in the genes that manage the growth and function of the brain, can result in intellectual disabilities. Down syndrome is an example of such a disability that can be caused by these gene mutations. Similarly, DNA replication errors also cause genetic disorders that are passed down from parents to their children. Examples of such inherited genetic disorders include cystic fibrosis, sickle cell anemia, and Huntington's disease. On the other hand, in same living organisms, when mutations occur during DNA replication, they can interfere with normal cellular functions, which can eventually lead to dysfunction and death of the affected cell. Mutated cells can form abnormal tissue growth, which can cause tissue damage and potentially lead to the development of tumors. DNA replication errors that occur in genes that regulate immune system function can increase the risk of autoimmune diseases and cancers.

To better understand most crucial biological processes including reproduction, gene expression, one must understand the nature of both nucleic acids (DNA, RNA). Living organisms DNA has two different components, coding DNA and non-coding DNA. Only around 2% of the entire DNA corresponds to coding DNA which contains the sequences that encode for proteins. Whereas, a major part around 98% of entire DNA corresponds to non-coding DNA which contains the sequences that do not encode for proteins but perform regulatory roles. The percentage of coding and non-coding DNA varies across different living organisms. In fact, the proportion of non-coding DNA varies greatly among the species. However, across most eukaryotic species, only a small chunk of DNA is coding and a major chunk is non-coding. Contrarily, in prokaryotes, coding DNA has a major portion whereas only about 20% portion corresponds to non-coding DNA.

Coding sequences are mainly present inside the coding regions of genes. The coding region is comprised of the sequences called exons that represent the portion of the genes containing genetic code needed for the production of unique proteins. Besides exons, there are non-coding sequences within the genes called introns which do not encode for proteins but help to determine when and how to produce the mRNA molecule that later translates to proteins. Genes can significantly vary in terms of length, from fewer than 100 nucleotides to thousands and even millions of nucleotides. For instance, a gene related to human dystrophin, which is responsible for strengthening the muscle fibres and protecting them from injuries, is based on 2.4 million nucleotides. Furthermore, genes are in huge numbers across all multi-cellular living organisms. For example, humans have more than 20,000 genes which showcase unique expression patterns to produce RNA and protein molecules. Taking the variety of protein-coding genes and their biological importance into account, researchers have classified protein-coding genes into various categories based on their expression patterns and functions.

On the other hand, considering, non-coding DNA does not have the genetic instructions needed for protein synthesis. Non-coding DNA was once assumed to be "junk," with no known functions and received very little attention. However, with the advancements of research in molecular biology, different types of non-coding DNA has been identified that perform unique roles. These types include introns, repetitive DNA, regulatory DNA, etc. Introns are noncoding DNA segments found within genes that do not code for proteins. Introns perform a variety of regulatory activities such as enhancing the gene expression or providing alternative splicing possibilities that enable the generation of multiple protein variants from a single gene. Repetitive DNA sequences are those kind of non-coding DNA sequences which are repeated several times throughout the genome. Repetitive DNA sequences are of different kinds and major types include telomeres, tandem repeats, and interspersed repeats. Telomeres guard the ends of chromosomes by acting as protective caps and play a role in genome stability and accurate genome replication. Tandem repeats play crucial roles in chromosomal organisation and gene control. They also serve as genetic markers which help to distinguish the individuals and perform evolutionary investigations. Interspersed repeats are non-coding DNA sequences that have move around the genome through the evolution and facilitated genetic differences among the individuals and species.

Another kind of non-coding DNA sequences operate as regulatory elements, dictating when and where genes are switched on and off. Such elements mainly provide the places for specific proteins (e.g. transcription factors) to attach and either activate or suppress the critical gene expression process by which information from genes is translated into proteins. For instance, promoters are non-coding DNA sequences located at beginning of genes that help to initiate the process of converting coding DNA into coding RNA, a critical first phase of gene expression also called transcription. Enhancers are also DNA sequences located at beginning or end of genes which increase the transcription of genes by binding to particular proteins and enhancing the activity of the promoter. Silencers regions are similar to enhancers but instead reduce or "silence" gene expression. Insulators facilitate the binding sites for proteins that regulate transcription in a variety of ways. Some of the insulators known as enhancer-blocker insulators mainly restrict the enhancers which eventually reduce their assistance in transcription. Others insulators known as barrier insulators prevent the structural changes in the DNA that suppress the expression of genes. Few insulators can act as an enhancer blocker as well as a barrier depending on the unique cellular conditions.

Other segments of non-coding DNA facilitate instructions for the production of particular kinds of RNA molecules. Few example of non-coding RNAs molecules produced from non-coding DNA include ribosomal RNAs (rRNAs), transfer RNAs (tRNAs) which help to assemble the protein building blocks called amino acids into a chain that forms a specific protein. MicroRNAs (miRNAs) are short length sequences that block the process of protein production, long non-coding RNAs (lncRNAs) are long RNA sequences that play a variety of rules in controlling gene activity. Based on cellular functionality, variation in sequence length, unique structure, physical, chemical properties [32], non-coding RNAs can be segregated

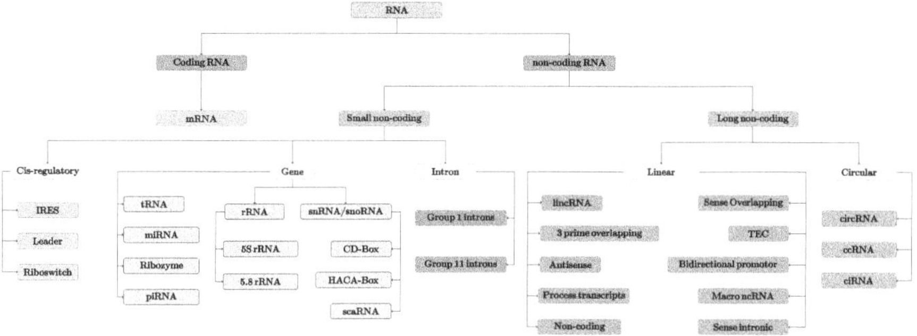

Fig. 1.7 Hierarchical representation of RNA classes

into different sub-classes, a taxonomy of which is depicted in Fig. 1.7. Accurate discrimination of non-coding RNAs from coding RNAs and identification of its sub-type can lay the foundation of demystifying the core function and biological roles of non-coding RNAs, their involvement to suppress the mechanism [597] underlying complex human diseases [67, 432] or to develop effective treatments and optimize therapeutics [700, 1057].

Considering coding and non-coding regions of both nucleic acids (DNA, RNA) are based on well ordered sequence of nucleotides and any kind of variation in underlay nucleotide distribution will have consequences on their molecular activities and eventually on the health of living organisms. Researcher are applying Artificial Intelligence (AI) approaches to identify genes, predict gene function, infer gene variations, forecast drug impact on gene, distinguish genes from regulatory elements, discriminate various kinds of regulatory elements, and predict the impact of non-coding RNAs on the activation or suppression of gene expression [24, 580, 1155].

1.2.2 *Proteins*

Proteins are known as the workhorses of living organisms because they are vital molecules that perform almost all essential functions of cells. Specifically, proteins are involved in diverse types of biological functions including catalyzing chemical reactions, providing support to cellular structure, immune system and energy metabolism. Apart from supporting, protein provide energy to various biological processes of living organisms. In our daily routine activities, we consume a lot of proteins, even we cannot breathe without burning proteins. Proteins can be classified into different types based on various criteria such as function, structure and composition.

Humans have 0.2 to 2 million different proteins. The important question is what type of proteins at what time will perform which kind of activities. All this information is present in amino acid sequences. Mainly, repetitive patterns of 20 unique amino acids make a polypeptide that is typically between 50 and 2000 amino acids in length. A single or multiple polypeptides fold into a specific three-dimensional structure to form a functional protein. Proteins that are made up from folding of single polypeptide are known as monomeric proteins. Similarly, multiple polypeptides fold together to make oligomeric proteins. The unique amino acid sequences primarily determine three-dimensional structures of proteins, which in turn decide proteins functions. Hence, any irregularity or change in amino acid sequence will effect the structure as well as function of a protein.

The biosynthesis of proteins having unique structures, functions and other characteristics is a core biological process in all living organisms. Almost all proteins are produced by transcribing the DNA into mRNA and translating the mRNA into one or more chains of amino acids which fold and mature before becoming the proteins. After synthesis, proteins undergo various modifications called post-translational modifications (PTMs) that significantly affect their nature. Some of the most common PTMs are Phosphorylation, Acetylation, Ubiquitylation, etc. that alter the structure and maturity of proteins [833]. These modifications can improve the activity of proteins in various cellular processes such as, regulation of genetic expression, activation of genes, DNA repair and cell cycle progression, different physical processes such as signal transduction, chromatin stability, protein–protein interaction and nuclear transport [833]. Contrarily, some modifications such as Nitration, S-nitrosylation and S-palmitoylation can adversely affect the activity of proteins in various aforementioned processes [833]. To date, more than 620 different types of PTMs have been identified, that influence the biology of the cell by affecting the functional diversity of the variety of proteins [833].

Furthermore, functions of the proteins are also dependent on their abilities to interact with other molecules including enzymes, proteins, non-coding RNA etc. These interactions are also dictated by amino acid sequences, protein structures, and PTMs. To decode various biological processes, initiation and progression of diseases, and guide the process of drug development through comprehensive analysis of protein structures, functions, interactions, and post-translational modifications. Researchers are applying AI approaches to predict protein family, protein-protein interactions, protein sub-cellular localization patterns. Also they are applying AI to infer protein structure, protein properties (e.g. stability), protein function, protein post-translational modifications, protein interaction with other biomolecules such as non-coding RNA, and protein domains or sub-structure which are responsible for specific interactions, function or biological behaviour [547]. Furthermore, AI approaches are assisting to manipulate existing protein's amino acid sequence, its 3D structure, to create a protein with improved characteristics like increased stability, specificity, or activity [247, 421, 747]. AI approaches are assisting to guide the process of designing and synthesizing artificial proteins (Protein engineering) for sake of optimizing overall well being or the management and treatment of particular disease [247, 421, 747].

1.2.3 Lipids

Lipids molecules are insoluble in water (hydrophobic) and are soluble in nonpolar solvents. Cells require lipids for a variety of important functions such as to store energy, to maintain temperature of body and to perform signaling. Based on working paradigm within the cells, lipids can be categorized into different classes such as fats, phospholipids, waxes, and sterols. Unlike other macromolecules, lipids are not the polymers of any single monomers, rather they are composed of many distinct components. Oils, fats, and phospholipids are made up of fatty acids and glycerol backbone, similarly waxes are based on fatty acids attached to long-chain alcohols, sterols have four fused carbon-rings. The typical arrangement of inherent components eventually determines the characteristics and functions of lipids.

Lipids store energy in a tiny package since they have a great deal of carbon-hydrogen connections. You may be aware that fats contain 9 calories per gram, but carbohydrates and proteins include just roughly 5 calories per gram. Clearly, lipids are the energy efficiency champion among macromolecules. This is the reasons why people gain more weight when they increase food intake and decrease intensive exercise where extra amount of energy turns into fat. Beside, energy storage, some lipids such as phospholipids and cholesterol control the structure as well as function of cell membrane. This membrane encloses and protects the contents of a cell and act as a strong physical barrier between the outside environment and inside of the cell. Furthermore, in humans, fat actually works as a shock absorber between the organs as well as an insulator around the very sensitive nerve cells (similar to the coating of plastic on the copper wires. Several animals make use of fat for insulation such as whales, seals, and polar bears manage to tolerate extremely cold temperatures. Lipids also assist the plants and water birds in waterproofing. Plants use various waxes to protect all the leaves from various kind of damages, as well as water birds use waxes to maintain the dry feathers. Lipids are also capable to cause significant developmental changes in humans, such as testosterone and estrogen lipids levels are instrumental in sexuality as well as fertility. They are responsible for the pregnancy, puberty, menopause, menstruation, sex drive, sperm production, etc.

Considering the biological significance of lipids, researchers are actively applying diverse AI approaches to identify lipid species, explore their structures, functionalities, and various characteristics such as chain length, double bonds, inherent fatty residues, etc. They are also applying AI to determine lipid-RNA interactions which significantly mediate ribozyme activity, lipid interactions with intrinsically disordered regions of proteins which affects lipid binding affinity and result in different diseases. Another promising area which is extensively being explored is finding the association of lipid dysregulation with the development of various diseases, and designing novel biomarkers for the effective treatments [509, 631, 853].

1.2.4 Carbohydrates

Carbohydrates are water soluble molecules and are comprised of small building blocks of sugar monomers that contain carbon, hydrogen, and oxygen atoms at their chemical level. It is evident in Fig. 1.2, among the four essential macromolecules (nucleic acids, proteins, lipids, and carbohydrates), carbohydrates have the most hydroxyl groups (-OH) linked to their carbon atoms. Broadly there are two kinds of carbohydrates: Monosaccharides and Polysaccharides. Monosaccharides are single sugars which predominantly have two hydrogen atoms and one oxygen atom for every carbon atom. Glucose and fructose are most commonly known monosaccharides that you may be aware of. Glucose is a naturally occurring sugar found in many foods that serves as the major source of energy for the body's cells. Fructose is a naturally occurring sugar that is found mostly in fruits and vegetables and is often used as a sweetener in processed foods and beverages. One might assume that two monosaccharides with the same number of carbon, hydrogen, and oxygen atoms will behave in a same manner. However, it is not the case because they will have substantially distinct characteristics mainly due to different organization of atoms.

For example, Glucose and galactose are nearly identical in terms of molecular formula and structure, however with one difference in atom arrangement, they behave substantially differently in cells. Glucose is directly used for energy production, while galactose must be converted to glucose in the liver before it can be used for energy. Hence, the way atoms are linked together is critical to the structure and function of carbohydrates. Polysaccharides are complex carbohydrates made up of long chains of monosaccharides. These monosaccharides are bonded together via condensation, a process in which a water molecule is eliminated in order to form a covalent connection between the monosaccharides. Most commonly known examples of polysaccharides include starch and cellulose. Starch and cellulose are both polymers of glucose, with starch found in bread, potatoes, rice, and pasta, while cellulose is a major component of the fiber in the human diet. Like monosaccharides, polysaccharides also behave differently depending on the way their inherent monosaccharides are linked together. Despite the fact that starch, cellulose, and glycogen are all constituted entirely of glucose, they all react extremely differently in the body. In the human digestive tract, starch and glycogen are easily broken down. However, cellulose, a kind of fibre can not be broken down and exits as a waste.

Carbohydrates are probably best known for providing fast-acting energy for nearly all cells on Earth. However, they perform a plethora of key functions for cells such as storing energy for later use. Furthermore, carbohydrates are crucial cellular identity markers. Glycoproteins, protein molecules with an attached sugar, mark the surfaces of cells and distinct cells have distinct glycoproteins on their surfaces. This cell signaling helps the cells to recognize liver cells as liver cells, heart cells as heart cells, etc., and communicate with one another effectively. Carbohydrates enables the bacteria to attach to the surfaces, hence the recognition of such carbohydrates help the immune system to destroy or fight against them. Carbohydrates can affect the folding and stability of proteins, which can impact their function and activity.

A critical analysis of carbohydrates can provide significant insights into various diseases, such as cancer, autoimmune disorders, and infectious diseases. It can also assist in the discovery of drugs and designing carbohydrate-based therapeutics. Carbohydrate interacts with a wide range of protein families, including lectins, antibodies, sugar transporters, and enzymes to mediate intercellular signaling, cellular recognition, cellular adhesion, protein folding, subcellular localization, and ligand recognition [104]. Researchers are applying AI approaches to predict protein-carbohydrate binding site, protein-carbohydrate interactions, human immunogenicity, bacterial pathogenicity and virulence that refer to the ability of an organism to infect the host and cause a disease. Also, Researcher are forecasting the taxonomic origin, structure, functions, and families of complex carbohydrates to better understand various biological processes [104].

In a nutshell, these four different biomolecules (nucleic acids, proteins, lipids, carbohydrates) build the macromolecular machines essential for life. These macromolecular machine are responsible to carry various functions, copy the key biological information present in nucleic acids to generate functional proteins, protect the cell from intruders, as well as to utilize the environment energy in best possible manner. After understanding the biological significance of most common macromolecules and the way AI approaches are being used in context of each of them. Let's understand the overall genetic flow of information essential for the survival, reproduction, and functioning of living organisms.

1.3 The Central Dogma of Molecular Biology: DNA to RNA to Protein

In previous section, we have briefly described the nucleic acids (DNA, RNA) and proteins. However, we are left with questions of how RNA gets genetic information from DNA and guides the process of protein production? Also, how information is transformed from a static sequence of fundamental nucleotides into a very dynamic and necessary biological activity referred as gene expression to guide how the life at core level of an organism proceeds?

The answer of these questions lies under the hood of central dogma of molecular biology. The central dogma of molecular biology describes the flow of genetic information in cells that is graphically illustrated in Fig. 1.8. The dogma was proposed by Francis Crick in 1958 [206] and states that, master plan of cell life is comprehensively encoded in the DNA. However the implementation and execution of this master plan happens through the expression of various genes into their respective functional forms. These functional molecules include messenger RNAs (mRNA) which carry the information that is later translated into proteins. In contrast, non-coding DNA is transformed into non-coding RNA functional

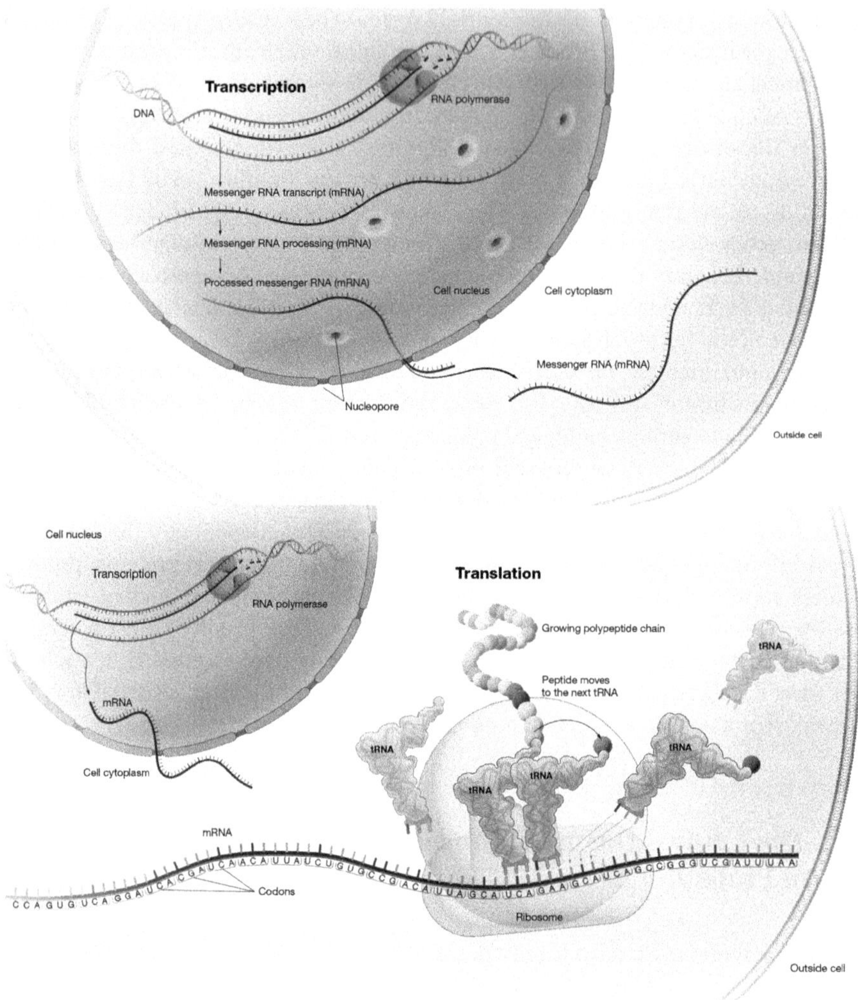

Fig. 1.8 The workflow of transcription and translation processes [815]

molecules which are not directly translated into proteins, but they overall regulate the process of gene expression. The functional molecules produced by coding and non-coding DNA are the main functional, structural, and regulatory components of cell, responsible for all the reactions which harvest the energy, make cellular components, aid the import or export of different materials, and ultimately implement all the cellular functions.

More precisely, it is evident in Fig. 1.8, the main idea of central dogma of molecular biology is that genetic information flows from DNA to RNA to protein. The process starts with transcription, where DNA is transcribed into RNA. This process occurs in the nucleus of eukaryotic cells and in the cytoplasm of prokaryotic cells. The second step in central dogma is translation, where RNA is translated into a sequence of amino acids that make up proteins. Molecular machinery including Ribosomes, and transfer RNA work together to translate mRNA message and assemble the corresponding amino acids into a protein. The central dogma describes the one-way flow of genetic information from DNA to RNA to protein. However, there are some exceptions to this flow, such as reverse transcription, where RNA is used as a template to create DNA. For example, Viruses like HIV inject their genetic information in RNA, so reverse transcription enables such viruses to copy their genetic material in the host's cells by converting RNA into DNA. Scientists are trying to explore and understand the process of forward and reverse transcription so that they can design better therapies to improve immune response and develop more efficient drugs for viral diseases.

After taking a brief look into central dogma of molecular biology, now let's more precisely explore the process of transcription and translation.

1.3.1 Transcription Is the Initial Phase in Gene Expression for the Generation of Coding and Non-coding RNAs

Transcription is the very first phase of gene expression in which DNA is copied into RNA. Figures 1.8 and 1.9 illustrate the process of transcription. When the cells require to construct proteins to perform various functions, they recognize the genes present in the coding region of DNA through promoter sequences that are located at the very beginning of genes. Afterwards, cells hire RNA polymerase as special worker that reads the genes and copies the genetic information into RNA molecule.

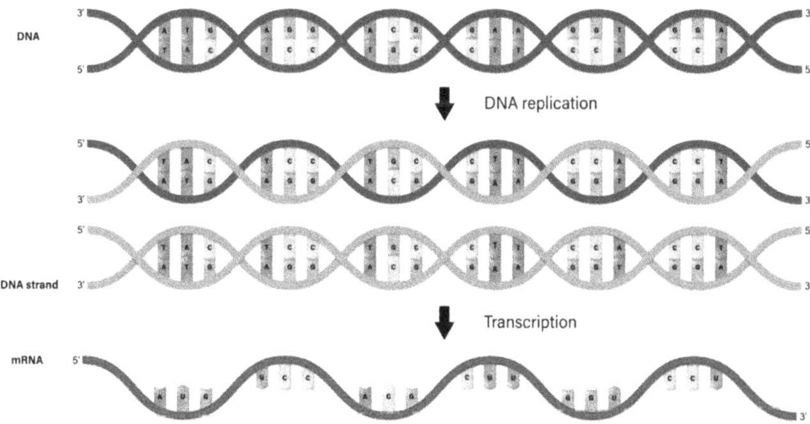

Fig. 1.9 An illustration of first phase of central dogma of molecular biology, DNA transcription to RNA

RNA polymerase requires assistants to recognize the promoters and initiate transcription. These assistants are regulatory proteins that fully control the transcription process. In eukaryotes, these regulatory proteins are called transcription factors that aid RNA polymerase binding to promoters. After binding, RNA polymerase starts reading and transcribing the DNA information into an RNA molecule. RNA polymerase opens up the double helix structure of DNA and only one strand of DNA is used as a pattern for the construction of a RNA molecule. The DNA strand that is read by RNA polymerase is referred as template strand and the strand which is not utilized by RNA polymerase is referred as non-template strand. To copy the genetic code, RNA polymerase primarily follows RNA base pairing rules, Adenine (A) pairs with Uracil (U), and Guanine (G) pairs with Cytosine (C). It brings the nucleotides in RNA in such a manner that they match up with nucleotides of DNA. As a result, it builds up a RNA molecule that is complementary to DNA strand. It catalyzes the creation of covalent bonds among the nucleotides of growing RNA molecule. When RNA polymerase connects RNA nucleotides with template strand DNA nucleotides, the nucleotides should be turned antiparallel with respect to each other. Hence, RNA polymerase interprets the template strand of DNA from 3' to 5' to construct the new RNA molecule in 5' to 3' direction.

Just like the promoter sequences that mark the beginning of genes, transcription terminators are the sequences in DNA that mark the end of genes. Transcription terminators primarily trigger the RNA polymerase to completely release the DNA and halt the process of transcribing. A transcription terminator has the code which construct a unique kind of RNA that completely fold back on itself to formulate the hairpin loop. This irregular kind of RNA effectively detaches the RNA polymerase from DNA and end the transcription of particular gene. However, it is important to mention that RNA polymerase does not stop there, it goes on to transcribe several more genes using the same process. Another important thing is that the process of producing RNA molecule through transcription is a perfect instance of chemical reaction which constructs complex molecules from the smaller pieces. Considering such chemical reactions require energy, cells make use of different energy-rich molecules to fulfil energy needs. Cellular respiration produces adenosine triphosphate (ATP) that empowers the transcription machinery. Furthermore, transcription energy is acquired from high-energy phosphate bonds present in the ribonucleotide triphosphates (NTPs), that are used as the building blocks for the expanding RNA molecule.

At this stage, one important question may arise: Is the RNA produced by transcription ready for the production of proteins? The answer of this question is a big no. Even after the transcription of eukaryotic genes, RNA is not yet ready for translation into proteins. In fact, the RNA produced by the transcribed eukaryotic genes is referred to as a pre messenger RNA (pre-mRNA) or primary RNA transcript just to showcase that the transcript is not yet complete. One issue that must be addressed before translation is that eukaryotic genes do not only contain the code for protein production. More specifically, the protein code has the short stretches of nucleotides which break up the protein code known as introns. Before the mRNA can be translated, the introns in the pre-mRNA must be removed. Exons are the parts of the pre-mRNA that contain the information for the protein. Only Exons are retained in the finished mRNA and expressed in the final protein.

Cell use small RNA and protein particles to delete introns from pre-mRNA. The particles are known as small nuclear ribonucleoproteins (snRPs). Several snRPs come together to create spliceosomes, which clip the pre-mRNA at boundaries among exons and introns, delete the introns, and then build bonds between the exons. This approach of deleting introns is known as RNA splicing and it is quite similar to the practise of clipping off array elements using any renowned programming language and then resealing the remaining elements to construct a new version of array. The spliceosome enzymes which trim and reseal the pre-mRNA appear to be small nuclear RNA (snRNA) that build up the snRPs. Most enzymes are proteins, but snRNA in snRPs is yet another example in an expanding list of RNA molecules acting as enzymes. In addition to splicing, the cell performs two further changes to the pre-mRNA to prepare it for translation: A protective cap known as the 5' cap is introduced to the 5' end of the pre-mRNA. The 5' cap identifies the completed mRNA as an RNA that has to be translated. Enzymes cut the 3' end of the pre-mRNA and replace it with an RNA fragment containing 100 to 250 adenine nucleotides. This poly-A tail prevents the cell from breaking down the final mRNA. After RNA processing, final mRNA is obtained and used by the cells to produce proteins.

At DNA transcription stage, AI approaches are being used to perform various tasks. More specifically, AI approaches are identifying which genes are activated or de-activated in a particular cell or tissue. AI approaches are assisting the researchers to predict the location of regulatory regions such as promoters and enhancers that control gene expression. AI approaches are inferring the binding sites of transcription factors which bind to DNA and regulate gene expression. Also, AI approaches are simplifying the identification of genetic variants that may contribute to disease or other traits. Some researchers are using AI approaches to design gene editing tools which can alter specific regions of genome with great accuracy. AI approaches are also actively predicting RNA stability and RNA splicing sites. Overall, AI is playing an increasingly important role in advancing our understanding of the genome and its regulation, which has important implications for fields such as medicine, agriculture, and biotechnology.

1.3.2 Transcription Is Followed by Translation that Interprets Coding RNA to Synthesise Functional Proteins

When a cell wants to make a protein, RNA polymerase reads the genetic code from the coding DNA and copies it into an mRNA molecule. The mRNA is subsequently transported from the nucleus to the cytoplasm, where translation machinery operates on the mRNA code to construct the protein. When the protein is finally created, it performs dedicated function in a particular type of cell. mRNA transcripts are

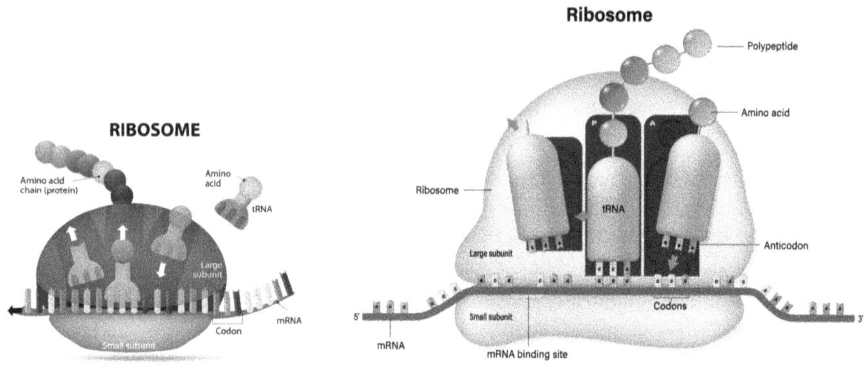

Fig. 1.10 A workflow of RNA translation into proteins based on ribosome machinery

mainly encoded by the sequence of four different linearly organized nucleotides (C, G, A, and U) that appear in the strand. The size of the strand, the proportion of every nucleotide within the strand, as well as the order of nucleotides can all vary between mRNA molecules. For instance, the mRNA that has the code for the protein called insulin, would be very different from the mRNA which includes the code for the protein called collagen. This is because insulin is a kind of globular signaling protein as compared to collagen which is a kind of fibrous structural protein. Furthermore, their main structures, sequence of amino acids are also distinct. The changes in the fundamental structure of the protein are determined by differences in the code in the mRNA molecules.

It is indicated in Fig. 1.10 that, in the process of translation, the reading of a group of three nucleotides called codon at a time mainly defines which of the 20 possible amino acids must be integrated at particular point inside the protein. Unique sequences within mRNA determine the initial and ending sites for the generation of proteins, whereas the successive codons in between mainly encode the chain of amino acids which make up the ultimate protein product. All the codons produce a specific outcome, either the integration of a particular amino acid or act as a stop site to halt the process of translation, hence codons are truly non-ambiguous and their behaviors are deterministic. A pie chart summarizing the association of various codons with respective amino acids in shown in Fig. 1.11. Translation is mainly carried out by a macro-molecular machine called ribosome that comprises of both molecules, RNA and proteins. The functional molecule which operates at the heart of translation is known as transfer RNA (tRNA) which serves as a physical link among codons and their respective amino acids. Figures 1.10 and 1.11 demonstrate the way tRNAs operate as decoders which makes the translation feasible. A collection of tRNAs reads the codons sequentially on mRNA template, recruiting the amino acids to the expanding protein chain in the exact order specified by the mRNA. The ribosome is a structure that facilitates this process. The resulting protein product is often not fully functional and requires additional processing. This post-translational processing involves removing certain portions of the protein and adding other chemical groups.

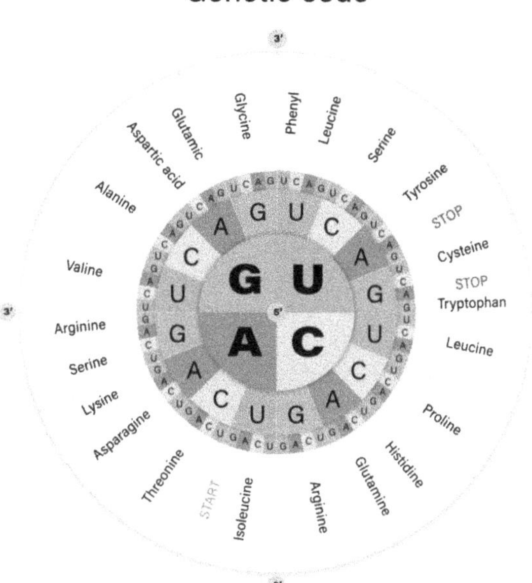

Fig. 1.11 Association of codons with respective amino acids

1.3.3 The Functional Repository of Cell Is Way More Complex than Entire Genome

In humans, inside the compartments of cells at a particular time, hundreds and thousands of genes express to produce mRNAs which have the codes to produce tens of thousands of different proteins. To control overall flow of genetic information from DNA to RNA to Protein production, diverse types of interactions between DNA, RNA, and proteins happen inside the cell. Like microRNAs (miRNAs) are small non-coding RNA molecules that bind to messenger RNAs (mRNAs) to either prevent translation or promote degradation of the mRNA, leading to hamper protein synthesis. Long non-coding RNAs (lncRNAs) interactions with mRNAs regulate gene expression by affecting mRNA stability, splicing, localization, translation, or epigenetic modifications. LncRNAs interact with other molecules including RNA binding proteins or chromatin modifiers and effect gene expression.

Proteins interact with different types of RNAs and control diverse types of biological processes such as mRNA stability, localization, splicing, editing, and translation. Ribosomal proteins interact with mRNA to initiate translation and ensure accurate and efficient protein synthesis. Standalone proteins are unable to perform several biological processes such as signal transduction, DNA replication and transcription, protein folding, and degradation. In fact, multiple proteins interact with each other to perform these types of complex functions. Proteins also interacts with viruses and play a key role in the life cycle of viruses by serving as carriers for viral particles. When a virus infects a host cell, it can hijack the cellular machinery to produce viral proteins and replicate its genetic material. Some viral proteins act as structural components of the virus, while others have specific functions that help the virus to enter and exit host cells, evade the immune system, and hijack cellular processes for its own purposes. We have only mentioned few kinds of interactions, however, several other interactions also happen inside the cell. To better understand the working paradigm of these diverse types of interactions and their biological impacts, AI aided programs are being used to find interactions among different kinds of RNAs, RNAs with proteins, protein-protein interactions and protein virus interactions.

Furthermore, DNA, RNA, and protein molecules undergo different kind of modifications. This is why, although the human genome only encodes approximately 20,000 to 25,000 protein-coding genes, the number of functional RNAs and proteins produced by human genes is projected to be around two million. It's very likely that this projection is truly an underestimate. Furthermore, is considered that some categories of functional molecules have remained unnoticed due to low gene expression levels as well as small sizes which are hard to identify. Nevertheless, technological advances have led to the development of more systematic gene identification techniques, resulting in the discovery of several diverse classes of small sized functional RNAs. These techniques facilitate a compelling example regarding how the gene count is expected to significantly fluctuate in coming years. It is the precise control over the synthesis of an increasingly great number of cellular components that allows the researchers, scientists, and clinicians to observe the amazing complexity in living organisms.

In a nutshell, the central dogma of molecular biology provides a basic platform for understanding the flow of genetic information in cells and the relationship between DNA, RNA, and proteins. Researchers are using AI approaches at different levels of central dogma of molecular biology with an aim to optimize diagnosis and treatment of diseases. AI approaches are being used to analyze large datasets of genomic data and identifying patterns in these datasets in order to elucidate the mechanisms that underlie gene expression and regulation. AI approaches are comparing the genetic information of a seemingly healthy individual with disease-inflicted individual using known disease-associated mutations of public databases. Furthermore, AI approaches are serving as simulation and modelling tools where they help to predict the behaviour of biological systems. For example, they assist to model the interactions between transcription factors, RNA polymerase, and DNA to understand how genes are regulated.

More specifically, at the DNA level, AI algorithms are identifying genetic variations or mutations that are associated with certain diseases. For example, AI approaches are analyzing genomic data to identify single nucleotide polymorphisms (SNPs) that may be associated with increased risk of developing Cancer or heart disease. At the RNA level, AI approaches are helping the scientists to identify aberrant splicing events or changes in RNA expression levels that may be indicative of a disease. AI approaches are assisting to distinguish normal changes from disease associated changes in RNA sequences. At the protein level, AI approaches are predicting the impact of mutations on protein structure and function. This can be done using machine learning algorithms that are trained on known protein structures and functional data. These algorithms can then be used to predict the functional impact of novel mutations in proteins.

After understanding the central dogma of molecular biology and getting an idea how AI approaches are being used at various levels of central dogma of molecular biology. Now, we will dive into the biological significance of gene expression regulation and how AI approaches are being used in this context.

1.4 Gene Expression Regulation

The process of gene expression is tightly regulated and involves many different molecular mechanisms that control when and where specific genes are expressed. At any particular compartment and time, cells only express a small fraction of the genes from their genome. Cells may alter the collection of genes they are using depending on various signals or alterations in the environment. As the genetic information is copied from DNA to RNA through transcription and then translated into proteins through translation. Hence, when cells alter the collection of genes for expression, a very different functional molecules (RNA, Proteins) are produced inside the cells. Almost all powerful changes inside the cell occur primarily through the precise control of genes that are activated and deactivated inside the cell.

Each phase in the process of gene expression including, transcription of DNA, processing of RNA, translation of RNA, modification of proteins exposes an opportunity for precise regulation and control of genes. Gene expression regulation is the most critical process of molecular biology because it permits which gene must be expressed at what time, location, and amount to produce the target functional molecules needed by living organism. Figure 1.12 illustrates different levels where gene expression can be regulated in context of central dogma of molecular biology.

There are many factors that affect gene expression regulation at a certain level. Broadly, these factors are divided into two classes: internal and external factors. Internal factors include mutations in the DNA sequence which can alter the function of genes and their regulatory regions, leading to changes in gene expression. Transcription factors are proteins that bind to DNA and can activate or repress the expression of genes. Mutations in the genes that encode transcription factors can alter their function and lead to changes in gene expression. Epigenetic modifications are chemical modifications that alter the structure of DNA or the histone proteins around which DNA is wrapped, leading to changes in gene expression. Also, chromatin structure such as the way DNA is packaged into chromatin affect the accessibility of genes for transcription.

RNA interference in which small RNA molecules bind to messenger RNA (mRNA) and either degrades it or prevents it from being translated into proteins. Cell types also impact gene expression regulation because different cell types express different sets of genes, allowing them to perform specific functions. There exist pathways called signal transduction pathways through which cells

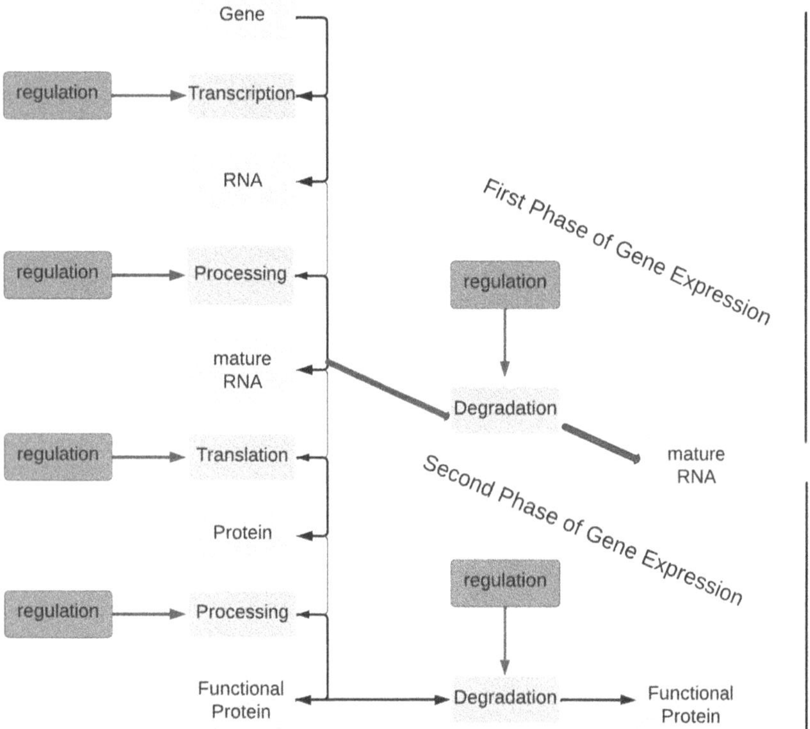

Fig. 1.12 The regulation of gene expression at different levels

communicate with each other, and they can lead to changes in gene expression. Furthermore, the expression of certain genes can change as an organism develops from embryo to adult, so development stage also plays a key role in gene expression regulation. Hormones are special chemical messengers which affect gene expression by binding to particular receptors on the surface of cell. Also, hormones impact the activity of transcription factors, and stability of mRNA molecules, which in turn affects the amount of protein production.

External factors that affect gene expression regulation include environmental factors like exposure to certain chemicals (e.g. ultraviolet radiation) or conditions. Toxins and drugs affect gene expression by damaging DNA or changing the activity of the regulatory proteins. Pathogens including viruses hijack cellular machinery of living organisms, alter the gene expression patterns in a manner that facilitate their survival and propagation. Overall, gene expression regulation is a complex and dynamic process that is positively or negatively influenced by a wide range of factors. Irregularities in gene expression due to malfunction of internal factors or negative effects of external factor lead to the development of various diseases. For instance, cancer cells often have altered gene expression profiles that contribute to their uncontrolled growth. For example, cancer cells often exhibit altered gene expression profiles that contribute to their uncontrolled growth. Therefore, it is important to understand the non-organic factors that lead to irregularities in gene expression and their negative effects in order to detect diseases, develop drugs, and optimize therapeutics.

AI approaches are helping the researchers to analyze large and complex datasets, identify useful patterns in raw genetic sequences to better understand various factors that influence gene expression regulation. AI approaches are predicting the effects of genetic mutations on gene expression and ultimately health of living organisms. AI approaches are identifying both known as well as novel regulatory or non-coding regions of DNA which control gene expression. AI approaches are helping the researchers to develop the gene expression models that can be used to infer the expression of a gene in response to specific stimuli. Overall, AI approaches are providing new insights into the complex process of gene expression regulation. They are paving ways to optimize disease initiation, progression, and treatments strategies.

As mentioned earlier, many factors contribute in the regulation of gene expression at different levels. Let's explore some of the major levels at which regulation of gene expression is most critical and also has greatest impact on cellular activities for living organisms.

1.4.1 Gene Expression Regulation by Controlling the Amount and Degradation of Functional Molecules

Figure 1.12 depicts one of many ways through which gene expression can be regulated. Specifically, this mechanism involves controlling the amount of functional molecules that are produced by the cell through low or high level expression of specific genes. The promotion of transcription increases the expression of genes, generation of mRNA molecules and ultimately translation of protein molecules. Whereas, repression of transcription decreases the expression of genes, generation of mRNA molecules and finally reduces the production of protein molecules. In this way, by promoting or avoiding the generation of the RNA molecule at the transcription level or by regulating the translation of a functional protein, gene expression can be regulated. Additionally, gene expression can be controlled by promoting or avoiding the degradation of RNA transcript or protein. To better understand these processes, let's consider an example.

Figure 1.13 illustrates the importance and effects of low as well as high level expression of a particular gene called guardian of the genome, denoted as p53. The expression of P53 gene in right amount is very crucial for avoiding the development of deadly cancer. It mainly serves as the tumor suppressor by closely monitoring the cells health and triggering the cells death when it detects any kind of abnormalities. It is shown in Fig. 1.13, the expression of P53 gene that produces p53 protein

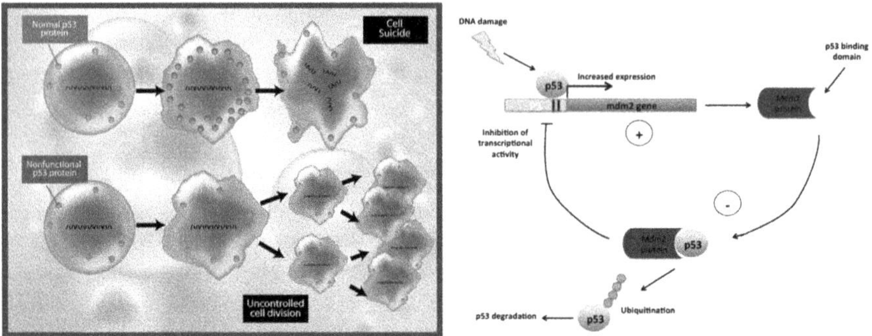

Fig. 1.13 The workflow of high level expression and activation of P53 protein that leads to uncontrolled cell division and huge cell death

has normally low levels. Hence, it must be very carefully regulated because any irregularity in expression of this gene such as uncontrolled high level of expression causes uncontrolled cell division and huge cell death that can eventually lead to permanent tissue damage or even cancer.

Furthermore, it is also shown in Fig. 1.13, activation and function of p53 proteins are controlled through number of post-translational modifications, including phosphorylation, acetylation, and ubiquitination and negative feedback loop. At post-translational stage, when p53 protein levels are very high, negative feedback loop activates expression of a unique gene named as MDM2 that causes P53 protein degradation. This degradation breaks p53 protein into individual amino acids that are later used by the cells in different metabolic processes or to create new proteins. On the other hand, when p53 protein levels are low, MDM2 gene is never activated, enabling P53 gene to trigger the cell death only when necessary.

Researchers are using AI approaches to edit the genes and control their expressions and replace the faulty genes with therapeutic genes. Also, AI approaches are assisting to design as well as optimize proteins with certain functions, and to develop synthetic pathways to produce certain cellular material or control specific biological process [95, 215, 499, 702, 927].

1.4.2 Regulating Gene Expression in Space and Time Dimensions

The variations in expression of multiple genes with respect to time and location define the cellular identity and behaviour inside living organisms. For instance, different tissues are characterized by the distinct cells from which they are generated, and distinct cells are characterized by the specific subset of genes they express. Hence, a liver cell will surely express a very different set of genes as compared to the

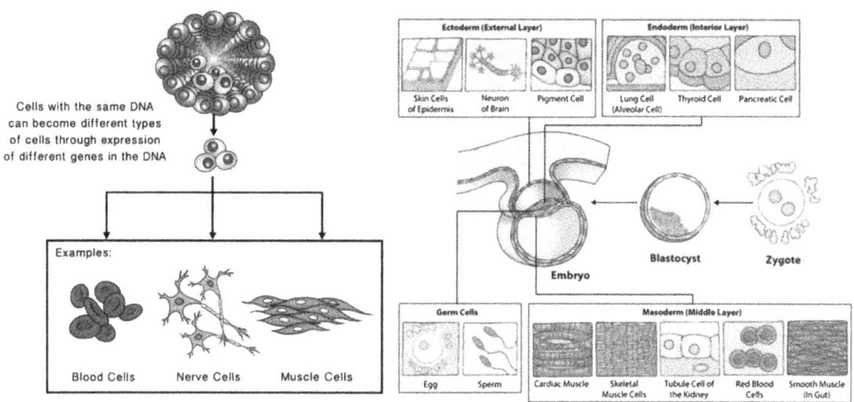

Fig. 1.14 The phenomenon of cell specialization initiates and depends on gene expression regulation

nerve cell. These variations in gene expression imply that both cells are biologically very different.

Figure 1.14 illustrates gene expression regulation across time and space dimensions gives birth to cell specialization in living organisms. To better understand cell specialization, an example of developing embryo is given in Fig. 1.14 where gene expression is controlled in a highly systematic and coordinated manner. During the development of embryo, distinct collection of genes are activated or de-activated at particular times as well as in certain regions, which lead to the formulation of various cell types and tissues.

More specifically, although all cells of developing embryo have same genome. However, expression of specific genes lead the cells present in one segment of the embryo to become specialized muscle cells, while in other regions, such expression causes the cells to become nerve cells, and so on. Gene expression regulation is crucial for effective development and differentiation of cells, tissues, and organs. Irregularities in precise control of gene expression across time or space dimension lead to various development abnormalities and birth defects. Gene expression regulation also plays a key role in avoiding the uncontrolled growth and division of cells. Irregularities in gene expression allow the cells to grow as well as divided uncontrollably that leads to cancer. Gene expression regulation plays a crucial role in preserving the appropriate functionality and interconnectivity of neurons within the brain. Irregularities in gene expression regulation can contribute to the development of neurodegenerative disorders such as Alzheimer's, Parkinson's, and Huntington's diseases. The appropriate differentiation and activation of immune cells depend on accurate gene expression regulation. Dysregulation of gene expression hampers the proper functioning of immune system and leads to autoimmune disease where the immune system mistakenly attacks the body's own tissues. Gene expression regulation is also essential for the proper regulation of different metabolic pathways and to balance the energy within the body. Dysregulation of gene expression can lead to different metabolic disorders like obesity, diabetes.

In a nutshell, dysregulation of gene expression has sever consequences on the well-beings of living organisms. Hence, fully understanding the factors and mechanisms that properly regulate gene expression or improperly regulate the gene expression is an active area of molecular biology. In this regard, AI approaches are helping the researchers to perform a comprehensive analysis of gene expression regulation across time and space dimensions to optimize disease detection and treatment [95, 215, 499, 702, 927].

1.4.3 Critical Functional Molecules Are Regulated at Multiple Levels

Many functional molecules of gene expression that are critical to cell functioning are regulated at multiple levels to ensure their proper usage. Messenger RNAs, and proteins which assist the cells to take critical decisions about development, metabolism, etc. as well as get oftenly mutated in different diseases like cancer, are some of the examples of functional molecules where gene expression levels are very tightly regulated at all the possible levels. For instance, E. coli protein levels which mainly control the expression of all genes essential for surviving the stress, is regulated at various levels. At transcriptional level, its expression is regulated on the basis of availability of carbon sources. The mRNA translation is regulated on the basis of oxygen levels, cell membrane changes, and temperature. Finally the activity and stability of the protein produced by the translation of mRNA is regulated on the basis of availability of magnesium and phosphate.

In another example which we have discussed and illustrated earlier in Fig. 1.13, the tumor suppressor gene p53 is also regulated at different levels to ensure proper functioning. P53 gene expression is regulated at the transcriptional level through the help of regulatory MDM2 protein. At the post-transcriptional level, mRNA molecule which encodes for p53 protein is regulated by non-coding RNAs which bind to mRNA and avoid its translation into P53 protein. The activity and stability of the P53 protein is controlled by various post-translational modifications, such as ubiquitination. Also, environmental factors like exposure to radiation or chemicals that cause DNA damage, do affect the p53 gene expression. P53 gene trans-localization to various cellular compartments including mitochondria and nucleus also change its expression and response to cellular stress. In a nutshell, multiple factors regulate the expression of genes at different levels and a comprehensive gene expression regulation analysis needs to be performed to take a tiny yet worthy step forward in the field of molecular biology. Researchers are using AI approaches to predict various factors involved in regulating gene expression at different levels. They are also using AI approaches to find the association of gene expression patterns with various diseases, identify potential drug targets, develop novel therapeutics, and quantify the effects of drugs on underlay genetics [95, 215, 499, 702, 927].

1.5 Cellular Organization Facilitates Chemical Reactions and Gene Expression Regulation

The cell, despite its minuscule size, harbors a vast and intricate world within it. We can think of cell as a factory. Factory has different operational units where each operational unit produces special product. Like factory, cell also has different compartments that carry out unique biological tasks. Figure 1.15 graphically illustrates the internal compartments of the cell. This section precisely dives into how the critical process of gene expression, which involves retrieving information from genome and expressing it to produce functional molecules is mainly accommodated inside the cell. Specifically, it explores the subcellular compartments in which all of these processes actually take place.

1.5.1 Eukaryotic Cell vs Prokaryotic Cell

Living organisms are broadly classified into two categories: eukaryotic and prokaryotic due to the structural and functional differences. Some well known eukaryotes include humans, plants, animals, and prokaryotes include Bacteria, Archaea. In general, eukaryotic cells are larger and way more complex than prokaryotic ones. Eukaryotic cells have a clearly defined nucleus that houses the genetic material, whereas prokaryotic cells contain a nucleoid region that houses the genetic material. In eukaryotic cells, DNA is packaged into chromosomes with the use of histone proteins, but in prokaryotic cells, DNA is packaged into a single circular chromosome.

While most eukaryotic cells lack a cell wall, prokaryotic cells have one that facilitates structural support and protection. In prokaryotes, capsule act as a protective layer exterior to the cell wall and allows the cell to completely evade host immune system, cytoplasm contains diverse molecules needed for the cellular metabolism, ribosomes synthesize proteins. Plasmids in prokaryotes are tiny circular DNA molecules which are separate from main chromosomes and usually carry the genes that enable them to survive and reproduce more effectively in specific environments. More specifically, plasmids may carry the genes that promote resistance to antibiotics, increase the production of toxins that can kills several cells, or enable the cells to more efficiently utilize different nutrients in order to adapt to challenging environments. Flagellum helps in movement, Fimbriae and Pili help the prokaryotic organisms to attach to different surfaces or to other cells, even Pili can also aid bacteria to transfer genetic material between cells.

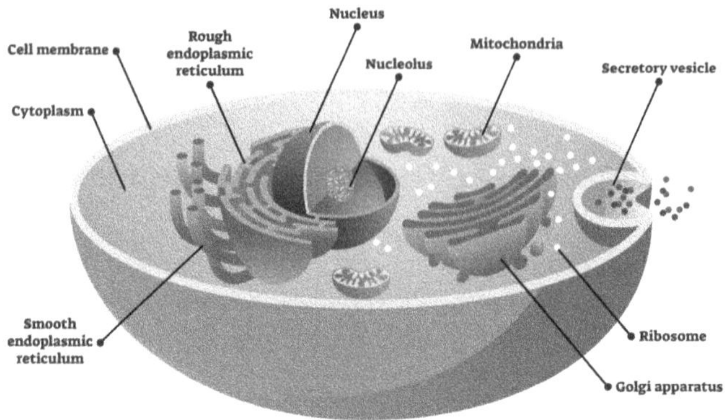

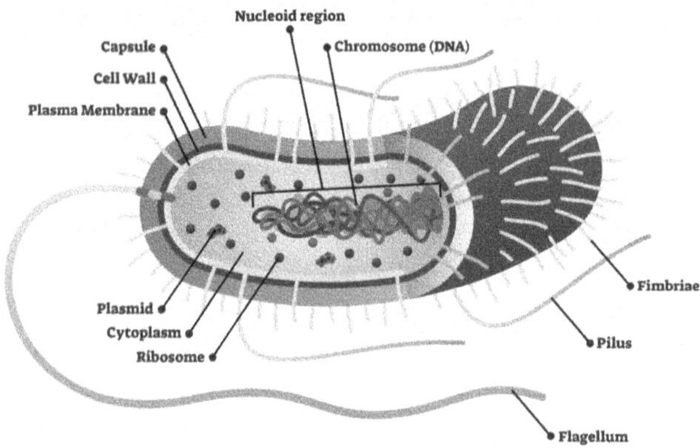

Fig. 1.15 A brief comparison of eukaryotic and prokaryotic cell

Furthermore, eukaryotic cells mainly reproduce through mitosis or meiosis, whereas prokaryotic cells primarily reproduce through binary fission. Eukaryotic cells have a way more complicated metabolism with a broader spectrum of metabolic pathways than prokaryotic cells. Eukaryotic cells are assumed to have evolved from the prokaryotic cells via a process called endosymbiosis. In this process, one cell encompass another cell and form a symbiotic relationship which ultimately resulted in the formation of membrane-bound organelles and the nucleus in the eukaryotic cells. Eukaryotic cells have a more complicated internal structure that includes membrane-bound organelles namely mitochondria, endoplasmic reticulum, Golgi apparatus, chloroplasts, whereas prokaryotic cells do not have such organelles. Mitochondria are usually called powerhouse of eukaryotic cell as they have the efficient macromolecular machinery for extracting the chemical energy from glucose and fatty acids derived from the food consumed by living organisms. Cell utilize this energy to perform different tasks.

Chloroplasts are mainly found in the plant cells and they have the machinery for deriving energy from the sunlight and turning it into usable molecules such as sugars within the cells. The endoplasmic reticulum (ER) is a network of flattened tubes and sacs located in the cytoplasm of cells. There are two types of ER: rough endoplasmic reticulum (RER) and smooth endoplasmic reticulum (SER). The RER is studied with ribosomes, which play key role in protein synthesis. The SER plays crucial role in lipids synthesis and drug detoxification. Both types of ER significantly contribute in the production of materials and components that are eventually exported from the cell, such as proteins and lipids. The Golgi apparatus is an organelle that processes and modifies the proteins and lipids in order to export them to other parts of the cell. The Nucleolus is a structure within the nucleus where ribosomes are produced that are mainly responsible for protein synthesis.

In context of cellular organization, AI approaches have numerous applications. Researchers are using AI approaches to analyze pictures of cells and their internal structures which assist the researchers in accurate identification and classification of various cell types, and enable them to explore the spatial organization of all cellular components. AI approaches are modelling and simulating cellular activity to acquire deeper understanding of functionality induced by the interplay of different cellular components [92, 798]. Furthermore, AI approaches are quantifying the effects of different medications on cellular processes and assisting to identify novel therapeutic targets which are based on unique spatial arrangement of cellular components. Another promising area in which AI approaches has shown great promise includes the creation and optimization of synthetic cells and cellular components. These synthetic components are not only deepening scientists biological knowledge but also offer tremendous applications in personalized healthcare [92, 798].

1.5.2 Compartmentalization Facilitates Chemical Reactions

Compartmentalization creates a special environment which allows the cells to carry out chemical reactions more efficiently. Biomolecules and regulatory components involved in each chemical reaction are retained in their own separate compartments, enabling the simultaneous occurrence of several chemical reactions at distinct parts of the cell without any kind of interference. It enables the cells to better control the delivery of raw materials that can be kept in same compartment or transported to particular compartment when they are needed. It enables the cells to eliminate waster products produced during the chemical reactions. For example, cellular respiration is a biological process in which cells decompose the glucose and other kinds of nutrients to release the energy essential to power diverse cellular processes. This biological process includes a series of chemical reactions that happen in mitochondria of eukaryotic cells and cytoplasm of prokaryotic cells. The raw materials needed for cellular respiration are glucose and oxygen, and the waste products are carbon dioxide and water.

Furthermore, compartmentalization enables the coupling of unfavorable chemical reactions with the energetically favorable chemical reactions. This allows cells to use energy in the most efficient manner because unfavorable chemical reactions need energy, so they use the energy released by favorable chemical reactions. Researchers are using AI approaches to model cellular chemical reactions with an aim to understand various components involved and their roles in the chemical reactions. In addition, they are using AI approaches to study how cells process different substances under various conditions in order to distinguish normal cellular behavior from pathogenic behavior.

1.5.3 Compartmentalization Is Essential for Regulating Gene Expression in Eukaryotic Cells

Based on the fact that DNA in eukaryotic cells is contained in a membrane-bound compartment, called nucleus, a significant degree of regulatory control over gene expression has evolved as a natural result of the physical separation of the nucleus from the other components of the cell. Transcription of DNA occurs in the nucleus, whereas translation of RNA occurs in the cytoplasm. This variation in location provides a rapid means of regulation: just as one can control access from one drive to another drive in a computer by locking or unlocking the drive through various privacy software. The physical barrier provided by the nuclear membrane facilitates the ability to regulate the flow of information from the nucleus to the cytoplasm (and vice versa), just as flow can only occur when certain conditions are met.

For instance, we are aware of the fact that in the eukaryotic cells, mRNAs produced in the nucleus are substantially modified through RNA processing after the process of transcription. We also know that this modification happens before the transcript is transferred from nucleus to cytoplasm, where it gets translated. In reality, transcripts cannot be transferred from nucleus until they have been entirely processed. This characteristic allows the cell to perform quality control such as blocking the transit of the transcript from nucleus to cytoplasm through the nuclear membrane and preventing improperly processed mRNAs from meeting the translational machinery. In bacterial cells, the genetic material or DNA is not stored within a distinct nucleus, as is the case in eukaryotic cells. Instead, the DNA is located in the cytoplasm, where it is almost immediately available for transcription and translation. This proximity between the DNA and the ribosomes allows for spatial coupling of transcription and translation in bacteria. This means that the ribosomes can directly translate the mRNA transcript as it is being produced, without any barriers or regulations to control the process. This spatial coupling of transcription and translation in bacteria results in unique forms of control on gene expression. Since there is no physical separation between transcription and translation, the process of gene expression in bacteria is more streamlined and efficient, but it also requires unique mechanisms for regulation.

An extra complexity encountered by gene expression is the creation of proteins destined for various locations, such as for cytoplasm, for cellular membranes, or for the sites that are beyond the scope of the cell. How could proteins be targeted to their right locations? Proteins targeted for membrane or for cell export, have physical features that are fully incompatible for moving through membrane without great assistance. In bacteria, due to the absence of internal organelles, proteins destined for the purpose of secretion are directly targeted to extracellular membrane. Eukaryotic cells seem to have refined this strategy where rather than directly targeting proteins to exterior of cell, translation takes place on the membrane of a large organelle system known as the endoplasmic reticulum. This organelle is actually a network of many tubules through which membrane as well as secreted proteins are translated, folded, and changed before sending them on their ways.

Golgi apparatus is the organelle that alters almost completed proteins with the sugars before packing the protein for ultimate export, and hence it is the final stop in maturation of several proteins from genes. Proteins are transported from the endoplasmic reticulum to the Golgi apparatus for export, and then from the Golgi to cell membrane. In a nutshell, it is crucial to realize how significant cellular structure is for all types of biological processes (e.g. gene expression). In this regard, researchers are using AI approaches to predict the sub-cellular localization of different molecules including RNA, Protein along with their impact on various biological functionalities inside living organisms [8, 1180]

1.6 Genome Expression

In the previous sections, we have examined the expression of specific genes. However, genes do not function in isolation. Just as modules in a software program that interact with other components and influence the behavior of other components. Genes also operate as full-fledged networks, where the activity of a particular gene can often influence the activity of another gene. Here, we go beyond the expression of single genes to investigate how whole communities and networks of genes, which are actually called genomes, are primarily expressed.

1.6.1 The Genome Output Depends on Interplay of Genes

The physical characteristics of an organism are determined by the expression of key information present within the genome. The visible characteristics and traits

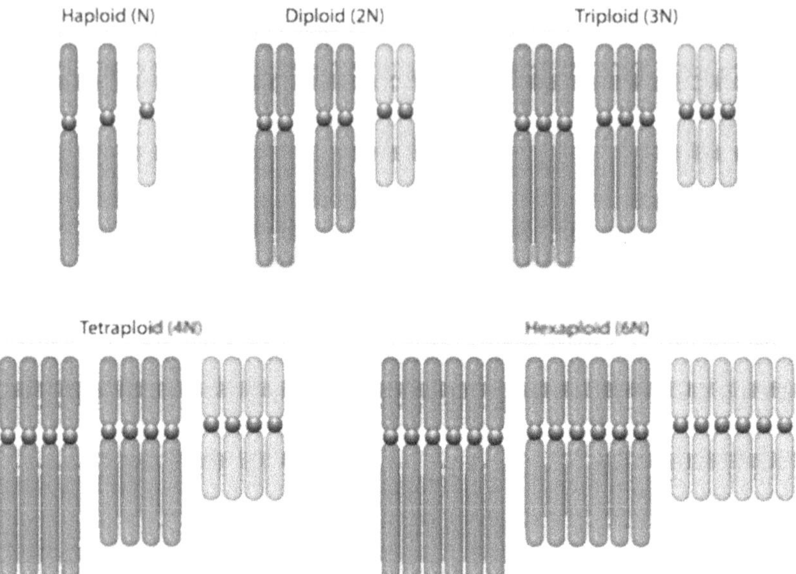

Fig. 1.16 Organisms have one or more copies of genes, haploid has only one copy, diploid has two copies, and so on

of an organism are known as phenotype, whereas the collective genetic sequence (DNA) that decides the phenotype is called the genotype of an organism. Phenotype is mainly determined by which set of genes are expressed, at what point in time, at what levels, and at what place in the cell. Hence, every living organism is the result of a well-coordinated process governed by a large number of genes, working, interacting, and influencing together.

Furthermore, Fig. 1.16 shows that organisms can have one or more than one copy of discrete genes. For instance, yeast may survive forever like a haploid organism that has a solo replica of its genome and express only one version of the genes. On the other hand, highly complex multi-cellular living organisms often possess two replicas of their genome, and hence called diploid. In Diploid, a mixed collection of functional molecules are produced through the expression of two separate replicas of the gene. Few living organisms like frog species namely Xenopus laevis oftenly used in biochemical test, have four replicas of every gene and hence called as tetraploids. Similarly, living organisms which have multiple copies of the genes are called polyploids.

The important question which must be answered is what is the significance of having multiple copies of every gene, called ploidy? The simple answer of this question is that the more the copies the better, Why so? This is because in polyploids, the risk of reducing the functional molecules due to the defects and flaws in the genes is almost negligible due to the availability of multiple copies. More specifically, if one version of a particular gene is defective in a manner that a fully functional product of the gene is not generated when the gene is completely expressed (or functional molecule is produced in little amounts), a diploid living organism has the benefit of a second replica of the gene to fully compensate for the faulty one. Whereas, a haploid living organism lacks the advantage of a 'back-up' gene, placing itself at a higher risk of having phenotypic problems if its single copy of a particular gene is altered. Researchers are employing AI approaches to analyze the expression of multiple genes at various locations along with their phenotypic impacts [767].

1.6.2 Genetic Analysis Is the Key to Decode the Organic Function of Every Gene

We have seen in earlier sections how different genes are expressed to produce distinct RNA as well as protein molecules, and how the coordinated action of all the genes within a genome regulates the growth, function, and survival of a living organism. However, how do we establish what single genes in reality do? How could we say with certainty that particular gene (m) is the driving force behind the process (n)?

Most of our understanding and findings of biology originates from genetic approaches, which investigate the organic function of single genes by comparing the wild-type meaning normal properties and common forms with non-wild type properties and forms, meaning the mutant ones. In traditional genetic approaches, researchers and scientists observed the living organisms under the umbrella of diverse environmental conditions after inflicting DNA damage, seeking for apparent phenotypic changes. When such kind of mutant (variant) living organisms were discovered, the particular mutations in the genome that defined the difference among mutant and wild-type organisms were investigated to facilitate key information on the potential function of the gene. To simply put, they mainly investigated the effects of various genes being turned inactive or changed.

For instance, consider a plant specie known as Hibiscus Syriacus that normally produces white, pink, blue, and purple coloured flowers. However, a defect in a specific gene (x) generated a mutant where half flower is in blue-violet and other half of the flower is white. One can easily deduce from the premises that the functionality of that specific gene (x) was to decide the color of the flower in some way. Hibiscus Syriacus flowers with normal blue-violet color and mutant blue plus white color are shown in Fig. 1.17. In research, this general strategy of finding genetic basis of phenotypic changes is also called forward genetics and these two different versions of the gene are also referred as wild-type and mutant allele. With a better understanding of genomes and their whole composition, as well as powerful gene manipulation technologies, it is now possible to specifically disrupt a desired gene in some organisms and analyse its phenotypic effects. This sophisticated approach, known as reverse genetics, has been extensively utilized to decipher the gene function.

Fig. 1.17 Color mutation in hibiscus syriacus flower: half flower of the plant is blue-violet and other half is white

It is important to mention that it may seem like that forward and reverse genetics are same things, however, forward genetics mainly involves observing a phenotype and identifying its underlying genetic cause. Whereas reverse genetics is all about disrupting a gene and observing the resulting phenotypic changes to understand the gene's function. In this regard, AI approaches are not only being used to predict the organic function of genes, distinguish normal gene from faulty gene. But they are also being used in genetic engineering where the ultimate aim is to manipulate the genes to introduce novel traits or modify the existing traits. Researchers are also using AI approaches in gene replacement therapy which exchanges a faulty gene with a healthy gene [215, 927] and extend the lifespan of living organisms.

1.6.3 A Proximal Relation Between Genotype and Phenotype

In the research of forward genetics, the phenotype of specific mutant organism is generally tracked by researchers and scientists with an aim to find the particular changes inside DNA (mutation) which resulted in phenotype being observed. This kind of study mainly reveals the sensitivity of specific tiny genetic changes which eventually result the changes in physical properties of living organisms. There are

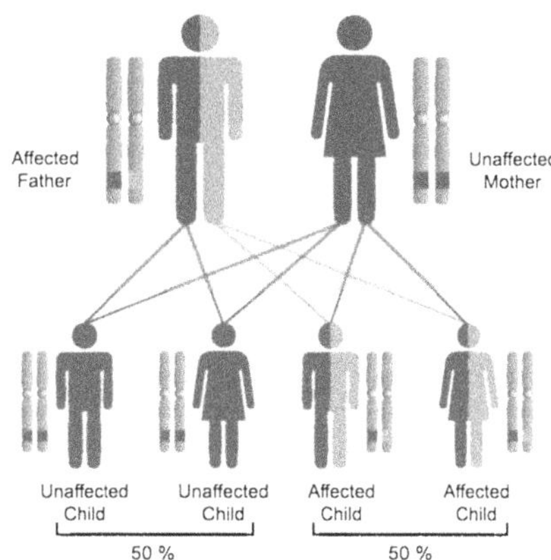

Fig. 1.18 A comparison of dominant and recessive inheritance in humans

different kinds of mutations which is why scientists have classified them in different classes. Mutations which cause the differences from typical wild-type organism are classified as recessive or dominant. To better understand these terms, one shall remember that living organisms do not all the time have single copy of every gene. Consider a diploid organism, if phenotype of the product of mutant gene is fully masked mainly by the appearance of phenotype of wild-type gene product, then this mutations is called recessive mutation.

A simple example of recessive mutation is alteration in enzyme which decreases its activity and overall impact, however it is compensated by the wild-type version which facilitates sufficient activity. On the other hand, if the product of a mutant gene has pretty obvious and non-conceivable phenotypic even in the presence of wild-type gene product then this kind of mutation is called dominant mutation. A simple example of dominant mutation is that an alteration in a primary structure of protein eventually disrupts the higher order structure of protein regardless of whether wild-type protein is available or not available. Figure 1.18 demonstrates dominant and recessive mutations impact on human inheritance. It is easy to understand that in case of recessive inheritance, the chances to transfer the condition or disease in offspring very low as compared to dominant inheritance where the chances of transferring a condition of disease to offspring is high.

It is important to understand what exactly is physical nature of these mutations which can trigger such phenotypic changes in living organisms. Figure 1.19 demonstrates how mutations can just be the result of a tiny alteration in gene such as change of a basic nucleotide, also called point mutations, re-arrangement of small, medium or large size regions of chromosome, duplication of certain region of DNA, insertion of novel segment of DNA, or deletion of small, medium or large segments of DNA. These type of modifications have the power to directly alter the product encoded by the underlay gene or disrupt the overall expression of underlay gene without changing the actual identity of the corresponding gene product.

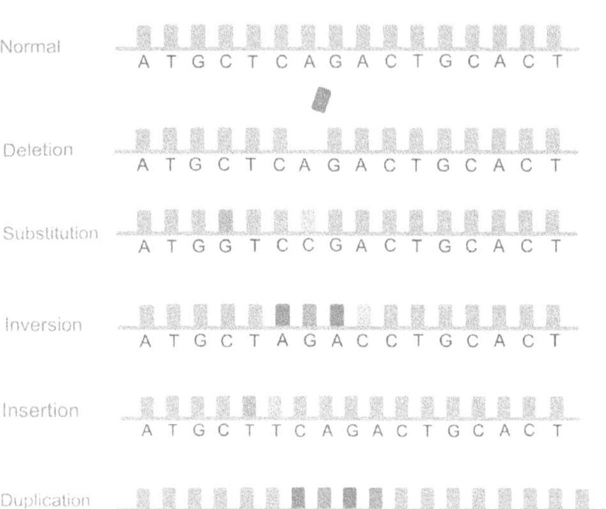

Fig. 1.19 Different kinds of DNA mutations

Mutations which remove the functionality of a gene are referred as null mutations or loss of function mutations. These kinds of mutations are usually associated either with an alteration in product that fully remove the function or a very large disruption within genetic sequence that remove the expression of underlay gene. Single nucleotide alterations within genetic sequences can change the identity of the product encoded by the underlay gene, either at RNA level or protein level, as demonstrated in Fig. 1.20. It is evident in Fig. 1.20 that an alteration of one nucleotide in protein coding gene is good enough to alter the corresponding codon and ultimately respective amino acid. Such alterations in identify of amino acids are called missence mutations. If the substitution of amino acid eventually result in the elimination of gene function then missence mutation can also be called null mutation. On the other hand, other substitutions can result in generation of partially functional genetic products.

Figure 1.20 also illustrate how in few cases, there can be an alteration in genetic sequence which introduces the pre-mature stop signal and as a result merely truncated species of the protein products are produced. These types of alterations are called nonsense mutations. Also, few times, there can be alteration in genetic sequence like an alteration in codon encoding a specific amino acid to other encoding the exact same type of amino acid that obviously do not produce phenotypic changes. Such kind of changes are usually called silent mutations. Beside DNA level, such mutations also occur on RNA and protein level too which are also explained in Fig. 1.20. In this regard, AI approaches are being employed to infer the functional impact of genetic changes at various levels of gene expression,

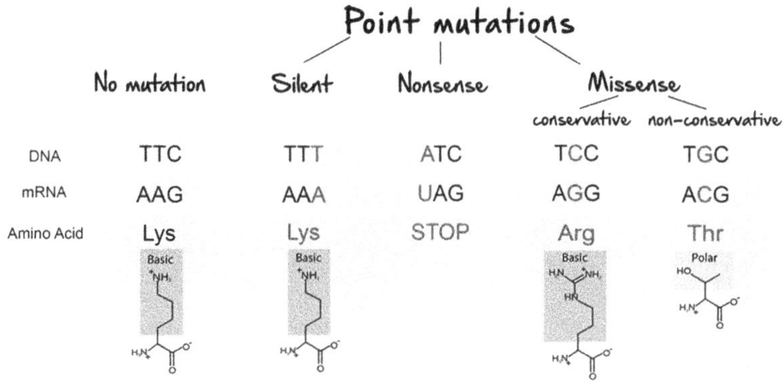

Fig. 1.20 Different kinds of point mutations

distinguish healthy genes from disease causing genes, predict phenotypic effects of different genetic variants, identify the potential drug targets based on unique genetic variants which are linked with specific phenotypes or diseases [499]. Also, AI approaches are being used to develop accurate personalized treatment plans by taking individual genomic as well as phenotypic characteristics [499] into account.

1.6.4 Epigenetic Modifications Can Lead to Phenotypic Changes

We have already seen how very large genomic DNA is actually packaged into a very compact kind of structure with the help of chromosomes and histone proteins. We have also discussed the presence of specialized mechanisms which alter the opening of compacted DNA. These unique modifications in the chromosome packaging usually do not alter the DNA sequence, however still can be transferred from generation to generation. Few times, these modifications do bring phenotypic changes, for instance, expression of particular gene can be activated (turned on) or de-activated (turned off or silenced) through increasing the tight compaction of DNA, eventually leading to phenotypic alteration in a living organism without altering the DNA sequence.

These heritable modifications in chromosome packaging are termed as epigenetic modifications, literally meaning changes which are in addition to genetics. In this regard, AI approaches are being employed to identify various kind of epigenetic modifications such as DNA methylation, histone modifications like acetylation, methylation, and phosphorylation, and chromatin structure analysis. Researchers are also applying AI approaches to analyze the expression, function, and sub-cellular localization of non-coding RNAs, their role in gene regulation, disease development, bio-marker identification, and many other important biological processes [453].

1.6.5 Organism Genetic Mutations Can Be Traced to Find the Roots of Diseases

In the complex multi-cellular organisms, there exist two general kinds of cell, somatic cells and germline cells. Somatic cells primarily make up the fundamental building blocks of organisms bodies and transfer their genetic information to the daughter cells present within same organism. Whereas germline cells play key roles in production of offspring and transferring the genetic information to next generations. After understanding these two kinds of cells, it is easy to comprehend that gene mutations happen in germline cells significantly impact the offspring of organism, and gene mutations happen in somatic cells significantly impact the same organism. A highly concerning drawback of gene mutation is mainly the development of a disease which disrupts the organic biological function in one or other way.

Few disease are the result of alterations in an individual gene, and therefore are called monogenic. A renowned monogenic disease example is Cystic fibrosis cause by the mutation in specific CFTR gene which encodes a protein that controls the transfer of ions across the cell membrane. This mutation leads to the generation of defective protein which causes the generation of very thick and sticky mucus inside the pancreas, lungs, and some other organs, causing the people digestive and respiratory problems like difficulty in breathing. In contrast, some diseases are the result of alterations in multiple genes and hence called polygenic or multi-factorial diseases. Almost all of the prevalent diseases including heart disease, diabetes, Alzheimer's disease are classified as polygenic diseases.

However, it is important to mention that presence of certain mutations do not always cause a certain disease, like alterations in genes namely BRCA1, BRCA1 are linked with breast cancer, however, not all people who inherit these genes develop cancer. In this regard, penetrance of the mutation is defined, that is just the percentage of people who carry a certain mutant genotype which exhibit particular mutant phenotype. Hence, the threat of developing a certain mutant phenotype while carrying the particular mutant genotype is estimated by genetic makeup of living organism as well as environmental factors including diet plans, medications, exposure to diverse environmental conditions, etc.

The extent up to which genotype (nature) as well as environment (nurture) impact phenotype of organism is not a new topic of debate at all as it has been discussed for years and still remain a controversial subject. In case of BRCA1, BRCA2 mutations, it is pretty evident that despite carrying the mutations, many people did not develop cancer. This indicates that there are other genetic factors too which influence whether cancer or any disease will be developed in people or not, and environmental factors like diet as well smoking habits can significantly impact the penetrance of such mutations. A deep understanding of genetic and non-genetic (epigenetics) basis of complicated diseases is essential to provide key information to various stakeholders especially clinicians. In this regard, researchers are applying AI approaches to predict functional consequences of different gene mutations, early detection of diseases, discover and analyze the impact of various chemical compounds on disease specific targets [95, 702]. Furthermore, they are using AI to guide the process of designing and analyzing clinical trials by identifying the patient subgroups which are most likely to respond to certain treatment [95, 702].

1.7 Obtaining a Complete Genome Raw Sequence

The very first genome which was entirely sequenced in 1995 belonged to Haemophilus influenzae. Since then, hundreds and thousands of raw sequences of various genomes belonging to different organisms have been published. The large availability of whole genome raw sequences has greatly revolutionized the field of molecular biology. Because, it has allowed the scientists and clinicians to study various organisms in unprecedented detail. However, an important question is how scientists obtain and use organism's genome raw sequences? In order to obtain and use raw sequences of the genome of a specific organism, two approaches are needed: DNA sequencing and computational analysis.

DNA sequencing is a process that determines the order of core building blocks like nucleotides of complete DNA. Computational analyses are required to make sense of the data and uncover valuable aspects of the genome by searching, recognizing, and extracting key patterns and relationships using different algorithms. Both of the approaches have significantly evolved over the years. The continuous improvements in DNA sequencing technologies over the years with technological breakthroughs has allowed for more efficient and quicker sequencing of genomes (e.g. high-throughput DNA sequencing technologies). Likewise, advancements in computational analysis methods has increased the pace of acquiring useful insights from humongous raw sequences. The prime focus of this section is not to dwell into various DNA sequencing technologies and computational analysis techniques. But to only discuss as simply as possible, the common underlying strategy used by advanced DNA sequencing and computational analysis technologies.

1.7.1 A Unified Whole Genome Sequencing Strategy: Analyzing Several Overlapping DNA Fragments

The process of sequencing an entire genome involves analyzing many small overlapping fragments of DNA. This is because we currently lack the technology to sequence an entire chromosome from one end to the other. Genome sequencing methods are quickly evolving, however, they are particularly unified by the common underlying strategy called "Shotgun". To understand this strategy intuitively first, consider a computer program called "shredder" that takes a large text file such as a novel in which text is completely out of order and jumbled up.

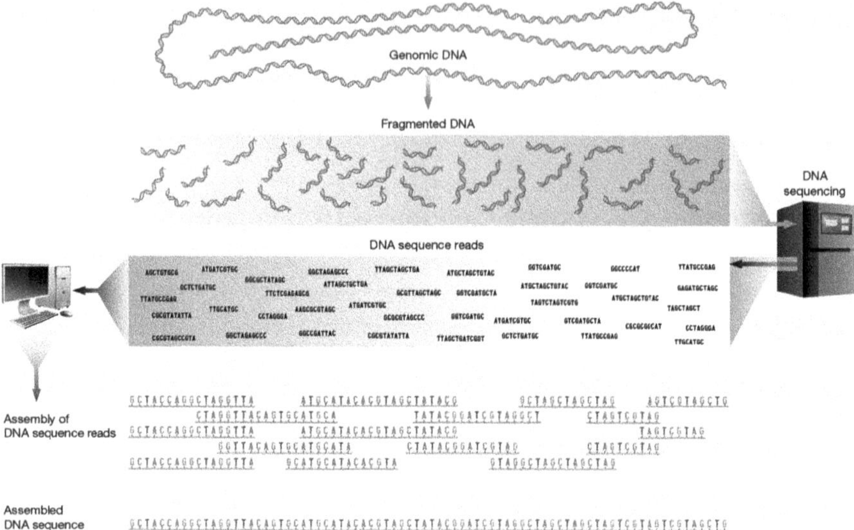

Fig. 1.21 Assembly of a genome sequence from the sequences of Multiple overlapping fragments. (Adapted from National Human Genome Research Institute, 2023 [293])

It will first segregates the text file into several tiny fragments and then analyzes every fragment for keywords or patterns to understand its content. Once the shredder has successfully analyzed all small fragments, it can reconstruct the original file in right order. Like shredder, the general approach of complete genome sequencing is to get a large collection of tiny overlapping DNA fragments which together represent the whole genome. Scientists are using various laboratory techniques to generate these random fragments such as Polymerase Chain Reaction (PCR), however, the important part is that, these random fragments do not need to be in order at the start and it is not essential to identify where each fragment is coming from the genome. By estimating the base sequence of every fragment as well as accurately comparing the sequences of several overlapping regions, whole sequence of particular genome can be assembled in right order with the help of a computer as explained in Fig. 1.21.

1.7.2 Genome Sequencing Requires Over-Sequencing

To acquire a whole genome sequence, it is essential to sequence several more base pairs than just the overall length of the genome due to two reasons. First of all, because DNA fragments are totally sequenced randomly, several distinct overlapping fragments should be sequenced in order to make sure that all areas of particular genome have been considered at least once. Theoretically, in order to make the coverage around 99.99%, clones correlating to eight genome equivalents shall be sequenced. For instance, to get a fairly complete sequence of E. coli chromosome, having the size of 4.6 Mbp using the shotgun sequencing strategy, it is essential to sequence around 37 Mbp of DNA, equivalent to eight genome. However, practically, even greater coverage is generally sought, because it comes with the additional benefit of boosting the correctness of the final sequence. This is primarily due to the idea that any kind of error in the sequence of a solo fragment can be identified and rectified when several independent clones with respect to same area of genome are sequenced.

Another reason that a full genome sequencing necessitates over-sequencing is that a substantial quantity of random overlapping DNA segments are required to establish the precise order wherein overlapping sequences should be aligned in order to build the whole genome. The presence of repeated DNA in many genomic regions complicates the hand on task since they make it hard to uniquely align fragments having similar repetitive sequences. The hand on task becomes almost infeasible if the areas of the genome having several repeats are longer than the sequenced fragments produced using shotgun strategy. Computer science technology will not be able to build this segment of genome from the overlapping sequences as the repeat would have homology in several different areas. This underlay limitation of shotgun technique is the core reason why earliest genome sequences of different organisms such as humans had missing regions. But, increasingly advanced computer assembly methods and the addition of genetic linkage data are making it easier to assemble regions having repeated DNA sequences. Furthermore, improved sequencing methods allow for the sequencing of larger stretches, and attach unique sequences with repeated sections in order to effectively distinguish them from one another.

1.8 Commonalities and Differences of Living Organisms

Life is incredibly diverse and can be found in all sorts of environments, from a tiny bacteria breathing in the super-heated environment present in the depths of the ocean to a huge polar bear wandering around the very cold arctic circle. Some living organisms are very small and made up of just a single cell, while others, like whales, have trillions of cells (more than 10^{15}). Although certain conditions, like very high acidity or saltiness, can be deadly for living organisms, life has found a way to thrive in many different types of environments. Aside from cellular differences, there is a huge variety of ways in which living organisms gather nutrients, protect themselves, move, reproduce, and care for their offspring. Even at the molecular level, there is a lot of diversity among living organisms. For example, proteins with similar structure and function may have very diverse chemical make-ups, just as creatures with similar appearances may have very different genetic make-ups. The genome size varies from 5000 elements in quite simple organisms such as viruses to all the way up to 10^{11} elements in plants, whereas humans genomes are comprised of almost 3×10^{11} elements.

Despite all this diversity, most living organisms have the same living parts like cells, membrane, and mechanisms at work inside their cells, which are essentially tiny bags of chemicals that carry out specific sequences of reactions called core metabolic pathways. These reactions involve about a thousand different substances that are quite similar across all living organisms. Furthermore, the genetic information of every living organism, ranging from humans to viruses, is written in the universal language of DNA or RNA sequences, and the ways most multicellular organisms grow, develop, and mature over time are also similar. With the advancement of sequencing and biochemical technologies, the genetic sequences of any organism's genome can now be easily and readily determined, making it feasible to catalogue, characterise, and compare multiple living organisms by sequence analysis. From such comparisons and analyses, it has become easy for researchers to mention the place of every organism within the family of living organisms called the "tree of life". This remarkable diversity of life forms and their underlying commonalities has led scientists to develop systematic ways of classifying and organizing living organisms. Through careful study of both visible characteristics and molecular evidence, researchers have established a comprehensive classification system that groups all known life forms into three main kingdoms. This fundamental organization of life, known as the tree of life, provides a framework for understanding how different organisms are related to each other and how they may have evolved over time (Fig. 1.22).

Tree of Life: Three Main Kingdoms of Living Organisms, Eukaryotes, Bacteria, and Archaea

There are a vast number of different types of living organisms in the world, with estimates ranging from 5 million to 50 million species. There exist more than 300,000 beetles with unique characteristics, and roughly more than 53,000 classes of the tropical-trees. Most familiar types of animals and plants cover a very tiny percentage of such huge number of species, roughly around 20%, where vertebrates, or animals having backbones, birds, mammals, amphibians, and fish reptiles species constitute just 3% of all species in the planet.

The traditional method of classifying living organisms is by comparing their outward appearances. For instance, we can observe that a fish has jaws, backbone, brain, eyes, etc. like human have but worms do not have. Likewise, we can observe that a rosebush is very much similar to the apple tree as compared to the grass. However, classification based on physiology becomes difficult when comparing organisms with great differences, such as how one can determine whether a particular fungus is more related to an animal or a plant. Microscopic organisms like bacteria are even more harder to classify, as they can look very similar to each other. Furthermore, most of our understanding of the microbial world has traditionally been limited to species that can be separated and cultured in the experimental laboratory. However, direct DNA sequencing of microbial populations in their natural environments—such as soil, sea water, or even human mouth, has shown us that the great majority of bacteria cannot be simply cultivated in the laboratory. They frequently thrive in the wild as core components of complex ecosystems and

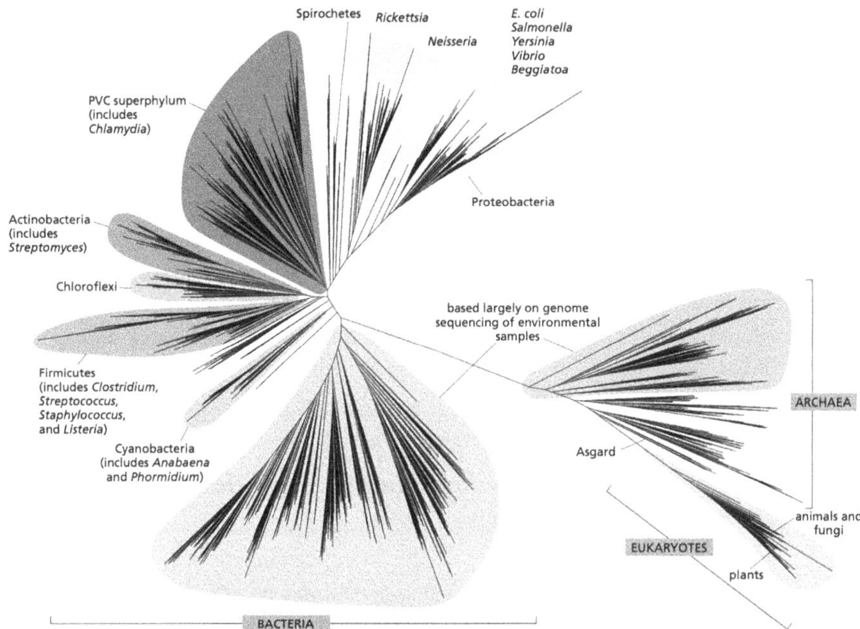

Fig. 1.22 Major kingdoms of the living organisms. The colored branches lengths are directly proportional to the differences between genomes found by making use of common genes which can be identified as well as compared across a plethora of species. Only few organisms are indicated. From the three kingdoms of life (eukaryotes, archaea, and bacteria), Bacteria have the greatest diversity, which corresponds to their ability to occupy practically every ecological niche on the earth. Currently, several novel bacterial species are being identified using DNA sequencing and naming all of them is a great challenge. Although eukaryotes especially humans are the prime focus of this book, they represent only a small portion of the global diversity. (Adapted from C.J. Castelle and J.F. Banfield, Cell 172:1181–1197, 2018.)

therefore can live when separated from their naturalistic environments. Prior to the development of modern DNA sequencing, we knew little about these species, especially those that live in very extreme environments like very deep Earth's crust or water thousand of miles below ocean surface. Genome analysis has provided us with a straightforward, effective, and simple method for determining evolutionary relationships. An organism's whole DNA sequence describes its nature with nearly perfect precision in great detail. Genome analysis has provided us with a straightforward, effective, and simple method for determining evolutionary relationships. An organism's whole DNA sequence describes its nature with nearly perfect precision in great detail.

In addition, DNA specification in form of string of letters also exist in a digital form which can be inserted into the computer, analysed and compared with corresponding information of other living organisms. Because DNA is susceptible to random mutations that accumulate over extended periods of time, the amount of differences between the DNA sequences of two living organisms can facilitate an objective and quantitative measure of evolutionary distance among them. For generating an accurate tree of life, it is essential that you begin with a piece of DNA that is readily identifiable in the genomes of all living organisms. To perform various biological functions, cells make use of same fundamental technique to translate a DNA sequence into RNA sequence and an RNA sequence into protein sequence through a decoding machine called ribosome. Ribosomes are basically similar in all creatures, and the well-preserved component of them is mainly RNA molecules that constitute their core. Although the exact ribosomal RNAs (rRNAs) sequence varies between organisms, they are sufficiently similar to be used as a measuring stick to determine up to what extent two species are related. Using this approach, scientists have discovered that the living world consists of three major domains: eukaryotes, bacteria, and archaea, details of which are given in following subsections.

1.8.1 Eukaryotes: Most Familiar Kingdom of Life

Eukaryotes make up the vast majority of the living organisms we observe in our environment. The name is derived from the Greek and means "really nucleated" (from the words eu, "well" or "truly," and karyon, "kernel" or "nucleus"), reflecting the fact that the DNA of these organisms is encapsulated in a membrane-bound organelle called the nucleus.

This characteristic, which can be seen under ordinary light microscopy, was utilized in the early twentieth century to categorize living organisms as either eukaryotes (those containing a nucleus) or prokaryotes (those without a nucleus). We are aware now that prokaryotes are two of the three major kingdoms of life, along with bacteria and archaea. Eukaryotic cells are often much larger compared to those of bacteria and archaea, and in addition to a nucleus, they specially contain a number of membrane-bound organelles that are absent in prokaryotes. Genomes of eukaryotes are typically substantially larger than those of bacteria or archaea, comprising of more than 20,000 genes for corals and humans, as opposed to 4000–6000 genes for the conventional bacterium or archaea. In addition to animals and plants, eukaryotes also include fungi (like mushrooms and yeasts used to make bread and beer), as well as a large variety of typical single-celled, microscopic life forms.

1.8.2 Bacteria: Most Diverse Group of Living Organisms on the Basis of Genome Analysis

When scientists constructed sophisticated trees of life using genome information, they found that bacteria are much more evolutionarily diverse than eukaryotes. This diversity reflects the fact that bacteria appeared much earlier in the planet's evolutionary history as compared to eukaryotes. Bacteria are invisible to unaided eyes, spherical or rod-shaped, and very small, having the size of a few micrometers in linear dimension. Unlike most multi-cellular living organisms, they typically live as independent entities or in loosely connected communities. They have a thick protective coat known as cell wall, under which there is a plasma membrane enclosing one cytoplasmic compartment (cytoplasm) containing DNA, RNA, proteins, as well as many tiny bio-molecules necessary for the life.

Bacteria live in a broad spectrum of environment, and they have diverse biochemical capabilities, including the ability to utilize virtually any kind of organic molecule as food or exploit light energy in different ways for various purposes, just like plants use it for photosynthesis and release oxygen as an important by-product. Bacteria can have an impact on human health, causing diseases like the bubonic plague and tuberculosis. Still, thousands of different bacterial species reside in our gut as well as on our skin, often proving beneficial. The study of bacteria has helped the scientists to comprehend basic biological processes, including the way gene are expressed or suppressed in organisms to regulate different biological functions. Furthermore, bacteria are being used to generate biofuels, human pharmaceuticals, and various other high worth chemical products.

1.8.3 Archaea: Most Mysterious Kingdom of Life

Archaea is the least understood domain of the three kingdoms of life. Most archaea have been recognized solely by their DNA, and just a few have been thoroughly investigated in the laboratory. They are tiny, like bacteria, and lack the internal, membrane-bound organelles found in eukaryotes. Yet, they differ from bacteria in a variety of areas, such as the chemistry of their cell walls, the types of lipids in their membranes, and various biochemical reactions they can conduct. Another amazing finding came from comprehensive genome comparisons is that: despite archaea resemble bacteria in terms of outward appearances, their genomes are more strongly related to eukaryotes as compared to bacteria. Archaea were originally considered to only exist in severe conditions like salt lakes, volcanoes, cattle stomachs, and acid hot springs, but they are now found in soils, saltwater, and on human skin. Distinct species of archaea have very diverse chemistries and they mainly play important roles in the recycling of carbon and nitrogen in ecosystem.

After briefly discussing diversity of multicellular organisms, in the following section, we will delve deeper into the functioning of these complex organisms and provide a more detailed examination of the components of eucaryotic cells.

1.9 Molecular Biology Progress Is Based on the Study of Model Organisms: A Look Back into the Future

The phrase "Molecular Biology" was primarily coined to put great emphasis on molecules that perform a plethora of biological processes. In order to comprehend how something happens within the cell, one shall identify all the molecules that contribute to the biological processes and determine their ultimate functions. A very powerful and efficient approach for determining all components needed in biological processes along with macromolecule functionalities, is mainly to replicate the entire process with purified components in-vitro (literally meaning within a glass or artificial vessel). An mRNA copy of a certain gene can be produced within a test tube by mixing DNA with RNA polymerase, magnesium, nucleoside triphosphate, and some other buffers. As a result of the mixture of test tubes fulfilling the minimum requirements of RNA synthesis inside the cell, RNA synthesis will occur without the presence or intervention of any other cellular components. However, it is important to mention that in-vitro experiments have several limitations such as when a specific biological process is modelled in a test tube, many proteins which regulate the process might still be missing or a significant post-transnational modification might be missing in the synthetic protein used for the experiment.

The explosion of nucleic acids, proteins, and key molecules information in every organism, whole genome sequencing technologies, the use of advanced computational methods and technologies, and improvements in microscopic techniques have made it feasible to perform new types of experiments with far more effectiveness and efficiency. For instance, the in-vivo (literally meaning within the living organism) experiment in which analysis of biological processes and macromolecules is performed within living organisms. This section is not meant to compare in vitro and in vivo experiments, but to provide you with sufficient knowledge about which organisms from different species have served as models for studying and exploring biology in all kinds of cells? What is the rationale behind the selection of a handful of organisms from millions of species? What properties are shared among these organisms? And what are some of the most common organisms from different kingdoms of life that have been extensively studied?

The number of species on Earth is very difficult to estimate because of several factors such as inaccessible habitats, location, tiny size, and disagreements among biologists about what constitutes a species. It is estimated that there are 5.3 million to 1 trillion species on Earth, of which 8.7 million are eukaryotic species. However, the thing that can be said without a doubt is that so far scientists have been able to identify around 1.2 million species. It is not possible to deeply study these millions of species and scientists extensively use only a few dozen species, called model organisms. Model organisms are not studied because of economical significance, or their disease rate. Instead, they are very easy to propagate, and they can facilitate generic insights into key biological processes, molecular processes, and macromolecules. They facilitate insights that can be applied to other living organisms and diseases considering several molecules and their processes are conserved in the flow of evolution. For instance, Arabidopsis, a flowering plant is not horticulturally significant, and Caenorhabditis elegans is not a medically or agriculturally significant roundworm. However, both organisms are studied and used as laboratory models because they provide scientists with insights into fundamental biological processes. Similarly, most forms of Escherichia coli species are not classified as human pathogens, however, they are still very similar to bacteria in so many ways, hence they are also extensively studied. These carefully selected model organisms have proven invaluable to biological research not by chance, but because they possess specific characteristics that make them ideal for laboratory studies. Understanding these shared properties helps explain why certain species, from the simple E. coli to more complex organisms like Arabidopsis, have become the cornerstones of molecular biology research. By examining the common traits among these diverse model organisms, we can better appreciate their collective contribution to our understanding of fundamental biological processes.

Model Organisms Share Properties and Offer Different Advantages

The most frequently used model organisms have a number of biological traits in common that make them the gold standard for research. A model organism must have relatively straightforward and, ideally, well-defined nutritional needs as well as a well-understood life cycle that enables it to be grown in large quantities in the lab without risk. Short generation time, compact size, and the capacity to store or preserve organism stocks for several years are additional advantages. Almost all significant model organisms have undergone in-depth genetic analysis research. This indicates that mutations that decrease or change gene functionality are simple to introduce, propagate, recognise, and characterise. Since the advancement of recombinant DNA technology, it has also become necessary to introduce modified genes into various organisms to study the effects of any type of mutation. Most model organisms' genomes have been sequenced, and there are large communities of researchers examining these organisms who have also developed huge databases having the most recent data on gene expression, biomolecule interactions, biomolecules roles in various diseases, etc. The availability of such humongous biological data has increased the development of data-driven computational approaches which have significantly accelerated the study of molecular biology, rate of discovery, and transformed the ways to identify diseases in early times, their progression, and most effective treatments.

Model organisms originate from different kingdoms and phylogenetic trees and represent various biological and evolutionary properties, as shown in Fig. 1.23. Single-celled eukaryotes, bacteria, small animals, flowering plants, mammals, and lower vertebrates are all well-studied model organisms. The key question is how do scientists select an experimental model to investigate a particular biological process? The answer depends on the question asked and the experimental instruments and techniques readily available at the time. For instance, almost all of scientists' knowledge regarding DNA transcription, translation or even DNA replication originated from the studies and exploration of bacteria. To some extent, this is due to the fact that bacteria were some of the first model organisms available. This is because it is relatively easy to replicate a very large amount of bacterial cells as well as to mutate them. We now understand that several features of the processes characterised specifically in bacteria are pretty much universal to all organisms, validating and justifying the use of bacteria as a model organism.

Eukaryotic cells shall be used to study phenomena that are specific to eukaryotes. The internal workings of a cell's nucleus can be studied easily in single-celled

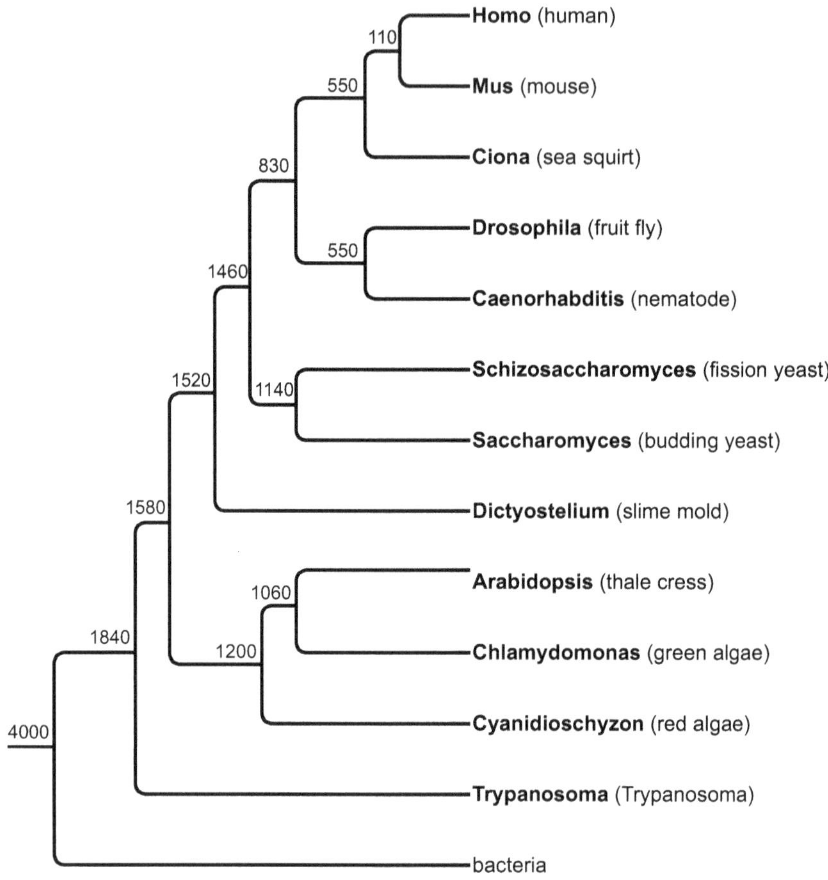

Fig. 1.23 A phylogenetic tree of common model organisms. The time of divergence among model organisms in terms of millions of years is indicated at the node of phylogenetic tree

eukaryotes like yeast, whereas the processes through which a multicellular organism fully develops from a tiny fertilised egg shall be studied in a multi-cellular eukaryote like a nematode or fruit fly. Similarly, phenomena specific to mammals or vertebrates shall be investigated in a most appropriate representative of that very class of the organism. The selection of appropriate model organisms is crucial to make quick experimental and industrial progress. It is important to mention that the model organisms we have illustrated and discussed in the text only represent a small subset of the most informative and important model organisms. In molecular biology, there is a dedicated field called phylogenetics which investigates the evolutionary relationships between different species. It involves the utilization of molecular data, and other biological data to recreate the evolutionary history of life on Earth. AI approaches are increasingly being employed in phylogenetics to enhance precision, and scalability of phylogenetic tree creation and analysis. These approaches are being used to automate tree construction, incorporate new data sources, and forecast evolutionary events. AI is assisting scientists to gain a better understanding of the evolutionary links between various species and subgroups of organisms [66, 955].

1.10 Potential and Benefits of Computational Biology over Experimental Methods

To study the structure, function, and interactions of biomolecules such as proteins, nucleic acids, and lipids, scientists have created a number of experimental methodologies and instruments [135, 1100]. X-ray crystallography, Nuclear Magnetic Resonance (NMR) spectroscopy, cryo-electron microscopy (cryo-EM), biochemistry methods such as enzyme assays, protein purification, and protein-protein interaction assays, and molecular cloning techniques such as polymerase chain reaction (PCR) and site-directed mutagenesis are among the most widely used techniques. The 3D structure of biomolecules is determined via X-ray crystallography by studying the diffraction patterns of X-rays flowing through a crystal of the biomolecule. The magnetic characteristics of certain nuclei are used in NMR spectroscopy to identify the 3D structure of a biomolecule. An electron microscope is used in cryo-EM to scan a biomolecule that has been flash-frozen in a thin coating of vitreous ice [135, 1100]. Biochemistry methods allow scientists to study chemical reactions and interactions at the molecular level; enzyme assays can be used to measure the activity of a specific enzyme, protein purification methods can isolate and purify specific proteins for further study; and protein-protein interaction assays can be used to identify and study interactions between different proteins.

Scientists may modify and analyze individual genes and proteins using molecular cloning techniques. These experimental procedures, however, have several limitations. Large quantities of pure protein are required for X-ray crystallography and NMR spectroscopy, which may be limited to tiny proteins or specialized circumstances. Cryo-EM has a lesser resolution than X-ray crystallography and NMR spectroscopy, and it also necessitates the use of pure proteins. Biochemistry procedures demand extremely specialized circumstances to get accurate findings, and studying individual proteins might be challenging. The modification of genes and proteins is required for molecular cloning procedures. In recent years, the rising use of artificial intelligence-based computational approaches such as deep learning has been presented as a strong alternative to experimental methods in molecular biology. These computational tools can evaluate enormous volumes of data, make predictions, and identify patterns that experimental methods cannot. Protein architectures, protein-protein interactions, drug-target interactions, and prospective drug candidates may all be predicted using these methods. They may also be used to evaluate high-throughput data and simulate complicated systems, and they can be applied to a wide range of issues that are difficult or impossible to answer experimentally [1155].

Chapter 2
DNA, RNA and Protein Structures

The central dogma of life hinges on three fundamental molecules i.e., DNA, RNA, and proteins. These molecules serve crucial functions in living organisms i.e., DNA contains the genetic information of an organism, RNA translates genetic information from DNA into proteins. In addition, proteins carry out a myriad of intercellular and intracellular functions. Within cells, they act as enzymes catalyzing biochemical reactions, receptors transmitting signals, and structural components providing support. Outside the cells, they play crucial roles in immune responses, cell signaling, and as transporters moving molecules across cellular membranes. The function of these molecules is controlled by both their sequence and structure. While the previous chapters have explained sequence details and their related prediction tasks, we focus solely on the details of DNA, RNA, and protein structures.

Each of these molecules encompasses diverse types of structures such as DNA molecule adopts diverse types of 3-dimensional structures/conformations including A-DNA, B-DNA, Z-DNA, H-DNA, G-Quadruplex, and I-Motif. Similarly, RNA molecule adopts tertiary (riboswitches, ribozymes) and different secondary structures including loops and stacking. Protein molecules also contain diverse types of structures including primary, secondary, tertiary, and quaternary.

These structures provide insights into fundamental processes of life by deciphering genetic codes and unraveling the complexities of cellular mechanisms. Particularly, they are useful for understanding disease mechanisms and designing targeted drug therapies. Additionally, they also enable the exploration of genetic variations and their potential implications to support advancements in personalized medicine. In the biotechnological field, DNA, RNA, and protein structure information is used to guide the engineering of novel biomolecules with tailored functions. Similarly, in the synthetic biology field, structural information is utilized to design bioengineered proteins and genetic circuits for various applications, including biopharmaceuticals and biofuel production. In essence, the ability to predict DNA, RNA, and protein structures from raw sequences is transformative

M. Nabeel Asim et al., *Artificial Intelligence for Molecular Biology*,
https://doi.org/10.1007/978-3-031-90450-9_2

and enables breakthroughs in diverse scientific disciplines and applications, from understanding the foundations of life to developing innovative solutions for medical and industrial challenges.

In essence, predicting the structure of RNA, DNA, and proteins is transformative across various fields. These capabilities can revolutionize personalized medicine by enabling treatments tailored to individual genetic profiles. Furthermore, they hold the potential to advance biotechnology significantly, from developing new therapeutics to creating innovative bioengineering solutions. Considering the importance of structure prediction, a diverse set of structure prediction tools have developed for DNA, RNA, and proteins. In terms of DNA structure prediction, a diverse set of AI-mediated applications have been developed until now. For instance, Mu et al. [730] proposed DL models to predict single and double-stranded DNA structures. Beknazarov et al. [82] proposed DL applications for the prediction of Z-DNA, and Cherednichenko et al. [181] explored the potential of generative adversarial networks (GANs) for the generation and prediction of non-B-DNA structures such as Z, G4, and H-DNA.

Approximately, more than 20 ML and DL approaches have been developed for the prediction of 2D RNA structures [494, 880]. These tools include NeuralFold [18], UFold [320], ExternalFold [1083], LinearFold [448], MXFold [17], MXFold2 [882] and so on. These tools vary in their capabilities. Some tools specialize in predicting a single RNA secondary structure i.e., Justyna et al. [494], benchmarked 15 different ML, DL, and shallow learning tools to predict canonical, non-canonical, and pseudoknot-associated base pairs. In contrast, more advanced tools can predict multiple potential structures, which enables them to capture inherent uncertainty in RNA folding and provide a broader perspective on possible conformations. For instance, Sato et al. [882] proposed MXFold2 that can predict RNA structures except for pseudoknots by utilizing the potential of nucleotides folding scores, convolutional neural networks (CNNs), long short-term memory network (LSTM), and skip connections.

Similar to DNA and RNA structure prediction tools, more than 15 ML and DL approaches have been developed for protein structure prediction. These tools include AlphaFold [493], RoseTTAFold [545], trRosetta [545], and DeepMind's AlphaFold2 [493]. These tools vary in their capabilities and methodologies. Some specialize in predicting specific structural features, such as secondary structures or contact maps, while others, like AlphaFold2, can predict highly accurate three-dimensional structures. For instance, AlphaFold2 has revolutionized the field by using advanced neural network architectures to achieve unprecedented accuracy in protein folding predictions. Similarly, RoseTTAFold combines different network types, including convolutional and recurrent neural networks, to improve its predictive performance. These advancements have greatly enhanced our ability to understand protein folding mechanisms and their functions in various biological processes.

2.1 DNA Structures

DNA structures are categorized based on the arrangement of nucleotide strands i.e., single, double, or more strands. Double-stranded DNA (dsDNA) forms a stable double helix with complementary base pairing, which is essential for genetic information storage and replication. Single-stranded DNA (ssDNA) is more flexible, forming various secondary structures critical for replication, transcription, and repair processes. The following subsections comprehensively discuss the details of single and double-stranded DNA structures.

2.1.1 Double Helix Structure

In 1953, a significant milestone in modern biology was reached when Watson and Crick [1082] unveiled the fundamental structure of DNA, known as the B form. In a double helix structure, two strands act as the backbone, and an internal ladder-like structure refers to the base pairs between the two strands. These base pairs exhibit significant pseudo-symmetry (Fig. 2.1), where the base pairs not only have similar dimensions but also identical external shapes relative to the backbone. This pseudo-symmetry is crucial for the uniform helical structure, as it ensures that the distance between $C_1^{'}$ atoms across pairs remains consistent for both AT and GC base pairs.

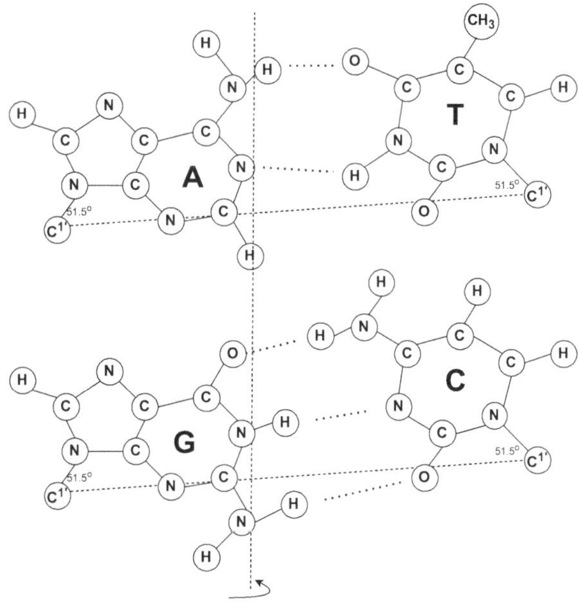

Fig. 2.1 DNA complementary pairs

However, DNA's ability to function efficiently within the cell extends beyond this singular form. The molecule's versatility is a result of its capacity to adopt various structural conformations in response to distinct environmental and physiological conditions. Factors such as hydration levels, ionic strength, and mechanical stress can induce conformational changes in the DNA helix. Under different conditions, DNA can transition from its most common structure to other forms that are more suited to specific cellular needs. For instance, changes in hydration can compact the DNA, while variations in ionic concentrations can induce twisting and untwisting of the helix. These structural variations enable DNA to protect genetic information, regulate gene expression, and facilitate interactions with other molecules. The uniformity in base-pair geometry, which creates an angle of 51.5° with the $C_1^{'}$-$C_1^{'}$ line for both AT and GC base pairs, allows the DNA to seamlessly transition between different forms without disrupting the helical backbone.

Understanding the dynamic nature of DNA structures sheds light on its biological significance. The ability of DNA to transition between different forms ensures that it can maintain genomic stability and function under various conditions. This structural plasticity is essential for processes such as replication, transcription, and repair, allowing DNA to interact effectively with proteins and other nucleic acids. The interplay between DNA's structure and its environment highlights the molecule's complexity and underscores its critical role in the life of the cell. The following subsections embody a brief description of four different DNA structural configurations namely: (1) B-DNA, (2) A-DNA, (3) Z-DNA, (4) B'-DNA.

2.1.1.1 B-DNA

B-DNA, or B-form DNA, is the most common and stable form of DNA found in biological systems. It is a right-handed double helix, characterized by its helical structure where the two strands twist around each other every 10.5 base pairs, creating a helical repeat of about 3.4 nm. The B-DNA helix has a wide major groove and a narrow minor groove, which play crucial roles in protein-DNA interactions. These grooves allow proteins to recognize and bind to specific DNA sequences, facilitating processes like transcription, replication, and repair.

The structure of B-DNA is composed of nucleotides, each consisting of a phosphate group, a deoxyribose sugar, and one of four nitrogenous bases: A, T, C, or G. A pairs with T through two hydrogen bonds, while C pairs with G through three hydrogen bonds. This complementary base pairing is essential for the double-helix structure and ensures the accurate replication of genetic information. The anti-parallel orientation of the two strands, with one running 5' to 3' and the other 3' to 5', is another critical feature of B-DNA.

B-DNA's structure is ideal for its biological functions i.e., allows for efficient packing within the cell nucleus, where DNA is wound around histone proteins to form nucleosomes, further compacting into chromatin. This compaction is essential for fitting the long DNA molecules into the relatively small nuclear space. Furthermore, the specific recognition sites on B-DNA's major and minor grooves enable regulatory proteins to bind precisely, controlling gene expression and ensuring proper cellular function.

While B-DNA is the most stable and predominant form under physiological conditions, DNA can adopt other conformations depending on environmental factors such as ionic strength and humidity. For example, A-DNA and Z-DNA are alternative helical forms that DNA can transiently assume. B-DNA's stability is attributed to its optimal hydrogen bonding and base stacking interactions, which make it the preferred structure in vivo. However, the ability of DNA to switch between forms is crucial for its biological versatility, allowing it to participate in various cellular processes that require structural flexibility.

2.1.1.2 A-DNA

A-DNA, or A-form DNA, is one of the alternate structural forms of DNA that occurs under specific conditions. Unlike the more common B-DNA, A-DNA is a right-handed double helix with a more compact and dehydrated structure. It is characterized by 11 base pairs per helical turn, resulting in a shorter and wider helix compared to B-DNA as show in Fig. 2.2. The helical repeat of A-DNA is approximately 2.8 nm. The major groove of A-DNA is deep and narrow, while the minor groove is wide and shallow, differing significantly from the grooves found in B-DNA.

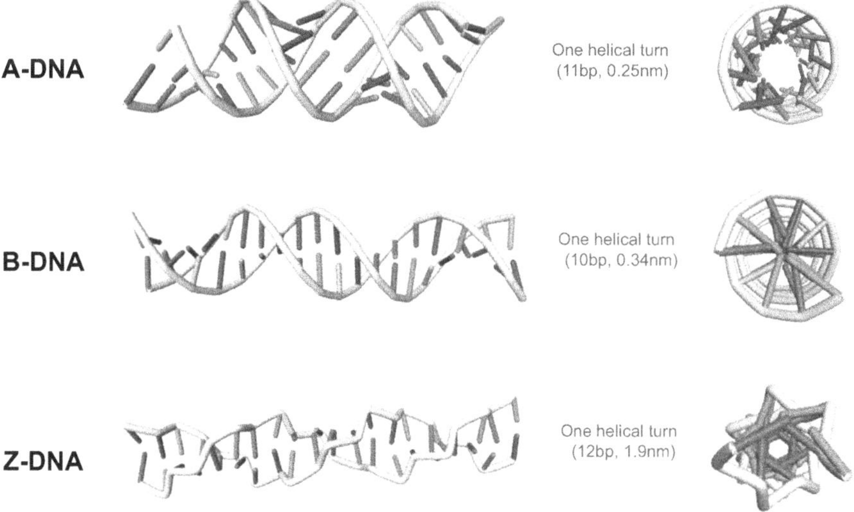

Fig. 2.2 Different DNA structural forms

Similar to B-DNA, the structure of A-DNA is composed of nucleotides, each with a phosphate group, a deoxyribose sugar, and one of the four nitrogenous bases. The base pairing in A-DNA follows the same principle as in B-DNA, however, the base pairs in A-DNA are tilted relative to the helical axis, which results in a central hole along the axis of the helix, contributing to its unique structural properties.

A-DNA typically forms under conditions of low humidity and high salt concentrations. It is less common in vivo compared to B-DNA but can be observed in certain DNA-protein complexes and during specific stages of cellular processes like sporulation in bacteria. The formation of A-DNA is believed to protect DNA in extreme environments, such as desiccation, by adopting a more compact and stable configuration. This structural adaptability underscores the importance of A-DNA in providing resilience to environmental stress.

2.1.1.3 Z-DNA

DNA conformation in Z form differs significantly as compared to A and B forms [62]. Figure 2.2 depicts left-handed double helix, Z-DNA, which consists of two base pairs. In Z-DNA, helical repeat occurs every 12 base pairs per turn, with each base pair rising about 3.7 Å, whereas in B-DNA, it's every 10.5 base pairs per turn with a rise of about 3.4 Å. Unlike the smooth path followed by backbone in B-DNA, backbone in Z-DNA takes a zigzag path. Because of zigzag backbone path, some phosphate groups are closer and electrostatic repulsion between them is greater than in B-DNA. Therefore, Z-DNA is stabilized by high salt concentrations or polyvalent cations that shield inter-phosphate repulsion better than monovalent cations. Hence, electrostatic interactions play a crucial role in Z-DNA formation and other factors contribute to Z-DNA stability. In each strand of Z-DNA, glycosyl angle swaps between syn and anti conformations, with one base in anti conformation while other adopts syn conformation for each base pair. Sugar conformations alternate along each polynucleotide chain and adopt $C_2^{'}$-endo and $C_3^{'}$-endo conformations corresponding to anti and syn nucleotide conformations, respectively. Rotation of helix between adjacent base pairs, also known as winding angle, rotates between -9° to -51° for different base pairs contacts.

The presence of Z-DNA is associated with specific functions in the cell. It is believed to play a role in the regulation of gene expression, DNA recombination, and chromatin organization. Certain proteins specifically recognize and bind to Z-DNA, influencing the structural dynamics of the genome. Additionally, Z-DNA formation can impact DNA repair mechanisms and the stability of genetic sequences. Its unique structure and properties make it an important, though less common, form of DNA that contributes to the complex regulation of genetic information and cellular processes.

2.1.1.4 B'-DNA

B'-DNA is a specific DNA conformation that is commonly adopted by poly(dA)poly(dT) sequences under various solution conditions. It introduced substantial intrinsic curvature into DNA molecules, particularly at junctions between B and B' helices. This intrinsic curvature, often induced by repeated adenine (A) or thymine (T) sequences, enables B'-DNA to participate in various DNA structure and function tasks, influencing processes such as DNA packaging, protein binding and gene regulation. Despite its similarity with B form, it exhibits narrower minor groove and significant positive propeller twist of base pairs [23]. Studies revealed that any sequence with pattern d($A_m T_{nm}$) adopts this conformation for $n \geq 4$ [388]. $B^{'} - B$ induces significant inherent curvature within DNA molecules that mainly appear at stacks between B and B' helices [388, 539]. As for a sequence of six consecutive adenines, total bend reaches approximately 19° [539, 684]. When these bends align with DNA helix natural pattern, they create significant natural curve in double helix. This bending phenomenon has also been observed in kinetoplast DNA segments which contain organized A-tracts made up of 5 to 8 nucleotides [696]. However, *X-ray* related studies of A-tracts initially failed to detect DNA bends in solutions. Later, it was discovered that in solutions of *2-methyl-2, 4-pentanediol*, the dehydrating agent commonly used in crystallization of oligonucleotides, intrinsic curvature linked to the A-tracts nearly vanished [944].

Besides the canonical base pairs, there are several other base pairs that have been observed experimentally [926, 1099]. Some of them have been observed in pairs formed by modified bases or in DNA-protein complexes. Figure 2.3 graphically illustrates Hoogsteen and reverse Hoogsteen that are characterized in triple helices and considered particularly significant.

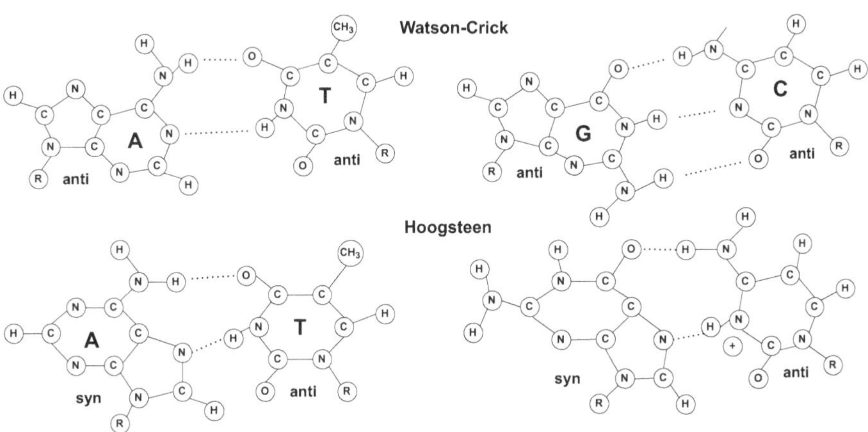

Fig. 2.3 Watson-Crick and Hoogsteen Base Pairing (SS)

2.1.2 Single-Stranded DNA Structure and Conformation

Single-stranded DNA (ssDNA) is composed of a deoxyribose sugar, a phosphate group, and nitrogenous bases—adenine A, T, C, and G. These nucleotides are linked by phosphodiester bonds, which create a backbone with a 5' phosphate end and a 3' hydroxyl end, giving the ssDNA a distinct polarity. The foundation of various DNA structures is established by ssDNA through its non-covalent interactions, such as hydrogen bonding. These interactions are crucial in forming complex DNA secondary structures like hairpins, loops, pseudoknots, and quadruplexes. ssDNA consists of a backbone of repeating units (sugar and phosphate group) and nitrogenous bases attached as side chains, acting as a linkage between two strands of DNA, which then coil to form the double helical structure of DNA.

In solution, ssDNA often assumes a random coil conformation due to the flexibility of its backbone. However, it can form secondary structures such as hairpins, where complementary bases within the same strand pair up to create a stem-loop formation, as well as bulges and internal loops caused by mismatched or unpaired bases. Guanine-rich regions can form G-quadruplexes, stabilized by Hoogsteen hydrogen bonds and cations like potassium. The conformation of ssDNA is influenced by its nucleotide sequence, environmental factors such as ionic strength, pH, and the presence of binding proteins, as well as temperature. Higher temperatures can disrupt secondary structures, leading to a more relaxed coil.

Figure 2.6 illustrates the clear directional flow of the ssDNA backbone, with the numbering of the sugar's carbon atoms indicating either a 3'-5' or 5'-3' orientation. Typically, ssDNA follows a 5'-3' orientation in its base sequence. This directional arrangement is crucial for various biological processes, including DNA replication and transcription, as it provides a standardized framework for interpreting genetic information. In ssDNA, the β-glycosidic bond links the base and sugar in nucleosides, contributing to the molecule's flexibility. The rotation angle (χ) between the $O_1^{'}–C_1^{'}$ bond and the $N_1–C_6$ bond in pyrimidines (C and T) and the $O_1^{'}–C_1^{'}$ and $N_9–C_8$ bonds in purines (A and G) defines two main rotational configurations: anti-conformation (approximately 210°) and syn-conformation (approximately 0°), as shown in Fig. 2.7. These configurations, along with the slight variations in bond lengths and angles, contribute to the structural flexibility of the ssDNA backbone. However, potential steric clashes within the repeating unit restrict these rotation angles.

The deoxyribose ring in ssDNA also exhibits versatility by adopting various conformations, which further influences the overall structure. Typically, four of the five atoms in the ring are in the same plane, with the fifth atom deviating. This ring can assume four different conformations: C_2-endo, C_2-exo, C_3-endo, and C_3-exo, based on whether atom C_2 or C_3 lies outside the plane. Figure 2.8 shows that in the C_2-endo conformation, C_2 is above the plane and parallel to both the base and the 5'-carbon. The C_3-exo conformation is similar but with C_3 deviating more from the plane compared to C_2. In the C_3-endo conformation, C_3 is above the ring, while the C_2-exo conformation mirrors the C_3-endo conformation. These structural features of ssDNA—directional orientation, glycosidic bond flexibility, and deoxyribose ring conformations—are integral to its biological functions and interactions.

The roles of ssDNA in biological processes are manifold. During DNA replication, ssDNA serves as a template for synthesizing a complementary strand. In transcription, regions of dsDNA unwind to expose ssDNA, allowing RNA polymerase to synthesize RNA. ssDNA is also an essential intermediate in homologous recombination and DNA repair processes. Some viruses, such as parvoviruses, have ssDNA genomes, which must be converted to dsDNA within host cells for replication and transcription.

2.1.2.1 Hairpin and Stem Loops

A hairpin loop forms when a single strand of DNA folds back on itself, creating a double-stranded stem with an unpaired loop at the end. This structure arises from sequences within the strand that are complementary to each other, typically palindromic sequences where one side of the loop is the reverse complement of the other. Hairpin loops are crucial in various biological processes, including the regulation of gene expression and the initiation of DNA replication. They are also a common feature in RNA secondary structures, influencing RNA stability and function.

Similar to a hairpin loop, a stem loop consists of a paired stem and an unpaired loop, but the loop region can be larger and more variable in size. This structure forms when stretches of complementary bases are separated by non-complementary sequences that form the loop. Stem loops play significant roles in RNA biology, including RNA splicing, translation, and regulation of gene expression. They are also important in the folding of RNA molecules and the formation of complex three-dimensional structures.

2.1.2.2 Bulge and Internal Loops

A bulge loop occurs when an unpaired nucleotide or group of nucleotides is on one side of a double-stranded stem. This structure forms due to extra bases in one strand that do not have complementary partners on the opposite strand. Bulge loops can affect the binding of proteins and other molecules to the DNA, influencing the stability and flexibility of the DNA structure. They are involved in various cellular processes, including the regulation of transcription and the recognition of DNA by proteins.

An internal loop is formed when there are unpaired nucleotides on both sides of a paired region, creating a loop within the double-stranded stem. This structure occurs due to mismatches or gaps between complementary base pairs. Internal loops introduce flexibility into the DNA or RNA molecule, affecting its three-dimensional conformation and interactions with proteins. They are often found in functional RNA molecules and play roles in RNA splicing, translation, and the regulation of gene expression.

2.1.2.3 Pseudoknot

A pseudoknot is a complex structure where bases in a loop pair with complementary bases outside the loop, creating an interwoven configuration involving multiple stem-loop regions. Pseudoknots form when sequences from different parts of the strand come together to create additional base-paired regions. These structures are important in the folding of RNA and have roles in regulating gene expression, ribosomal frameshifting, and viral replication. Pseudoknots add stability and functional diversity to RNA molecules.

2.1.2.4 G-Quadruplexes Structure

Quadruplex structures, particularly G-quadruplexes, are formed by guanine-rich sequences where four guanine bases form a square planar structure known as a G-tetrad. In G-Quadruplexes structures, four guanine bases are linked together through Hoogsteen hydrogen bonding as illustrated in Fig. 2.4. G-quadruplex architecture is comprised of stacks of multiple G-quartets which can vary in size and complexity depending on the sequence.

Studies related to fiber diffraction of poly(G) suggested that quadruplexes is a right-handed helix structure with rise of 0.34 nm and 30-degree helical twist, where all glycosidic angles are in anti conformation and all four strands are parallel [1248]. Oligonucleotides with sequence d(TTAGGG) adopts similar structural configuration and has lowest free energy among G-quadruplex structures in the absence of loop constraints [1047]. However, further research revealed the possibility of numerous four stranded structure based on G-quartet [556, 807, 1096]. Moreover, strands polarity is a crucial factor in such structures and all possible strand polarity combinations are depicted in Fig. 2.5.

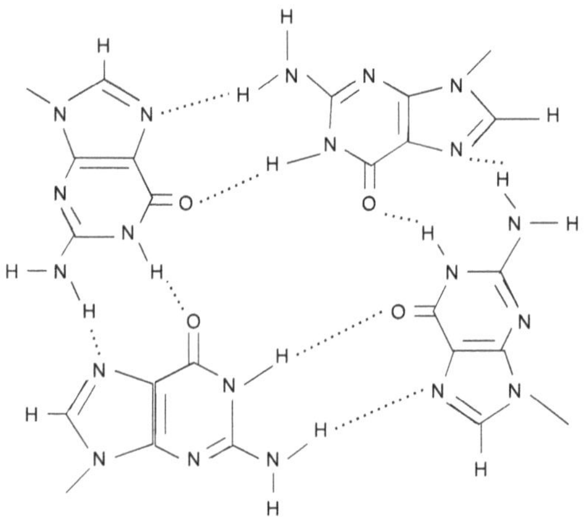

Fig. 2.4 The pattern of hydrogen bonds in the G-quartet. Guanines form a stable quartet by acting as hydrogen bond donors and acceptors. The quartet is further stabilized by the presence of K+ or, to a lesser extent, Na+. These cations coordinate at the center quartet, interacting with electronegative carbonyl oxygens

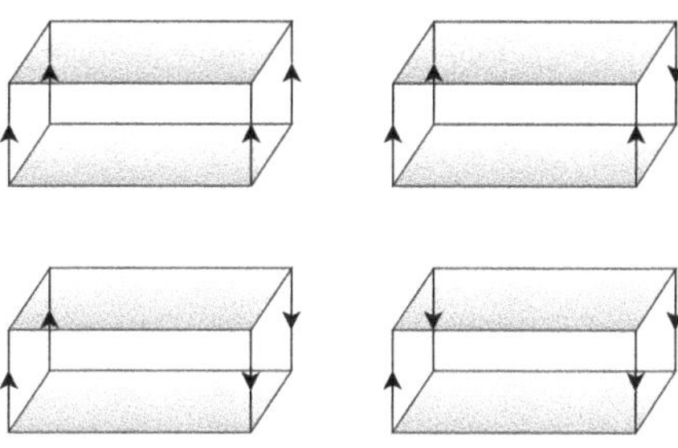

Fig. 2.5 Possible orientations of DNA strands in the G-quadruplexes

Further studies revealed that G-quadruplex structures are intramolecular oligonucleotides with telomere-like sequences. In these structures, DNA strands form loops whose shapes depends strand segments polarity. Although loop formation influences structural preferences, multiple loop types have been observed in G-quartet structures. Oligonucleotide d(GTGGTGGGTGGGTGGGT), also known inhibiting HIV integrase [1106], forms a parallel-stranded quadruplex structure with single-nucleotide loops [595, 732]. This ability to form small loops separates Q-quadruplexes apart from hairpins with B-form stems, where loops must contain at least three nucleotides.

There is no easy rule to predict structure created by a particular oligonucleotide with G-repeats. Although there are 26 potential arrangements of loops in intramolecular G-quadruplexes, only few has been experimentally identified through X-ray or NMR analysis [807]. Stable quadruplex structures can be created with only two G-quartets [682]. Guanines glycosyl angles within G-quartets can adopt either anti or syn conformations depending on directionality of strand.

Intramolecular quadruplexes folding results in a relatively small energy gain, typically less than 10 kcal/mol at 37 °C, particularly when structure contains three G-quartets. This energy gain is significantly lower than energy utilized in forming duplex using two complementary oligonucleotides of same length, which can reach up to 30 kcal/mol [556]. Likelihood of G-quadruplex formation within regular double-stranded DNA (dsDNA) is low, as its formation requires unwinding of the duplex. G-quadruplexes formation can potentially be facilitated by negative supercoiling, which reduces energy associated with supercoiling due to unwinding of corresponding DNA segment. However, experimentation revealed that this reduction in supercoiling energy only favors quadruplex formation at superhelix density of approximately −0.06, a level of torsional stress which is rarely achieved in vivo and difficult to replicate in vitro. Hence, negative supercoiling alone may not be sufficient to induce quadruplex formation by unwinding dsDNA segment. Moreover, quadruplex formation may occur through proteins that specifically bind to this DNA structure. Despite these considerations, G-quadruplexes formation through unwinding dsDNA segments has not been observed, even in laboratory settings.

Single-stranded overhangs at ends of eukaryotic chromosomes are highly favorable for quadruplex formation. These structures exhibit diverse topologies, including parallel, anti-parallel and hybrid forms, as well as varying loop arrangements

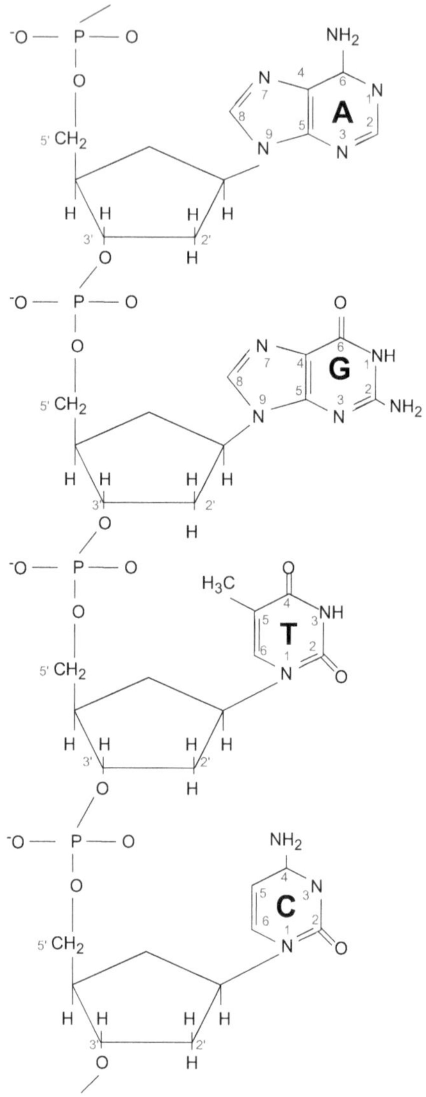

Fig. 2.6 Single-stranded DNA (ssDNA) chemical structure includes 2-deoxyribose repeating units and negatively charged phosphate residues forming backbone which is made up of four bases: adenine (A), guanine (G), thymine (T), or cytosine (C). Adenine and guanine are purines, while thymine and cytosine are pyrimidines, being smaller in size. Carbon and nitrogen atoms in both backbone and bases are labeled with red digits to indicate their numbering systems

between the G-quartets. The specific arrangement of loops and strands contributes to the stability and functionality of G-quadruplexes. For example, the number and composition of loops can influence the thermal stability and folding kinetics of the structure.

The time it takes for quadruplex formation depends on the reaction order. When it comes to quadruplex intramolecular folding, this process usually occurs within tens of milliseconds, whereas typically time can span many days for four oligonucleotides structures [556].

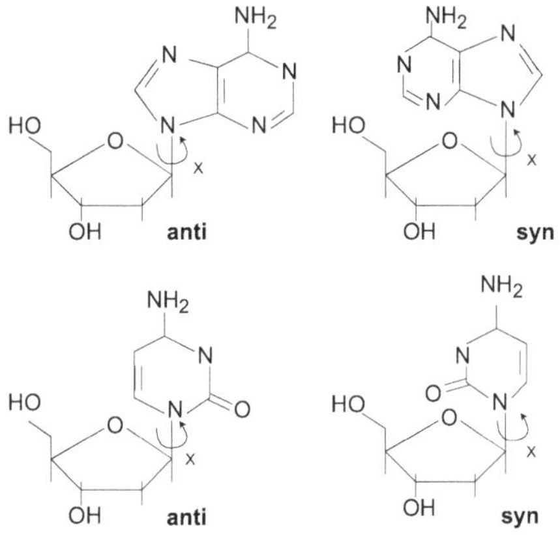

Fig. 2.7 Anti-conformation (anti) and syn-conformation (syn) refer to different alignments of purines and pyrimidines which varies with 210° angle χ

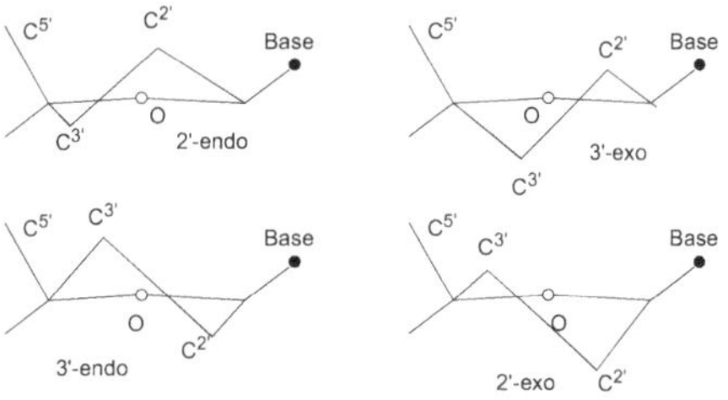

Fig. 2.8 Deoxyribose ring major conformations

2.1.3 Triple Helix Structure

Triple helix structure is comprised of three strands of nucleic acids winding around each other in a helical manner. Unlike more common double helical DNA structure which consists of two intertwined strands, triple helix involves association of three strands. In triple helical structure, two strands typically form Watson-Crick base-paired double helix, similar to structure of DNA and third strand binds to major groove of B-DNA through Hoogsteen or reverse Hoogsteen base pairing. Overall, triple helix structure provides unique ways to interact and form stable complexes that enables DNA to perform various tasks such as gene regulation, molecular recognition and potential therapeutic applications.

Although double helix structure varies slightly during triple helix formation but it retains *B-DNA* essential features. Specifically, glycosyl angle corresponds to anti conformation while sugars maintain C_2*-endo* conformation and triple helix repeat is around 12 triads per turn [829]. Moreover, specific interactions between bases of third strand and double helix facilitates triple helix stability. These interactions involve hydrogen bonding and other molecular forces that help maintaining triple helix structure. Third strand in triple helix must form hydrogen bonds with purines of double helix [316]. However, only sequences with homopurine-homopyrimidine segments in *B-DNA* can form triple stranded helix due to structural constraints. Hydrogen bonding patterns between purines in *B-DNA* and bases of third strand involve Hoogsteen and reverse Hoogsteen base pairs. Based on these patterns, the triple helices can be divided into two groups: (1) Hoogsteen base pairing triplexes, (2) Reverse Hoogsteen base pairing triplexes. Following subsections provides a brief description of triple helices groups.

2.1.3.1 Hoogsteen Base Pairing Triplexes

Specifically in Hoogsteen base pairing triplexes, third strand is composed entirely of pyrimidines and forms Hoogsteen base pairs, namely: (1) TA base pair, (2) CG base pair. Two strands must run parallel to connect properly for Hoogsteen base pairs while third strand should be anti-parallel to pyrimidine strand of Watson–Crick duplex and its sequence must be reverse of that particular strand. Furthermore, cytosine participating in Hoogsteen pairing must be protonated in N^3 position. Figure 2.9 graphically illustrates hydrogen bonding patterns in Hoogsteen base pairing triplexes.

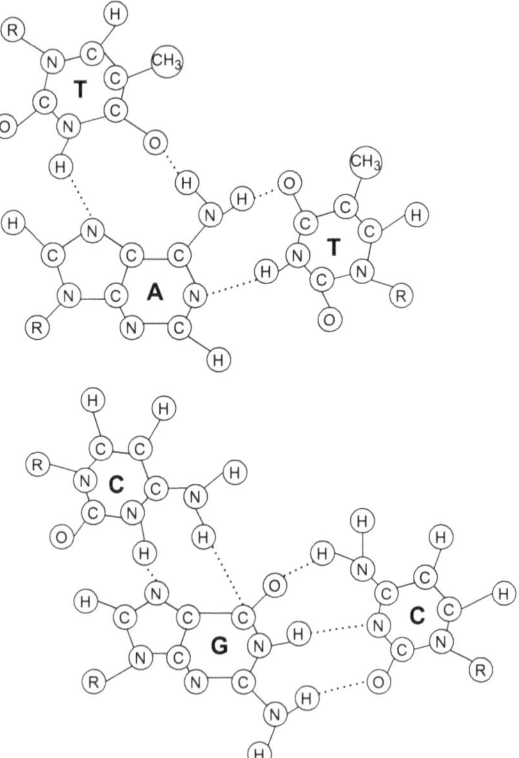

Fig. 2.9 Base triads in the Hoogsteen triplexes. The cytosine of the third strand is protonated in the N^3 position

2.1.3.2 Reverse Hoogsteen Base Pairing Triplexes

It was assumed that triplexes generated with reverse Hoogsteen base pairing must required purines to create triads namely: (1) A*AT triad, (2) G*GC triad. Later, it was discovered that T*AT triad can also enhances thermal stability of triplexes [80]. Figure 2.10 graphically illustrates hydrogen bonding patterns in reverse Hoogsteen triplexes triads. The two linked DNA strands in these triplexes must be anti-parallel to each other and any deviation from the specified sequence requirements significantly reduces triplexes stability.

2.2 RNA Structures

RNA (ribonucleic acid) primary structure is a linear sequence of four basic nucleotides (A,C,G,U) and each nucleotide consists of three components: a phosphate group, a ribose sugar, and one of four nitrogenous bases. The RNA primary structure is determined by the specific order in which these nucleotides are covalently linked through phosphodiester bonds. The RNA primary structure is fundamental for understanding the genetic information encoded in RNA molecules. It serves as a template for the synthesis of proteins during the process of translation and plays essential roles in various cellular functions, including gene expression, regulation, and catalysis in certain types of RNA molecules.

RNA secondary structure represents local spatial arrangement and interactions between nucleotides within a single RNA molecule [422, 699]. Unlike the primary structure, which describes the linear sequence of nucleotides, the secondary structure provides information about how these nucleotides fold and pair with each other. RNA secondary structure is important for understanding the functional aspects of RNA molecules. It plays a significant role in determining how the RNA molecule interacts with other molecules, such as proteins or other nucleic acids, and how it contributes to biological processes like translation, splicing, and catalysis.

RNA tertiary structure represents overall three-dimensional arrangement of an entire RNA molecule, including its secondary structural elements as well as the interactions and spatial relationships between different regions of the molecule [422, 699, 1091]. While the secondary structure focuses on local interactions like base pairing and stem-loop structures, the tertiary structure provides a broader perspective on how the entire RNA molecule folds and adopts a specific 3D shape. RNA tertiary structure provides insights into how the molecule achieves its unique shape and how this shape relates to its biological activity.

These higher-order structures are crucial for the diverse functions of RNA which influence its interactions with proteins, small molecules, and other cellular components. Understanding the secondary and tertiary structures of RNA is pivotal in unraveling the molecular mechanisms underlying processes like translation,

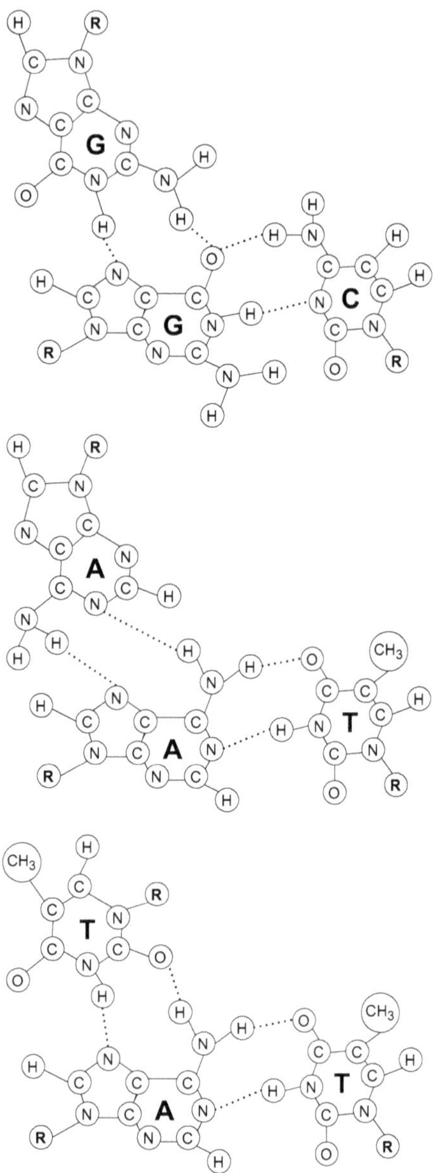

Fig. 2.10 Base triads in reverse Hoogsteen triplexes

transcriptional regulation, and RNA catalysis. For example, RNA structures like riboswitches can act as molecular switches which can alter their conformation in response to specific signals and modulate gene expression accordingly. Table 2.1 provides a high-level comparison between RNA secondary and tertiary structures and the following subsections provide more insights about RNA secondary and tertiary structures.

Table 2.1 A comparison of secondary and tertiary RNA structures

Property	Hairpin loop	Internal loop	Bulge loop	Stacking	External loop	Multibranch loop	RNA pseudoknot	Riboswitch	Ribozyme	tRNA cloverleaf structure
Formation	Sequence folds back on itself, creating a loop	Unpaired bases disrupt helical structure within a dsRNA region	Single-stranded segment interrupts dsRNA region	Stacking of adjacent base pairs	Unpaired bases in a loop region external to the double-stranded RNA	Junction of three or more helices, creating multiple branches	Base-pairing between distant regions creates a knot-like structure	Adopts different conformations in response to ligand binding	Catalytic RNA molecule capable of enzymatic functions	Distinct cloverleaf shape with acceptor, anticodon, D, and TC loops
Function	Regulatory roles, recognition sites for proteins	Influences overall folding and stability of RNA	Contributes to structural diversity, molecular interactions	Contributes to overall stability of RNA structure	Influences local RNA structure, may participate in interactions	Creates complex RNA structures, often involved in RNA folding	Diverse roles including ribosomal frameshifting, viral replication	Gene expression regulation based on ligand binding	Catalyzes enzymatic reactions, e.g., self-cleavage	Adapter molecule in protein synthesis, ensures accurate amino acid pairing
Role in RNA World	Commonly found in RNA secondary structures	Essential for RNA folding, stability, and dynamics	Adds diversity to RNA structures, molecular recognition	Contributes to overall stability and stacking interactions	Impacts local RNA structure, participates in interactions	Essential for creating complex RNA structures	Plays roles in regulatory mechanisms, structural complexity	Molecular switches for sensing and responding to environmental signals	Demonstrates RNA's catalytic capabilities	Exemplifies RNA's adaptability in adopting specific structures for specialized functions

(continued)

Table 2.1 (continued)

Property	Hairpin loop	Internal loop	Bulge loop	Stacking	External loop	Multibranch loop	RNA pseudoknot	Riboswitch	Ribozyme	tRNA cloverleaf structure
Examples	Commonly found in mRNA and regulatory RNAs	Occurs in tRNA, rRNA, and mRNA, influencing RNA folding	Found in various RNA molecules, influencing local structure	Contributes to overall stability of RNA secondary structures	Often found in structured RNA elements, such as ribosomal RNA	Junctions in RNA secondary structures, such as in ribosomal RNA	Found in untranslated regions of mRNA, modulating gene expression	Examples include hammerhead, group I intron	Examples include thiamine pyrophosphate (TPP) riboswitch, flavin mononucleotide (FMN) riboswitch	Examples include hammerhead, group I intron

2.2.1 RNA Secondary Structures

In the mid-twentieth century, initial breakthroughs were made through pioneering work on transfer RNA (tRNA), where researchers deciphered the cloverleaf secondary structure that characterizes tRNA molecules [976]. The discovery of tRNA secondary structures laid the foundation for broader investigations into RNA folding patterns [993]. As experimental techniques such as chemical probing and enzymatic probing emerged, researchers explored RNA secondary structures with higher precision [542]. In the 1980s, the advent of computational methods, particularly thermodynamic algorithms, facilitated the prediction of RNA secondary structures by analyzing sequence-structure relationships.

The exploration of ribosomal RNA (rRNA) secondary structures further expanded the understanding of RNA's structural diversity [764]. High-resolution structural studies, including X-ray crystallography and cryo-electron microscopy, revealed the secondary structures within the large ribosomal RNA molecules which provided crucial insights into their functional roles in protein synthesis.

The late twentieth century and early twenty-first century witnessed a surge in efforts to catalog and annotate RNA secondary structures [130]. Projects like the Comparative RNA Web (CRW) database became invaluable resources that compiled experimentally validated and computationally predicted RNA secondary structures from various organisms [130].

The formation of RNA folding is caused by the complementary base-pairing interactions between adenine (A), uracil (U), guanine (G), and cytosine (C). These hydrogen bonding interactions lead to the establishment of stable helical and other most commonly known RNA secondary structures such as, Stem-Loop (Hairpin) Structure, Internal Loop, Bulge Loop, Multi-Loop (Multibranch Loop), and Pseudoknot. These RNA secondary structures differ based on their distinct arrangements of base pairing and unpaired regions: Stem-Loop forms a short double-stranded stem and a loop; Internal Loop features unpaired bases between paired stretches; Bulge consists of a single unpaired nucleotide; Multi-Loop connects multiple unpaired loops; Pseudoknot shows complex interwoven base pairing; Junction is a point where multiple helices meet; Base Pair Stack represents a continuous helical sequence; Tertiary Interaction involves non-local interactions shaping the overall 3D structure. Figure 2.11 graphically represents the formation of different RNA secondary structures. A more comprehensive detail of these structures is given in the following subsections:

2.2.1.1 Hairpin Loops

Hairpin loops are distinctive RNA structural motifs characterized by a sequence that folds back on itself and forms a stem-loop structure [968]. In this arrangement, the stem comprises base-paired nucleotides, while the loop consists of unpaired nucleotides. This structural feature plays a pivotal role in various regulatory

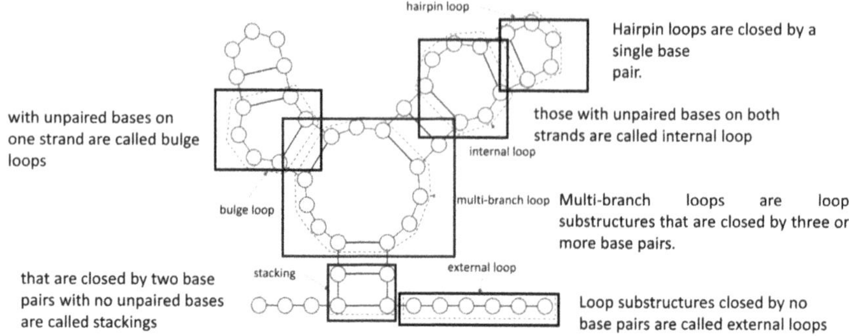

Fig. 2.11 Description of RNA secondary structures

processes, including transcription termination. Additionally, hairpin loops serve as recognition sites for proteins and other molecules, influencing RNA-protein interactions and contributing to the dynamic functional diversity of RNA molecules. The flexibility of hairpin loops makes them integral components of RNA secondary structures, showcasing the adaptability and regulatory significance of these structures in cellular processes.

2.2.1.2 Internal Loop

Internal loops are essential structural elements in RNA characterized by unpaired bases disrupting the helical structure within a double-stranded region [675]. These loops contribute to the overall folding and stability of RNA molecules. The unpaired bases in internal loops introduce structural flexibility, allowing RNA to adopt diverse conformations. Furthermore, internal loops play crucial roles in mediating RNA interactions with other molecules, such as proteins or additional nucleic acids.

2.2.1.3 Bulge Loop

Bulge loops represent unique RNA structural motifs where a single-stranded segment interrupts a double-stranded region [993]. These structural features contribute to the diversity of RNA structures and can participate in molecular interactions that influence RNA function and stability. Bulge loops are particularly relevant in RNA-protein interactions and may act as recognition sites for binding partners. Their presence introduces local distortions in the RNA structure, impacting its overall conformation. Studying bulge loops provides valuable insights into the versatile ways RNA molecules can adopt specific shapes, highlighting their significance in various cellular processes (Fig. 2.11).

2.2.1.4 RNA Pseudoknot

RNA pseudoknots are intricate structural motifs formed by base-pairing interactions between distant regions of a single-stranded RNA molecule, creating a knot-like structure [881]. These pseudoknots play diverse roles in RNA-mediated processes, including ribosomal frameshifting and viral replication. The complex folding patterns of pseudoknots contribute to their functional versatility. Their involvement in regulatory mechanisms underscores their significance in modulating gene expression and other essential cellular functions. The dynamic nature of RNA pseudoknots adds an additional layer of complexity to the understanding of RNA structures and their multifaceted roles in cellular biology.

2.2.1.5 tRNA Cloverleaf Structure

The transfer RNA (tRNA) cloverleaf structure is a hallmark of tRNA molecules, featuring an acceptor stem, anticodon loop, D-loop, and TC loop [969]. This intricate structure allows tRNA to function as an adapter molecule in protein synthesis, facilitating the accurate pairing of amino acids with the corresponding codons on messenger RNA during translation. The unique features of the cloverleaf structure ensure the precision of this molecular matching process, highlighting the crucial role tRNA plays in the fidelity of protein synthesis. The tRNA cloverleaf structure is a testament to the remarkable adaptability of RNA in adopting specific conformations tailored to perform specialized functions within the cellular context.

2.2.2 RNA Tertiary Structures

The discovery of RNA's 3D structures began with pioneering efforts to elucidate the overall components of ribosomal RNA (rRNA) in the 1960s and 1970s through X-ray crystallography. In 1974, the first crystallographic structure of the 30S subunit of the bacterial ribosome was determined by Ada Yonath et al. [197]. This breakthrough provided valuable insights into the complex organization of rRNA within the ribosome. Over subsequent years, advancements in crystallography techniques and technology allowed researchers to unveil higher-resolution structures of both bacterial and eukaryotic ribosomes and RNA which contribute to the understanding of the molecular mechanisms underlying protein synthesis. Simultaneously, nuclear magnetic resonance (NMR) spectroscopy emerged as a powerful tool for studying the structure of biological macromolecules in solution.

The 3D structure of RNA unveils a complex and dynamic arrangement of its constituent molecules. It is mainly composed of nucleotide units featuring a sugar (ribose), a phosphate group, and one of four nitrogenous bases—adenine (A), guanine (G), cytosine (C), and uracil (U) and phosphodiester bonds. This structural framework is foundational to the complex 3D structure that governs RNA's various biological functions.

The formation of RNA tertiary structures is driven by various interactions, including:

Base Pairing Distant regions of the RNA sequence can form base pairs, bringing different parts of the molecule closer together. This can include interactions between loops, bulges, or even non-contiguous sequences.

Base Stacking Stacking interactions occur between adjacent bases within the same strand or between different strands, contributing to the overall stability of the three-dimensional structure.

Hydrogen Bonding Hydrogen bonds form between complementary bases and other functional groups, stabilizing the interactions between different regions of the RNA.

Metal Ion Coordination Metal ions, such as magnesium, can coordinate with specific sites on the RNA, influencing the folding and stability of the tertiary structure.

Long-Range Interactions Long-range interactions involve the interaction of distal regions of the RNA molecule, often facilitated by specific structural motifs or elements. Pseudoknots, for example, are tertiary structures formed by base-pairing interactions between non-adjacent regions.

Protein Interactions Proteins, such as RNA-binding proteins, can bind to specific regions of the RNA molecule and influence its folding. These protein-RNA interactions play a crucial role in the formation and stabilization of tertiary structures.

RNA molecules can adopt complex tertiary structures shaped by long-range interactions. These interactions involve distant parts of the RNA chain that contribute to the molecule's overall conformation. Helical regions, similar to the double helix in DNA are prevalent in RNA structures, but they are typically more dynamic and often contain non-canonical base pairs that enhance structural complexity.

Specific functional sites within the RNA 3D structure play crucial roles in various biological functions. For example, in ribosomal RNA (rRNA), functional sites are involved in the binding of transfer RNA (tRNA) and catalyzing the synthesis of proteins. The following subsections describe 3D RNA structures that contribute to the overall function of RNA.

2.2.2.1 Riboswitches

Riboswitches represent sophisticated RNA structures capable of adopting different conformations in response to ligand binding which allows them to regulate gene expression [726]. These molecular switches play crucial roles in sensing cellular conditions and modulating RNA structure accordingly to control biological processes. Riboswitches are often found in the untranslated regions of messenger RNAs, where they act as sensors for small molecules [726]. The conformational changes triggered by ligand binding influence downstream gene expression, highlighting the regulatory capabilities of RNA structures beyond their role as mere carriers of genetic information.

2.2.2.2 Ribozymes

Figure 2.12 shows 3D structure of a riboswitch. Ribozymes are catalytic RNA molecules with the unique ability to perform enzymatic functions, such as self-cleavage or ligation [264]. This intrinsic catalytic activity sets ribozymes apart from the traditional view of RNA as a passive information carrier. Ribozymes play crucial roles in RNA processing, demonstrating the diverse functions that RNA can perform beyond its canonical role [264]. Their involvement in cleaving and joining RNA sequences underscores their significance in cellular processes and has led to the exploration of ribozymes in biotechnological and therapeutic applications (Fig. 2.13).

Fig. 2.12 3D sample Lysine riboswitch representation

Fig. 2.13 A sample ribozyme 3D structure

2.3 Protein Structures

The protein primary structure refers to the linear sequence of amino acids that make up a polypeptide chain [1094]. It is the specific order in which amino acids are covalently linked through peptide bonds and form the backbone of the protein molecule. The primary structure is a fundamental level of protein organization and serves as a basis for higher-order structural complexities in proteins [1094].

2.3.1 Secondary Structure

Protein secondary structure refers to the local 3D arrangements of amino acid residues within a protein molecule [549]. It primarily involves the regular folding patterns and structural motifs that emerge as a result of hydrogen bonding interactions between nearby amino acids in the polypeptide chain. The two most common types of protein secondary structures are alpha helices and beta sheets [237, 513]. In an alpha helix, the polypeptide chain twists into a right-handed helical structure, stabilized by hydrogen bonds between the carbonyl oxygen of one amino acid and the amide hydrogen of an amino acid three or four residues away. In a beta sheet, neighboring strands of the polypeptide chain align side by side and are held together by hydrogen bonds between the carbonyl oxygen of one strand and the amide hydrogen of an adjacent strand.

2.3.2 Tertiary/Quaternary Structures

The tertiary structure of proteins refers to the 3D arrangement of AAs and folding of the entire polypeptide chain into a specific and unique spatial conformation [842]. This structure arises from interactions between amino acid residues that are not in close proximity with each other. Protein quaternary structure refers to the spatial arrangement of AAs and interaction between multiple polypeptide chains. In other words, it describes the organization of individual protein subunits and their interactions to form a functional, biologically active macromolecular complex. Not all proteins have quaternary structures; it specifically applies to those composed of more than one polypeptide chain. Figure 2.14 graphically illustrates the 3D structure of gap junction beta-2 (GJB2) protein.

Understanding the 3D (tertiary/quaternary) structure of proteins holds paramount importance in deciphering their roles within biological processes. The spatial arrangement of amino acid residues in proteins directly influences their functions and dictates the specificity of molecular interactions and catalytic activities crucial for cellular life. For enzymes, the precise folding in three dimensions is imperative

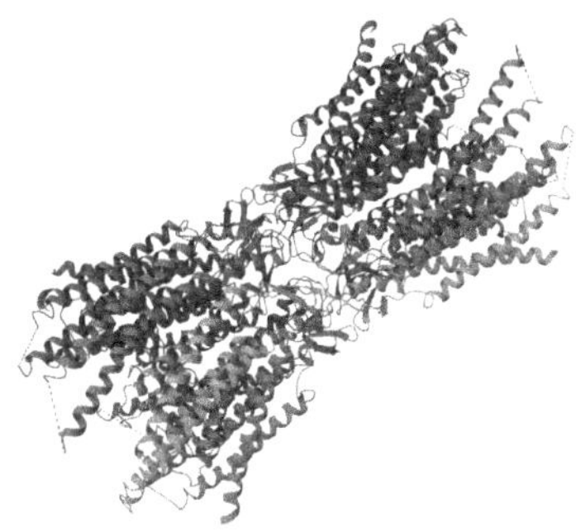

Fig. 2.14 3D structure of gap junction beta-2 (GJB2) protein

for substrate recognition and catalysis. Structural insights into receptors aid in comprehending signaling pathways, while the conformation of transport proteins determines their efficiency in molecular trafficking. Additionally, protein structure information is useful for understanding and finding protein-protein interactions that are crucial for cellular regulation and communication and are tied to their structural conformations. In diseases, structural anomalies often underlie dysfunctional protein behavior [842]. Therefore, a comprehensive understanding of protein structures provides a foundation for advancing knowledge of biological processes, targeted drug design, and therapeutic interventions.

Diverse experimental methods like X-ray crystallography, nuclear magnetic resonance (NMR) spectroscopy, Cryo-EM, and mass spectrometry have traditionally been employed to determine and explore protein structures. These methods are expensive, laborious, and time-consuming with limited scalability. Considering these challenges, computational methods have emerged as powerful tools for understanding and predicting protein structures. These methods leverage principles of physics, bioinformatics, and AI to predict the folding patterns of amino acid chains that constitute proteins.

Computational methods can be placed in two different categories i.e., conventional bioinformatics approaches, and modern AI-based tools. In the conventional bioinformatics-based category two prominent methods are template-based modeling (TBM) [498] and ab initio (or de novo) modeling [783]. TBM relies on known protein structures (templates) to predict the structure of a target protein based on sequence similarity. This method is particularly effective when the target protein shares homology with proteins of known structures. On the other hand, ab initio modeling aims to predict protein structures without relying on template information. It involves exploring the vast conformational space of a protein and identifying the most energetically favorable fold. While TBM excels in cases where suitable templates are available, ab initio modeling is essential for addressing proteins with unique folds or limited homologous templates. Integrating these approaches, along with experimental data, presents a holistic strategy to enhance the accuracy of protein structure predictions.

Chapter 3
Exploration of AI-Driven Genomic and Proteomic Sequence Analysis Landscape

History of bio-informatics began hundred years ago with the contributions of Gregor Mendel, an Austrian monk who is widely recognized as the "Father of Genetics" [282]. In the primary work, he cross-fertilized different colors of the same species of flowers to discover different types of genetic relations [282]. Similar to the classical problem whether chicken came first or egg, there exist multiple theories whether RNA, Protein or Metabolism discovered first in origin of life. Overall, most evolutionary biologist believe that first discovered genetic material was RNA. However, a minority of scientist advocate protein as earliest molecule of life [461]. Interestingly a recent history uncovers that development of biological databases started with protein data [870].

Prior to 1950, the prevailing view was that proteins were shapeless and had no specific chemical composition. Due to this perspective, researchers did not put significant efforts to further explore shape and composition of proteins [872]. In 1951, a British biochemist Frederick Sanger revolutionized this view and proved that proteins have defined shape and chemical compositions [870, 871]. He claimed that most of the information about the shape and composition of proteins is encapsulated in the protein sequence [870, 871]. He used Edman degradation method to extract raw sequence of bovine insulin protein, consisting of only 51 residues [870, 871]. This finding opened new horizons of exploring and understanding shape and composition of proteins and later in 1958, Sanger received Noble Prize in chemistry. Afterwards, in 1965, the very first nucleic acid sequence of yeast alanine tRNA consisting of 77 bases, was discovered by Robert Holley's team at Cornell University [433]. He was awarded with Noble Prize in Physiology or Medicine in 1968. They employed chemical and enzymatic methods (ribonuclease digestion) to break down the RNA molecule into smaller fragments and analyzed each segment to determine its sequence [433].

Considering the need of a comprehensive approach for extracting DNA, RNA or Protein sequences across different species, in 1977, at the newly established Laboratory of Molecular Biology in Cambridge, Sanger et al. [946] developed the

M. Nabeel Asim et al., *Artificial Intelligence for Molecular Biology*,
https://doi.org/10.1007/978-3-031-90450-9_3

very first DNA sequencing technique namely "the dideoxy chain-termination or Sanger method". At the same time, with an aim to produce human proteins with therapeutic potential, the very first genetic engineering company "Genetech" was launched. Sanger's Noble Prize winning technique has been extensively used for sequencing various model organisms and entire human genome. Specifically, in 1977, the first living organism, a single-stranded DNA virus namely Phage $- X174$ was sequenced using Sanger technique [946]. The prime motivation behind the sequencing of Phage $- X174$ was to acquire a better comprehension of its genetic makeup, functions, gene expression, regulatory components, and pathogenesis of viruses [946]. Building on this work, in 1995, complete genome sequence of bacterium namely Haemophilus influenzae along with associated information such as genes and functional elements inside genome were published. Afterwards, several eukaryotic and prokaryotic living organisms have been sequenced such as Mycoplasma genitalium in 1995, Saccharomyces cerevisiae in 1996, Escherichia coli around 1997, Saccharomyces cerevisiae and Caenorhabditis elegans around 1998, Arabidopsis thaliana around 2000, and human genome around 2001 [44, 334].

In a nutshell, there was a marathon of sequencing diverse organisms and discover key insights about basic principles of genome organization, replication, expression, and host interaction. Genetic sequence analysis technologies that are competent in analyzing a massive amount of DNA molecules in parallel manner are labeled as next-generation sequencing (NGS) technologies [40, 357]. NGS technologies enable the exploration of hundreds and thousands of genes to demystify the associations of genetic variations with different diseases and biological phenomenon [40, 357]. NGS technologies provide the basis to control the regulation of gene expression and have revolutionized the development of new applications in clinical and genomic research, reproductive health, environmental, agricultural and forensic science [132, 241, 330, 417, 551, 917, 1191].

The high throughput paradigm of NGS technologies has given birth to exponentially increasing Genomics and Proteomics data which is of great significance [224]. The humongous Genomics and Proteomics data analysis is useful to comprehensively understand diverse biological processes, gene expression patterns and their associations with the initiation, progression and treatment of different diseases through optimizing therapeutics [781]. An accurate exploration of genetic data will not only unlock unimaginable insights of life sciences, but it will also turn personalized healthcare into reality [781].

This chapter provides a comprehensive overview of diverse spectrum of DNA, RNA, and protein sequence analysis tasks. Additionally, it establishes a foundational stage for the subsequent seven chapters by presenting a high-level overview of AI-driven genomics and proteomics sequence analysis pipeline. Specifically, it offers essential biological knowledge and motivations for the development of AI applications across 47 DNA, 45 RNA, and 62 protein sequence analysis tasks. To streamline the application of AI methodologies within genetic sequence analysis landscape, it aligns DNA, RNA, and protein sequence analysis tasks with fundamental AI paradigms such as classification, regression, and clustering.

Additionally, the chapter presents a comprehensive architecture for an end-to-end AI-driven genomics and proteomics sequence analysis pipelines. Within this framework, it illustrates various strategies for benchmark datasets development and provides a high-level overview of contemporary sequence representation learning techniques that effectively transform raw genetic sequences into rich numerical representations. It delves into the intricacies of feature engineering methodologies and a comprehensive knowledge of machine learning and deep learning models along with associated evaluation metrics. The chapter concludes with a holistic overview of existing genomics and proteomics sequence analysis frameworks available to the research community. Partial content of this chapter is published in [57–60].

3.1 Genomics and Proteomic Sequence Analysis Landscape

Genomics and Proteomics sequence analysis are scientific studies for understanding the complex languages of Deoxyribonucleic acid (DNA), Ribonucleic acid (RNA), and Protein. Deoxyribonucleic acid (DNA) contains instructions for the development, operation, growth, and reproduction of all living organisms [28]. Organisms grow from a single cell, such as a fertilized egg, to a multi-cellular adult and replace worn-out or damaged cells with new ones over time to maintain the health of tissues and organs through cell division. When a cell divides, each new cell requires an exact copy of the DNA to function correctly [28]. DNA replication and repair processes ensure that each daughter cell receives the same genetic information as the parent cell, which is essential for the survival and proper functioning of all living organisms. However, DNA is continuously exposed to various internal and external factors that can cause damage. Internal factors include harmful molecules produced within the cell, spontaneous chemical changes, errors during DNA replication, and natural cell chemicals. External factors include UV and ionizing radiation, harmful chemicals, environmental pollutants, and lifestyle choices like diet and smoking. Alterations in genetic information due to internal or external factors can lead to complex diseases and disorders such as cancer [28]. To detect susceptibility, initiation, and progression of such diseases at early stages, scientists perform large-scale DNA sequence analysis. Through DNA sequence analysis, scientists can decode the intricate genetic data by uncovering the origins of genetic mutations and disorders. Additionally, this analysis is crucial for the development of targeted therapies and the advancement of personalized medicine [28].

Similarly, Researchers are actively exploring RNA sequence data to understand the distinct roles of RNAs in living organisms (e.g. protein synthesis and gene regulation) and their associations with various diseases, including cancers, genetic disorders, and neurodegenerative conditions [543]. protein sequence analysis objective is to unlock diverse types of a wealth of knowledge about biological processes and genetic disorders. Researchers are gaining deep understanding about biological processes in which proteins are involved, such as enzyme activity, cell signaling, and immune responses. Researchers are also gaining understanding about genetic disorders by pinpointing mutations that alter proteins functionalities. It helps in forecasting diseases susceptibility by finding unique protein signatures, or biomarkers that are linked to particular disease states. Specifically, this analysis enables researchers to identify individuals at higher risk for developing certain diseases before symptoms even appear. Following sub-sections provides high level overview of different tasks that are being performed at DNA, RNA, and protein levels.

3.1.1 *Biological Foundations of DNA Sequence Analysis Goals and Tasks*

With an aim to find molecular basis of diseases initiation and progression, their effective detection at early stages, and development of potent drugs, researchers are trying to understand DNA sequence language by performing a variety of sequence analysis tasks. Every unique DNA sequence analysis task aims to enhance the understanding of one specific aspect of DNA and a bunch of tasks can enhance the understanding of specific major biological goal. To summarize the biological background of 44 distinct DNA sequence analysis tasks, we have categorized them into 8 major biological goals. Figure 3.1 depicts the biological categorization of 44 unique DNA sequences analysis tasks into 8 different goals including genome structure and stability, gene expression regulation, gene analysis, gene network

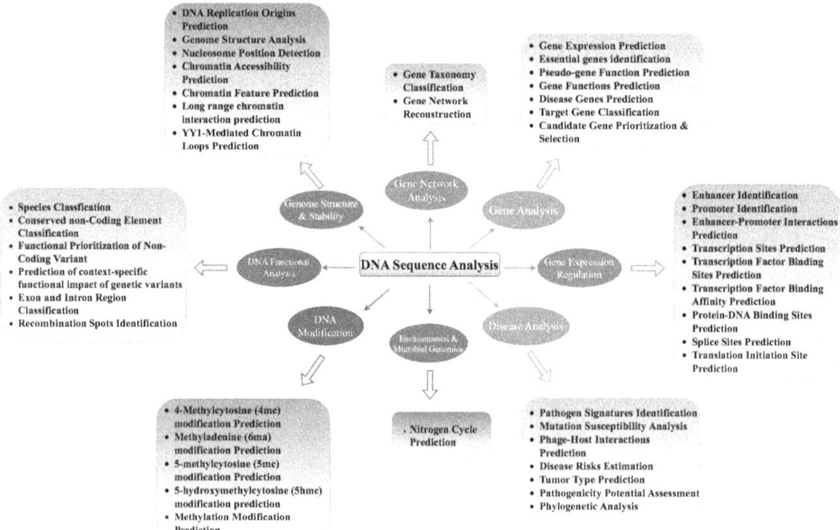

Fig. 3.1 DNA sequence analysis tasks and goals

analysis, DNA modifications prediction, DNA functional analysis, environmental and microbial genomics, and disease analysis.

In living organisms, DNA is packaged at multiple levels to condense vast genetic information into a well-organized structure within the cell nucleus [28]. At the first level, DNA is wrapped around histone octamers also known as nucleosomes. These nucleosomes further assemble into chromatin, which then folds and condenses into an even more compact structure known as the genome [28]. The exploration of genome structure and stability is pivotal in understanding the biological intricacies and potential therapeutic avenues. Genome structure can affect how genes are accessed and used. Disruptions in this structure, like missing or misplaced DNA sections, or changes in how tightly DNA is wrapped around histone octamers, or irregularities in nucleosomes positions can lead to genes being turned on or off at the wrong times or in the wrong amounts [28]. This can cause various diseases and biological disorders. DNA is an instruction manual that control biological functioning within living organisms. If genome gets unstable, the manual gets messed up such as typos, missing sections, etc. It can lead to cells uncontrolled growth (cancer) and genes improper working (many diseases) [28]. In a nutshell, a stable genome possesses clear, complete instruction manual, essential for keeping biological functions working smooth. To better understand genome structure and stability, it is essential to explore various tasks such as DNA Replication Origins Prediction [1036, 1166], Genome Structure Analysis [275, 616], Nucleosome Position Detection [302, 396], Chromatin Accessibility Prediction [306, 376, 1214], Chromatin Feature Prediction [981, 1147, 1214], Long-range Chromatin Interaction Prediction [326, 1012], and YY1-Mediated Chromatin Loops Prediction [2, 1216]. These tasks are crucial for comprehending the intricate mechanisms governing genetic information processing and regulation within cells [165].

DNA replication origins prediction is fundamental as accurate replication of the genome is vital for maintaining genomic stability [1036]. The prediction of replication origins involves calculating DNA structural properties to identify sites crucial for initiating DNA replication [1036]. Understanding where these sites are located and how they are specified is essential for comprehending DNA replication and ensuring genome integrity [757]. Genome structure analysis plays a pivotal role in deciphering the organization and arrangement of genetic material within the cell [275]. By analyzing the structural features of the genome, researchers can gain insights into the functional and spatial organization of chromosomes, aiding in the identification of genomic elements involved in gene regulation and phenotypic variations [275, 491]. Furthermore, nucleosome position detection is essential for understanding how nucleosomes, the basic units of genome, are arranged along the DNA strand [302, 719]. This information is crucial for elucidating gene regulation mechanisms and chromatin dynamics within the cell [302, 719]. Chromatin accessibility prediction is a key task that involves determining the regions of chromatin that are accessible for transcription factors and other regulatory proteins to bind [306, 376, 1214]. Prediction of chromatin accessibility across different cellular contexts provides valuable insights into gene regulation and chromatin dynamics [306, 376, 1214]. Chromatin feature prediction complements accessibility prediction by identifying specific chromatin features and epigenetic markers that influence gene expression and regulatory processes [753, 981, 1147, 1214]. These features include transcription factor (TF) binding sites, DNase I-hypersensitive sites (DHS), and histone marks (HM). By understanding these features, researchers can unravel the mechanisms underlying chromatin regulation and gene expression [1147]. Long-range chromatin interactions make bridges between distant enhancers and promoters. These interactions enable interactions between enhancers and promoters by bringing them closer to each other [326, 1012]. YY1-mediated chromatin loops prediction provides comprehensive understanding about gene regulation [2, 223, 1216]. YY1 is a protein that makes loop between enhancers and promoters. These loops are essential for gene regulation and by predicting these loops, we can see which genes can be controlled through YY1 protein [2, 223, 1216]. This knowledge is valuable for understanding diseases where gene regulation goes wrong. To sum up, only through multi-dimensional exploration of genome structure and stability, researchers can discriminate healthy cellular processes from malfunctioned processes, find the root causes of diseases, and develop potent therapies.

Another major goal of molecular biologists behind is gene expression regulation. Gene expression regulation provides fundamental insights into how genes are activated or repressed in response to various cellular cues [580]. Specifically, researchers are trying to unravel the intricate mechanisms that control when and up to what extent specific genes are turned on or off in different cells and tissues [580]. This knowledge forms the basis for understanding the functional behavior of genes in different biological contexts and sets the stage for further analyses. Hence it holds immense promise for scientists and pharmaceutical industries. This helps scientists to detect irregularities in normal gene expression regulation, the way diseases develop at the molecular level, and identify potential drug targets [580]. Furthermore, this understanding can assist pharmaceutical industries to develop improved diagnostic tools, innovative personalized therapies, and targeted interventions, which will ultimately contribute to advancements in personalized healthcare [580]. Additionally, it can provide a deeper understanding of biological systems which can lead to breakthroughs in biotechnology [580]. For better understanding of gene expression regulation, researchers are performing 9 different DNA sequence analysis tasks including enhancer identification [515], promoter identification [613], enhancer-promoter interactions prediction [717], transcription site prediction [196], transcription factor binding sites prediction [342], transcription factor binding affinity prediction [1128], protein-DNA binding sites prediction [637], splice sites prediction [637], and translation initiation site prediction [1207]. Enhancers [45, 338, 411, 465, 482, 515, 560, 563, 571, 591, 620, 646, 653, 655, 656, 704, 718, 779, 1043, 1067, 1137, 1156] and promoters identification [562, 603, 613, 974, 1149, 1215, 1215], along with their interactions [434, 756, 931, 975, 1246] prediction are important to decipher a complex control panel for gene expression [515, 613, 717]. Enhancers are known as distant switches of genes, while promoters are the landing sites where gene activation starts. Identification of these elements and predicting how they loop together, provide a comprehensive understanding of gene regulation, including which genes are activated or repressed, the intensity of their expression, and the specific cell types involved [643, 1212]. This knowledge reveals the intricate regulatory code that governs gene expression and offers valuable insights into the mechanisms underlying normal cellular function as well as the dysregulation that may contribute to various diseases.

Furthermore, prediction of different genomic sites including transcription sites [196], transcription factor binding sites [35, 195, 481, 723, 907], transcription factor binding site affinity [1128], Protein-DNA-binding site [496, 637, 673, 735], splice site [214, 481, 495, 662, 897], and translation initiation site [196, 614] provide deep insights into gene expression regulation. A transcription site refers to the specific location on the DNA where the process of transcription takes place. Transcription is the synthesis of RNA from a DNA template, and the transcription site represents the region where the RNA polymerase enzyme binds and initiates the transcription process. Whereas, transcription factor binding sites are specific DNA sequences where transcription factors (proteins), that regulate gene expression, bind. These binding sites are typically located near the transcription start site and are recognized by transcription factors to control the initiation or repression of transcription. In contrast, transcription factor binding site affinity refers to the strength or affinity with which a transcription factor binds to its specific binding site on DNA. It represents the likelihood of a transcription factor binding to its target site and influencing gene expression. A protein-DNA binding site refers to any region on the DNA where a protein binds. This can include transcription factors, as mentioned earlier, as well as other proteins involved in various cellular processes such as DNA replication, repair, and chromatin remodeling. Splice sites are specific sequences within a gene's DNA that mark the boundaries of introns and exons. During the process of RNA splicing, introns are removed from the pre-mRNA molecule, and exons are joined together to form the mature mRNA. Splice sites are essential for the accurate and precise splicing of RNA. Translation initiation site (TIS) is the specific location on the mRNA molecule where the process of translation begins. TIS prediction seems like a RNA sequence analysis task, however, in molecular biology research, in order to study gene expression, researchers are synthesizing complementary DNA (cDNA) data from messenger RNA (mRNA) template through a process called reverse transcription. In the context of cDNA data, the translation initiation site (TIS) represents the position where the ribosome, the cellular machinery responsible for protein synthesis, binds to the mRNA to initiate translation. The TIS is typically identified by the presence of specific start codons, such as AUG, which serve as signals for the ribosome to start protein synthesis.

To better understand gene functions and their roles in disease initiation, researchers are exploring various aspects such as gene expression prediction [20, 841], identification of essential [564, 676, 885, 902, 1115, 1211, 1211] and disease-specific genes [769], gene functions prediction [295, 443], pseudo-gene function prediction [295], target gene classification [42], and candidate gene prioritization [998]. Overall together these tasks provide a comprehensive platform for disease diagnosis and development of treatment strategies by uncovering disease mechanisms, identifying potential therapeutic targets, and organizing genes into functional categories. Specifically, gene expression prediction provides useful information about the level of gene activity in different cells or tissues [12]. This task is vital for understanding the molecular mechanisms underlying complex diseases like cancer and identifying potential therapeutic targets. Essential genes identification is another critical task in gene analysis that helps researchers pinpoint genes that are crucial for an organism's survival and development [455]. This task is particularly important in understanding gene function and the genetic basis of various disorders. Gene functions prediction elucidates the roles of genes in different pathways and biological processes and provides valuable insights into disease mechanisms and potential therapeutic interventions.

Apart from gene functions prediction, pseudo-gene function prediction has gained a lot of attention as a critical task in gene analysis [295]. Pseudogenes were once thought to be useless DNA because they can't code for proteins due to mutations that happened over time. However, recent studies have shown that pseudogenes actually play important roles in controlling genes, especially in cancer. For instance, the pseudogene PTENP1 helps to regulate the tumor suppressor gene PTEN in various cancer conditions, showing that pseudogenes can have important functions. Pseudogenes function prediction offers numerous advantages, including better understanding of gene regulation, disease mechanisms, evolutionary biology, and the potential for new biomarkers and drug targets. In addition, disease gene prediction is a pivotal task in gene analysis that focuses on identifying genes associated with specific diseases or disorders [243]. By pinpointing disease-related genes, researchers can unravel the genetic basis of diseases, discover novel biomarkers for diagnosis and prognosis, and develop targeted therapies. This task is instrumental in precision medicine approaches, where understanding the genetic underpinnings of diseases is crucial for personalized treatment strategies. Target gene classification involves categorizing genes based on their functions, interactions, or regulatory mechanisms [1189]. By classifying target genes, researchers can better understand gene networks, signaling pathways, and biological processes. This task is essential for deciphering the complex relationships between genes and their roles in health and disease. Candidate gene prioritization and selection are critical tasks in gene analysis that aim to identify genes with the highest likelihood of being involved in a particular biological process or disease [211]. By prioritizing candidate genes, researchers can focus their efforts on studying genes that are most likely to have significant effects, accelerating the discovery of novel gene functions and disease mechanisms. This task is crucial for efficiently allocating research resources and maximizing the impact of genetic studies. Aforementioned 7 DNA sequence analysis tasks are essential for advancing our understanding of genes and their roles in health and disease. By leveraging these tasks, researchers can unravel the complexities of the genome, uncover novel gene functions, and pave the way for innovative diagnostic and therapeutic strategies in various fields of biology and medicine.

Furthermore, gene network analysis is a promising goal that seeks to comprehend the intricate interactions and relationships between genes within a biological system.

Two primary tasks within Gene Network Analysis are Gene Taxonomy Classification and Gene Network Reconstruction. Gene Taxonomy Classification [724, 911, 1017] involves categorizing genes based on their evolutionary relationships and functional similarities, providing a structured framework for organizing genetic information. Gene Taxonomy Classification plays a crucial role in gene network analysis by offering a foundational structure for understanding the evolutionary history and functional relationships between genes. By classifying genes into taxonomic groups based on shared characteristics and evolutionary relatedness, researchers can infer valuable insights into the origins and evolutionary trajectories of genes within a network [199]. This classification allows for the identification of core genes that have remained conserved throughout evolution, providing a basis for inferring phylogenetic relationships and understanding the fundamental building blocks of gene networks. Moreover, Gene Taxonomy Classification enables researchers to utilize existing knowledge about gene functions and evolutionary relationships to guide Gene Network Reconstruction. By categorizing genes into taxonomic groups, researchers can pinpoint gene clusters with similar functions or evolutionary origins, facilitating the identification of modules within gene networks that exhibit coordinated activity [200]. This classification serves as a roadmap for exploring the functional roles of genes within a network and understanding how these roles have evolved over time. On the other hand, Gene Network Reconstruction [273, 810, 811] involves creating a detailed map of the interactions and regulatory relationships between genes within a cell or an organism. The primary input for gene network reconstruction is gene expression data obtained through high-throughput techniques like RNA sequencing (RNA-seq) or microarrays. This task is pivotal for understanding how genes work together to control various biological functions and processes [557]. By reconstructing gene networks, researchers can uncover key regulatory hubs involving highly connected genes, clusters of closely interacting genes, pathways, and interactions that steer cellular functions and responses to external stimuli [947].

DNA modifications prediction is also a crucial goal where researchers aim is to decipher how tiny tweaks to the DNA code can lead to big changes in cellular functions. In DNA modifications, distinct chemical groups are added to specific locations on the DNA molecule. These additions do not change the actual sequence of nucleotides (A, C, G, T) but can alter the physical properties of DNA sequence. Understanding these modifications, such as 4-Methylcytosine (4mc) [300, 516, 618, 755, 861, 1121, 1124, 1160, 1188, 1254], Methyladenine (6ma) [1, 297, 449, 605, 750, 914, 1004, 1233], 5-methylcytosine (5mc) [948, 1000], 5-hydroxymethylcytosine (5hmc) [488, 1174, 1186, 1247], and methylation modifications [478, 488, 750, 1072, 1174, 1186, 1247], are essential for advancing our comprehension of epigenetic regulation [534, 678, 941]. Specifically, methylation modifications that occur due to the addition of methyl groups to DNA molecules, play a pivotal role in regulating gene expression and maintaining genomic integrity. Similarly, Methyladenine modifications, such as DNA N6-methyladenine (6mA) occurs due to the addition of a methyl group to the adenine base of DNA. DNA

6mA modifications dynamically influence DNA thermal stability, curvature, and transcription factor interactions, impacting gene expression in a heritable manner. Understanding the prediction of 6mA sites are pivotal for both basic and clinical research as it aids in the identification of gene expression patterns and potential epigenetic changes induced by environmental factors. These predictions enhance our ability to study the role of 6mA modifications in diseases and could lead to improved therapeutic strategies, highlighting the relevance of accurate prediction methods in unraveling the complexities of DNA modifications. Moreover, 5-Methylcytosine (5mc) modification occurs due to the addition of a methyl group to the cytosine base of DNA. Whereas, 5-Hydroxymethylcytosine (5hmc) modification is an oxidized derivative of 5mc, where an additional hydroxyl group (-OH) is added to the methyl group of 5mc. Prediction of 5-Methylcytosine (5mc) and 5-Hydroxymethylcytosine (5hmc) modifications is essential for decoding their roles in gene regulation, developmental processes, and disease states. These critical epigenetic modifications are dynamically regulated by enzymes and influence gene expression crucial for neuronal differentiation and cellular proliferation. Abnormal levels of these modifications have been linked to diseases like cancer. Precise prediction of 5mc and 5hmc sites is useful for development of targeted therapies and improved prognostic assessments.

Functional genomics is also a critical goal that encompasses multiple sub-tasks including species classification [753], conserved non-coding elements (NCEs) classification [344], functional prioritization of non-coding variants [1147], prediction of context specific functional impact of genetic variants [326], exon and intron region classification [14], and recombination spots identification [261]. Each of these tasks plays a vital role in unraveling the complexities of genetic regulation and molecular mechanisms within the genome. In biomedical research, understanding the genetic similarities and differences between humans and other species is crucial for modeling diseases and studying genetic disorders. Majority of the genome is conserved across different species which makes it difficult to distinguish humans and non-human species. Despite very high genetic similarity across species (<10% sequence divergence), small differences are extremely valuable and they have significant biological implications. Specie classification determines the source species of genetic sequences based on such differences and pave way for better modeling diseases, and studying genetic disorders [753]. Conserved non-coding elements classification is another critical task in functional genomics that focuses on identifying and understanding non-coding regions of the genome that are evolutionarily conserved across different species [344]. It is essential for advancing our understanding of gene regulation, evolutionary biology, and the genetic basis of diseases. By elucidating the functions of these non-coding regions, researchers can gain insights into the intricate regulatory networks that govern gene expression and cellular processes, and contribute to the development of targeted therapies.

Functional prioritization of non-coding variants [1147] is another crucial task for making sense of the vast amount of genetic data generated by modern sequencing technologies. By identifying which variants have significant biological impacts, researchers can gain a deeper understanding of the genetic architecture of complex diseases, uncover novel therapeutic targets, and advance the field of precision medicine. This prioritization is essential for translating genomic research into practical health benefits and ultimately improving patient outcomes and advancing our knowledge of human biology [1147]. As functional prioritization of non-coding variants task involves identifying which non-coding variants among millions are likely to have functional consequences, it does not account for the specific context in which these variants might exert their effects. Whereas, prediction of context-specific functional impact of genetic variants aims to provide a detailed understanding of how specific variants influence gene function in different contexts (e.g. specific tissue) [326]. This is particularly important for genetic studies that seek to uncover the mechanisms by which variants contribute to disease phenotypes. Unlike functional prioritization of non-coding variants task which only filters the variants that are most likely to have functional significance. Prediction of context-specific functional impact of genetic variants provides a finer level of detail by predicting the actual effect of a variant on gene expression or other functional outcomes in specific tissues. This granularity is essential for precisely understanding the specific biological mechanisms and for developing targeted therapies [326].

Exon and intron region classification is crucial for understanding gene structure and function within the genome. Exons are coding regions that are translated into proteins, while introns are non-coding regions that are spliced out during mRNA processing. By classifying exons and introns, researchers can describe gene boundaries, identify functional elements, and elucidate the mechanisms of gene expression regulation [938]. This task is essential for deciphering the genetic code and unraveling the complexities of gene regulation in health and disease. Recombination spots identification is a pivotal task in functional genomics that focuses on mapping regions of the genome where genetic recombination events occur. Genetic recombination is a natural process where DNA segments are exchanged between two chromosomes during cell division. Recombination plays a vital role in generating genetic diversity, ensuring proper chromosome segregation, and driving evolution [791]. By identifying recombination hot spots, researchers can gain insights into the mechanisms underlying genetic diversity and genome evolution, shedding light on the processes that shape genetic variation and adaptation in populations. In conclusion, the tasks related to functional genomics, including species classification, conserved non-coding element classification, functional prioritization of non-coding variant, prediction of context-specific functional impact of genetic variants, exon and intron region classification, and recombination spots identification, are essential for advancing our understanding of genetic regulation, molecular mechanisms, and disease pathogenesis. By delving into these tasks, researchers can unravel the complexities of the genome, decipher the genetic basis of diseases, and pave the way for precision medicine and personalized healthcare interventions tailored to an individual's genetic profile.

Another goal of researchers is to study overlap between two distinct fields namely environmental science and Microbial genomics [275]. This interdisciplinary study enables researchers to explore how environmental factors such as pollution, climate change, and agricultural practices affect on function and diversity of microbial communities [275]. A key area of focus in this field is the nitrogen cycle prediction. By examining the genomes of microbes involved in nitrogen fixation, nitrification, and denitrification, scientists can predict how these processes might respond to environmental changes [600]. This prediction provides understanding about potential impacts of environmental shifts on ecosystem health [530] and nitrogen availability, which are essential for plant growth and overall biogeochemical cycles [874].

From all 8 different biological goals, disease analysis goal has received huge attention in scientific community as it aims to understand, diagnose, and treat various illnesses. Within this field, several tasks play a vital role in enhancing our comprehension of diseases. One such task is Pathogen Signatures Identification [746], which involves identifying specific markers or characteristics of pathogens that can aid in their detection and classification [1236]. By pinpointing these signatures, researchers can develop targeted diagnostic tools and therapies, ultimately improving disease management and control. Mutation Susceptibility Analysis [1165] is another essential task in disease analysis. This task focuses on investigating the genetic variations that make individuals more prone to developing certain diseases [177]. Understanding mutation susceptibility can aid in personalized medicine approaches, where individuals at higher risk can be identified early for preventive interventions or closer monitoring. Phage-Host Interactions Prediction [793, 822, 1074] is a task that delves into the relationships between bacteriophages and their host bacteria [284]. By predicting these interactions, researchers can gain insights into how phages influence bacterial populations, which is crucial for developing phage-based therapies to combat bacterial infections and antibiotic resistance. Disease Risks Estimation [723] is a fundamental aspect of disease analysis that involves assessing the likelihood of an individual developing a particular condition based on various factors such as genetics, lifestyle, and environmental exposures [399]. Accurately estimating disease risks enables healthcare providers to offer targeted interventions and counseling to high-risk individuals, potentially preventing the onset or progression of diseases. Tumor Type Prediction [784] is a significant task in disease analysis that focuses on identifying the specific type of tumor a patient may have based on various characteristics such as genetic markers, imaging features, and histopathological findings [341]. Predicting tumor types is essential for determining the most effective treatment strategies and prognostic outcomes for patients with cancer. Pathogenicity Potential Assessment [275] is a critical task that involves evaluating the ability of pathogens to cause disease in a host. By assessing the pathogenicity potential of different microorganisms, researchers can prioritize the development of interventions against the most virulent pathogens, thereby improving disease prevention and control strategies. Phylogenetic Analysis [846] is a key component of disease analysis that involves studying the evolutionary relationships between different strains of pathogens or tumor cells. Phylogenetic analysis provides insights into the origins, spread, and diversification of diseases, aiding in the development of targeted interventions and understanding disease transmission dynamics.

3.1.2 Biological Foundations of RNA Sequence Analysis Goals and Tasks

This section offers a high-level overview of RNA sequence analysis world. Specifically, it focuses on 47 distinct tasks that researchers are employing to unlock the secrets of RNA molecules. Scientists aim behind analysis of these tasks is to gain a deeper understanding of RNA's diverse biological roles within living organisms. Additionally, these tasks play an important role in understanding potential associations between RNA and various diseases. To facilitate a more organized comprehension of these 47 tasks, we have categorized them into 10 distinct goals. These goals represent the core objectives that researchers aim to achieve through RNA sequence analysis. A visual representation of these 10 major goals, along with the associated tasks that contribute to them, is presented in Fig. 3.2. This graphical illustration serves as a valuable roadmap to provide a clear overview of the multifaceted nature of RNA sequence analysis.

The central dogma of molecular biology states that genetic information from Deoxyribonucleic acid (DNA) is transcribed into Ribonucleic acid (RNA) and RNA is translated to proteins [136]. Only 3% region of DNA is based on protein coding information, whereas, 97% region of DNA is transcribed into different non-coding RNA molecules. Initially, scientists thought that non-coding RNAs were useless or just noise, but recent studies have unveiled their crucial roles in gene regulation

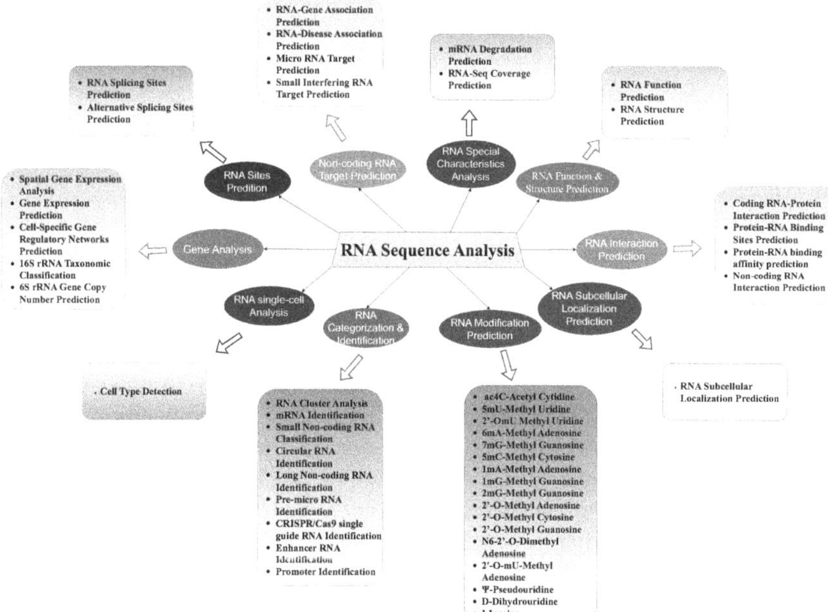

Fig. 3.2 RNA sequence analysis tasks and goals

and cellular development. These groundbreaking discoveries have spurred scientific researchers to delve deeper into the functions of non-coding RNAs and their potential links to a broad range of diseases [105]. A critical first step towards deciphering functional landscape of non-coding RNAs lies in their comprehensive categorization, including the differentiation of non-coding RNAs from coding RNA molecules [245, 889]. Furthermore, according to region of interest, ncRNA classification is performed at different levels such as ncRNA classification into small non-coding RNA [50] and Long-non coding RNA [989]. Small non coding RNA further classification into 13 distinct classes including miRNA, ribozymes, 5S rRNA, 5_8S_rRNA, HACA-box, CD-box, tRNA, scaRNA, IRES, Intron_gpI, Intron_gpII, riboswitch, and leader. Moreover, non coding RNA classification into circular RNAs [761, 953] and long non-coding RNAs (lncRNAs) [212]. Similarly, non coding RNA classification into pre-cursor microRNA [379] and microRNA. Apart from categorization of these different types of RNAs, identification of three special types of RNAs including single guide RNA (sgRNA) [1245], enhancer RNA [1196], and Promoter RNA [1035] is also very crucial.

Each type of RNA classification serves a specific purpose in enhancing our understanding of the diverse regions and functions of RNA molecules. For instance, discrimination of coding RNAs from non-coding RNAs [889] helps to unravel the regulatory mechanisms controlling gene expression and protein synthesis [587]. Small non-coding RNA classification assists to decode ncRNAs functions like post-transcriptional gene regulation, and involvement in regulatory pathways. Circular RNA classification helps to uncover their roles in cellular processes and disease pathogenesis to offer new insights into RNA biology and potential biomarkers for various conditions [761]. LncRNAs classification assists to better understand chromatin remodeling, transcriptional regulation, and cellular signaling [989]. Pre-microRNAs undergo processing to generate mature microRNAs that target specific mRNAs for degradation or translational repression [378]. Pre-microRNAs classification assists to understand the biogenesis and regulatory roles of microRNAs in gene expression and cellular processes [378].

Furthermore, sgRNA identification is critical for gene editing which has revolutionized the field of biology by enabling the scientists to make precise changes to the genetic material, such as adding, deleting, or modifying specific DNA segments. In a powerful CRISPR/Cas9 gene editing system, sgRNAs guide the Cas9 molecular scissor to target specific DNA sequences. Precise sgRNAs identification can help the scientists to select sgRNAs with higher editing efficiency to reduce the risk of unintended off-target effects [1245]. Enhancer RNAs are non-coding enhancer RNAs (eRNAs) transcribed from enhancer regions of the genome [1196]. ERNAs identification helps to understand the activation of target genes and formation of chromatin loops that bring enhancers and promoters into close proximity [1196]. Also, it provides insights into gene regulation mechanisms and cell identity to offer new avenues for understanding transcriptional control in development and disease [1196]. Promoter RNA molecules are transcribed from promoter regions of genes, which play roles in transcriptional regulation and gene expression control [1035]. Promoter RNAs identification helps to decipher the regulatory networks governing gene expression and cellular processes [1035]. With the passage of time, researchers are discovering new classes of RNAs to further extend its classification tree. RNA cluster analysis is useful to discover new RNA classes as it involves the grouping of RNA molecules based on similarities in sequence, structure, or function [16]. It also enables the researchers to discover the functions, interactions, regulatory networks of new types of RNA molecules on the basis of their similarities with existing types of RNA molecules [16].

Apart from RNA categorization, another important RNA sequence analysis goal mentioned in Fig. 3.2 is RNA subcellular localization prediction. RNA subcellular localization refers to the specific regions within a cell where different RNA molecules are found. This targeted positioning allows them to interact with the right machinery and carry out their functions effectively [52, 639, 1187, 1219]. For example, mRNAs localization in the nucleus regulates gene expression by removing defective RNAs and influencing the expression levels of ncRNAs [639, 1187, 1219]. mRNA localization in the cytoplasm provides quantitative and spatial control over protein production and contributes to cell membrane maintenance and nutrient metabolism control [639, 1187, 1219]. miRNA localization in the nucleus is vital for cell division and promoting organism growth by replacing worn-out cells [52]. Whereas, miRNAs localization in the cytoplasm induces disease gene silencing by binding to mRNA molecules [52]. snoRNAs localization in the nucleus plays a crucial role in post-transcriptional regulation by guiding RNA modifications [52]. LncRNAs localization in nucleus controls gene expression through chromatin remodeling [52]. Besides this, its localization in the cytoplasm prevent mRNA degradation and suppress regulatory effects of miRNAs on mRNAs [52]. Dysregulated localization of RNA molecules disturbs the production and distribution of proteins inside cells which directly impacts cellular function and cellular response to environmental changes [52]. It is associated with diverse complications like Cancer and Huntington's disease. Researchers are determining single compartment as well as multi-compartment subcellular localization of RNAs to better comprehend RNAs functional paradigms, relevance to diseases, therapeutic significance, and potential as drug targets [52, 53, 639, 1187, 1219].

In addition to RNA classification and subcellular localization, RNA sequence analysis delves into the intricate network of interactions among RNA molecules. It investigates both RNA-RNA interactions and RNA-protein interactions. Grasping these interactions is crucial for understanding how RNA molecules orchestrate complex biological processes. Inter-RNA interactions enable communication between different types of RNAs which physically bind together inside the cells. This interplay can influence gene expression by either activating or repressing target RNAs, assist in RNA processing through modifications or assembly, and direct RNA molecules to specific cell locations for specialized functions. Understanding these interactions is crucial for elucidating the RNA's functional networks and regulatory pathways.

Considering every unique interaction contributes differently to the comprehension of gene regulation, disease mechanisms, and therapeutic development. Researchers are predicting interaction between non-coding RNAs like lncRNA-miRNA [1226], circRNA-miRNA [374], coding RNA and proteins [394] like mRNA-protein, tRNA-protein, and Viral RNA-protein. LncRNAs can act as miRNA "sponges," to push miRNAs away from their mRNA targets. Predicting lncRNA-miRNA interactions [1226] unveils another layer of complex gene regulatory networks by exposing how lncRNAs contribute to the fine-tuning of gene expression. LncRNA and miRNA are involved in various diseases including cancer, cardiovascular, and neurodegenerative diseases. Due to extensive involvement, lncRNA-miRNA interactions can serve as biomarkers for diagnosis, prognosis, and treatment response [1226]. Likewise, circRNAs are highly stable due to their circular structure. This structure makes them efficient miRNA sponges. CircRNA-miRNA interactions [374] prediction helps to identify these stable regulatory elements that can influence miRNA availability and activity over extended periods. Also, circRNAs have emerged as potential therapeutic agents and targets due to their ability to modulate miRNA activity. CircRNA-miRNA interaction prediction can lead to the development of circRNA-based therapies, which could offer advantages in stability and specificity over linear RNA-based approaches [374].

Moreover, researchers are also predicting interactions between non-coding RNAs and proteins [394] like lncRNA-protein, rRNA-protein, snoRNA-protein, snRNA-protein, and miRNA-protein to expand their understanding of gene expression regulation. For example, LncRNAs perform their functions by interacting with the RNA binding proteins. Hence, lncRNA-protein interaction prediction [394] provides useful insights into transcriptional and post-transcriptional gene regulation. Similarly, prediction of other non-coding RNAs and proteins provides valuable insights into RNA-protein complexes, regulatory roles, disease associations, therapeutic targets, and systems-level understanding of cellular processes. In addition, to decipher the intricacies of RNA-protein interactions, researchers are also determining RNA-binding-proteins (RBPs) bindings sites on RNAs [1146]. RBPs selectively bind to specific regions on RNA molecules and form complexes that control various cellular processes. Identifying these binding sites helps to understand the way RBPs interact with RNA [1146]. RBPs-RNA interaction plays a key role in regulating various cellular processes, including RNA splicing, transport, stability, and translation [394]. Aberrations in RBPs functions or their interactions with RNA can lead to dysregulation of these processes which contribute to the development and progression of diseases, such as cancer, neurodegenerative disorders, and genetic diseases. By understanding and predicting RBP-RNA interactions, researchers can gain insights into the molecular mechanisms underlying diseases, identify potential biomarkers for diagnosis and prognosis, and develop targeted therapies [394]. For instance, targeting specific RBP-RNA interactions can modulate the expression of disease-related genes and offer new avenues for therapeutic intervention. Likewise, Protein-RNA binding affinity prediction [905] focuses on accurately determining the strength of interactions between proteins and RNA molecules. It aids in understanding the specificity and selectivity of protein-RNA interactions and sheds light on the regulatory networks and pathways involved [905]. Prediction of binding affinities can uncover key residues and motifs involved in the binding interface and facilitate the design of experiments to validate and manipulate these interactions. Dysregulation of protein-RNA interactions is associated with various diseases, including cancer, neurodegenerative disorders, and viral infections [905]. Accurate prediction of binding affinities can aid in identifying disease-associated mutations, potential therapeutic targets, and designing RNA-based therapies. It can also contribute to drug discovery and development by predicting the binding affinities of small molecules or drug candidates targeting RNA-binding proteins.

RNA target prediction is another major goal of RNA sequence analysis which is very similar to RNA interaction prediction with one major difference. Unlike RNA interaction prediction where interacting partners are non-coding RNAs or proteins, it focuses on the interactions of microRNA (miRNAs) with messenger RNA (mRNA), coding transcript sequences (CTS), and non-coding transcript sequences [1220], small interfering RNAs interactions with genes [1130], and non-coding RNA interactions with different diseases (miRNA-Disease, circRNA-disease, and lncRNA-disease). MicroRNA (miRNA) target prediction is vital as miRNAs are known to modulate gene expression by binding to specific mRNA targets, leading to mRNA degradation or translational repression [72, 287]. MiRNA target prediction helps to unravel the intricate regulatory networks within cells [634] and elucidate the roles of miRNAs in various biological processes, including development, differentiation, and disease progression [325, 1220]. Likewise, siRNA target prediction helps to identify the target genes that can be effectively silenced by siRNAs [1130]. siRNAs are synthetic double-stranded RNA molecules that induce sequence-specific gene silencing by targeting complementary mRNA sequences for degradation [122]. Predicting siRNA targets helps to understand functional genomics studies, and develop potential therapeutic applications capable to control gene expression levels [1130]. Non-coding RNAs (ncRNAs) including miRNAs, circular RNAs (circRNAs), and long non-coding RNAs (lncRNAs), have emerged as key players in the regulation of gene expression and the development of diseases [325, 921]. Non-coding RNAs-disease association prediction can provide valuable insights their roles in disease pathogenesis and offer potential disease biomarkers, and therapeutic targets [921]. By accurately predicting RNA targets, researchers can uncover novel regulatory pathways, identify potential therapeutic targets for various diseases, and design RNA-based therapeutics with enhanced specificity and efficacy [970].

Another RNA sequence analysis goal that enhances the understanding of gene expression regulation is RNA Site prediction. To explore the functional implications of different RNA sites, RNA splice prediction [158] and alternative splicing site prediction [787] are two most widely studied areas. RNA splicing is a fundamental post-transcriptional process in eukaryotic gene expression that involves removing introns from precursor messenger RNAs (pre-mRNAs) and joining exons to form mature mRNAs [158]. This process ensures the removal of non-coding regions and the proper arrangement of coding regions for translation into functional proteins. Accurate RNA splicing prediction enhances our understanding of gene function, genetic variant interpretation, disease mechanisms, splicing regulation, and therapeutic target identification. It provides valuable insights into the complexity of gene expression and contributes to various areas of biomedical research and clinical applications [158]. In contrast, alternative splicing refers to the phenomenon where different combinations of exons within a pre-mRNA can be selected and joined together which result in the production of multiple mRNA isoforms [787]. These isoforms can have different exon compositions that lead to the production of distinct protein variants from a single gene. Alternative splicing greatly increases the diversity of proteins that can be generated from a limited number of genes and allows the generation of different protein isoforms with unique functions or regulatory properties [787]. It plays a crucial role in various biological processes, including tissue development, cell differentiation, and response to environmental cues. Many human diseases, such as cancer, neurodegenerative disorders, and genetic diseases, are associated with abnormalities in alternative splicing patterns. By accurately predicting alternative splicing site, researchers can gain insights into disease mechanisms, identify potential biomarkers, and develop targeted therapies. For example BRCA1 gene is associated with an increased risk of breast and ovarian cancer [475, 1123]. Alternative splicing of this gene produces multiple protein isoforms, some of which have been linked to cancer susceptibility and response to treatment. Predicting alternative splicing patterns in the BRCA1 gene can aid in identifying individuals at higher risk and developing personalized treatment strategies [475, 1123].

Furthermore, one more promising RNA sequence analysis goal is to identify and understand the complex landscape of post-transcriptional modifications in RNA molecules. These modifications are chemical alterations that occur to different functional classes of RNA, including ribosomal RNA (rRNA), transfer RNA (tRNA), messenger RNA (mRNA), and long non-coding RNA (lncRNA). Accurate prediction of these modifications is essential as they play significant roles in fine-tuning gene expression, regulating RNA stability, splicing, translation, and protein binding [1231]. For example, N6-methyladenosine (m6A) methylation is reversible and dynamically controlled modification that impacts RNA processing, nuclear export, translation, cell differentiation, sex determination, and stress responses. M6A modification prediction offers insights into RNA degradation regulation, subcellular localization, splicing, and RNA transcript conformational changes. Similarly, ac4C-Acetyl Cytidine modification plays a crucial role in various biological processes, including gene expression regulation, RNA structure, and interactions with other molecules. Ac4C-Acetyl Cytidine prediction helps to comprehend intramolecular interactions, mRNA stability, translational efficiency, and RNA-protein interactions. N1-methyladenosine (m1A) modification is involved in regulating RNA structure and function. Prediction of m1A modification contributes to our understanding of RNA modification-mediated modulation of protein binding and other regulatory mechanisms. Similarly, 5-methylcytosine (m5C) modification is associated the development and progression of several diseases. m5C modification prediction helps in identifying potential disease biomarkers and understanding the regulatory mechanisms of RNA modifications in disease processes. Also, 5-methyluridine (m5U) modification is involved in RNA structure stabilization and RNA-protein interactions. M5U modification prediction enhances our understanding of the interplay between RNA and proteins and their functional implications. m6Am (N6,2'-O-dimethyladenosine) modification prediction helps in studying the regulatory effects of this modification on RNA processing and metabolism. Pseudouridine modification plays vital roles in stabilizing RNA structure and facilitating RNA-protein or RNA-RNA interactions. Pseudouridine modification prediction contributes to understanding the impact of this modification on RNA structure and function. In addition, prediction of different 2'-O-methyl modifications such as Am (2'-O-methyladenosine), Cm (2'-O-methylcytosine), Gm (2'-O-methylguanosine) help in studying their roles in RNA structure and function, as well as their potential regulatory effects on RNA-protein interactions. Prediction of Um modification (2'-O-methyluridine) enhances our understanding of the functional implications of this modification in RNA structure and function. D-Dihydrouridine modification is involved in stabilizing RNA structure and has been implicated in neural and brain development. D-Dihydrouridine modification Prediction contributes to understanding its impact on RNA structure and function. Inosine modification has diverse functions, including RNA editing and regulation of gene expression. Inosine modification prediction helps in studying its roles in RNA function and gene regulation.

Beyond post-transcriptional modifications, RNA function prediction is another promising sequence analysis goal that allows to understand specific biological activities and mechanisms associated with the RNA molecule. Researchers are predicting the functions of different RNAs including MicroRNA (miRNA), rRNA, mRNA, bpRNA-1m, and bpRNA to comprehend cellular regulation and disease development. For instance, miRNA function prediction helps to understand their roles in post-transcriptional regulation, gene expression, and disease development. It aids in uncovering the complex networks of miRNA-mRNA interactions, their implications in cellular processes, and potential therapeutic targets. Similarly, as rRNA is a fundamental component of protein synthesis machinery called ribosome, rRNA function prediction helps in understanding its role in ribosome structure and function. It contributes to our knowledge of translation processes and the regulation of protein synthesis. Also, mRNA function prediction is crucial for understanding its role in gene expression and protein synthesis. A special RNA called bpRNA-1m is actually a 1-methyladenosine modification in base-pairing RNA. bpRNA-1m and bpRNA function prediction help to understand the impact of these modifications on RNA structure, stability and interactions. RNA functions are intimately connected with their structures as the conformational dynamics of RNA molecules influence their functional activities. With an aim to unravel the complex mechanisms of gene expression regulation [259], researchers are deciphering complex folding patterns of RNA molecules through structures prediction. Transfer RNAs (tRNAs) are essential for protein synthesis, and accurate prediction of their structures aids in understanding translation mechanisms [380]. Similarly, predicting the structures of ribonuclease P (RNase P) and various ribosomal RNAs (rRNAs) like 23S rRNA, 5S rRNA, and 16S rRNA is crucial for deciphering their roles in catalysis and ribosome function [380]. Moreover, predicting the structures of other RNAs like tmRNA, telomerase RNA component (TERC), signal recognition particle RNA (SRP), and pseudoknot RNAs provides insights into their functions in translation, telomere maintenance, protein targeting, and gene regulation [380].

Another promising goal of RNA sequence analysis is studying specific features or properties of RNA molecules to bridge the gap between RNA sequence information and functional implications. In this regard, two tasks are being most commonly studied: prediction of degradation rates of mRNA molecules [412] and prediction of coverage or read counts of RNA-seq experiments [632]. Understanding the special characteristics of RNA, such as degradation rates and coverage patterns, is essential for unraveling the mechanisms of post-transcriptional regulation and gene expression control. Specifically, accurate prediction of mRNA degradation rates provides insights into the stability and turnover of mRNA molecules [412]. It helps in understanding the dynamics of gene expression and the impact of mRNA decay on protein synthesis. mRNA degradation prediction aids in studying post-transcriptional regulation and the control of gene expression [412]. Similarly, prediction of RNA-Seq coverage allows precise interpretation of RNA-seq data, which provides information about gene expression levels [632]. It helps in identifying regions of high or low coverage to better indicate differential expression or regulatory events [632]. RNA-Seq coverage prediction aids in the detection of rare or low-abundance transcripts that may have functional significance. Also, it enables the identification of differentially expressed genes and regulatory elements across different conditions or cell types [632].

Another promising RNA sequence analysis goal is gene analysis that is a fundamental aspect of understanding biological processes. Researchers are performing diverse tasks to decode the dynamics of genes and their impact on cellular processes. For example, spatial gene expression analysis is crucial for investigating gene expression patterns in specific locations within tissues or cells [939]. This analysis helps researchers unravel the spatial organization of gene activity and provides insights into tissue-specific functions and regulatory networks [939]. Identification of active genes in specific regions aids in understanding the molecular basis of various physiological processes and diseases [939]. Gene expression levels prediction is another essential task in gene analysis that helps forecast gene expression under different conditions or in various cell types [1231]. This prediction is valuable for understanding how genes respond to stimuli, environmental changes, or genetic modifications [1231]. Anticipating gene behavior in different scenarios aids in identifying key regulatory elements and pathways governing gene expression [1231]. Prediction of gene expression is crucial for elucidating regulatory mechanisms controlling gene activity and predicting gene functions in diverse biological contexts [1231]. Similarly, prediction of cell-specific gene regulatory networks is a significant endeavor in gene analysis that focuses on deciphering interactions among genes and regulatory elements within specific cell types [939]. By the prediction of these networks, researchers can describe regulatory relationships governing gene expression in distinct cell populations [939]. Understanding cell-specific regulatory networks is vital for unraveling cell identity, differentiation processes, disease mechanisms, cellular functions and dysregulation [939]. This prediction is essential for elucidating gene regulatory networks orchestrating cellular processes and phenotypic traits [939]. Taxonomic classification of microbial species based on 16S rRNA gene sequences is a critical component of gene analysis [1089]. This classification is essential for characterizing microbial communities, assessing biodiversity, and understanding the ecological roles of microorganisms [1089]. Inferring the taxonomic composition of microbial populations through 16S rRNA sequences enables the study of microbial diversity, community dynamics, and ecological interactions [1089]. This classification is crucial for microbial ecology, epidemiology, and biotechnological applications relying on accurate identification of microbial taxonomy [1089]. Prediction of copy number of 16S rRNA genes in microbial genomes is a significant task in gene analysis [701]. Estimation of these copy numbers is important for normalizing microbial abundance data, assessing microbial growth rates, and understanding the metabolic potential of microbial communities [701]. Accounting for variations in gene dosage across different microbial species enables more accurate quantification of microbial populations and functions [701]. Prediction of 16S rRNA gene copy numbers is essential for interpreting metagenomic data, studying microbial physiology, and inferring the ecological roles of microorganisms in diverse environments [701].

Single-cell multi-omics analysis is a promising goal in biological research as it provides a deep insight into cellular diversity and molecular mechanisms at a single-cell level. This goal integrates various omics data types, including genomics, transcriptomics, epigenomics, metabolomics, and proteomics, to offer a comprehensive holistic view of complexities of biological systems, cellular function and regulation [823]. Technological advancements like single-cell RNA sequencing (scRNA-seq) and mass spectrometry have enabled the simultaneous profiling of multiple omics layers within individual cells and transformed the way of studying biological systems [823]. A fundamental task of single-cell multi-omics analysis is clustering analysis, which groups cells based on similarities in their omics profiles to identify distinct cell populations, characterize cell types, and comprehend cellular heterogeneity within a sample [1178]. By applying clustering algorithms to multi-omics data, researchers can reveal hidden patterns and relationships that may not be evident when analyzing individual omics datasets separately. This leads to the discovery of novel cell subtypes and critical regulatory networks for various biological processes [680]. Another essential task in single-cell multi-omics analysis is Gene Ontology (GO) analysis, which aims to clarify the functional significance of genes within identified cell clusters. Through annotating genes with specific biological functions and pathways, researchers can gain insights into the molecular processes driving cellular diversity and behavior [1239]. Integrating GO analysis with multi-omics data enhances the understanding of the biological mechanisms underpinning cell identity and function [153].

Furthermore, survival analysis is another task which holds significant importance in single-cell multi-omics investigations, especially in the realm of cancer research and personalized medicine. By associating omics profiles with patient outcomes, researchers can pinpoint molecular signatures linked to disease progression, treatment response, and overall survival [209]. This process aids in the identification of biomarkers for patient stratification and the development of tailored therapies based on individual molecular profiles. This enhances clinical decision-making and patient outcomes [209]. Genetic perturbation prediction stands out as another critical task in single-cell multi-omics analysis. It focuses on predicting gene regulatory networks and the impact of genetic changes on cellular phenotypes. Through integrating multi-omics data with perturbation information, researchers can forecast how specific genetic modifications or interventions may influence cellular behavior and function [134]. This task is pivotal in grasping the molecular underpinnings of diseases, spotting potential drug targets, and unraveling intricate gene-protein-metabolite interactions within cells [209]. Batch correction emerges as a foundational preprocessing step in single-cell multi-omics analysis that is essential for eliminating technical variations introduced during data generation and processing. Batch effects can confound biological signals to produce misleading outcomes that affect subsequent analyses and interpretations [1132]. By employing batch correction algorithms on multi-omics datasets, researchers can standardize data from different experimental batches or platforms to accurately capture and analyze true biological differences [1132]. In a nutshell, each task within single-cell multi-omics analysis plays a distinctive and critical role in enriching our comprehension of cellular diversity, disease mechanisms, and therapeutic targets. It perfectly sets the stage for groundbreaking discoveries in precision medicine and individualized healthcare.

3.1.3 Biological Foundations of Protein Sequence Analysis Goals and Tasks

Proteins sequences are made up from repetitive patterns of 20 unique amino acids namely alanine, arginine, asparagine, aspartic acid, cysteine, glutamic acid, glutamine, glycine, histidine, isoleucine, leucine, lysine, methionine, phenylalanine, proline, serine, threonine, tryptophan, tyrosine, and valine [216]. Within proteins sequences, arrangements of these amino acids represent diverse types of information such as protein's structure, function, and interactions of proteins [22]. Irregularities or mutations in the arrangement of amino acids can lead to various biological disorders and diseases, such as cystic fibrosis, sickle cell anemia, Huntington's disease, Tay-Sachs disease, and different forms of cancer [281, 850]. Specifically these mutations can leads toward defective protein function, misfolding, or loss of stability. These disruptions initiate and propagate Alzheimer's disease and muscular dystrophy [819, 847]. With an aim to understand roles of proteins in diverse types of biological functions and their associations with genetic disorders and diseases, researchers are exploring the realm of proteins from various perspectives [728]. To equip AI researchers with biological foundations of diverse types of tasks, we have categorized 63 distinct protein sequence analysis tasks into 11 distinct biological goals namely Protein Identification, Properties Prediction, function and structure prediction, Modification prediction, Interaction Prediction, Sub-cellular location

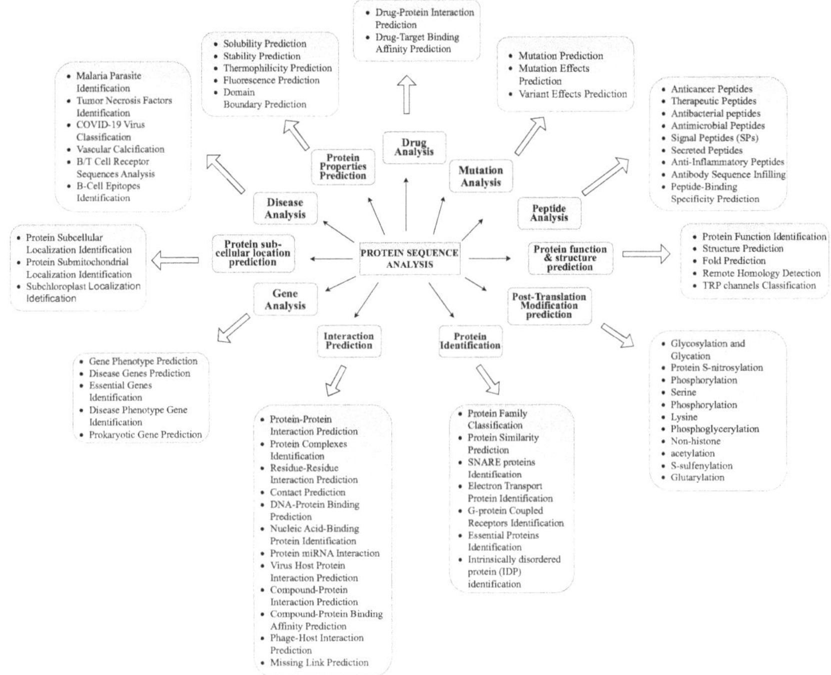

Fig. 3.3 Protein sequence analysis tasks and goals

prediction, Peptide Analysis, Gene Analysis, Mutation Analysis, Disease Analysis, and Drug Analysis. A graphical illustration of all 11 goals and their associated tasks is shown in Fig. 3.3.

Within cellular environment, proteins act as an essential workhorses. Each protein possesses a unique function (e.g., enzyme activity, structural support) and structure such as 3_{10}-helix (G), α-helix (H), π-helix (I), β-strand I, bright (B), turn (T), bend (S) and others (C) [115, 292, 414, 1133]. Proteins functions holds valuable information about biological activities, such as catalyzing biochemical reactions [858, 992], providing structural support [837, 913], and facilitating cellular communication [1182] and transport [1167, 1182]. Structural information reveals how a protein interacts with other molecules. A comprehensive information about function and structure of a protein is useful for understanding cellular machines working paradigm in cellular environment [116].

Living organisms contain millions of proteins in simple cells and billions in complex organisms [774]. However, due to alternative splicing and post-translational modifications, accurate estimation of exact number of proteins is difficult [351]. To thoroughly explore the distinct functionalities and properties of proteins, considering the fact proteins within the same family share similar characteristics, scientists are used to study them at family level rather than individually [1181].

This family-level exploration and analysis require proteins classification into various families, such as kinases, phosphatases, G-protein coupled receptors (GPCRs) [1181], immunoglobulins, heat shock proteins (HSPs), cytochromes, proteases, transcription factors, transporters, and structural proteins [743]. Protein family classification facilitates valuable clues about the structure and function of uncharacteristic proteins in the cell on the basis of known structures and functions of family members [459]. This is especially useful when no experimental evidence is available and scientists may have to annotate the structure and functions of proteins in a newly sequenced genome [459]. Moreover, it can significantly aid the researchers to design drugs which specifically target proteins of one particular family involved in disease processes, while minimizing effects on other proteins [1119]. Researchers can identify patterns of dysfunction within a particular protein family that may contribute to the development or progression of a disease [1119].

Protein family classification is performed at different levels such as their classification into four major classes namely enzyme (E), G-protein coupled receptor (GPCR), ion channel (IC) and nuclear receptor (NR) [459]. G-protein coupled receptors proteins are further classified into 5 sub classes namely Peptide, Calc-Sense, GlutaMeta, GABA, and cAMP [1181]. Similarly, these 5 classes proteins are further classified into 86 different classes which can be further classified into 108 sub classes [1181]. Researchers are discovering new sub classes of proteins, so the number of classes in which protein are usually classified are being change with the passage of time. Apart from proteins family classification into multiple classes, researchers are also utilizing classification paradigms to identify proteins with unique functionalities and properties, such as identification of essential proteins [467, 615, 666, 862, 1163], SNARE proteins [514, 561], electron transport proteins [426], G-protein coupled receptors [522], and intrinsically disordered proteins (IDPs) [1133]. From hundreds and thousands of proteins, only few proteins actively involved in fundamental biological processes and pathways. These proteins are known as essential proteins and can serve as biomarkers for disease diagnosis, prognosis, and monitoring [862]. Dysfunction or absence of essential proteins can severely hamper the proper functioning of living organisms [728]. Many diseases, including genetic disorders and cancers, are associated with mutations or alterations in essential proteins [420, 1090]. Essential proteins identification is important for development of diagnostic tests that can detect diseases at earlier stages [1135]. Pharmaceutical companies are also interested in deeply analysing essential proteins [1026]. Drugs that can modulate the activity of these proteins have the potential to treat diseases caused by their dysfunction [1026]. With essential proteins identification, drug discovery efforts can be more targeted and efficient which leads to the development of therapies that are more effective and have fewer side effects [1201]. Similar to essential proteins, Electron transport proteins (ETPs) are key players of cellular respiration process [31]. Cells utilize this process for transforming nutrients, particularly glucose, into energy and for releasing waste products [31]. Energy is essential for performing cellular functions such as maintaining the cellular structure, repairing any damage, and synthesizing molecules (DNA, RNA, Proteins) [668]. Energy is also crucial for active transport as it allows the cells to move molecules across their membranes [668]. Moreover, cells require energy to generate electrical signals, such as nerve impulses, and to support the immune system in defending against pathogens [1081]. Researchers are identifying ETPs to gain insights into the mechanisms underlying mitochondrial dysfunction, which is associated with various diseases including neurodegenerative disorders, metabolic disorders, and cancer [426]. Also, ETPs identification and understanding are paving way for the development of targeted therapies aimed at modulating mitochondrial function to treat or prevent diseases associated with mitochondrial dysfunction [426].

Another group of proteins called Soluble N-ethylmaleimide-sensitive factor Activating Protein Receptors (SNARE), are considered essential components in the cellular machinery of eukaryotic organisms [514]. SNARE proteins play a vital role in cellular processes by facilitating the fusion of membranes, which is necessary for the transportation and delivery of materials within eukaryotic cells [1009]. In essence, SNARE proteins ensure that cargo is delivered precisely within cells, promoting the integration and communication of cell membranes [1009]. This mechanism is crucial for the proper functioning of cells and, consequently, the overall well-being of an organism [1009]. SNARE proteins play key role in cell growth, cell division, and release of neurotransmitters from nerve cells for intercellular communication in the nervous system [1009]. The identification of SNARE proteins is extremely important in scientific research, particularly in relation to human diseases like cancer [514, 561]. Studies have indicated that SNARE proteins are linked to a wide range of human diseases. Identification and understanding of functions and mechanisms of SNARE proteins can provide valuable insights into the pathology of these diseases. For example, disruptions in the normal functioning of SNARE proteins can disturb cellular communication and transportation, potentially leading to uncontrolled cell growth and disease progression. By unraveling the intricacies of SNARE proteins, researchers can pave the way for novel therapeutic strategies and interventions in the field of medicine [7].

Within protein identification landscape, protein similarity prediction groups proteins into clusters based on the distribution of amino acids in their sequences [1069]. This approach aids in discovering new classes of proteins with unique characteristics [1069]. Moreover, protein similarity prediction facilitates the annotation of newly sequenced proteins by inferring functions based on similarity to known proteins [1069]. Thereby it accelerates the discovery of biological pathways and processes [1069]. Scientists benefit from insights into evolutionary biology because it helps to elucidate how proteins evolve and diversify across different species [1069]. For the pharmaceutical industry, protein similarity prediction is invaluable in drug discovery and development. Identifying proteins similar to known drug targets can reveal new therapeutic opportunities [528].

In addition to protein classification, scientists are performing protein subcellular localization prediction to gain insights into proteins roles in different cellular compartments [670]. Primarily, proteins core biological activities are strictly linked with their presence in different cellular compartments such as Cell junction, Cell membrane, Cell projection, Cytoplasm, Golgi apparatus, Lysosome,

Mitochondrion, Nucleus, Secreted, Endoplasmic reticulum, Plastid, Extracellular, Signal, Chloroplast, Lysosome/Vacuole and Peroxisome [1077]. Within a cell different compartments provide distinct microenvironments [1077]. For instance, in mitochondria compartment proteins perform metabolism related activities, while proteins get localized to nucleus are responsible to carry DNA replication process [345]. Without proper localization, cellular chaos would lead to large scale dysfunction [345]. Subcellular localization prediction ensures that proteins are positioned correctly to perform their functions in an organized and efficient manner [733]. It also helps in understanding cellular processes, such as DNA replication, protein synthesis, and energy production [733]. A deep understanding of Subcellular localization association with irregularities in different cellular processes helps in identifying disease mechanisms [733]. This knowledge can be utilized to develop drugs or therapies that target affected proteins within a specific compartment and restore normal cellular function [733].

Protein modification prediction is important for understanding protein's functional landscape including its stability and activity in various biological processes and diseases [307]. Proteins undergo diverse kinds of post-translational modifications (PTMs) including methylation [162], glycosylation [440], acetylation [710], phosphorylation [1131], and ubiquitination [827]. PTMs modify various properties of proteins such as their structure, electrophilicity, and interactive capacity which enable them to take part and regulate a variety of cellular processes [833]. With over **200/400** diverse types of post-translational modifications, the challenge for researchers lies in efficiently identifying and understanding their role within the protein's life-cycle [833]. Accurate identification of post-translational modifications facilitates a deeper understanding about their roles in various processes such as signaling pathways, immune responses, metabolic regulation [833]. Specifically, researchers are putting efforts to identify S-sulfenylation [260, 1217] and Glutarylation sites in proteins [1034]. S-sulfenylation involves the reversible bonding between a cysteine residue in a protein and a sulfenic acid group [260, 1217]. It plays key role in signal transduction, regulation of protein activity, function, and interactions in cellular environment [260, 1217]. Identification of S-sulfenylation sites provides valuable insights into the transmission of physical cues, such as blood flow or muscle contraction, into biochemical signals that regulate multiple processes like metabolism, inflammation, and vascular tone [260, 1217]. Also, it helps in understanding how cells respond to changes in their environment [260, 1217]. Dysregulated S-sulfenylation has been linked to the development of vascular diseases like aortic aneurysms, atherosclerosis, and thrombotic diseases [260, 1217]. It has also been associated with oxidative stress and the progression of diseases such as cancer and neurodegenerative disorders [260, 1217]. Accurate discrimination of regular S-sulfenylation sites from irregular S-sulfenylation sites can help to detect disease at early stages. On the other hand, glutarylation refers to the addition of a glutaryl group to specific lysine residues in proteins [260, 1217]. Identification of glutarylation sites aids in unraveling the role of this PTM in cellular metabolism, particularly in energy production and mitochondrial function [260, 1217]. Also, Glutarylation can modulate the activity of enzymes, histones, and transcription factors, thereby affecting gene expression and various cellular processes [1034]. When glutarylation is dysregulated, it creates metabolic disorders such as glutaric aciduria and lysine degradation [1034]. Additionally, it can contribute to pathologies like neurodegenerative diseases and cancer [508]. Accurate prediction of glutarylation is essential to detect various disease and develop potent therapeutics [1034]. In a nutshell, identification of different types post-translational modifications have significant implications for protein function and drug development [833].

Similar to protein classification, modification detection and sub-cellular location prediction, protein interaction prediction landscape offers valuable insights about how proteins work together within biological systems, interactions role in cellular communication, and how irregularities in interactions contribute to disease mechanisms and affect biological processes [833]. Protein interactions are the cornerstone of nearly all cellular processes because they mediate signaling pathways, genetic expression, and cellular machinery functions [219]. The protein interactions landscape includes the interactions between proteins and various molecules including DNA/genes, viruses, RNAs and compounds [219, 274, 504, 1190]. Each type of interaction prediction yields unique insights such as Protein-gene interactions are pivotal in understanding gene regulation and expression process [1190]. The journey from a naive cell to a specialized one is governed by the intricate interplay between genes and the proteins [472]. Protein-protein interaction prediction enables the scientists to understand proteins dynamic roles in various biological processes, unknown protein functions, structure-function relationships, and complex signaling networks [55, 1190]. Protein-RNA interaction prediction helps in understanding gene regulation, potential therapeutic targets identification, and uncovering mechanisms of diseases [393, 592]. Furthermore, virus-host protein interactions (VHPIs) prediction sheds light on how viruses hijack the host's cellular machinery through an intricate network of protein interactions [137, 685, 1120, 1153]. The importance of these interactions is paramount across different stages of the viral life cycle, as they orchestrate a virus's ability to penetrate the host cell, replicate its genetic material, subvert the immune defenses, and eventually progenate [1190]. Investigating virus-host protein interactions illuminates the pathways of viral infections and immune responses, which is critical for developing effective antiviral strategies [137].

Protein interaction prediction landscape can be broadly categorized into four key areas:

- Interaction Prediction: Scientists determine whether a molecule will interact with a protein or not. This knowledge provides a high-level overview of potential partnerships within a biological system [1190].
- Binding Affinity Prediction: Scientists determine detailed information about the strength of these interactions. This refined information offers important insights into the nature and functionality of protein-molecule complexes [541].
- Missing link Prediction: Scientists determine missing links within complex networks encompassing all possible interactions between proteins and other molecule [71, 352]. Missing links identification can reveal unknown functional relationships and can provide valuable insights into biological pathways and

cellular mechanisms [748, 800]. Furthermore, this network-centric approach aids in development of new therapies by studying diseases [304, 691].

- Module prediction: Module prediction is a multi-scale network analysis that involves simultaneously analyzing the interactions of proteins with other proteins as well as various molecules such as genes and disease-related molecules [1075]. Considering the availability of diverse molecular networks including gene-gene, protein-protein, disease-gene, disease-protein, researchers are performing complex identification or module identification to analyze how molecules interact and influence each other [190, 1039, 1075]. This analysis highlights tightly interconnected groups of molecules such as proteins, genes, or metabolites that work together to perform specific biological functions [1039]. Furthermore, module identification can identify key molecular players, pathways, and regulatory mechanisms involved in specific diseases or physiological conditions [1039]. It helps in identifying potential biomarkers for various diseases, including atherosclerosis and Parkinson's disease [190]. Identifying and understanding the modules affected by diseases can help researchers gain insights into disease mechanisms and identify points of intervention for treatment [190]. Additionally, module detection can contribute to personalized medicine by identifying patient-specific variations in molecular networks. This allows the development and application of tailored treatments based on an individual's unique molecular profile [1075].

In the realm of drug discovery and personalised medicine, similar to Protein-DNA [637], and Protein-RNA interactions [393], protein-drug interaction [1229] and affinity [1111] prediction holds immense significance [637, 1111]. Protein-drug interaction knowledge empowers scientists to pinpoint promising therapeutic targets and assess drug efficacy and safety [1229]. This knowledge streamlines drug development and tailoring therapies having fewer side effects [1229]. Drug-protein interactions play an essential role in the development and discovery of drugs, as well as understanding the life-cycle and positive or negative impacts of drugs on living organisms [637, 1111, 1229]. Without comprehension of drug-protein interactions, it is not feasible to determine the effectiveness, safety, and implications of drugs [100]. The fundamental purpose of any drug is to have a therapeutic and healing effect on the body. This is normally accomplished by the drug interacting with a particular protein within the body, which can be a receptor, enzyme, or transporter protein [637, 1111, 1229]. The interaction between the drug and the protein can either suppress or activate the function of protein, resulting in a therapeutic effect. For example, many antiviral medications suppress the viral proteins and prevent them from replicating and harming the human bodies. A deep comprehension of drug-protein interactions can assist in the prediction and mitigation of adverse medication reactions. When a drug interacts with unanticipated proteins known as off-target interactions, it can cause undesired severe side effects [895]. By detecting potential off-target interactions quite early in the drug development process, pharmaceutical scientists can change the medication structure or other properties to limit off-target interactions and significantly improve the drug safety profile.

Drug-protein interactions also impact pharmacokinetics which include the way drug is absorbed, transported, metabolized, and excreted from human body [1139]. For instance, several drugs are primarily broken down or metabolized by cytochrome P450 enzymes inside the liver [1139]. Therefore, if a drug undesirably suppress cytochrome P450 enzyme then this may lead to toxicity due to the increase impact of other drugs that are normally metabolized by cytochrome P450 enzymes. Furthermore, how a particular drug impacts the human body, also called pharmacodynamics is mainly determined through its interaction with specific proteins [895]. For example, if a particular drug activates a receptor protein, it may lead to increase the blood pressure. In addition, the identification of drug-protein interactions is also valuable for advancing personalized medicines. All individuals have different versions or isoforms of proteins in their bodies mainly because of genetic variations. These isoforms could interact differently with the drugs, resulting variations in responses to drug across the individuals [637, 1111, 1139]. By comprehending such interactions, pharmaceutical scientist can develop very effective and customized treatment plans for the individuals based on their unique genetic makeup. Understanding drug-protein interactions can also help with drug re-purposing initiatives, which involve the efforts to find new usages for existing medications. Like, if a protein becomes non-functional due to a disease and a known drug interacts with that protein, researchers may investigate if that drug could be used to treat the disease and revive the functionality of protein.

Proteins are built from small building blocks called peptides which are short chains of amino acids linked together [729]. With an aim to more deeply understand peptides landscape, researchers have categorized them based on diverse types of properties like stability [804], bio-availability [893], efficacy [363], action mechanisms [81], involvement in cellular processes [844], utilization in drugs [1177] and therapies [1041]. The field of peptide research is constantly expanding, with the discovery of novel peptide types encompassing unique properties. The most well-characterized and diverse peptide types include anticancer peptides [15, 246, 964], antibacterial peptides [903, 1169], antimicrobial peptides [1113, 1125], signal peptides (SPs) [986], secreted Peptides [1074], and anti-inflammatory peptides [517, 840]. Each type of peptide has unique importance such as anti-inflammatory peptides (AIPs) are valuable in modulating the immune system, inflammatory response, and defense against chronic inflammation [517, 840]. They have potential as therapeutics for conditions like rheumatoid arthritis, inflammatory bowel disease, asthma, and drug-resistant pathogens [517, 840]. AIPs also aid in wound healing and tissue regeneration while reducing inflammation at the site of injury [517, 840]. Some AIPs exhibit neuroprotective effects and are being explored for treating neurodegenerative diseases such as Alzheimer's [517, 840] and Parkinson's [517, 840]. Similarly, high efficacy, low toxicity, and lack of drug resistance properties have made anti-cancer peptides (ACPs) ideal candidates for the development of cancer therapies [15, 246, 964]. ACPs have the ability to kill cancer cells through various mechanisms [15, 246, 964]. ACPs can directly bind to membranes of cancer cells and initiate their destruction by damaging their membranes. They can also add a kill tag to cancer cells so they become recognizable to immune system for destruction

[15, 246, 964]. Anti-bacterial peptides (ABPs) protect the host from harmful bacteria [903, 1169]. These peptides eliminate bacteria by either blocking their internal activity or by disrupting their cell membrane [903, 1169]. Because of the growing concern over antimicrobial resistance, researchers are identifying ABPs to develop effective drugs as an alternative to traditional antibiotics [903, 1169]. With an aim to more deeply study functional paradigms of different kinds of peptides to develop modern drugs, precise identification of type-specific peptides is an important task [903, 1169].

Another goal of scientists is to predict a variety of protein properties, including solubility [1031], stability [1031], thermophilicity [409], fluorescence [1070], and domain boundaries [408]. Protein solubility refers to the ability of a protein to dissolve and remain in solution without aggregating or precipitating out [409]. Various factors, such as amino acid composition, hydrophobicity, and interactions of different molecules within cellular environment influence protein solubility. It facilitates valuable information about the functioning and behaviour of proteins in living organisms [1031]. Proteins need to fold into their correct three-dimensional structures to perform their dedicated biological functions. Unfolded or misfolded proteins can generate protein aggregates that are linked with neurodegenerative diseases such as Parkinson's or Alzheimer's [1031]. Protein solubility prediction assists to identify proteins that are more prone to misfold or aggregate, enabling the scientists to comprehend the underlying mechanisms of complex diseases and develop potential interventions. Furthermore, protein solubility prediction is essential for drug discovery [1031]. Poorly soluble, or insoluble proteins are not only hard to extract from a complete mixture but they are also less suitable as drug targets [1031]. Protein solubility prediction can assist the scientists to optimize purification protocols and experimental conditions in order to extract suitable proteins that can serve as drug targets. Protein's stability prediction facilitates researchers to engineer proteins with improved stability for various applications, such as enzymes for industrial processes or therapeutic proteins. This information is also useful to determine optimal storage conditions for proteins used in research or therapeutic applications [1031]. Thermophilicity information facilitates the identification of enzymes from extremophilic organisms or the engineering of enzymes suited for high-temperature environments [409]. These enzymes hold immense potential for biotechnological applications, including bioremediation and geothermal energy production processes. Protein fluorescence is a process in which a protein absorbs light at a specific wavelength and subsequently emits light at a longer wavelength. This property is useful for studying protein interactions and designing biosensors [409, 1031]. Protein domain boundaries define the regions within a protein where distinct structural or functional units are separated [408]. These boundaries prediction helps in understanding the modular nature of proteins, protein engineering, and functional annotation. Accurate prediction of domain boundaries aids in interaction sites identification, designing stable and functional protein constructs, and understanding the protein's role in various biological processes [408].

In the realm of disease analysis, protein-centric investigations focus on parasite identification [410], characterization of immune factors [980], viral classification [922], and analysis of protein sequences associated with specific pathological conditions [510], such as vascular calcification [144] and B/T cell receptor sequences [785]. With the influx of diverse types of diseases, researchers are putting tremendous efforts to identify proteins relations with diverse types of diseases. For instance, malaria is a disease caused by parasites transmitted through the bites of infected mosquitoes. In this disease two different proteins namely secretory and non-secretory proteins are important in the malaria parasite's life cycle [510]. Secretory proteins enable parasites to enter and infect red blood cells and assist the parasite in avoiding detection and attack of host's immune system [901]. They also help parasites in acquiring essential nutrients from the host [901]. Non-secretory proteins maintains parasite's cellular structure and organization and enable them to resist antimalarial drugs [501]. By identifying and distinguishing between these two types of proteins, scientists can more deeper understand how the parasite invades human cells and survives [501].

Likewise, in another task called vascular calcification [144], calcium builds up in the blood vessels and make them stiffer which leads to various cardiovascular complications such that it act as a silent killer for populations with chronic kidney disease (CKD), diabetes mellitus (DM), and those of advanced age (older than 65 years) [144]. Adverse micro-environment hormones, uremic toxins and epigenetic mechanisms including microRNAs and methylation abnormalities, contribute to the progression of vascular calcification. To address this problem, researchers are identifying compounds that prevent or reverse the calcification by targeting the responsible pathways of involved compounds [144]. Furthermore, global pandemic caused by SARS-CoV-2 has underscored the importance of rapidly identifying potential treatments for new viral diseases. By understanding the proteins involved in the virus's life cycle and how it interacts with human cells, scientists are identifying existing drugs and compounds that could act as inhibitors. Such inhibitors block these interactions and stop the virus from replicating. In a nutshell, identification of disease related proteins, compounds, and inhibitors are accelerating drug repurposing [144]. This method is much faster and cost-effective than developing new drugs from scratch, as these compounds have already been tested for safety in humans. It offers a rapid response to emerging diseases and providing immediate treatment options while new vaccines and specific antiviral drugs are being developed [144].

Humans are exposed to various pathogens over the time, however, their bodies are protected from getting infected again due to the adaptive nature of immune system [887]. This nature enables the immune system to remember the pathogens for an extended period of time and respond more effectively on account of future encounters [1095]. The immune system achieve this through two kinds of receptors: T-cell [1095] and B-cell [887]. While T-cells bind to antigen fragments presented on the surface of infected or abnormal cells [1095]. B-cells take up antigens directly by the production of antibodies [887]. Researchers are performing identification of T-cells and B-cells to gain a deeper understanding of how the immune system recognizes and responds to pathogens and diseases. This knowledge can then be applied to disease diagnosis, monitoring disease progression, and evaluating treatment effectiveness [1095]. Additionally, B/T cell receptor analysis provides valuable insights into vaccine development by allowing scientists to study the immune response of individuals with strong vaccine reactions [785]. This information can be used to design more effective vaccines. Furthermore, analyzing B/T cell receptors contributes to the development of personalized medicine approaches by enabling scientists to tailor treatments based on individual receptor sequences [785]. This optimization can lead to improved therapeutic outcomes. Lastly, B/T cell receptor analysis aids in drug discovery by identifying potential targets for intervention and designing therapies that specifically target these receptors [785].

In the realm of protein science, another important area of study named mutation analysis encompasses several key sub-tasks to predict, compare and understand the implications of genetic variations [1011]. This multifaceted task three key sub-tasks are Mutation Prediction [1008], Mutation Effects Prediction [707, 894, 954, 1078], and Variant Effects Prediction [698]. Mutation prediction highlights different areas of sequences that might undergo genetic changes [1008]. Specifically, it identifies changes of amino acids within protein sequences [1008]. Mutation effects prediction focuses on determining genetic changes functional consequences. Specifically, scientists find how these mutations affect protein stability, folding dynamics, and interactions with other biomolecules [707, 954, 1078]. Overall, this analysis offers insights into impact of mutations on cellular processes [707, 954, 1078]. Variant Effects Prediction provides insights into how mutations influence disease susceptibility or treatment response [698]. This task holds immense potential for advancement of precision medicine strategies. This knowledge not only empowers researchers to elucidate the molecular mechanisms of disease but also paves the way for the development of targeted therapeutic strategies [698].

In the realm of gene analysis, most important tasks include gene phenotype prediction [497], disease genes prediction [686], essential genes identification [1064], and prokaryotic gene prediction [1005]. Specifically, gene phenotype refers to physical characteristics and traits of an organism which results from specific gene expression or set of gene [497]. These physical characteristics and traits include height, eye color, hair color, blood type, blood pressure, heart rate, enzyme activity, or metabolic processes [497]. A group of proteins also exhibits phenotypic functional features namely conditional phenotypes, cell cycle defects, mating and sporulation defects, auxotrophies, carbon, and nitrogen utilization defects, cell morphology and organelle mutants, stress response defects, carbohydrate and lipid biosynthesis, nucleic acid metabolism defects, sensitivity to amino acid analogs and other drugs, sensitivity to antibiotics, and sensitivity to immunosuppressants [175]. These phenotypic genes often leads to chronic diseases such as sickle cell anemia, multiple sclerosis, Huntington's disease, type 2 diabetes, heart disease, and many forms of cancer [497]. For instance, an autoimmune disease namely multiple sclerosis (MS) disrupts myelin and axons and causes brain and spinal cord inflammation, and identifying its associated genes can elucidate its molecular mechanisms [659]. Identification of such genes will effectively contribute to discovering the inner molecular mechanisms of multiple sclerosis as a disease [659].

Identifying essential genes helps researchers understand vital genetic components for survival and discover new drug targets [774, 1064]. Prokaryotic genes are crucial for understanding bacterial biology, pathogenicity, and antibiotic resistance, with significant applications in medicine and biotechnology [1005]. Understanding gene phenotypes and essential genes is fundamental for advancing medical research and developing targeted therapies [774, 1005, 1064].

3.2 A Look on DNA, RNA and Protein Sequence Analysis Tasks from the Perspective of Computer Scientists

With the influx of biological data and rise of AI, researchers are increasingly applying AI in diverse areas of molecular biology. Development of large scale AI applications requires comprehensive understanding of diverse sequence analysis tasks. However, there exist a huge knowledge gap between computer scientists and molecular biologists. Molecular biologists know the need, biological importance, and pharmaceutical worth of different sequence analysis tasks. However, they lack expertise to determine which machine or deep learning models can effectively replace or complement experimental work. Conversely, computer scientists excels at designing AI-driven predictive pipeline for specific data type, however, they have limited knowledge about nature of biological sequence analysis tasks. For instance, DNA sequence analysis tasks such as gene function prediction, gene network reconstruction, gene expression prediction, and disease risk estimation can be challenging for computer scientists to grasp. However, a comprehensive literature review that explains the basics of these tasks can bridge the knowledge gap between molecular biologists and computer scientists. For example, gene function prediction is a multi-label classification tasks, gene expression prediction is a regression task, while gene network reconstruction and disease risk estimation are binary classification tasks. With this foundational understanding, computer scientists can more effectively develop predictive pipelines for these complex binary, multi-label classification, and regression tasks. To empower AI experts, we have mapped 44 DNA, 47 RNA and 62 Protein sequence analysis tasks into core AI tasks like classification, regression and clustering. Figures 3.4, 3.5, and 3.6 graphically represent mapping of all three macro molecules tasks into core AI tasks. A high level analysis of all three figures reveal that nature of DNA, RNA and protein sequence analysis tasks can be categorized into three primary types: regression, clustering, and classification where classification can be further subdivided into three secondary types: binary classification, multi-class classification and multi-label classification. Let's mathematically formulate the possible natures of sequence analysis tasks.

Binary classification involves predicting the outcome of input sequence as a binary variable (0 or 1). Given a dataset with sequences $X \in \mathbb{R}^{nxd}$, binary labels $y \in 0, 1$, and training dataset $(x_1, y_1), (x_2, y_2), \ldots,$ prime objective is to learn a decision function $f : X \rightarrow Y$ that maps inputs to binary outputs 0, 1 on the basis of hypothesis function $h(x)$ learned from the training data.

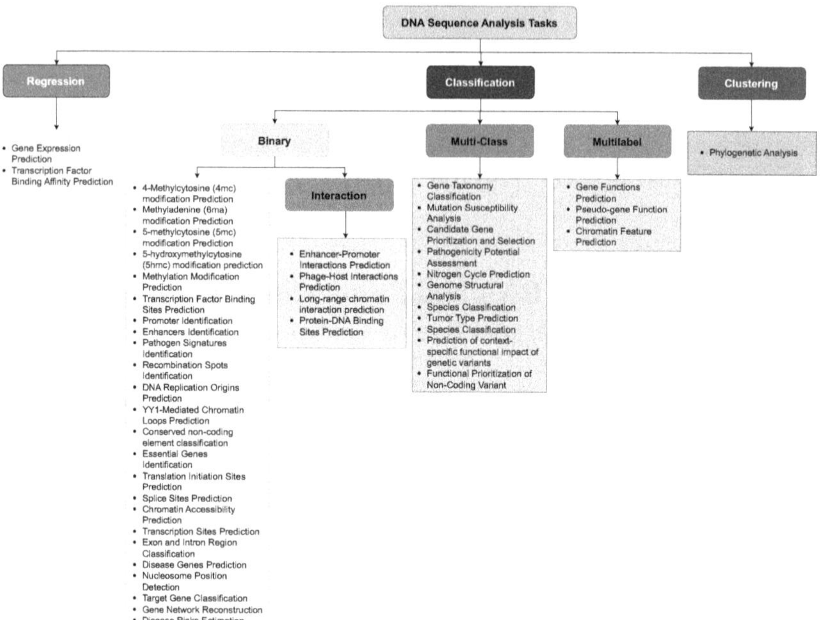

Fig. 3.4 DNA sequence analysis tasks and their corresponding AI paradigms

$$f(x) = \begin{cases} 1 & if h(x) \geqslant 0.5 \\ 0 & otherwise \end{cases} \tag{3.1}$$

Multi-class classification involves predicting outcome of input sequence from a set of three or more classes. Specifically, given a dataset with sequences $X \in \mathbb{R}^{nxd}$, labels $y \in 1, 2, \ldots, K$ where K is the number of classes, and training dataset $(x_1, y_1), (x_2, y_2), \ldots, (x_n, y_n)$ where $x_i \in X$ and $y_i \in Y$, prime objective is to learn a decision function $f : X \rightarrow Y$ that assigns inputs to one of the predefined classes.

$$f(x) = argmax_k h_k(x) \tag{3.2}$$

where $h_k(x)$ is the hypothesis function for class k learned from the training data. Contrarily, in multi-label classification, each input sequence can be assigned to multiple classes simultaneously. Given a dataset with sequences $X \in \mathbb{R}^{nxd}$, labels $y \in 1, 2, \ldots, K$ where K is the number of classes, and training dataset $(x_1, y_1, y_2, ..), (x_2, y_1, y_4, \ldots), \ldots, (x_n, y5, y_n, \ldots.)$ where $x_i \in X$ and $y_i \in Y$, prime objective is to learn a decision function $f : X \rightarrow 0, 1^K$ that assigns inputs to multiple classes simultaneously using hypothesis function $h_k(x)$ for class k learned from the training data.

$$f(x) = (h_1(x), h_2(x), \ldots, h_K(x)) \tag{3.3}$$

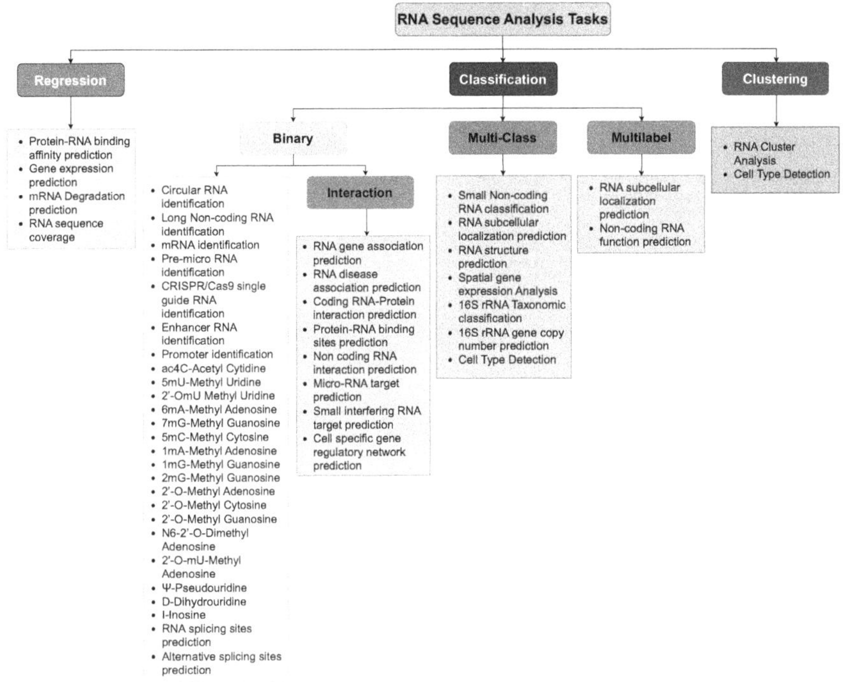

Fig. 3.5 RNA sequence analysis tasks and their corresponding AI paradigms

Furthermore, regression aims to predict a continuous outcome variable for an input sequence. Given a dataset with sequences $X \in \mathbb{R}^{nxd}$, labels $y \in \mathbb{R}$, and training dataset $(x_1, y_1), (x_2, y_2), \ldots, (x_n, y_n)$ where $x_i \in X$ and $y_i \in Y$, prime objective is to learn a function $f : X \rightarrow \mathbb{R}$ that predicts continuous outcomes using hypothesis function $h(x)$ learned from the training data.

$$f(x) = h(x) \tag{3.4}$$

Clustering aims to group similar sequences into same clusters. Given a dataset with sequences $X = x_1, x_2, \ldots, x_n$, where each $x_i \in \mathbb{R}^d$, prime objective is to find a partition of the data into clusters $C = C_1, C_2, \ldots, C_K$. This is accomplished based on distance metric $d(x, \mu_c)$ between sequences x and the centroid μ_c of cluster c.

$$f(x) = \text{argmin}_c d(x, \mu_c) \tag{3.5}$$

3.3 AI-Driven Genomic and Proteomic Sequence Analysis

Considering the aptitude of Artificial Intelligence (AI) approaches to automatically extract important hidden patterns, these approaches have been extensively utilized

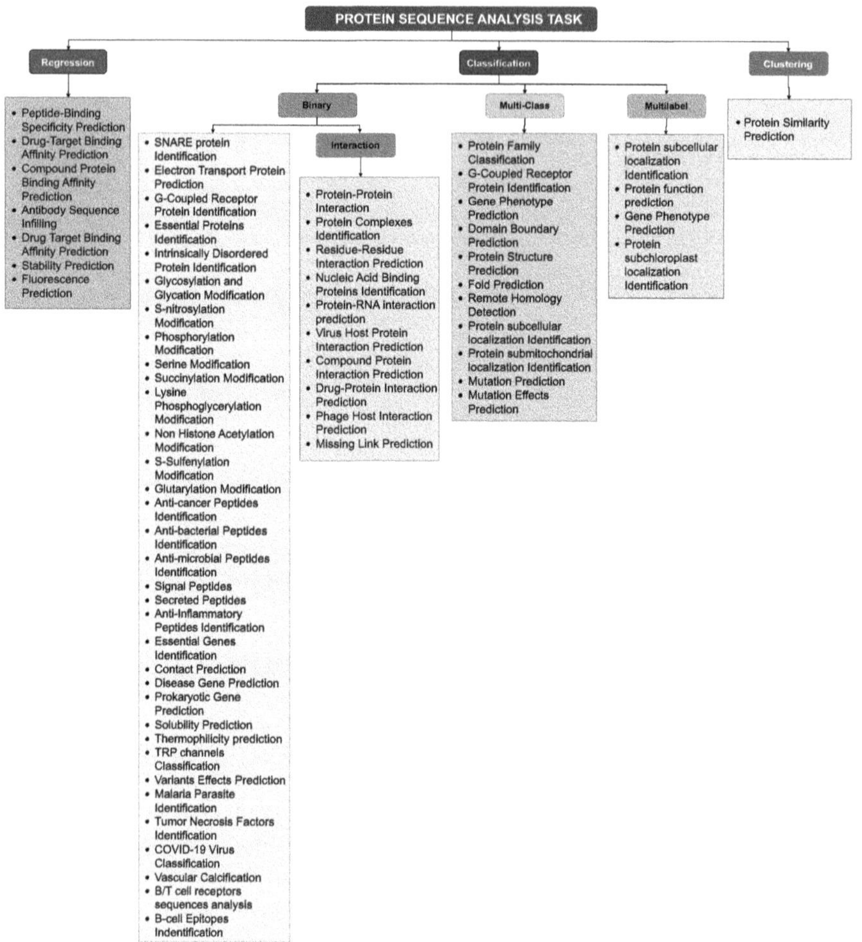

Fig. 3.6 Protein sequence analysis tasks and their corresponding AI paradigms

to explore the hidden potential of biological sequences for the establishment of economical large-scale Genomics and Proteomics sequence analysis landscape [255, 694]. Within this landscape, a closer look at various Genomics and Proteomics sequence analysis tasks reveals that most of the tasks fall under three different paradigms: (1) Classification, (2) Clustering and (3) Regression. In classification, the primary goal of predictive pipelines is to forecast pre-defined classes such as families of biomolecules, possibility of biomolecules interaction, localization of biomolecules within the cell, and discrimination of normal cells from disease cells [54, 361, 1155], etc. In clustering, the main focus of predictive pipelines is to group together sequences that have similar characteristics [361, 1155]. Whereas, in regression, the focus of predictive pipelines is to estimate continuous numerical

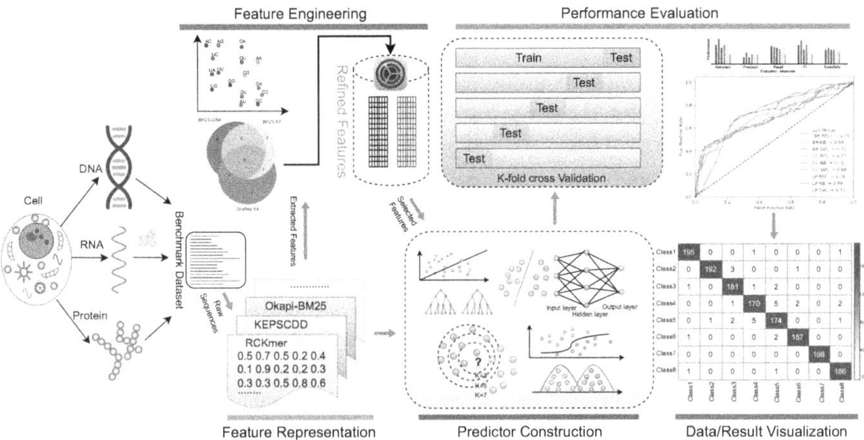

Fig. 3.7 Functional scope of AI-driven Genomics and Proteomics sequence analysis frameworks

values such as prediction of disease risk, quantifying the drug response and estimating DNA copy number variations [361, 1155].

Figure 3.7 illustrates generic workflow of AI predictive pipelines for Genomics and Proteomics sequence analysis tasks. To facilitate large scale genomic and proteomic sequence analysis, development of AI predictive pipelines hinges entirely on the quality and quantity of available data. Public biological databases house vast amount of data related to different sequence analysis tasks and species. Researchers are leveraging these databases for acquiring data for development of AI-empowered applications tailored for various genomic and proteomic sequence analysis tasks. Furthermore, apart from public databases, preprocessed datasets specifically tailored for distinct sequence analysis tasks are also publicly accessible. Chapter 4 provides a detailed description of diverse data formats used in biological databases and thorough descriptions of various data retrieval systems. It provides an extensive information of biological sequence databases, including 47 peptide, 35 DNA, 64 RNA, 68 protein, and 12 CRISPR system related databases. Moreover, Chapter 7 and Chaps. 3 to 6 (see Vol. 2) provide detailed descriptions of publicly available datasets related to a wide array of DNA, RNA, protein, peptide, and CRISPR system- related tasks. After data collection, in any Genomics and Proteomics sequence analysis task next step is data preprocessing in which usually sequence samples lengths are standardized. Two most widely used lengths standardization methods are sequence padding and sequence truncation. In sequence padding, shorter sequences are extended to match the length of the longest sequence. Alternatively, sequence truncation shortens longer sequences to align with the length of the shortest sequence. Sequence truncation can lead to the loss of nucleotide or amino acid patterns due to removal of essential segments of the sequence. On the other hand, sequence padding method preserves the original sequence information but may introduce bias in sequences due to repetition of added character.

After generating fixed-length sequences, the next step is to transform raw sequences into numerical representations, or vectors, as machine and deep learning models inherently operate on numerical data. The performance of these models is directly linked to the inclusion of informative and discriminative nucleotide or amino acid patterns within the generated vectors. Consequently, even simple machine learning models can outperform complex models when furnished with feature vectors enriched with informative and discriminative nucleotide or amino acids patterns. Conversely, complex models may under-perform when provided with less informative representations. A high level overview of diverse types of statistical representation generation methods is given in Sect. 3.4.

Afterwards, statistical vectors of sequences are either passed to deep learning or machine learning predictors. Deep learning predictors possess the capability to autonomously perform feature engineering while machine learning predictors require explicit feature engineering to remove redundant or noisy features. Consequently, statistical vectors are directly passed to deep learning predictors, while traditional machine learning predictors require a preceding feature engineering step. Moreover, it is widely accepted that not all input features significantly contribute in accurate classification of hand on task [386]. Feeding traditional machine learning models with highly informative features significantly assist them to find useful correlations for accurate predictive modeling. It is widely acknowledged that even a basic predictor performs better when it is fed with a relevant and informative subset of input features. Contrarily, even a complex predictor under-performs on account of noisy features. Generally, there are two approaches to select the most informative subset of features. One approach is to apply feature selection that removes irrelevant and redundant features and retain the most informative features from original set of features. Other approach is to apply dimensionality reduction approaches that transform original feature space into reduced feature space by eliminating redundant correlations of features. A comprehensive details of a variety of feature selection and dimensionality reduction algorithms is summarized in Sect. 3.5.

It can be seen in Fig. 3.7, in AI based genomic and proteomic sequence analysis predictive pipelines, after feature engineering stage next step is to pass optimized statistical vectors to machine learning predictors. Section 3.6 summarizes details of diverse types machine and deep learning predictors. In AI-driven Genomic and Proteomic sequence analysis landscape, the final step involves performance evaluation of predictive pipelines. To accurately gauge the effectiveness of predictors, a diverse array of evaluation metrics is employed. Section 3.7 provides a detailed description of various evaluation metrics.

3.4 Sequence Representation Learning Methods

Sequence encoding or representation learning methods are key players of AI-driven Genomic and Proteomic Sequence analysis predictive pipelines. These methods capture nucleotides or amino acids distribution patterns within raw sequences and include these patterns into statistical vectors. There is a marathon for development of powerful sequence encoders capable of generating informative distribution patterns aware statistical vectors of raw sequences. To date, more than **100** sequence encoding methods have been developed that can be broadly classified into three categories: (1) domain specific methods, (2) neural word embedding, and (3) language models. Within domain specific methods, a sub-category of methods named physico-chemical properties based methods generate statistical vectors of raw sequences using pre-computed physical and chemical values of nucleotides or amino acids. Similarly, another sub-category named statistical methods rely on position and occurrence frequencies of individual or group of nucleotides or amino acids within raw sequences. Neural word embedding methods learn distributed representations of nucleotides or amino acids in the continuous vector space. These methods capture the syntactic and semantic similarities of individual or group of nucleotides or amino acids by mapping them to vectors in a high-dimensional space. Primarily, these methods produce similar representations of individual or group of nucleotides with similar contexts. Language models based predictive pipelines make use of a pre-trained language model or train a large language model from scratch on huge biological sequence data. The pretrained language models poses rich syntactic and semantic patterns of nucleotides or amino acids.

All three types of sequence encoding methods are powerful but also carry some inherited limitations that can lead to performance degradation for specific tasks. Physico-chemical properties based and statistical methods are widely used for genomics and proteomics sequence analysis. These methods capture the intrinsic characteristics of biological sequences, such as amino acid composition, hydrophobicity, and molecular weight. However, these methods lack to capture complex relationships of residues such as long range interactions of residues in the sequences [609]. Also, these methods may not fully capture the semantic and functional similarities between sequences [609]. Neural word embeddings methods are primarily used for diverse Natural Language Processing (NLP) tasks and have been adapted for genomics and proteomics sequence analysis. These methods efficiently capture semantic and contextual information of residues. However, these methods lack to efficiently handle different contexts of same nucleotide or amino acid. Likewise, language models are competent in capturing complex relationships of nucleotides or amino acids, however there is a deficiency of pretrained language models for sequence data. Moreover, language models pretraining require large amount of sequence data for training and hyperparameters optimization. In a nutshell, utilization of a sequence encoding method primarily depend on the characteristics of sequence analysis problem at hand. There is no generic sequence encoding method that can be applied to any sequence analysis task.

Table 3.1 illustrates diverse array of sequence representation learning methods across all three categories. Specifically, first three columns illustrates names of domain specific representation learning methods capable of transforming DNA/RNA, protein, and all three types of sequences into statistical vectors. A detailed description of domain specific DNA/RNA sequence encoding methods is given in Chap. 5 and details of domain specific protein sequence encoding methods

Table 3.1 Genomic and Proteomic sequence representation learning methods

DNA/RNA encoding methods	Protein encoding methods	DNA/RNA/protein encoding methods	Embedding methods	Language models
Nucleotide chemical property (NCP) [170]	Adaptive skip dipeptide composition (Protein_ASDC) [1086]	Basic Kmer (Kmer) [93, 571]	Word2Vec [715]	ALBERT [1186]
Reverse compliment kmer (RCKmer) [381, 762]	Kmer dipeptides composition (Protein_DPC) [93, 878]	Tri-Peptide Composition (TPC) [93, 878]	DeepWalk [805]	AlphaFold [851]
Pseudo dinucleotide composition (PseDNC) [650, 651]	Dipeptide deviation from expected mean (Protein_DDE) [878]	Enhanced Amino Acid Composition (EAAC) [157, 1232]	ELMo [806]	AlphaFold2 [851]
Pseudo k-tupler composition (PseKNC) [650, 651]	PseAAC of distance-pairs and reduced alphabet (Protein_DistancePair) [97, 649]	Accumulated Nucleotide Frequency (ANF) [170]	FastText [716]	BERT [250]
Parallel correlation pseudo dinucleotide composition (PCPseDNC) [650, 651]	Conjoint triad (Ctriad) [906]	Pseudo KNC (PseudoKNC) [650, 651]	GATNE [1175]	BigBird [212]
Parallel correlation pseudo trinucleotide composition (PCPseTNC) [650, 651]	Conjoint k-spaced Triad (KSCTriad) [157, 1232]	Binary (binary) [159, 164]	GEMSEC	ELMo [806]
Series correlation pseudo dinucleotide composition (SCPseDNC) [650, 651]	Enhanced amino acid composition (EAAC) [157, 1232]	Electron-ion interaction pseudopotentials value (MappingClass_eiip_fourier) [553, 740]	GloVe [802]	ELECTRA [194]
Series correlation pseudo trinucleotide composition (SCPseTNC) [650, 651]	Weighted Sparse Representation based Classification Global (WSRC_global) [536]	MappingClass_integer_fourier (MappingClass_integer_fourier) [553, 740]	GraRep [133]	ESM-1 [1146]
Dinucleotide-based auto covariance (DAC) [265, 375, 651]	Weighted Sparse Representation based Classification Local (WSRC_local) [536]	Term Frequency Inverse Document Frequency (TFIDF) [389, 834]	Graph embedding	ESM-2 [1133]
Dinucleotide-based cross covariance (DCC) [265, 375, 651]	Weighted Sparse Representation based Classification Local+Global (WSRC_local_global) [536]	Okapi-BM25 [1092]	MetaGraph2Vec [276]	GPT [1186]

Dinucleotide-based auto-cross covariance (DACC) [265, 375, 651]	Enhanced Grouped amino acid composition (EGAAC) [157, 1232]	Mono Mono K-Gap (monoMonoKGap) [731]	HAKE	Graph Transformer Network [582]
Trinucleotide-based auto covariance (TAC) [651]	Grouped amino acid composition (GAAC) [157, 1232]	Mono Di K-Gap (monoDiKGap) [731]	HIN2Vec [244]	Heterogeneous Graph Transformer [644, 667, 1175, 1250]
Trinucleotide-based cross covariance (TCC) [651]	Grouped tripeptide composition (GTPC) [157, 1232]	Mono Tri K-Gap (monoTriKGap) [731]	HOPE [786]	IgFold [709]
Trinucleotide-based auto-cross covariance (TACC) [651]	Grouped dipeptide composition (Protein_GDPC) [157, 1232]	Di Mono K-Gap (diMonoKGap) [731]	Hyper2Vec	LongFormer [84]
Electron-ion interaction pseudopotentials of Trinucleotide (PseEIIP) [553, 740]	Composition of k-spaced amino acid group pairs (CKSAAGP) [157, 1232]	Di Di K-Gap (diDiKGap) [731]	Laplacian eigen maps [83]	pLM-TAPE (BERT)
MappingClass_binary_fourier [107, 108]	Pseudo-amino acid composition (PAAC) [188, 189]	Di Tri K-Gap (diTriKGap) [731]	LINE [979]	RNAFormer [313]
MappingClass_zcurve_fourier [107, 108]	Amphiphilic PAAC (APAAC) [188, 189]	Tri Mono K-Gap (triMonoKGap) [731]	Locally linear embedding [854]	RoBERTa [640]
MappingClass_real_fourier [107, 108]	Pseudo K-tuple reduced amino acids composition (PseKRAAC type 1 to type 16) [1255]	Tri Di K-gap (triDiKGap) [731]	Mashup	T5 [795, 1179]
MappingClass_complex_number [107, 108]	Protein_binary_6bit [97, 1049]	Composition of k-spaced Nucleic Acid Pairs (CKSNAP) [157]	Node2Vec [369]	Transformer [560]
MappingClass_atomic_number [107, 108]	Protein_binary_5bit_type_1 [97, 1093]	Complex Network (complex_network) [107, 108]	OPA2Vec [935]	Transformer-XL [196]

(continued)

Table 3.1 (continued)

DNA/RNA encoding methods	Protein encoding methods	DNA/RNA/protein encoding methods	Embedding methods	Language models
classifical_chaos [107, 108, 731]	Protein_binary_5bit_type_2 [97, 1093]	Enhanced Complex Network (enhanced_complex_network) [107, 108]	Random Watcher-Walker (RW2)	ULMFiT [441]
frequency_chaos [107, 108, 731]	Protein_binary_3bit_type_1 [1086]	Position-specific trinucleotide propensity based on single-strand (PSTNPss) [210, 413]	RandomWalk [369]	Vision Transformer [268]
zCurve [328, 731]	Protein_binary_3bit_type_2 [1086]	Position-specific trinucleotide propensity based on double-strand (PSTNPds) [210, 413]	RotatE [131]	XLNet [1186]
gcContent [107, 108, 731]	Protein_binary_3bit_type_3 [1086]		RWR [990]	
cumulativeSkew [107, 108, 731]	Protein_binary_3bit_type_4 [1086]		SDNE [1032]	
atgcRatio [107, 108, 731]	Protein_binary_3bit_type_5 [1086]		SocDim [977]	
spectrum [107, 108]	Protein_binary_3bit_type_6 [1086]		Struc2Vec [1126]	
orf [107, 108]	Protein_binary_3bit_type_7 [1086]		SVD [397]	
fickett_score [107, 108]	Overlapping property features (Protein_OPF_10bit) [1086]		Topo2Vec	
	Overlapping property features (Protein_OPF_7bit_type_1) [1086]		TransE	
	Overlapping property features (Protein_OPF_7bit_type_2) [1086]		tSNE	
	Overlapping property features (Protein_OPF_7bit_type_3) [1086]		DANE	
	Learn from alignments (Protein_AESNN3) [97, 629]		Graph2vec [745]	
	BLOSUM62 (BLOSUM62) [572]			
	ZSCALE (ZSCALE) [167]			
	AAINDEX (AAINDEX) [1007]			

	Composition (CTDC) [125, 126, 277, 278, 395]			
	Transition (CTDT) [125, 126, 277, 278, 395]			
	Distribution (CTDD) [125, 126, 277, 278, 395]			
	Sequence-order-coupling number (SOCNumber) [187]			
	Quasi-sequence-order descriptors (QSOrder) [187]			
	Moran [305, 627]			
	Geary [937]			
	NMBroto [437]			
	auto_covariance [265, 375, 651]			
	auto_cross_covariance [265, 375, 651]			
	bi_auto_covariance [265, 375, 651]			

is available in Chap. 6. A detailed description of word embedding methods and large language models is given in Chapter 1 and 2 (Vol. 2), respectively.

3.5 Feature Engineering Methods

In feature engineering domain, researchers have always been striving to develop innovative techniques capable of selecting relevant and more appropriate features. In this marathon, a variety of feature selection and dimensionality reduction approaches have been proposed. Following sub-sections briefly describes different feature selection and dimensionality reduction approaches.

3.5.1 Feature Selection Methods

The working paradigm of feature selection approaches can be categorized into three main classes: filter [386], wrapper [533] and embedded [552]. The filter based feature selection techniques filter the corpus features based on their general properties, such as correlation with the dependent variable. It is considered the fastest and the best approach particularly for corpus with large number of features. Most widely employed filter based feature selection methods are Pearson correlation [386], mutual information [386], uni-variate [386], constant [552], quasi constant [552] and duplicate feature removal [386] encoders.

Uni-variate [386], constant [552] and quasi constant [552] feature selection methods eliminate duplicate features. Pearson correlation retains subset of features that are significantly correlated with the target but having minimum inter-feature correlation. Univariate feature selection (ANOVA) selects an informative subset of features by making use of the Gaussian distribution to compute linear connections between the input and output variables. The mutual information technique measures the reduction in uncertainty in one variable 'X' when the variable 'Y' is known. Mutual information assigns a score to each feature by utilizing information of input and output variables. Higher mutual information values imply that the target 'Y' has a low uncertainty given the predictor 'X'.

Unlike filter based methods, wrapper based methods rely on predictors to select the important features. Although this approach is computationally expensive, however, it is considered better than filter-based feature selection methods in terms of performance. Most commonly used wrapper techniques for sequential feature selection include forward feature selection, backward feature selection, recursive feature selection and exhaustive feature selection. These approaches iteratively choose the most informative subset of features from the feature space. Forward feature selection starts with the feature that performs best against the target. Then additional features are added in manner that maximize performance when combined with previously selected features. This procedure is repeated until the predetermined criterion is satisfied [1016]. Backward feature selection, also known as backward elimination, operates in the exact opposite way as forward feature selection. This technique starts with complete feature set and builds a model around them. At each iteration, it removes least impactful feature from the feature set until the stopping criterion is met. Recursive Feature Elimination (RFE) employs a greedy search method to find the best feature subset. It builds models iteratively, identifying which features perform best or worse in each iteration. It continues to develop models based on the features that are left in the feature space until all of them have been examined. The features are then graded according to how likely they are to be removed. Exhaustive feature selection evaluates all possible combinations of features, to find the optimal subset of features. It is computationally expensive for larger datasets.

Embedded feature selection methods inherently integrate feature selection within the model building process. This technique include regularization techniques and tree based methods. In this category Lasso regularization method mitigates linear models overfitting by incorporating a penalty on the absolute value of coefficients. This strategy effectively eliminates less influential features. On the other hand, Ridge regularization method reduces the overall impact of irrelevant features by penalizing the sum of squared coefficients. Apart from this, ensemble methods including Random Forest, Decision Trees, Gradient Boosting, and Extreme Gradient Boosting assess feature importance based on their contribution to the purity of decision nodes (i.e. better discrimination of classes) within ensemble models.

3.5.2 Dimensionality Reduction Methods

The aim of dimensionality reduction procedures is to transform original high dimensional feature space into the low dimensional feature subspace, while preserving the essential features. Most widely employed 14 dimensionality reduction methods are shown in Table 3.2. These methods can be categorised into two different groups: linear and nonlinear.

Linear dimensionality reduction methods including non-negative Matrix Factorization (NMF) [221], Independent Component Analysis (ICA) [570], Principal Component Analysis (PCA) [439], Truncated SVD [397], Factor Analysis (FA) [856] and linear discriminative analysis (LDA) transform high-dimensional feature space into a low-dimensional feature space as a linear combination of the original variables. The low-dimensional feature space retains the intrinsic structure of statistical sequence vectors and manages to capture the essential sequence features, while significantly reducing number of parameters. NMF [221] decomposes statistical feature space into two non-negative matrices known as NMF matrix and coefficients matrix and original statistical feature space is transformed into reduced feature space through additive combination of vectors present in underlay matrix. ICA also generates reduced feature space by separating original statistical feature space into additive components. PCA [439] converts original statistical vectors into n principal

Table 3.2 Feature selection and dimensionality reduction methods

Method	Algorithm	Reference
Wrapper based feature selection approaches	Forward feature selection	[533]
	Backward feature selection	[533]
	Recursive feature selection	[533]
	Exhaustive feature selection	[533]
Embedded feature selection approaches	Lasso	[552]
	Redige	[552]
Tree based feature selection approaches	Random forest based feature importance	[705]
	Decision tree based feature importance	[705]
	Gradient boost based feature importance	[705]
	Extreme gradient boost based feature importance	[705]
Filter based feature selection approaches	Mutual information	[386, 386]
	Univariate feature selection	[386]
	Pearson	[386]
	Constant feature	[552]
	Quasi constant features	[552]
	Duplicate features	[386]
Dimensionality reduction	K-means	[398]
	T-SNE	[681]
	Principal Component Analysis (PCA)	[439]
	Kernel PCA	[891]
	Locally linear embedding	[883]
	Singular Value Decomposition (SVD)	[397]
	Non-Negative Matrix Factorization (NMF)	[221]
	Independent Component Analysis (ICA)	[570]
	Multi-Dimensional Scaling (MDS)	[476]
	Factor analysis	[856]
	Feature agglomeration	[898]
	Gaussian random projection	[232]
	Sparse random projection	[583]
Auto encoder	Auto encoder	[1055]

components that represent the most relevant information. Similarly, Truncated SVD factorizes the original statistical vectors into number of columns equal to truncation to retain only a few largest singular values, Factor Analysis finds factor to describe the covariance of correlated observed variables. LDA focuses on low dimensional feature space with maximum separability between the groups.

Contrary to linear methods, nonlinear methods are applied to original statistical vectors that contain a nonlinear relationship. These methods preserve the global as well as local features of high-dimensional feature space in low-dimensional feature space. Nonlinear dimensionality reduction methods including K-means, t-SNE, Kernel PCA [439], Isometric Mapping (Isomap) and multi-dimensional scaling (MDS) preserve the global features of original statistical vectors of sequences. K-means computes the cluster centers and sets the number of clusters equal to target dimensions of statistical feature space. The new statistical feature space is generated in which new features are actually the distances of each point with respect to each cluster center. The t-SNE algorithm constructs a probability distribution on the feature pairs in the higher dimensions in such a manner that similar features are assigned higher probabilities and dissimilar features are assigned lower probabilities. Kernel PCA projects the nonlinear inseparable statistical vectors onto a higher dimensional feature space where it becomes linearly separable. Isomap generates the neighborhood networks and preserves the geodesic distance in low-dimensional feature space. Similar to Isomap, MDS measures the similarities and dissimilarities between the observed variables.

Instead of preserving global features, some methods try to preserve only geometrical properties of local features of nonlinear original statistical vectors such that locally linear embedding (LLE), Hessian LLE and Laplacian eigenmap. These methods preserve the local features in low-dimension statistical vectors assuming that only the local distances are reliable in high-dimensional statistical vectors. LLE reduces the original feature space to lower embedding while preserving the embedding of original sequences. Laplacian eigenmap maps the embedding corresponding to nearest neighbor and represents the graph with its Laplacian matrix. Hessian LLE is an extension of LLE that first minimizes the curviness of high dimensional original statistical vectors and then transfers to low dimensional feature space to make low dimension feature space locally isometric. Feature Agglomeration is another nonlinear dimensionality reduction approach which groups various components that behave similarly using hierarchical clustering. This behavior can be achieved by clustering in the feature direction or clustering transposed feature space. Gaussian Random Projection reduces the dimensions of high-dimensional feature space by projecting the original input's dimensional space onto a randomly generated matrix. Sparse Random Projection transforms feature space to sparse random matrices. These matrices are the best alternative to the dense Gaussian random projection matrices as they generate similar quality feature space in less memory and allow faster computation on the new feature space.

AutoEncoder is another very efficient dimensionality reduction approach based on artificial neural network. It is based on encoder-decoder paradigm where encoder compresses the original feature space into lower dimensions using bottleneck layers and decoder produces the original feature space from compressed representation. By reducing the reconstruction loss, an effective compressed representation of statistical feature space is learned.

3.6 Artificial Intelligence Predictors

In the realm of computational biology, Artificial Intelligence applications inception can be traced back to the employment of statistical algorithms for diverse array of sequence analysis tasks including sequence alignment, phylogenetic analysis, and population genetics. Early pioneers of computational biology, such as Margaret Dayhoff and Walter Fitch, employed statistical algorithms to study protein sequences and evolutionary relationships. Later, researchers introduced the applications of Markov models, conditional random fields [457] algorithms in distinct types of sequence analysis tasks including gene prediction [321], protein secondary structure prediction [346], protein domain prediction [933] and sequence alignment [39]. Markov models make use of probabilistic paradigm to predict the next state based solely on the current state. These models are useful for sequential data but lack in capturing dependencies between non-linear data. On the other hand, Conditional Random Fields model also makes use of probability theory for finding relationships between input sequences and output labels.

In computationally biology landscape, for sequences clustering [821] task, the most widely used statistical algorithms are cosine similarity [1165], Louvain clustering [839], and coverage clustering algorithm [1110]. Prime objective of these algorithms is to split a dataset into discrete clusters, based on degree of similarity between dataset instances. The statistical models rely on strong assumptions of underlying data distribution, which may not always align with real-world genetic sequence analysis scenarios. Moreover, these algorithms also lack in capturing complex patterns and non-linear relationships. Additionally, they often require significant manual feature engineering, which can be time-consuming and prone to human error. These limitations paved the way for the subsequent rise of machine learning algorithms. This section provides a high level overview of machine and deep learning predictors that are being utilized to develop AI-driven Genomics and Proteomic sequence analysis predictive pipelines.

3.6.1 Machine Learning Predictors

Following drawbacks of statistical algorithms and to harness the potential of steadily increase of data, researchers have introduced diverse types of machine learning algorithms. The genomics and proteomics sequence analysis domain has witnessed extensive applications of a diverse array of machine learning algorithms in distinct types of sequence analysis tasks including essential gene identification [213, 548], DNA methylation modification [1000, 1160], essential protein identification [862, 1163], disease gene association prediction [769], recombination spots identification [261] etc. The most widely used traditional machine learning algorithms include decision trees (DT), random forest (RF) [64], adaptive boosting (AdaBoost)[909], gradient boosting machine (GBM) [1153], XGBoost [332], CatBoost [332], extra trees (ET) [213], LightGBM [1153], support vector machine [1006, 1029, 1087], Naive Bayes [459], k-nearest neighbors (kNN) [15], linear regression [1227], logistic regression [645, 1118, 1227], ridge regression, and ElasticNet regression [486]. Based on working paradigm these algorithms can be broadly categorised into 5 distinct categories namely memory based algorithms, likelihood estimator based algorithms, support vector machine, regularization based algorithm, and tree based algorithms. The following subsections provide detailed description of the algorithms within each of these categories.

3.6.1.1 Memory Based Algorithms

K-Nearest Neighbors (KNN) [15, 246] is a memory-based algorithm that utilizes similarity based approach to make predictions. Unlike other algorithms, KNN does not explicitly learn a predictive function during training. Instead, it stores the entire training dataset in memory and uses it directly during the prediction phase. Specifically, for new data instance, the algorithm identifies "k" closest neighbors from training data based on a distance metric. Equation 3.6 illustrates mathematical expressions of three distinct distance functions that are usually employed to find distance between train set instances and new instance.

$$Distance_{\mathrm{kNN}} = \begin{cases} Euclidean\ Distance = \sqrt{\sum_{i=1}^{k}(x_i - y_i)^2} \\ Manhattan\ Distance = \sum_{i=1}^{k} |x_i - y_i| \\ Minkowski\ Distance = \left(\sum_{i=1}^{k} (|x_i - y_i|)^q\right)^{\frac{1}{q}} \end{cases} \tag{3.6}$$

In Eq. 3.6, y_i indicates new instances or test set instances whereas x_i refers to training instances. Furthermore, in KNN algorithm value of "k" is a hyper-parameter. This hyperparameter decides the number of instances of train set also known as neighbours, whose class labels will be considered to decide test instance class label. A smaller value of "k" can lead to over-fitting while a larger value can potentially cause under-fitting. Generally the value of k is chosen after multiple experimentation's. KNN is primarily used for classification and regression tasks. For classification, output is based on majority class voting, while for regression predicted value is average of the neighbors' values.

To better understand the working paradigm of each algorithm, consider a hypothetical dataset presented in Table 3.3. In this table, the first column represents dataset partitioning into train and test sets. The second column lists 5 sequence IDs, the third column contains 5 DNA sequences. The fourth column displays the overlapping 2-mers derived from DNA sequences, and the last column shows the corresponding class labels.

Furthermore, the 2-mer sequences are converted into statistical vectors by computing occurrence frequencies of unique 2-mers within the 2-mers sequences. For instance, in 2-mer sequence sample having Sequence ID 1, 2-mer AT has

Table 3.3 Sample dataset

Dataset partition	Sequence ID	DNA sequence	Sequence 2-mers	Class label
Train set	1	ATCG	AT, TC, CG	0
	2	GCTA	GC, CT, TA	1
	3	CGAT	CG, GA, AT	0
	4	TCGC	TC, CG, GC	1
Test set	5	GCGAT	GC, CG, AT	0

Table 3.4 Statistical representation of sample dataset sequences using 2-mer term frequencies

Dataset partition	Sequence ID	AT	TC	CG	GC	CT	TA	GA	Class label
Train set	1	1	1	1	0	0	0	0	0
	2	0	0	0	1	1	1	0	1
	3	1	0	1	0	0	1	1	0
	4	0	1	1	1	0	0	0	1
Test set	5	1	0	1	1	0	0	0	0

occurrence frequency of 1 and 2-mers TC and CG also have occurrence frequencies of 1. Across the entire dataset, there are 7 unique 2-mers, so each 2-mer sequence sample is represented as a 7-dimensional vector. 2-mers sequence sample, having Sequence ID 1, contains 3 2-mers. Since this sequence sample does not include the other 4 unique 2-mers, their frequency values will be set to zero in the vector representation. This process ensures that all sequence samples are represented as vectors of the same dimensionality. Table 3.4 illustrates statistical representation of sample dataset sequences using 2-mers term frequencies.

To classify the test sequence first of all kNN computes distance of statistical vector of test sequence (sequence ID 5) with statistical vectors of all training sequences (ID 1, 2, 3, 4). In this example, distance is computed using the Euclidean distance formula. Following Eq. 3.6 mathematical expression of Euclidean distance, the distances between test instance and training instances is shown as D(1, 5), D(2, 5), D(3, 5), and D(4, 5) and computed as below.

- Distance between sequence 1 and 5:

$$\begin{aligned}
D(1,5) &= \sqrt{(1-1)^2+(0-1)^2+(1-1)^2+(1-0)^2+(0-0)^2+(0-0)^2+(0-0)^2} \\
&= \sqrt{0^2+(-1)^2+0^2+1^2+0^2+0^2+0^2} \\
&= \sqrt{0+1+0+1+0+0+0} \\
&= \sqrt{2} \\
&\approx 1.414
\end{aligned}$$

- Distance between sequence 2 and 5:

$$D(2,5) = \sqrt{(1-0)^2+(0-0)^2+(1-0)^2+(1-1)^2+(0-1)^2+(0-1)^2+(0-0)^2}$$

$$= \sqrt{1^2 + 0^2 + 1^2 + 0^2 + (-1)^2 + (-1)^2 + 0^2}$$
$$= \sqrt{1+0+1+0+1+1+0}$$
$$= \sqrt{4}$$
$$= 2$$

- Distance between sequence 3 and 5:

$$\begin{aligned} D(3,5) &= \sqrt{(1-1)^2 + (0-0)^2 + (1-1)^2 + (1-0)^2 + (0-0)^2 + (0-1)^2 + (0-1)^2} \\ &= \sqrt{0^2 + 0^2 + 0^2 + 1^2 + 0^2 + (-1)^2 + (-1)^2} \\ &= \sqrt{0+0+0+1+0+1+1} \\ &= \sqrt{3} \\ &\approx 1.732 \end{aligned}$$

- Distance between sequence 4 and 5:

$$\begin{aligned} D(4,5) &= \sqrt{(1-0)^2 + (0-1)^2 + (1-1)^2 + (1-1)^2 + (0-0)^2 + (0-0)^2 + (0-0)^2} \\ &= \sqrt{1^2 + (-1)^2 + 0^2 + 0^2 + 0^2 + 0^2 + 0^2} \\ &= \sqrt{1+1+0+0+0+0+0} \\ &= \sqrt{2} \\ &\approx 1.414 \end{aligned}$$

After computing Euclidean distance, kNN identifies k nearest number of test sequence based on value of k. Considering $k = 3$, 3 nearest neighbors for test sequences 5 are: sequence 1 (Class: 0), sequence 4 (Class: 1) and sequence 3 (Class: 0). Hence, test sequence has 2 neighbours from class 0 and 1 neighbour from class 1. Finally, based on majority vote of nearest neighbors, the KNN algorithm assigns 0 class to test sequence.

3.6.1.2 Likelihood Estimator Based Algorithms

Naive Bayes [459] makes use of Bayes' theorem to estimate the likelihood of new instances belonging to particular class. Naive Bayes leverages the conditional independence assumption to streamline the calculation of likelihoods. Classification through Naive Bayes algorithm involves the computation of three key probability distributions: (1) Prior Probability: This is the probability of each class occurring independently of the feature space. It represents the overall likelihood of each class before considering any specific features. (2) Likelihood: This is the probability of observing each feature given that the instance belongs to a specific class. Under the Naive Bayes assumption, these likelihoods are calculated for each feature independently, even though the features themselves might be correlated. (3) Posterior Probability: After calculating the prior probabilities and the likelihoods, the Naive Bayes classifier uses Bayes' theorem to compute the posterior probability of each class given the observed features. This posterior probability represents the likelihood of the instance belonging to a particular class after considering the specific features.

Mainly, Naive Bayes algorithm has 3 different variants: (1) Gaussian Naive Bayes, (2) Multinomial Naive Bayes, (3) Bernoulli's Naive Bayes. Multinomial and Bernoulli Naive Bayes are more suitable for discrete data whereas Gaussian Naive Bayes is more useful for continuous data. Working paradigm of all three Naive Bayes variants can be defined in 4 distinct steps including (1) prior probability computation, (2) likelihood computation, (3) posterior probability computation, (4) class prediction. Lets discuss in detail all four steps of NB variants:

- All three NB variants make use of Eq. 3.7 to compute prior probability of each dependent variable.

$$\begin{aligned}\text{Prior Probability} \\ &= P(y_i) \\ &= \frac{\text{Count of instances with Class } y_i}{\text{Count of total instances in dataset}} \quad \text{where } y_j \in \{y_1, y_2, \ldots, y_m\}\end{aligned} \tag{3.7}$$

- All three NB variants follow distinct criterion for likelihood computation. Gaussian Naive Bayes variant computes likelihood by estimating mean and variance of independent variables with respect to dependent variables.
  - First of all, it computes mean of independent variables using Eq. 3.8.

$$\mu_{y_i} = \frac{1}{N_{y_i}} \sum_{i=1}^{N} x_i \epsilon y_i \tag{3.8}$$

In Eq. 3.8, μ_{y_i} represents mean of independent variables x_i for a dependent variable y_i. Specifically,this mean is computed separately for each dependent variable y_i to describe the likelihood of observing a given independent variable x_i under that dependent variable. N_{y_i} denotes total number of instances that belong to class y_i. It serves as a normalization factor to ensure that the mean is computed as the average of all independent variables corresponding to the dependent variable y_i. The $\sum_{i=1}^{N} x_i \in y_i$ refers to the total sum of the independent variables x_i for all instances that belong to dependent variable y_i. The notation $x_i \in y_i$ implies that only those independent variable are summed that correspond to instances of class y_i. The fraction $\frac{1}{N_{y_i}}$ ensures that the sum of the independent variables is averaged by dividing by the total number of instances in class y_i. This gives us the mean independent variable with respect to that dependent variable.

- In the next step, it computes variance of independent variables using Eq. 3.9.

$$\sigma_{y_i}^2 = \frac{1}{N_{y_i}} \sum (x_i - \mu_{y_i})^2 \tag{3.9}$$

In Eq. 3.9, $\sigma_{y_i}^2$ represents the variance of independent variable x_i for a specific dependent variable y_i. This variance is computed separately for each dependent variable y_i and describes the spread or variability of the independent variable corresponding to a particular dependent variable. N_{y_i} refers to total number of instances that belong to dependent variable y_i. It serves as a normalization factor that ensures the variance is computed as the average of the squared deviations of independent variables from mean. $\sum (x_i - \mu_{y_i})^2$ represents total sum of squared deviations of independent variables x_i from the mean μ_{y_i} for all instances that belong to dependent variable y_i. The squared deviation $(x_i - \mu_{y_i})^2$ measures how far each independent variable x_i is from the mean μ_{y_i}, with squaring ensuring that deviations are treated as positive values. The fraction $\frac{1}{N_{y_i}}$ normalizes the sum of squared deviations by dividing by the total number of instances that belongs to dependent variable y_i. This gives us the average of the squared deviations, which is the variance for the dependent variable y_i.

- Afterwards, it makes use of mean and variance of independent variables and calculates likelihood using Eq. 3.10.

$$P(x_i|y_i) = \frac{1}{\sqrt{2\pi\sigma_{y_i}^2}} \exp\left(-\frac{(x_i - \mu_{y_i})^2}{2\sigma_k^2}\right) \tag{3.10}$$

In Eq. 3.10, $P(x_i|y_i)$ represents probability of observing a independent variable x_i given that instance belongs to a specific dependent variable y_i, μ_{y_i} is mean of independent variables x_i for dependent variable y_i, and $\sigma_{y_i}^2$ refers to variance of independent variable x_i for dependent variable y_i. $\frac{1}{\sqrt{2\pi\sigma_{y_i}^2}}$ is normalization factor of Gaussian distribution. It ensures that total area under probability density function equals 1. Factor $\sqrt{2\pi\sigma_{y_i}^2}$ is derived from standard deviation of the distribution, where σ_{y_i} is square root of variance, $\exp\left(-\frac{(x_i-\mu_{y_i})^2}{2\sigma_{y_i}^2}\right)$ represents probability density function of Gaussian distribution. The exponent $-\frac{(x_i-\mu_{y_i})^2}{2\sigma_{y_i}^2}$ measures how far the observed independent variable x_i is from the mean μ_{y_i}, scaled by the variance $\sigma_{y_i}^2$. The farther x_i is from the mean, the lower the value of this term, and thus the lower the likelihood.

On the other hand, Multinomial Naive Bayes computes likelihood in three different steps. It estimates the number of instances having independent variables with respect to dependent variables count(x_i, y_i), then counts the number of independent variables in class y_i, and finally likelihood is computed using these counts. Equation 3.11 illustrates mathematical expression for likelihood estimation.

$$P(x_j|y_i) = \frac{\text{count}(x_i, y_i) + \alpha}{\text{total\_count}(y_i) + \alpha \cdot \text{count of independent variables}} \tag{3.11}$$

The likelihood computation in Eq. 3.11 is adjusted with a smoothing parameter α. The term count$(x_i, y_i) + \alpha$ in the numerator ensures that even instances of independent variables that do not appear in the training set (i.e., with zero count) contribute a non-zero probability. The denominator, total_count(y_i) + α · count of independent variables, normalizes this probability, accounting for the total occurrences in class y_i and the effect of smoothing across all possible independent variables.

Bernoulli Naive Bayes variant estimates likelihood by computing the proportion of independent variables and compliment of this proportion. This proportion is a ratio of the count of instances where both the independent variable x_i and the dependent variable y_i occur to the total number of instances. Equation 3.12 provides the mathematical expression for determining this proportion.

$$P(x_i = 0|y_i) = \frac{\text{Count}(x_i = 0, y_i) + \alpha}{\text{Instances number } y_i + \alpha \cdot 2} \tag{3.12}$$

In Eq. 3.12, Count$(x_i = 0, y_i)$ represents number of instances where both the independent variable $x_i = 0$ and the dependent variable y_i occur together, and α is a smoothing parameter that is added to Count$(x_i = 0, y_i)$ to prevent the likelihood from becoming zero, especially when x_i does not occur with y_i in training data, also known as Laplace smoothing. Instances number $y_i + \alpha \cdot 2$ represents the total number of instances where y_i occurs, adjusted by the smoothing factor. The factor 2 accounts for the binary nature of Bernoulli Naive Bayes, where each feature can either be present or absent. Hence, this equation calculates smoothed likelihood of an independent variable $x_i = 0$ being present (or absent) given a specific class y_i, which is a fundamental step in the Bernoulli Naive Bayes classification process.

Afterward computing proportion, it computes likelihood by taking complement probability of this proportion. Mainly, it is probability that instance does not belong to the specified class. Equation 3.13 illustrates mathematical expression for likelihood computation.

$$P(x_j = 0|y_i) = 1 - P(x_j = 1|y_i) \tag{3.13}$$

- After computing likelihood, all three NB variants compute posterior probability that estimates the class label for each instance based on prior probabilities and likelihood information. Equation 3.14 depicts mathematical expression for computing posterior probability.

$$P(y_i|\mathbf{x_i}) \propto P(y_i) \cdot \prod_{i=1}^{n} P(x_i|y_i) \tag{3.14}$$

Naive Bayes Explanation Using a Hypothetical Dataset

To understand the working of Naive Bayes algorithm, lets reconsider the sample data given in Table 3.4.

In the sample dataset, the prior probabilities of both classes can be computed using Eq. 3.7. It can be seen in Table 3.4, out of 4 instances, 2 instances belong to class 1 and 2 instances belong to class 0. Hence, prior probability of each class can be calculated as follow.

$$P(y_i = 1) = \frac{2}{4}, \; P(y_i = 0) = \frac{2}{4}$$

$$P(y_i = 1) = \frac{1}{2} = 0.5, \; P(y_i = 0) = \frac{1}{2} = 0.5$$

In this example lets compute likelihood of independent variables using Bernoulli Naive Bayes approach. As mentioned before, while computing likelihood we need to decide a small value of alpha that avoids division with zero in Eq. 3.13. Lets set α value 1 and calculate total number of instances for dependent variable. Moreover, it determines count of instances where both independent and dependent variable occur together. By using these values, it computes likelihood as shown below.

Specifically, for independent variable AT, in class 0 2 instances contain independent variable AT, total number of instances of class 0 are also 2. Lets substitute these values in Eq. 3.13 to compute independent variable AT likelihood with respect to class 0 $P(\text{AT} = 1|y_i = 0)$ as follow.

$$P(\text{AT} = 1|y_i = 0) = \frac{2+1}{2+1 \cdot 2} = \frac{3}{4} = 0.75$$

Following similar criteria, it computes likelihoods of other independent variables with respect to class 0 as follows:

$$P(\text{TC} = 1|y_i = 0) = \frac{1+1}{2+1 \cdot 2} = \frac{2}{4} = 0.5$$

$$P(\text{CG} = 1|y_i = 0) = \frac{2+1}{2+1 \cdot 2} = \frac{3}{4} = 0.75$$

$$P(\text{GC} = 1|y_i = 0) = \frac{1+1}{2+1\cdot 2} = \frac{2}{4} = 0.5$$

$$P(\text{CT} = 1|y_i = 0) = \frac{1+1}{2+1\cdot 2} = \frac{2}{4} = 0.5$$

$$P(\text{TA} = 1|y_i = 0) = \frac{2+1}{2+1\cdot 2} = \frac{3}{4} = 0.75$$

$$P(\text{GA} = 1|y_i = 0) = \frac{1+1}{2+1\cdot 2} = \frac{2}{4} = 0.5$$

similarly, the likelihoods of all independent variables with respect to class 1 are computed as follows:

$$P(\text{AT} = 1|y_i = 1) = \frac{0+1}{2+1\cdot 2} = \frac{1}{4} = 0.25$$

$$P(\text{TC} = 1|y_i = 1) = \frac{2+1}{2+1\cdot 2} = \frac{3}{4} = 0.75$$

$$P(\text{CG} = 1|y_i = 1) = \frac{2+1}{2+1\cdot 2} = \frac{3}{4} = 0.75$$

$$P(\text{GC} = 1|y_i = 1) = \frac{2+1}{2+1\cdot 2} = \frac{3}{4} = 0.75$$

$$P(\text{CT} = 1|y_i = 1) = \frac{2+1}{2+1\cdot 2} = \frac{3}{4} = 0.75$$

$$P(\text{TA} = 1|y_i = 1) = \frac{1+1}{2+1\cdot 2} = \frac{2}{4} = 0.5$$

$$P(\text{GA} = 1|y_i = 1) = \frac{1+1}{2+1\cdot 2} = \frac{2}{4} = 0.5$$

By using these prior probabilities and likelihood, Naive Bayes algorithm estimates the posterior probability of a class for a new sequence, and it does not involve iterative optimization methods. To understand this, consider a test sequence {GCGAT} from Table 3.3 and employ Naive Bayes algorithm. In accordance to Table 3.4, {GCGAT} has 3 independent variables {CG, AT, GC}. Thus, it calculate the posterior probabilities for each class with respect to independent variables of test sequence.

For class 0, posterior probability can be computed as follow.

$$P(y_i = 0|x_i = \text{AT, CG, GC}) \propto P(y_i = 0) \times P(\text{AT} = 1|y_i = 0) \times P(\text{CG} = 1|y_i = 0) \times P(\text{GC} = 1|y_i = 0)$$

$$\mathbf{P(y_i{=}0|x_i{=}AT, CG, GC)} \propto 0.5 \times 0.75 \times 0.75 \times 0.5 \approx \mathbf{0.1406}$$

For class 1: posterior probability can be calculated as given below.

$$P(y_i = 1|x_i = \text{AT, CG, GC}) \propto P(y_i = 1) \times P(\text{AT} = 1|y_i = 1) \\ \times P(\text{CG} = 1|y_i = 1) \times P(\text{GC} = 1|y_i = 1)$$

$$\mathbf{P(y_i = 1|x_i = \text{AT, CG, GC})} \propto 0.5 \times 0.25 \times 0.75 \times 0.75 \approx \mathbf{0.0703}$$

After predicting the class probability, a classification decision is made by applying the argmax function. This function assigns class label to new instance based on the calculated probabilities for each class. The class with the highest probability is selected as the predicted class. In the special case where the predicted probability is exactly same, it indicates that the instance has an equal likelihood of belonging to either class. In this unique scenario, probability of class 0 is higher from both classes. Hence, it assigns test sequences class 0.

3.6.1.3 Support Vector Machine Algorithm

Support Vector Machine (SVM) is a supervised learning algorithm that determines an optimal hyperplane to separate the data points of different classes with maximum margin. A hyperplane is a n-dimensional plane that acts as a decision boundary between two or more classes. The optimal hyperplane is positioned at same distance from the closest data points of each class, known as support vectors. Equation 3.15 defines a hyperplane or linear decision boundary that separates data points into different classes.

$$y_i = \mathbf{w} \cdot \mathbf{x}_i + b \tag{3.15}$$

In Eq. 3.15, $\mathbf{w}$ is weight vector that determines the orientation of the hyperplane, b refers to bias term that controls the offset of the hyperplane from the origin, y_i represents the predicted class label for the ith instance.

Support Vector Machines (SVM) are efficient in designing optimal hyperplanes capable of discriminating data points into distinct classes when the data is linearly separable. However, real-world datasets often contain complex and non-linear patterns that cannot be separated into distinct classes using linear lines. To handle such complexity, SVM utilizes kernel functions that map the original data into a higher-dimensional space, where the data may become linearly separable. This transformation empowers SVM to effectively classify data that is not linearly separable in its original form. Most commonly employed kernel functions are polynomial, Gaussian radial basis function (RBF), and sigmoid.

$$K(\mathbf{x}_i, \mathbf{x}_j) = \begin{cases} Polynomial = ({x_i}^T x_j + c)^d, \quad \gamma > 0 \\ Gaussian\ RBF = e^{-\|\mathbf{x_i}-\mathbf{x}_j\|^2} \\ Sigmoid = tanh(\gamma.{x_i}^T x_j + c),\ \gamma > 0 \end{cases} \tag{3.16}$$

In Eq. 3.16,

- x_i, x_j: Statistical vectors of ith and j data points, respectively.
- d: Degree of the polynomial in the polynomial kernel.
- γ: A scaling parameter that controls the influence or spread of the kernel function.
- c: A coefficient that shifts the function, affecting the position and shape of the decision boundary.
- tanh: Hyperbolic tangent function, used in the sigmoid kernel.

The Support Vector Machine (SVM) incorporates the kernel matrix K within its decision function to predict the class labels. The mathematical expression of SVM decision function is shown in Eq. 3.17.

$$y_i = sign\ (\sum_{i=1}^{N} \mathbf{w} \cdot \mathbf{x}_\mathrm{i} \cdot \mathbf{K}(\mathbf{x_i}, \mathbf{x_j}) + b) \tag{3.17}$$

In Eq. 3.17,

- y_i: The predicted class label for the unseen data point $\mathbf{x}_i$.
- w: The learned weights (Lagrange multipliers) associated with the support vectors.
- $\mathbf{K}(\mathbf{x}_i, \mathbf{x}_j)$: The kernel function applied to pairs of data points $\mathbf{x}_i$ and $\mathbf{x}_j$.
- b: The bias term in the decision function.

To understand the working paradigm of SVM, lets begin with linear data and then expand the example for non-linear data to illustrate the concept of kernel functions. Table 3.5 illustrates a sample dataset consisting of 8 samples and each sample has 2 features. The samples belong to two distinct classes.

Table 3.5 Sample binary classification dataset with linearly separable data points

Sample ID	Feature 1 (x_1)	Feature 2 (x_2)	Class label
1	3	1	Positive
2	3	−1	Positive
3	6	1	Positive
4	6	−1	Positive
5	1	0	Negative
6	0	1	Negative
7	0	−1	Negative
8	−1	0	Negative

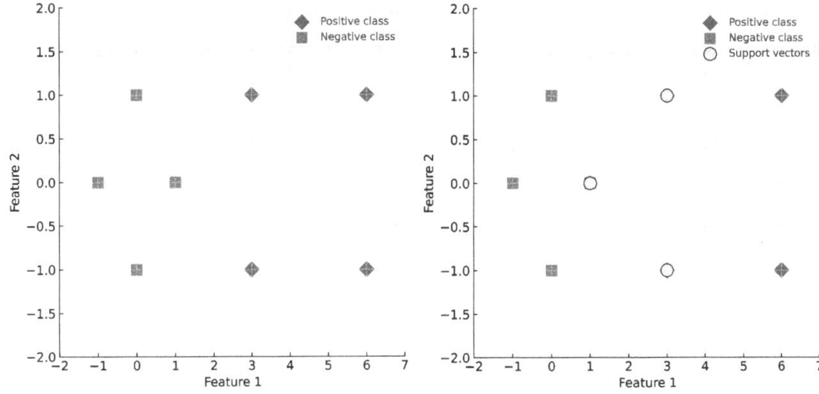

(a) Distribution of Sample data points in $\mathbb{R}^2$ feature space **(b)** Visualization of Support Vectors within a sample dataset

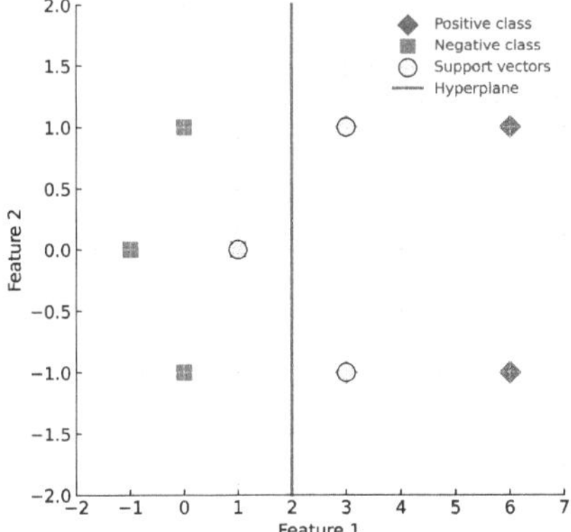

(c) Visualization of hyperplane within a Sample Dataset

Fig. 3.8 SVM model visualization: data points, support vectors, and decision hyperplane. (**a**) Distribution of Sample data points in $\mathbb{R}^2$ feature space. (**b**) Visualization of Support Vectors within a sample dataset. (**c**) Visualization of hyperplane within a Sample Dataset

Table 3.5 features and class labels are mapped in a two-dimensional space $\mathbb{R}^2$ as shown in Fig. 3.8a. In Fig. 3.8a, x-axis corresponds to feature 1 and the y-axis to feature 2. In this graphical representation, positive and negative class samples are distinguished using different colors. A closer analysis of this space shows that positive class samples 1 and 2, with feature values (3, 1) and (3, -1), respectively and negative class sample 5, with feature values (1, 0), are positioned close to each other. This proximity suggests that the optimal hyperplane should be positioned between these points, and as a result, the support vectors will likely be located near them.

Following analysis of Fig. 3.8a and aforementioned key points about support vectors, Fig. 3.8b highlights 3 support vectors. These support vectors are: $\{sv_1 = (1, 0), sv_2 = (3, 1), sv_3 = (3, -1)\}$. Moreover, it is evident from Fig. 3.8b that data points are linearly separable and linear hyperplane can be used to separate them.

SVM algorithm iteratively adjusts the weights associated with the support vectors to maximize the margin between the two classes while correctly classifying the training data. This process ultimately leads to the determination of the normal vector w, which defines the hyperplane. Equation 3.18 illustrates mathematical expression to compute an optimal normal vector w:

$$\mathbf{w} = \sum_i \beta_i \, \mathrm{sv}_i \tag{3.18}$$

In Eq. 3.18, β_i represents the weights or coefficients associated with the corresponding support vector sv_i. $\sum_i$ refers to sum up of support vectors. In this example, there are three support vectors ($\mathrm{sv}_1, \mathrm{sv}_2, \mathrm{sv}_3$) with corresponding weights ($\beta_1, \beta_2, \beta_3$). The equation would expand to:

$$\mathbf{w} = \beta_1 \cdot \mathrm{sv}_1 + \beta_2 \cdot \mathrm{sv}_2 + \beta_3 \cdot \mathrm{sv}_3 \tag{3.19}$$

The weights of support vectors are determined by solving an optimization problem that maximizes the margin between classes while correctly classifying data points. To achieve this objective, illustrated in Eq. 3.19, support vectors and variables can be mapped with their corresponding class labels as follows:

$$\begin{aligned} \beta_1 \cdot \mathrm{sv}_1 + \beta_2 \cdot \mathrm{sv}_2 + \beta_3 \cdot \mathrm{sv}_3 &= -1 \\ \beta_1 \cdot \mathrm{sv}_1 + \beta_2 \cdot \mathrm{sv}_2 + \beta_3 \cdot \mathrm{sv}_3 &= +1 \\ \beta_1 \cdot \mathrm{sv}_1 + \beta_2 \cdot \mathrm{sv}_2 + \beta_3 \cdot \mathrm{sv}_3 &= +1 \end{aligned} \tag{3.20}$$

Without a bias term, the hyperplane would always pass through the origin (0,0). This constraint severely limits the model's ability to fit complex datasets. By introducing a bias term, SVM allows the hyperplane to be shifted anywhere in the feature space. To design an optimal decision boundary, a bias term with value 1 is introduced in support vectors. The augmented support vectors are denoted by a tilde sign and are given as:

$$\{\tilde{\mathrm{sv}}_1 = (1, 0, 1), \tilde{\mathrm{sv}}_2 = (3, 1, 1), \tilde{\mathrm{sv}}_3 = (3, -1, 1)\} \tag{3.21}$$

The addition of bias into support vectors points will extended Eq. 3.20 into a new equation as follows:

$$\begin{aligned} \beta_1 \cdot \tilde{\mathrm{sv}}_1 + \beta_2 \cdot \tilde{\mathrm{sv}}_2 + \beta_3 \cdot \tilde{\mathrm{sv}}_3 &= -1 \\ \beta_1 \cdot \tilde{\mathrm{sv}}_1 + \beta_2 \cdot \tilde{\mathrm{sv}}_2 + \beta_3 \cdot \tilde{\mathrm{sv}}_3 &= +1 \\ \beta_1 \cdot \tilde{\mathrm{sv}}_1 + \beta_2 \cdot \tilde{\mathrm{sv}}_2 + \beta_3 \cdot \tilde{\mathrm{sv}}_3 &= +1 \end{aligned} \tag{3.22}$$

To find optimal position of the hyperplane, SVM computes dot product between support vectors. The dot product measures the similarity between vectors in the feature space. The higher values illustrates that the vectors are more aligned. This similarity plays a key role in determining the placement of the SVM decision boundary. The integration of the similarity between support vectors into Eq. 3.22 results in the following updated equation:

$$\begin{aligned}\beta_1\tilde{sv}_1\cdot\tilde{sv}_1+\beta_2\tilde{sv}_2\cdot\tilde{sv}_1+\beta_3\tilde{sv}_3\cdot\tilde{sv}_1&=-1\\ \beta_1\tilde{sv}_1\cdot\tilde{sv}_2+\beta_2\tilde{sv}_2\cdot\tilde{sv}_2+\beta_3\tilde{sv}_3\cdot\tilde{sv}_2&=+1\\ \beta_1\tilde{sv}_1\cdot\tilde{sv}_3+\beta_2\tilde{sv}_2\cdot\tilde{sv}_3+\beta_3\tilde{sv}_3\cdot\tilde{sv}_3&=+1\end{aligned}\tag{3.23}$$

Equation 3.24 demonstrates the substitution of $\tilde{sv}_i$ values from Eq. 3.21 into Eq. 3.23.

$$\begin{aligned}\beta_1(1,0,1)\cdot(1,0,1)+\beta_2(3,1,1)\cdot(1,0,1)+\beta_3(3,-1,1)\cdot(1,0,1)&=-1\\ \beta_1(1,0,1)\cdot(3,1,1)+\beta_2(3,1,1)\cdot(3,1,1)+\beta_3(3,-1,1)\cdot(3,1,1)&=+1\\ \beta_1(1,0,1)\cdot(3,-1,1)+\beta_2(3,1,1)\cdot(3,-1,1)+\beta_3(3,-1,1)\cdot(3,-1,1)&=+1\end{aligned}\tag{3.24}$$

In Eq. 3.24, the computation of the dot product results in the following equation:

$$\begin{aligned}2\beta_1+4\beta_2+4\beta_3&=-1\\ 4\beta_1+11\beta_2+9\beta_3&=+1\\ 4\beta_1+9\beta_2+11\beta_3&=+1\end{aligned}\tag{3.25}$$

Solving above expressions yield $\beta_1 = -3.5$, $\beta_2 = 0.75$, and $\beta_3 = 0.75$. Now these calculated β_i values are used to determine the discriminating hyperplane using the following expression.

$$\tilde{w}=\sum_i \beta_i\tilde{sv}_i$$

$$=-3.5\begin{pmatrix}1\\0\\1\end{pmatrix}+0.75\begin{pmatrix}3\\1\\1\end{pmatrix}+0.75\begin{pmatrix}3\\-1\\1\end{pmatrix}=\begin{pmatrix}1\\0\\-2\end{pmatrix}$$

Since vectors are augmented with a bias, last entry in $\tilde{w}$ is interpreted as hyperplane offset b, facilitating the expression of the separating hyperplane equation as $y = wx + b$ with $w = (1, 0)$ and $b = -2$. Plotting the line gives the expected decision boundary as illustrated in Fig. 3.8c.

To understand working of SVM with non-linear data points, consider sample data points in 2-dimensional space $\mathbb{R}^2$ as given in Table 3.6.

The objective is to identify an optimal hyperplane that discriminates instances into two classes. Table 3.6 features and class labels are mapped in a two-dimensional space $\mathbb{R}^2$ as shown in Fig. 3.9a. In Fig. 3.9a, x-axis corresponds to feature 1 and the

Table 3.6 Sample binary classification dataset with non-linear data points

Sample ID	Feature 1 (x_1)	Feature 2 (x_2)	Class label
1	2	2	Positive
2	2	−2	Positive
3	−2	−2	Positive
4	−2	2	Positive
5	1	1	Negative
6	1	−1	Negative
7	−1	−1	Negative
8	−1	1	Negative

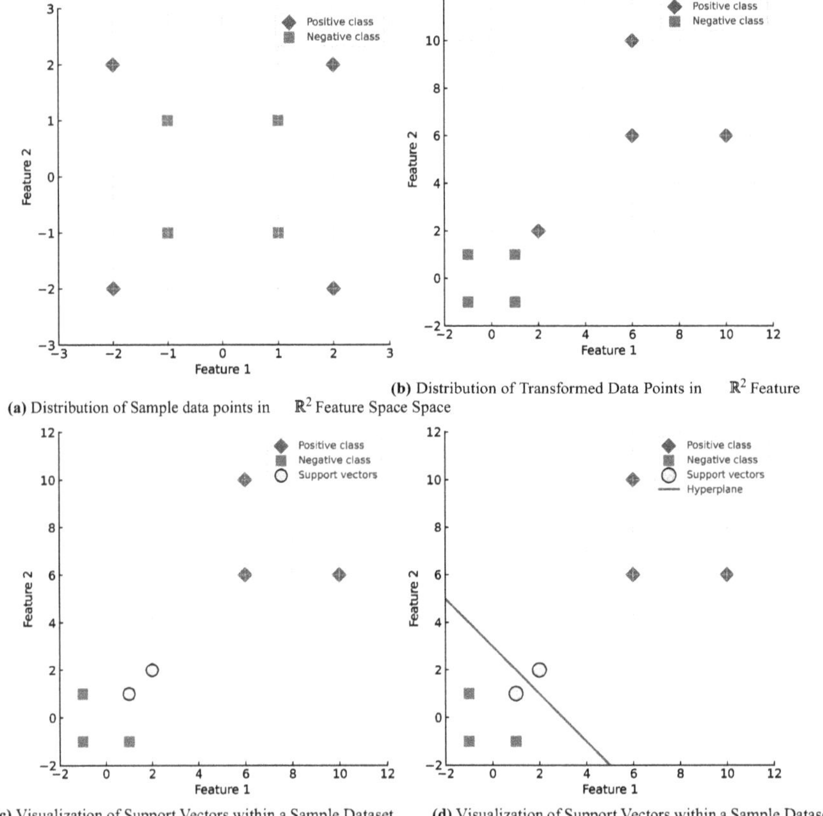

Fig. 3.9 Non-linear data samples based SVM model visualization: data points, support vectors, and decision hyperplane. (**a**) Distribution of sample data points in $\mathbb{R}^2$ feature space. (**b**) Distribution of transformed data points in $\mathbb{R}^2$ feature space. (**c**) Visualization of support vectors within a sample dataset. (**d**) Visualization of support vectors within a sample dataset

y-axis to feature 2. A close analysis of instances within this graphical representation illustrates that these instances are not linearly separable. In order to solve a non-linear problem, SVM transforms non linear data into linear data by using mapping function f. The mapping functions are usually designed according to number of input features. The example dataset (Table 3.6) encompasses two features, therefore, data can be transformed through a quadratic mapping function shown in Eq. 3.26.

$$f_1\begin{pmatrix}x_1\\x_2\end{pmatrix} = \begin{cases}\begin{pmatrix}4 - x_2 + |x_1 - x_2|\\4 - x_1 + |x_1 - x_2|\end{pmatrix} & \text{if } \sqrt{x_1^2 + x_2^2} > 2\\ \begin{pmatrix}x_1\\x_2\end{pmatrix} & \text{otherwise}\end{cases} \tag{3.26}$$

Equation 3.26 is a mapping function that transforms the input data points (x_1, x_2) into a new space. This transformation can help in making the data linearly separable in the new space, which is one of the core ideas behind using kernels in SVMs. The function has two cases, depending on the value of $\sqrt{x_1^2 + x_2^2}$, which represents the Euclidean distance of the point (x_1, x_2) from the origin. In first condition, transformation is applied when distance $\sqrt{x_1^2 + x_2^2}$ is greater than 2. The expressions $(4 - x_2 + |x_1 - x_2|)$ and $(4 - x_1 + |x_1 - x_2|)$ modify input data based on values of x_1 and x_2. This helps in creating a new feature space where a linear decision boundary may be more easily applied by SVM. In second condition, if distance $\sqrt{x_1^2 + x_2^2}$ is not greater than 2, function leaves data point unchanged, meaning it remains in its original form (x_1, x_2). In a nutshell, purpose of this mapping function is to introduce a non-linear transformation that can help SVM to better separate classes in data. In scenarios where data is not linearly separable in its original space, this function transforms data points into a new space where they might become linearly separable, allowing the SVM to effectively find a separating hyperplane.

In order to understand how this non-linear data is transformed into linear data, lets take a data point (2, 2), it first checks the conditions either these points are making a circle or not by substituting these values in $\sqrt{x_1^2 + x_2^2} > 2$ as given below:

$$\sqrt{2^2 + 2^2} = \sqrt{4 + 4} = \sqrt{8} = 2\sqrt{2} \approx 2.828 > 2$$

Since condition is fulfilled, it computes transformed values of x_1 and x_2 by substituting these values in $4 - x_2 + |x_1 - x_2|$ and $4 - x_1 + |x_1 - x_2|$ as follows.

Substitution of (2, 2) into $4 - x_2 + |x_1 - x_2|$:

$$4 - x_2 + |x_1 - x_2| = 4 - 2 + |2 - 2| = 4 - 2 + 0 = 2$$

Substitution of (2, 2) into $4 - x_1 + |x_1 - x_2|$:

$$4 - x_1 + |x_1 - x_2| = 4 - 2 + |2 - 2| = 4 - 2 + 0 = 2$$

Table 3.7 Transformed linear data from non-linear data points

Sample ID	Feature 1 (x_1^*)	Feature 2 (x_2^*)	Class label
1	2	2	Positive
2	10	6	Positive
3	6	6	Positive
4	6	10	Positive
5	1	1	Negative
6	1	−1	Negative
7	−1	−1	Negative
8	−1	1	Negative

Some data points does not fulfil the condition of making circle, so these points remain unmodified in transformed data. For instance, condition for data point (1, 1) is not full filled as given below:

$$\sqrt{1^2 + 1^2} = \sqrt{1+1} = \sqrt{1} = sqrt2 \approx 1.414 \not> 2$$

Thus, (1, 1) remained unmodified in transformed data.

The transformed data is shown in Table 3.7 and its visual representation is shown in Fig. 3.9b. A close analysis of Fig. 3.9b reveals that, both class instances are now separable with a linear line. As the data is now linearly separable in the transformed space, the SVM algorithm can proceed to determine the optimal hyperplane using the same methodology as applied in the linearly separable dataset.

3.6.1.4 Regularization Based Algorithms

Regularization-based algorithms (Ridge regression, Lasso regression, and Elastic Net) [486, 645, 1118, 1227, 1227] objective is to mitigate the overfitting phenomenon by introducing penalty terms to the learning process. In this category Linear regression [1227] is a foundational model that predicts dependent variables based on linear combinations of independent variables. Logistic regression algorithm [645] extends this foundational model concept to predict dependent variables by finding relationships between dependant and independent variables using sigmoid function. The other three algorithms (Ridge regression, Lasso regression, and Elastic Net) make use of linear and logistic regression working criteria with addition of penalty terms to the learning process.

Equation 3.27 depicts mathematical expression that linear regression [1227] algorithm utilizes to predict dependant variables by finding relationships between dependent and independent variables.

$$\hat{y_i} = \beta_0 + \beta_1 x_1 + \beta_2 x_2 + \ldots\ldots + \beta_n x_n + \epsilon \tag{3.27}$$

In Eq. 3.27, y_i refers to dependent variables and β_0 denotes intercept or constant term. Moreover, $\beta_1, \beta_2, \ldots, \beta_n$ are weights of independent variables and ϵ represents an adjustable parameter (bias term). The main goal of linear regression algorithm is to determine optimal values of $\beta_1, \beta_2, \beta_3, \ldots, \beta_n$. These optimal values are obtained by utilizing ordinary least square (OLS) loss function during training process. Equation 3.28 illustrates mathematical expression to compute ordinary least square (OLS) loss.

$$L_{OLS} = \frac{1}{n}\sum_{i=1}^{n}(y_i - \hat{y}_i)^2 \tag{3.28}$$

In Eq. 3.28, y_i indicates actual value and $\hat{y}_i$ represents predicted value of dependent variables. The computed loss value is utilized to update weights of independent variables during backpropogation process.

For a precise understanding of linear regression algorithm working paradigm, consider hypothetical dataset given in Table 3.4. First of all linear regression algorithm segregates train set data into independent variables matrix X and dependent variables vector Y as follows.

$$X = \begin{pmatrix} 1\,1\,1\,0\,0\,0\,0 \\ 0\,0\,0\,1\,1\,1\,0 \\ 1\,0\,1\,0\,0\,1\,1 \\ 0\,1\,1\,1\,0\,0\,0 \end{pmatrix} \quad \text{and} \quad Y = \begin{pmatrix} 0 \\ 1 \\ 0 \\ 1 \end{pmatrix}$$

In next step, it estimates weights of independent variables β by using mathematical expression of Eq. 3.29.

$$\beta = (X^T X)^{-1} X^T Y \tag{3.29}$$

In Eq. 3.29, X signifies independent variables matrix, X^T is the transpose of X, $(X^T X)^{-1}$ is the inverse of $X^T X$, and Y represents dependent variables vector. β values are estimated by first calculating $X^T X$ and its inverse $(X^T X)^{-1}$, and multiplying it with $X^T Y$. Using aforementioned matrix and vector y, Eq. 3.29 produces following weight vector $[-0.2, 0.4, 0.6, 0.1, 0.3, -0.1, 0.2]$. In weight vector, each entry corresponds to particular β_n, i.e., $\beta_1 = -0.2$, $\beta_2 = 0.4$, $\beta_3 = 0.6$, $\beta_4 = 0.1$, $\beta_5 = 0.3$, $\beta_6 = -0.1$, and $\beta_7 = 0.2$. Bias term ϵ is adjusted as 0.1. For training sequences in Table 3.3, it substitutes weights values (β), X and ϵ in Eq. 3.27 to obtain predicted Y vector as follows.

$$\hat{Y} = X \cdot \beta + \epsilon = \begin{pmatrix} 1\ 1\ 1\ 0\ 0\ 0\ 0 \\ 0\ 0\ 0\ 1\ 1\ 1\ 0 \\ 1\ 0\ 1\ 0\ 0\ 1\ 1 \\ 0\ 1\ 1\ 1\ 0\ 0\ 0 \end{pmatrix} * \begin{pmatrix} -0.2 \\ 0.4 \\ 0.6 \\ 0.1 \\ 0.3 \\ -0.1 \\ 0.2 \end{pmatrix} + \begin{pmatrix} 0.1 \\ 0.1 \\ 0.1 \\ 0.1 \end{pmatrix} = \begin{pmatrix} 0.9 \\ 0.4 \\ 0.6 \\ 1.2 \end{pmatrix}$$

After predicting the class probability $\hat{Y}$, a classification decision is made by applying the argmax function. This function assigns class label to an instance based on the calculated probabilities for each class. The class with the highest probability is selected as the predicted class. In the special case where the predicted probability is exactly same (e.g. 0.5 for binary classification), it indicates that the instance has an equal likelihood of belonging to either class. In this unique scenario, from both classes, a random class label is assigned to the instances. In the next step, it makes use of Y and $\hat{Y}$ vectors and computes loss by using Eq. 3.28 as follow.

$$L_{\text{OLS}} = \frac{(0-0.9)^2 + (1-0.4)^2 + (0-0.6)^2 + (1-1.2)^2}{4}$$

$$L_{\text{OLS}} = \frac{0.81 + 0.36 + 0.36 + 0.04}{4} = \frac{1.57}{4} \approx \mathbf{0.3925}$$

This loss value is utilized to compute gradient value that is backpropagated to update the β weights. This process is iteratively repeated to obtain optimal β weights.

Contrary to linear regression, logistic regression [645, 1118, 1227] predicts probability of an instance to a particular class by using a logistic (sigmoid) function. Equation 3.30 depicts mathematical expression to predict the probability of an instance to a particular class.

$$h_\alpha(y_i) = \frac{1}{1 + e^{-\alpha y_i}} \tag{3.30}$$

In Eq. 3.30, y_i indicates linear combination of independent variables $(x_1, x_2, \ldots, x_n)$ and $h_\alpha(y_i)$ refers to the predicted probability of dependent variable y_i. The cost function of logistic regression algorithm is derived from maximum likelihood estimation (MLE), also known as log loss. Equation 3.31 illustrates mathematical expression of cost function.

$$L_{Logistic} = -\frac{1}{n} \sum_{j=1}^{n} \left[y_i \log(h_\alpha(y_i)) + (1 - h_\alpha(y_i)) \log(1 - h_\alpha(y_i) \right] \tag{3.31}$$

In Eq. 3.31, n indicates total number of training samples, and $h_\alpha(y_i)$ refers to predicted probability of dependent variable y_i. Similar to linear regression, it calculates gradients of loss function which is backpropagate to update weights.

To understand the working paradigm of logistic regression, consider dataset illustrated in Table 3.3, in very first step, logistic regression set a random weight vector as $\beta = [0.5, -0.3, 0.2, 0.1, -0.5, 0.4, -0.2]$ and ϵ as 0.1. Moreover, it adjust α values as 1 and determines predicted labels of all training instances using Eq. 3.32 as given below.

$$\hat{Y} = X \cdot \beta + b \tag{3.32}$$

First of all, it computes $X \cdot \beta$ and adds bias values to output predicted $\hat{Y}$ vector as follow.

$$\hat{Y} = X \cdot \beta + \epsilon = \begin{pmatrix} 1\,1\,1\,0\,0\,0\,0 \\ 0\,0\,0\,1\,1\,1\,0 \\ 1\,0\,1\,0\,0\,1\,1 \\ 0\,1\,1\,1\,0\,0\,0 \end{pmatrix} * \begin{pmatrix} 0.5 \\ -0.3 \\ 0.2 \\ 0.1 \\ -0.5 \\ 0.4 \\ -0.2 \end{pmatrix} + \begin{pmatrix} 0.1 \\ 0.1 \\ 0.1 \\ 0.1 \end{pmatrix} = \begin{pmatrix} 0.5 \\ 0.1 \\ 1.0 \\ 0.1 \end{pmatrix}$$

Afterwards, it employs sigmoid function on $\hat{Y}$ to determine $h_a(\hat{Y})$ as follow.

$$h_a(\hat{Y}) = \begin{pmatrix} \frac{1}{1+e^{-0.5}} \\ \frac{1}{1+e^{-0.1}} \\ \frac{1}{1+e^{-1.0}} \\ \frac{1}{1+e^{-0.1}} \end{pmatrix} \approx \begin{pmatrix} 0.622 \\ 0.525 \\ 0.731 \\ 0.525 \end{pmatrix}$$

In the next step, it makes use actual and predictor dependent variable vector i.e., Y and $h_a(\hat{Y})$ to compute loss using Eq. 3.31. For 1st instances, it computes difference between actual and predicted dependent variable as follow.

$$Y_1 = [0 \cdot \log(0.622) + (1-0) \cdot \log(1-0.622)] = \log(0.378)$$

Similarly, it computes this difference for other instances Y_2, Y_3, and Y_4.

$$Y_2 = [1 \cdot \log(0.525) + (1-1) \cdot \log(1-0.525)] = \log(0.525)$$

$$Y_3 = [0 \cdot \log(0.731) + (1-0) \cdot \log(1-0.731)] = \log(0.269)$$

$$Y_4 = [1 \cdot \log(0.525) + (1-1) \cdot \log(1-0.525)] = \log(0.525)$$

Afterwards, it substitutes these values in Eq. 3.31 and computes loss as given below.

$$L_{Logistic} = -\frac{Y_1 + Y_2 + Y_3 + Y_4}{4}$$

$$L_{Logistic} = -\frac{\log(0.378) + \log(0.525) + \log(0.269) + \log(0.525)}{4}$$

$$L_{Logistic} = -\frac{-0.973 + (-0.648) + (-1.310) + (-0.648)}{4} \approx -\frac{-3.579}{4} \approx \mathbf{0.895}$$

Using this loss, it calculates gradients and backpropagates to determine optimal values of weights. Despite being efficient and computationally inexpensive, both linear and logistics regression have some drawbacks such that it only focuses on linear relationships between dependent and independent variables which may not hold in real-world data. Moreover, outliers and high correlations among independent variables can lead to unreliable weights estimation. To address these issues, regularization algorithms incorporate additional terms (θ, p, λ) into cost function to prevent overfitting and enhance model generalization. Most common regularization-based approaches are ridge regression, lasso regression, and elastic net regression.

Ridge regression [349, 1079], also known as L2 Regularization, is a variant of linear regression that incorporates a penalty term into cost function to deal with high correlation among dependent variables. Equation 3.33 presents mathematical expression to compute cost function for ridge regression.

$$L_{Ridge}(\hat{\beta}) = \sum_{i=1}^{n}(y_i - \hat{y}_i\hat{\beta}) + \lambda \sum_{j=1}^{m} \hat{\beta}_j^{\,2} = \|y - X\beta\|^2 + \lambda \|\beta\|^2 \tag{3.33}$$

Here $\lambda \|\beta\|^2$ refers to penalty term and λ is the regularization parameter that controls trade-off between fitting training data well and keeping weights small. At $\lambda = 0$, ridge regression aims to reduce ordinary least square loss and higher value of λ signifies shrinkage of weights values towards zero to reduce variance and avoid overfitting. Since ridge regression reduces variance, it increases biasness in model. Following the hypothetical example for linear regression, ridge regression adds penalty term (λI) in $X^T X$ as $(X^T X + \lambda I)$. Thus, Eq. 3.29 can be modified as follow.

$$\beta = (X^T X + \lambda I)^{-1} X^T Y \tag{3.34}$$

In Eq. 3.34, $X^T X$ refers to a gram matrix, which captures the relationships between the independent variables in X by computing dot product of independent variables vectors, λ is a regularization parameter used to control the strength of regularization. Primarily it reduces impact of multi-collinearity by penalizing large weights. Moreover, $X^T Y$ represents cross product of dependent and independent variables. Using Eq. 3.34, weight vector is computed as $[0.1, 0.3, 0.4, 0.2, 0.1, -0.1, 0.05]$. For similar train sequences, it substitutes weight vector and statistical vector of test sequence in Eq. 3.27 and determine predicted dependent variable vector $\hat{Y}$. Using actual and predicted values of dependent variables vector i.e., Y and $\hat{Y}$, it computes loss using Eq. 3.33 and calculates gradients which is backpropagate to update weights. Similar to linear regression, Lasso regression [201] algorithm determines dependent variables using Eqs. 3.27 and 3.29. The only difference is Lasso regression [201] algorithm adds penalty term $(\lambda\|\beta\|_1)$ to loss function and shrinks some weights to zero. Equation 3.35 depicts mathematical expression to determine loss function for lasso regression.

$$L_{Lasso}(\beta) = \sum_{i=1}^{n} \arg \min_{\beta} \left\{ \|\hat{y}_i - y_i\|^2 + \lambda\|\beta\|_1 \right\} \tag{3.35}$$

In Eq. 3.35, $\hat{y}$ and y_i indicates predicted and actual dependent variables, $\|\beta\|_1$ represents absolute weight vector, λ is a regularization parameter which controls the strength of penalty. A higher λ increases amount of shrinkage applied to weights. In a nutshell, $\lambda\|\beta\|_1$ is known as the L1 penalty.

Elastic Net regression algorithm [486] combines penalties of both lasso (L1) and ridge (L2) regression in loss function. Equation 3.36 embodies mathematical expression for computing cost function for Elastic Net regression.

$$\mathcal{L}_{\text{Elastic Net}}(\beta) = \|\hat{y}_i - y_i\|^2 + \lambda_1\|\beta\|_1 + \lambda_2\|\beta\|_2^2 \tag{3.36}$$

In Eq. 3.36, $\lambda_1\|\beta\|_1$ refers to L1 penalty, and $\lambda_2\|\beta\|_2^2$ indicates L2 penalty term. As Elastic Net combines both L1 and L2 penalties, and adjusts hyper-parameters values $\lambda_1 = 0.5$ and $\lambda_2 = 0.5$. By using these values, it formulated weight vector as $[0.15, 0.25, 0.45, 0.1, 0.05, 0, 0]$. By using Eq. 3.27 for similar train sequences, and weight vector, it determines predicted dependent variables vector. After predicting the class probability, a classification decision is made by applying the argmax function. This function assigns class label to an instance based on the calculated probabilities for each class. The class with the highest probability is selected as the predicted class. In the special case where the predicted probability is exactly same (e.g. 0.5 for binary classification), it indicates that the instance has an equal likelihood of belonging to either class. In this unique scenario, from both classes a random class label is assigned to the instance. Afterwards, it utilized Y and $\hat{Y}$ and computes loss using Eq. 3.36. Using this loss, it calculates gradients and backpropagates to determine optimal values of weights.

3.6.1.5 Tree Based Algorithms

Tree-based algorithms use the supervised learning paradigm to perform classification and regression tasks. In this paradigm, algorithms are trained to map input sequences to predefined class labels or continuous numerical values. Tree based category contains distinct algorithms including decision tree, random forest, Adaboost, gradient boosting machines, XGBoost, CatBoost, extra trees, and LightGBM. These all algorithms are derived from foundational algorithm named decision tree. Overall, based on working paradigm, these algorithms can be categorized into foundational, bagging and boosting classes. Bagging class includes random forest and boosting class includes Adaboost, gradient boosting machines, XGBoost, CatBoost, and LightGBM.

The primary objective of decision tree algorithm is to utilize independent variable values to map input data into a hierarchical structure comprising of root and leaf nodes. In this structure, root nodes represent decision making independent variables, while leaf nodes correspond to two distinct types: decision nodes and terminal nodes. Decision nodes can be further divided into sub-nodes, whereas terminal nodes indicate final label and do not undergo further splitting. To split decision nodes into sub-nodes, tree based algorithms utilize a specific set of splitting algorithms. The most commonly utilized splitting algorithms include: (1) ID3, (2) CART, (3) C4.5, (4) MARS. Equation 3.37 illustrates mathematical expressions of these four splitting algorithms.

$$Splitting\ Criterion = \begin{cases} ID3{:}\ Entropy\ (s) = \sum_{i=1}^{N} -p_i \log_2 p_i \\ CART{:}\ Gini\ Impurity\ (G.I) = 1 - \sum_{i=1}^{N} ({p_i}^2) \\ C4.5{:}\ Gain\ Ratio = \frac{Entropy(Before) - \sum_{j=1}^{K} Entropy(j,after)}{\sum_{j=1}^{K} w_j \log_2 w_j} \\ MARS{:}\ Variance = \frac{\sum (x_i - \bar{x}_i)^2}{n} \end{cases} \tag{3.37}$$

In Eq. 3.37, p_i refers to probability of ith independent variable, $Entropy(Before)$ is the entropy of data before split, K is number of nodes that can be generated, w_j is the weight of jth node and $Entropy\ (j, after)$ is entropy of jth sub-node after split. In addition, x_i indicates ith independent variable and n represents total number of instances.

Decision tree algorithm follows a series of steps to perform classification and regression tasks as given below.

- First step involves calculating entropy of entire dataset (S) by using any one of algorithm specified in Eq. 3.37.
- In the next step, it calculates the entropy for each independent variable (X) by using Eq. 3.38.

$$H(X) = \sum_{i=1}^{n} \frac{X_i}{X} H(X_i) \tag{3.38}$$

In Eq. 3.38, X_i refers to distinct values of independent variable X and $H(X_i)$ indicates the entropy of X_i. When computing the entropy of an independent variable, note that if it corresponds to only one dependent variable for all instances, its entropy will be zero.

- Afterwards, it computes information gain by using entropy of entire dataset and individual independent variables. Equation 3.39 is used to compute information gain for each independent variable.

$$I.G\,(S,\ X) = H(S) - H(X) \tag{3.39}$$

- Based on computed information gain values, it selects independent variable with highest information gain as root node.
- Subsequently, it creates branches based on distinct values of chosen independent variable (root node) and assigns data to each branch accordingly.
- Each branch repeats same process to determine the independent variables for its leaf nodes. This process continues until the entropy of branch data becomes zero.
- After constructing tree, this algorithm makes use of pruning to eliminate branches that do not contribute to the prediction.
- Finally, for a test instance, tree is traversed from the root to the leaf node by following the splitting rules at each step, and this algorithm outputs class label of the final leaf node as the predicted label.

In order to develop a comprehensive understanding of decision tree algorithm [954], consider a hypothetical dataset consisting of five instances, each instance contains four independent variables (f1, f2, f3, f4) and assigned to one of two binary labels (Yes or No), as illustrated in Table 3.8. To understand node splitting criteria lets play with given sample data and ID3 algorithm. First of all, it computes entropy of entire data by using ID3 algorithm as given below. For this, it computes probability of both classes. Probability of a particular class label is the ratio of number of instances in that class to total number of instances in dataset. In this example, 2 instances belongs to class No and 3 instances have class label Yes. Therefore, $P_{No} = \frac{2}{5}$ and $P_{Yes} = \frac{3}{5}$

Table 3.8 Sample dataset

Instances	f1	f2	f3	f4	Label
S_1	x	a	m	c	Yes
S_2	y	b	m	d	No
S_3	y	c	n	c	No
S_4	x	a	n	d	Yes
S_5	x	b	l	c	Yes

$$H(S) = P_{Yes} log_2 P_{Yes} + P_{No} log_2 P_{No}$$

$$H(S) = \left(\frac{3}{5}\log_2\frac{3}{5} + \frac{2}{5}\log_2\frac{2}{5}\right) \approx 0.971$$

In the next step, it computes entropy of each independent variable as follow.

- **Entropy of independent variable f1:** It can be seen in Table 3.3, independent variable f1 has two distinct values namely x and y. It first computes individual entropy of both values. Value x is present in three instances. Among 3 instances, 2 have class label Yes and 1 has a class label No. Entropy of value x can be computed using computation of these probability values: $P_{Yes|x} = \frac{3}{3} = 1$ and $P_{No|x} = \frac{0}{3} = 1$ and entropy can be computed as follow.

$$H(f1_x) = P_{Yes|x} \cdot log_2(P_{Yes|x}) + P_{No|x} \cdot log_2(P_{No|x})$$

$$H(f1_x) = 1 \cdot log_2(1) + 0 \cdot log_2(0) = 0$$

Similarly, for value y, $P_{Yes|y} = \frac{0}{2} = 1$ and $P_{No|y} = \frac{2}{2} = 1$ and entropy can be computed as given below.

$$H(f1_y) = P_{Yes|y} \cdot log_2(P_{Yes|y}) + P_{No|y} \cdot log_2(P_{No|y})$$

$$H(f1_y) = 0 \cdot log_2(0) + 1 \cdot log_2(1) = 0$$

In the next step, it computes sum of entropy of both x and y values of independent variable f1.

$$H(f1) = H(f1_x) + H(f1_y) = 0 + 0 = 0$$

In the similar manner it calculates the individual entropy for remaining 3 independent variables $f2$, $f3$, and $f4$ as follow.

- **Entropy of independent variable f2:**

$$H(f2_a) = 0, \quad H(f2_b) = -\left(\frac{1}{2}\log_2\frac{1}{2} + \frac{1}{2}\log_2\frac{1}{2}\right) = 1, \quad H(f2_c) = 0$$

$$H(f2) = \frac{2}{5} \cdot 0 + \frac{1}{5} \cdot 1 + \frac{2}{5} \cdot 0 = 0.2$$

- **Entropy of independent variable f3:**

$$H(f3_m) = -\left(\frac{2}{3}\log_2\frac{2}{3} + \frac{1}{3}\log_2\frac{1}{3}\right) \approx 0.918, \quad H(f3_n) = 1, \quad H(f3_l) = 0$$

$$H(f3) = \frac{3}{5} \cdot 0.918 + \frac{1}{5} \cdot 1 + \frac{1}{5} \cdot 0 = 0.751$$

- **Entropy of independent variable f4:**

$$H(f4_c) = -\left(\frac{2}{3}\log_2\frac{2}{3} + \frac{1}{3}\log_2\frac{1}{3}\right) \approx 0.918, \quad H(f4_d) = 1$$

$$H(f4) = \frac{3}{5} \cdot 0.918 + \frac{2}{5} \cdot 1 = 0.951$$

Afterwards, it calculates information gain for each independent variable. To compute entropy for independent variable f_i, it computes the difference between entropy of entire data and entropy of independent variable f_i.

- **Information gain for f1:** In first step, entropy of entire data is computed as H(S) = 0.971 whereas entropy of f1 is 0.

$$IG(S, f1) = H(S) - H(f1) = 0.971 - 0 = 0.971$$

- **Information gain for f2:**

$$IG(S, f2) = H(S) - H(f2) = 0.971 - 0.2 = 0.771$$

- **Information gain for f3:**

$$IG(S, f3) = H(S) - H(f3) = 0.971 - 0.751 = 0.220$$

- **Information gain for f4:**

$$IG(S, f4) = H(S) - H(f4) = 0.971 - 0.951 = 0.020$$

Since information gain of f1 independent variable is the highest, this algorithm selects it as root node and creates two branches based on its two distinct values: x and y. In the next step, it divides input dataset into two partitions such that S1, S4, S5 for x branch and S2 and S3 for y branch. Each branch of tree undergoes same process to select independent variables for its leaf nodes. This process is repeated until entropy of branch data reaches zero. To predict label for a test instance, it traverse from root to a leaf node, adhering to splitting rules at each step. Finally, it predicts the label of last node in this traversal for test sequence.

Despite being effective with both categorical and numerical data, decision trees have certain drawbacks. Decision trees fit perfectly on the training data, but even a small change in the data can result in an entirely new tree structure. This sensitivity leads to high variance, increased complexity, and potential bias in the model. Moreover, trees with more depth increase computational time of model, and results in poor predictive performance. To deal with these issues, ensemble approaches aggregate predictions of several algorithm to produce more accurate and reliable predictions. There are two types of ensemble approaches: bagging and boosting. Bagging algorithms generates multiple subsets of input data and train separate algorithm on each data and makes prediction on the basis of majority of votes whereas boosting algorithms combines multiple weak classifiers in sequential manner to develop a strong classifiers.

In bagging algorithms landscape, random forest (RF) [64] is the only algorithm that trains multiple decision trees on different subsets of training data. In the prediction phase, decision trees use majority voting to determine the final class label for classification tasks, while for regression tasks, they generate the final prediction by averaging the individual tree predictions. By aggregating outputs of multiple decision trees, random forest is less likely to over-fit but training multiple trees can be computationally expensive.

Beyond bagging algorithms, boosting algorithms build classifier in sequential manner, where subsequent classifier corrects the errors of previous ones. AdaBoost [909] is one of the earliest boosting algorithm which works by adjusting the weights of incorrectly classified instances so that subsequent classifiers focus on more difficult instances. AdaBoost classifier consists of multiple decision trees with one level or split, also called as stumps. It operates by combining multiple weak classifiers to construct a strong classifier for making predictions. Working paradigm of AdaBoost classifier consists of 7 different steps. In the very first step, it initializes the equal weights for each instance and first weak classifier is trained using these weights. In 2nd step, the weak classifier prediction error is computed using Eq. 3.40.

$$Err = \frac{\sum_{i=1}^{N} w_i \left(y_i \neq C_m\left(x_i\right)\right)}{\sum_{i=1}^{N} \left(w_i\right)} \tag{3.40}$$

In Eq. 3.40, w_i indicates weight of instance i, y is the actual label, $G_m\left(x_i\right)$ is the output of mth classifier, and N refers to total samples. On the basis of calculated error, Eq. 3.41 embodies mathematical expression to determine the performance of each classifier.

$$\alpha_m = \frac{1}{2} \log\left(\frac{1 - Err}{Err}\right) \tag{3.41}$$

In the subsequent step, it makes use of α_m and updates weights ($w_i = w_{i-1} * \alpha_m$). These weights are normalized and passed to next classifier which follows similar step to output final prediction. The final prediction is the sum of product of performance of each classifier and total number of classifiers. Equation 3.42 provides mathematical expression to determine final prediction.

$$C(x) = \sum_{m=1}^{M} \alpha_m C_m(x), \quad where\ C_m(x) \in \{-1, +1\} \tag{3.42}$$

The AdaBoost algorithm is particularly sensitive to outliers and noisy data. To achieve optimal performance, it is important to remove outliers during the preprocessing stage before feeding the data into the algorithm. In contrast to AdaBoost algorithm, gradient boosting machine (GBM) [1153] reaps the benefits of multiple decision trees having depth larger than 1 and number of leaf nodes ranging between 8 to 32. The prime objective of GBM is to optimize cost function by using residual weights and probabilities. Mainly, residual weight is the difference between actual and predicted values. In order to enhance this process, XGBoost [332], extension of GBM, calculates similarity score to update next residual value. Specifically XGBoost optimizes cost function by using second order differential to avoid over-fitting. LightGBM [1153] is another gradient boosting machine which constructs tree by using histogram based approaches. Specifically, LightGBM constructs histograms for each feature by counting how many data points fall into each histogram. Another gradient boosting algorithm is categorical boosting (CatBoost) [332], which integrates target statistics and other techniques to handle categorical variables. CatBoost introduces an ordered boosting technique to generate more sophisticated trees. These sophisticated trees are symmetric trees which adopt similar structure at all splits at a given level. Finally, Extremely Randomized Trees algorithm [213], also known as Extra Trees algorithm, works similar to Random forest algorithm. The only difference is that it sample data distinctly that every sample should be unique and there will be no overlapping of data. It selects a random subset of algorithms for each split. It makes a small number of randomly selected split points for each of selected algorithms from which it chooses a best split by using gini-index, entropy or information gain, as used in decision trees for splitting.

3.6.2 Deep Learning Based Predictors

Artificial intelligence field has witnessed a major breakthrough with the emergence and widespread adoption of deep learning architectures. Primarily, deep learning architectures are inspired by the cognitive processes of the human brain. Unlike, machine learning algorithms, which heavily relies on manually crafted features, deep learning architectures are capable of autonomously learning intricate patterns directly from raw data.

Genomics and Proteomics sequence analysis field has experienced the explosion of deep learning predictors, core architectures of which are mainly formed by deep feed forward neural networks [601], convolutional neural networks [601], recurrent neural networks [713] and hybrid networks that make use of both CNN and RNN layers [44, 953]. These architectures based predictive pipelines are comprises of an input layer, CNN/RNN layers and output layer. According to the nature of data, within these predictive pipelines researchers are also use to include pooling layers, dropout and batch normalization strategies [112].

Following subsections illustrates mathematical expressions of convolutional neural networks [601], recurrent neural networks [713] and hybrid networks

3.6.2.1 Input Layer

At input layer, raw genetic sequences are segregated into k-mers. Usually a fixed dimensional random embeddings are initialized to create numeric representation of corpus k-mers. In other words, given a sequence of n k-mers $K_1, K_2, K_3, \ldots\ldots.,$ K_n, for each corpus k-mer K_i, embedding vector e_i is generated by computing matrix vector product using embedding matrix $W \epsilon R^{d*|V|}$ where $|V|$ represents size of vocabulary and d associates to the dimension of real valued word embedding vector.

$$e_i = W v_i \tag{3.43}$$

In this way, each corpus sequence is represented in terms of several k-mer vectors containing real values in between 0 and 1 $e = e_1, e_2, e_3, \ldots\ldots., e_n$. These features are then fed to variety of feature extraction layers.

3.6.2.2 Convolution Layer

Convolutional layer performs feature extraction process by utilizing different sizes kernels. In case of symmetrical kernel, convolution operation certainly turns into a correlation operation [355]. Each kernel $w \in R^{kk}$ is applied upon a window containing h k-mers with certain stride size in order to generate fresh feature. For instance, a fresh feature c_i is produced from the widow of k-mers $x_{i:i+h-1}$

$$c_i = f(w.x_{i:i+h-1} + b) \tag{3.44}$$

In Eq. 3.44, $b \in R$ represents bias and f acts as a non-linear function like hyperbolic tangent. This kernel is executed over every possible window of k-mers present in a sequence $K_{1:h}, K_{2:h+1} \ldots\ldots..K_{n-h+1:n}$ to generate a feature map that can be represented as:

$$c = [c_1, c_2, c_3 \ldots\ldots\ldots\ldots c_{n-h+1}] \tag{3.45}$$

where $c \in R^{n-h+1}$. Model actually makes use of multiple kernels with diverse window and stride size to get collection of features. Convolution operation can be classified into distinct different types considering the size and type of filters, padding type, and convolution direction [112].

3.6.2.3 Pooling Layer

After extracting features from sequences, capturing their relative positions become an important task which is achieved by down-sampling or pooling. Pooling is a local operation that captures the dominant response of local region by aggregating similar neighbourhood information [568]. Pooling operation can be expressed as:

$$Z_l^k = g_p(F_l^k) \tag{3.46}$$

Here, Z_l^k refers to down-sampled feature map of lth layer for kth give feature map (F_l^k), g_p represents the kind of pooling operation. Pooling mainly assists to acquire feature combinations that are invariant to small distortions and translational shifts [836, 886]. Minimization of feature map alleviates the complexity of neural network and also assists in raise the generalization through reducing over-fitting. Variety of pooling operations are applied like average, min, max overlapping, L2, spatial pyramid, etc. [112, 1037]

3.6.2.4 Activation Function

It is a decision function and helps the network for learning complex patterns. Appropriate selection of activation function enhance the process of learning. For a feature map acquired through convolution operation, activation function can be written as:

$$T_l^k = g_a(F_l^k) \tag{3.47}$$

Here, (F_l^k) is produced by convolution, that is given to activation function represented as g_a. Activation function embeds non-linearity and yields output T_l^k for lth layer.

Critical analysis of literature shows that multifarious activation functions have been used like tanh, sigmoid, maxout, SWISH, ReLU, and its variants (LeakyReLu, PReLU, ELU) [371, 565, 830, 1037, 1127]. But, ReLU and variants of ReLU are mostly preferred by the researchers as they greatly assist in dealing with gradient vanishing issue [429, 772].

3.6.2.5 Batch Normalization

In order to resolve the issues related to covariance shift inside feature maps, batch normalization is widely used. Covariance shift refers to the change in distribution of network hidden units/values that significantly decreases convergence rate (by pushing learning rate to lower value) and demands watchful initialization of model parameters. For transformed feature map, batch normalization is given as follows:

$$N_l^k = F_l^k - u_b/\sqrt{\sigma_b^2 + \epsilon} \tag{3.48}$$

Here Eq. 3.48, for mini batch, N_l^k and F_l^k refer to normalized and input feature map, σ_b^2 and u_b correspond to variance and mean of a given feature map. To avoid zero division, ϵ is injected to add numerical stability.

Batch normalization standardizes distribution of values of feature map through setting them into unit variance and zero mean [468]. Also, it greatly flatten gradient flow and serve as a regularizing factor, through which network generalization is improved upto great extent.

3.6.2.6 Dropout

Dropout layer serves as a regularizer in neural network that eventually alleviates overfitting and improves generalization through randomly neglecting few connections or units with particular probability [423]. In neural networks, as several connections based on non-linear relation are co-adapted at times, therefore random dropping of few units creates multiple thinned deep architectures and afterward one optimal representative network architectures is opted with quite small weights. Then, this opted architecture is considered an approximation of all proposed networks [945].

3.6.2.7 Fully Connected Layer

The layer which is used at the end of neural network is called Fully connected layer. Unlike convolution and pooling, it can be classified as a global extraction. It takes the input from all feature extraction phases and analyzes the result of former layers [626]. As a result, it creates a non-linear association of selected features, that are used for sequence classification [838].

3.6.3 Recurrent Neural Network (RNN) And Its Variants (LSTM, GRU)

Researchers have extensively utilized recurrent neural network (RNN) for DNA, RNA, and Protein sequence analysis tasks [693, 966]. As RNN allocates more weights to former points and takes into account information of former nodes, hence it analyzes the structure of dataset in a more effective manner. Mostly, RNN makes use of Long Short Term Memory (LSTM) or Gated Recurrent Unit (GRU) that consists of embedding layer, hidden layers and output layer. This methodology can be expressed as:

$$x_t = f(x_{t-1}, u_t, \theta) \tag{3.49}$$

At time step t, here x_t represents state and u_t refers to the input. Using weights, it can be expressed as:

$$x_t = W_{rec}\sigma(x_{t-1} + W_{in}u_t + b) \tag{3.50}$$

Here W_{rec} represents recurrent weight matrix w_{in} is input weights, b refers to bias and σ implies the element wise operation.

RNN is highly vulnerable to exploding and vanishing gradient problems [86] when error of neural network is back propagated in the network. Due to these reasons, its variants LSTM and GRU are used mostly in experimentation, details of which are given below.

3.6.3.1 Long Short Term Memory (LSTM)

Long Short Term Memory (LSTM), a special type of RNN given by Hochreiter et al. [799] addressed the downfalls of RNN such as vanishing gradient issue through preserving long range dependencies in a very effective manner [799]. Although due to having chain like architecture it is quite similar to RNN but LSTM makes use of several gates in order to efficiently regulate information which is allowed for the state of each node.

$$i_l = \sigma(W_i[x_t, h_{t-1}] + b_i) \tag{3.51}$$

$$C_t = tanh(W_c[x_t, h_{t-1}] + b_c) \tag{3.52}$$

$$f_t = \sigma(W_f[x_t, h_{t-1}] + b_f) \tag{3.53}$$

$$C_t = i_t * C_t^{f_t C_{t-1}} \tag{3.54}$$

$$o_t = \sigma(W_o[x_t, h_t - 1] + b_o) \tag{3.55}$$

$$h_t = o_t(tanh(C_t)) \tag{3.56}$$

Here Eq. 3.51 refers to input gate, Eq. 3.52 refers to the value of candid memory cell, Eq. 3.53 represents forget gate activation, Eq. 3.54 computes value of fresh memory cell, and Eqs. 3.55, 3.56 describe the final yield of gate value. Moreover, b refers to bias, w represents the weight matrix, x_t represents the input at timestamp t, and indies i,c,f,o refer to input, memory of cell, forget, and final output gates in turn.

3.6.3.2 Gated Recurrent Unit (GRU)

Gated recurrent unit (GRU) is a more simplified version of LSTM architecture [192]. But unlike LSTM, it has two gates and does not have internal memory. In addition, it does not apply second non linearity [183].

$$z_t = \sigma_g(W_z * x_t + U_z * h_{t-1} + b_z) \tag{3.57}$$

Here z_t is the update gate representation of t, x_t is input vector, and W, b, U are parameter vectors. Activation function is either ReLu or sigmoid that can be formulated as:

$$r_t = \sigma_g(W_r * x_t + U_r * h_{t-1} + b_r) \tag{3.58}$$

Where z_t is the update gate representation of t and r_t is reset gate representation of t.

$$h_t = z_t \cdot h_{t-1} + (1 - z_t) \cdot \sigma_h(W_h * x_t + U_h * (r_t \cdot h_t - 1) + b_h) \tag{3.59}$$

For t, h_t is the final output vector where σ_h represents hyperbolic tangent operation.

3.7 AI-Driven Genomic and Proteomic Sequence Analysis Predictive Pipelines Performance Analysis

Predictive models training and evaluation undergo through two distinct experimental settings: (1) k-fold cross-validation [473, 1113] and (2) Train-test split [240, 928]. First strategy splits data into k same size folds (typically 5 or 10 folds), then reserves one fold for testing and trains predictor on remaining folds. In next iteration, from k-folds, another fold is used for test set and all other folds are used for model training. Following same process k number of times model is trained and tested. In this way model is evaluated on whole data. Specifically for deep learning predictors [240, 929], another set called validation set is created from training set (typically 10% of training data). This set is used to optimize hyper-parameters of predictor. On the other hand, train test split based experimental setting divides data into two distinct sets: (a) Train set, (b) Test set. Among both sets, train set is comprised of majority of data (usually 70% or 80%) while test set contains 20% or 30% data. Similar to k-fold setting, for deep learning models from train set here also a validation set is created.

Large-scale performance analysis of genomics (DNA, RNA) and Proteomics (protein) sequence analysis pipelines presents significant challenges due to sensitive nature of Genomic and Proteomic sequence analysis tasks. Unlike, Natural Language Processing, errors in this domain can pose a risk to millions lives. To rigorously evaluate the integrity, generalizability and applicability of AI-driven applications for genomics and proteomics sequence analysis tasks under the paradigm of classification, regression and clustering, diverse types of evaluation measures are available. Following sub-sections briefly describe the details of each AI-paradigm related evaluation measures for Genomics and Proteomics sequences analysis predictive pipelines.

3.7.1 *Binary or Multi-Class Classification Paradigm Predictors Evaluation Measures*

In this category most commonly used evaluation measures are Accuracy [51, 929], Precision [929], Recall [929], F1 score [777] and MCC [1200]. These evaluation measures are typically calculated using confusion matrix, which is comprised of four entities: true positives (TP), false positives (FP), true negatives (TN), and false negatives (FN). Figure 3.10 graphically represents confusion matrix and all 4 entities.

It can be seen in Fig. 3.10, TP and TN represent instances that are correctly predicted as positive and negative, respectively. Conversely, FP and FN denote instances that are incorrectly predicted as positive and negative, respectively. Figure 3.10 graphically illustrates confusion matrix and its entities namely TP, TN, FP, and FN. Equation 3.60 embodies mathematical expressions to compute the aforementioned measures.

$$f(x) = \begin{cases} Accurcay\ (Acc) = \frac{TP+TN}{TP+FP+TN+FN} \\ Precision\ (P) = \frac{TP}{TP+FP} \\ Recall\ (Rec) = \frac{TP}{TP+FN} \\ F1\text{-}Score = \frac{2*Precision*Recall}{Precision+Recall} \\ Sensitivity\ (Sn) = \frac{TP}{TP+FN} \\ Specificity\ (Sp) = \frac{TN}{TN+FP} \\ MCC = \frac{(TP\times TN)-(FP\times FN)}{\sqrt{(TP+FP)(TP+FN)(TN+FP)(TN+FN)}} \end{cases} \tag{3.60}$$

Fig. 3.10 Confusion matrix

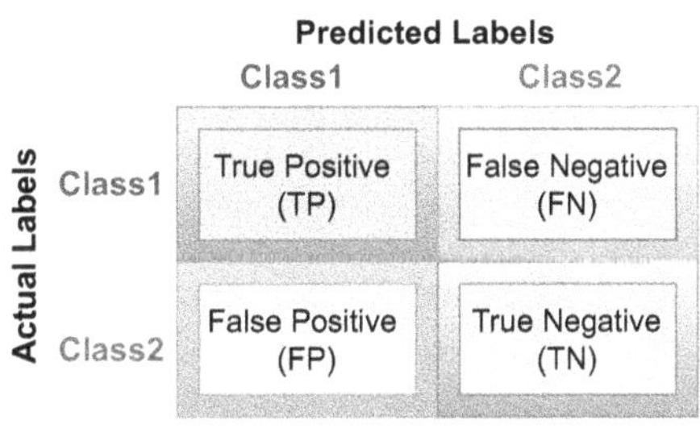

Predictive pipelines across balanced datasets are evaluated using diverse types of evaluation measures including F1-score, precision, accuracy, recall, specificity and Matthews correlation coefficient (MCC). On the other hand, micro, macro and weighted versions of these measures are used for evaluation of predictive pipelines across imbalanced datasets. To address class imbalance problem, weighted precision [185] determines proportion of precision and relative weight of each class. Precision of a class is ratio of true positives to total number of positives for that class, while relative weight of a class is proportion of instances of that class relative to total number of instances.

Similarly, weighted recall [185] and weighted F1-score [275] are computed by determining the weights, recall, and F1-score for each class. Macro precision [180] determines precision for each class independently and then computes average of these precision values. Macro recall [180] and macro F1-score [180] computes average recall and F1-score across all classes by treating each class equally regardless of its size, respectively. Unlike macro precision, micro precision [180] calculates precision globally by considering all true positives and false positives across all classes together. Micro recall [180] and micro F1-score [180] aggregates TP, FP and FN across all classes and provides fair and balanced evaluation of predictor's performance. Equation 3.61 embodies mathematical expressions for computing weighted, macro and micro scores for highly imbalance datasets.

$$f(x) = \begin{cases} Weighted\text{-}Precision = \frac{\sum_{i=1}^{n} Pr_i.w_i}{\sum_{i=1}^{n} w_i} \\ Weighted\text{-}Recall = \frac{\sum_{i=1}^{n} R_i.w_i}{\sum_{i=1}^{n} w_i} \\ Weighted\ F1\text{-}Score = \frac{\sum_{i=1}^{n} F1\text{-}score_i.w_i}{\sum_{i=1}^{n} w_i} \\ Macro\text{-}Precision = \frac{1}{n}\sum_{i=1}^{n} Pr_i \\ Macro\text{-}Recall = \frac{1}{n}\sum_{i=1}^{n} Rec_i \\ Macro\ F1\text{-}Score = \frac{1}{n}\sum_{i=1}^{n} F1\text{-}Score_i \\ Micro\text{-}Precision = \frac{\sum_{i=1}^{n} TP_i}{\sum_{i=1}^{n}(TP_i+FP_i)} \\ Micro\text{-}Recall = \frac{\sum_{i=1}^{n} TP_i}{\sum_{i=1}^{n}(TP_i+FN_i)} \\ Micro\ F1\text{-}Score = \frac{\sum_{i=1}^{n} 2.TP_i}{\sum_{i=1}^{n}(2.TP_i+FP_i+FN_i)} \end{cases} \tag{3.61}$$

Here Pr_i, R_i, and $F1\text{-}score_i$ refers to precision, recall, and F1-score of class i. w_i represents relative weight of class i, and n indicates total number of classes. In micro scores, TP_i, FP_i and FN_i refers to true positives, false positive and false negatives in class i, respectively.

3.7.2 Multi-Label Classification Paradigm Predictors Evaluation Measures

Performance evaluation of multi-label classification is challenging as compared to binary and multi-class classification. In binary or multi-class classification, each sample is assigned to only one class at time. Hence, predicted class label will either belong to true or false category. While in multi-label classification, a sample belongs to two or more labels simultaneously. A model may predict multiple labels where some are correct, some are incorrect, or all can be correct or incorrect. Therefore, output in multi-label classification can be partially correct, partially incorrect, fully correct and fully incorrect. The challenge arises from partial corrections, which commonly occur. The most widely used evaluation measures are accuracy [671], precision [671], recall [1113], coverage [671] and hamming loss [180]. Equation 3.62 demonstrates mathematical expression of these evaluation measures.

Accuracy assesses predictor performance by computing ratio between actual and predicted labels while precision computes performance by closely monitoring actual true labels from the set of labels that classifier predicted as true. Moreover, recall measures how many labels are correctly predicted from actual labels and F1-score is harmonic mean between precision and recall. The higher the value of accuracy, precision, recall and F1 the better will be the performance of classifier. Finally, hamming loss measures how many labels are wrongly predicted and how many labels remain unpredicted.

$$f(x) = \begin{cases} Recall\ (Rec) = \frac{1}{M}\sum_{i=1}^{M}\frac{|A_i \wedge P_i|}{|A_i|} \\ Accuracy\ (Acc) = \frac{1}{M}\sum_{i=1}^{M}\left|\frac{A_i \wedge P_i}{A_i \vee P_i}\right| \\ F1\text{-}Score = \frac{1}{M}\sum_{i=1}^{M}\frac{2*|Pre(n_i)*Rec(n_i)|}{|Pre(n_i)+Rec(n_i)|} \\ Precision\ (P) = \frac{1}{M}\sum_{i=1}^{M}\frac{|A_i \wedge P_i|}{|P_i|} \\ Hamming\ Loss = \frac{1}{ML}\sum_{i=1}^{M}\sum_{j=1}^{L}\left[|(A_i{}^j \neq P_i{}^j)|\right] \end{cases} \tag{3.62}$$

In these equations 3.62, M denotes total number of samples, n_i represents ith sample from M samples, A_i represents actual class label and P_i denotes predicted label of n_i sample, L represents length of sample, j represents class index, $\vee$ represents logical OR operator and $\wedge$ represents logical AND operator.

3.7.3 Regression Paradigm Predictors Evaluation Measures

Regression tasks differ fundamentally from classification tasks where model predicts continuous numerical values rather than discrete class labels. As binary/multi-class classification assigns each sample to a single class label and multi-label classification handles samples with multiple class labels, whereas regression tasks predict continuous values. Therefore, evaluation metrics for regression tasks include mean squared error (MSE), mean absolute error (MAE), mean bias error (MBE), mean absolute percentage error (MAPE), root mean square error (RMSE), R^2, relative mean absolute error (rMAE), relative mean square error (rMSE), relative root mean square error (rRMSE) and relative mean bias error (rMBE).

MAE assesses predictor performance by measuring absolute difference between predicted and actual values. MSE quantifies deviation by averaging squared differences between actual and predicted values. Similarly, RMSE calculates standard deviation of prediction errors, and demonstrates how tightly data points cluster around regression line. MBE assesses predictor performance in terms of under and over-fitting by enumerating average difference between predicted and actual value. MAPE calculates percentage variation between predicted and actual values. Smaller the values of MAE, MBE, MSE and MAPE, better will be predictor performance. Higher value of R^2 score signifies promising predictor performance as it measures proportion of variance in predicted dependent variable explained by independent variable to determine strength of relationship.

MAE, MSE, RMSE, and MAPE give average value of errors for N number of data points. Relative performance evaluation can improve quality of performance evaluation by reducing the noise from data. For relative performance evaluation, the percentage error of each metric is computed relative to the average of actual values. It facilitate in controlling factors that influence predictor performance by relatively calculating ratio of particular error with average of actual values. Since data continuously varies and produces varying predicted values at different time intervals, an overall percentage error is computed to obtain relative error of all data points. Equation 3.63 embodies mathematical expressions for aforementioned evaluation metrics.

$$f(x)_{regression} = \begin{cases} MAE = \frac{1}{M}\sum_{i=1}^{M}\left|P^i - A^i\right| \\ MSE = \frac{1}{M}\sum_{i=1}^{M}\left(A^i - P^i\right)^2 \\ RMSE = \sqrt{\frac{1}{M}\sum_{i=1}^{M}\left(A^i - P^i\right)^2} \\ MBE = \frac{1}{M}\sum_{i=1}^{M}\left(P^i - A^i\right) \\ MAPE = \frac{1}{m}\sum_{i=1}^{M}\left|\frac{P^i - A^i}{A^i}\right| \times 100 \\ R^2\ Score = 1 - \frac{\sum_{i=1}^{M}(P-A)^2}{\sum_{i=1}^{M}(A-avg(A))^2} \\ rMAE = \frac{MAE}{\bar{A}} \times 100 \\ rMSE = \frac{MSE}{\bar{A}} \times 100 \\ rMBE = \frac{MBE}{\bar{A}} \times 100 \\ rRMSE = \frac{RMSE}{\bar{A}} \times 100 \end{cases} \tag{3.63}$$

In Eq. 3.63, M denotes total number of samples, A^i represents actual value and P^i is predicted value where i denotes the sample number and $\bar{A}$ is the average of total actual values.

3.7.4 Clustering Paradigm Predictors Evaluation Measures

In contrast to first three categories explained, clustering tasks aim to group similar samples based on their features without predefined class labels. Prime objective is to use clustering algorithms and identify inherent patterns or structures of data. In these tasks, clusters of data samples with similar features are created, and predictors assigns new data points to appropriate clusters. A higher similarity to a cluster indicates that data sample belongs to that cluster. Over time, researchers have introduced various evaluation measures to assess predictor performance which accuracy, normalized mutual information (NMI), silhouette score (SS), Dunn index (DI) and Davies-Bouldin index (DBI).

Accuracy is the proportion of correctly predicted samples to total number of samples. NMI quantifies quality of predictor by measuring mutual information between predicted clusters and actual clusters. Mutual information refers to computed joint probability between predicted clusters and actual clusters. Silhouette score measures how similar data samples are within a cluster compared to other clusters. BDI evaluates average similarity ratio of each cluster with its most similar cluster. DI computes ratio of minimum inter-cluster distance to maximum intra-cluster distance. Equation 3.64 embodies mathematical expressions for these evaluations measures.

$$f(x)_{clustering} = \begin{cases} Accuracy = \underset{max}{m} \frac{\sum_{i=1}^{n} 1\{y_i = m(c_i)\}}{n} \\ NMI = \frac{I(y_i, c_i)}{\frac{1}{2}[H(y_i) + H(c_i)]} \\ SS = \frac{min\{d(y_i)\} - a(y_i)}{max\{min\{d(y_i)\}, a(y_i)\}} \\ DBI = \frac{1}{n} \sum_{i=1}^{n} \underset{j \neq i}{max} \left(\frac{\bar{S}_i + \bar{S}_j}{d(c_i, c_j)}\right) \\ DI = \frac{min_{1 \leq i < j \leq n} d(c_i, c_j)}{max_{1 \leq k \leq n} d'(c)} \end{cases} \tag{3.64}$$

In Eq. 3.64, y_i refers to predicted cluster, c_i and c_j indicates ith and jth clusters among n clusters. Moreover, $I(y_i, c_i)$ signifies mutual information, $H(y_i)$ and $H(c_i)$ shows entropy of predicted and actual clusters. $d(y_i)$ average distance from y_i to all points in other clusters, $a(y_i)$ is average distance of y_i to all points in that clusters. $d(c_i, c_j)$ represents inter-cluster distance between cluster i and cluster j, $\bar{S}_i$ represents mean distance from cluster mean for all observations in cluster i, while $\bar{S}_i$ denotes mean distance from cluster median for all observations in cluster j. An extensive analysis of existing literature reveals that most commonly used evaluation measures are accuracy and normalized mutual information.

3.8 A Brief Dive into AI-Driven Genomics and Proteomics Sequence Analysis Frameworks

Genomics and Proteomics sequence analysis field has witnessed the emergence of 7 open source AI frameworks namely BioSeq-Analysis [97], BioSeq-Analysis 2.0 [638], iLearn [157], iLearnPLus [156], iFeature [155], iFeatureOmega [151], and MathFeature [106]. Prime objective of these frameworks is to provide a pool of algorithms to develop end-to-end predictive pipelines for a wide range of biological sequence analysis tasks.

BioSeq-Analysis 2.0 [638] is a web-based platform that was launched in 2019 as an updated version of the original BioSeq-Analysis platform [97] established in 2017. BioSeq-Analysis 2.0 platform offers a larger pool of algorithms in five distinct modules: compatible data types, feature descriptor, feature analysis, predictor construction, and performance evaluation. The data type module contains functions for reading DNA, RNA, and protein sequences in plain text or FASTA format. The feature descriptor module provides 26 new sequence encoding methods. Among these methods, 7 methods are capable of transforming raw DNA sequences into statistical vectors, 6 methods for RNA sequences, and 13 for protein sequences. In total, BioSeq-Analysis2.0 now encompasses 116 sequence encoding methods. In feature analysis module, BioSeq-Analysis2.0 incorporates feature selection methods such as mutual information and the chi-square algorithm. Within an end-to-end predictive pipelines, objective of algorithms of this module is to remove irrelevant and redundant features. In the predictor construction module, BioSeq-Analysis2.0 facilitates five classification algorithms SVM, RF, optimized evidence-theoretic K-nearest neighbor (OET-KNN), covariance discriminant algorithm, and Conditional Random Field. Performance evaluation module encompasses two distinct experimental settings namely 5-fold cross-validation and independent tests to estimate different performance metrics such as sensitivity, specificity, accuracy, Matthews Correlation Coefficient, and AUC. The availability of BioSeq-Analysis2.0 as both a web server and a stand-alone package renders it a versatile tool for researchers to easily develop various predictors for biological sequence analysis tasks. The platform's ability to automatically carry out feature extraction, predictor construction, and performance evaluation steps makes it a unique and powerful tool in the field of computational biology.

In 2020, iLearn [157] appeared as an extension of iFeature [155] that was developed in 2018. iLearn offers a large pool of algorithms in 6 distinct modules: compatible data types, feature descriptor, feature analysis, predictor construction, performance evaluation, and visualization. Data type module contains different functions for reading DNA, RNA, and protein sequences. These functions expect

DNA, RNA, and Protein sequences in plain text or FASTA format. Feature descriptor module contains wide array of encoding methods for transforming raw sequences into statistical vectors. It offers 97 encoding methods including 53 encoding methods for protein sequences, 26 for DNA sequences, and 18 for RNA sequences. Feature analysis module contains 16 algorithms including 6 clustering algorithms (K-Means, Gaussian Mixture Model, Hierarchical Clustering, Mean Shift, DBSCAN, and Affinity Propagation), 5 feature selection algorithms (Chi-Square, Mutual Information, Information Gain, Pearson's Correlation Coefficient, and F1-score), 2 feature normalization algorithms, and 3 dimensionality reduction algorithms (Principal Component Analysis, Latent Dirichlet Allocation, and t-Distributed Stochastic Neighbor Embedding). Predictor construction module encompasses five machine learning algorithms namely SVM, RF, ANN, KNN, and LR. Performance evaluation module facilitates the selection of 2 experimental settings (k-fold cross validation, independent test) and most widely used 7 performance metrics namely accuracy, precision, recall, F1-score, sensitivity, specificity, Matthews Correlation Coefficient, Area Under the Receiver Operating Characteristic curve (AU-ROC), and Area Under the Precision-Recall Curve (AU-PRC). Visualization module offers different plot types for visualization, such as 2D and 3D scatter diagrams for clustering and dimensionality reduction results, and ROC and PRC curves for cross-validation results to perform comprehensive data interpretation and presentation. Cross-platform implementation provides iLearn in two formats: a web server and a python package along with command-line interface (CLI) to facilitate users with varying levels of computational expertise and different usage preferences. This framework supports the development of end-to-end predictive pipelines for different data types by taking distinct combinations of algorithms from each module for standalone or ensemble predictor construction.

iLearn is a comprehensive Python toolkit that integrates various features for analyzing DNA, RNA, and protein sequences. It combines feature extraction, analysis, predictor construction, model selection, ensemble learning, and performance evaluation. iLearn can calculate 53 different feature descriptors for proteins, 26 feature descriptors for DNA sequences and 18 feature descriptors for RNA sequences, respectively. Building on iFeature, iLearn incorporates six feature clustering algorithms (Meanshift, DBSCAN, K-means, Hierarchical Clustering, Spectral Clustering, Affinity Propagation), five feature selection methods (Chi-Square, Information Gain, F-Score, Mutual Information, Pearson Correlation Coefficient), and three dimensionality reduction techniques (PCA, LDA, tSNE). It supports four output feature formats, compatible with other tools. Additionally, iLearn offers five common machine learning algorithms: SVM (Support Vector Machine), RF (Random Forest), ANN (Artificial Neural Network), KNN (K-Nearest Neighbors), and LR (Logistic Regression). To enhance result interpretation, iLearn provides visualization options, including scatter diagrams for clustering and dimensionality reduction outcomes, and ROC and PRC curves for cross-validation results. These capabilities make iLearn a powerful and unique tool that significantly simplifies feature generation, analysis, model training, benchmarking, and predictions in sequence-based studies.

In 2021, an extension of iLearn namely iLearnPlus was released [156]. iLearnPlus framework consists of five distinct modules: compatible data types, feature descriptor, feature analysis, predictor construction, and performance evaluation. The platform supports DNA, RNA, and protein sequences in FASTA format. The feature descriptor module offers a total of 147 descriptors including 19 major classes of feature descriptors for protein sequences and 19 major classes for DNA and RNA sequences. Feature analysis module encompasses 10 feature clustering algorithms, 3 dimensionality reduction methods, 2 feature normalization approaches, and 5 feature selection techniques. The predictor construction module supports a variety of machine learning algorithms. This includes 12 conventional machine learning algorithms, 2 templates for ensemble predictors development, and 7 deep learning approaches. The platform enables automated parameter optimization and supports both binary and multi-class classification which eventually provides flexibility for diverse prediction tasks. Performance evaluation module includes K-fold cross-validation and independent test assessments. Furthermore, it provides 8 performance metrics for binary classification and an accuracy metric for multi-class classification. The platform also offers extensive data visualization options, including histograms, kernel density plots, heatmaps, scatter plots, boxplots, ROC curves, and PRC curves. Statistical analysis tools such as t-test and bootstrap test for comparing ROC/PRC curves are also included. iLearnPlus is designed with accessibility in mind, offering both a standalone software with a graphical user interface (GUI) available on GitHub (supporting Windows, macOS, and Linux) and a web server implementation running on Nectar cloud computing infrastructure. Its modular design, featuring iLearnPlus-Basic, iLearnPlus-Estimator, iLearnPlus-AutoML, and iLearnPlus-LoadModel components, caters to users with varying levels of expertise. This design allows for both rapid model development and sophisticated design protocols which makes iLearnPlus a versatile tool for biological sequence analysis and prediction.

MathFeature iFeatureOmega [151] appeared as an extension of iFeature [155] in 2021. It offers a large pool of algorithms in 4 distinct modules: compatible data types, feature descriptor, feature analysis, and data visualization. Data type module contains different functions for reading five distinct biological data types: DNA, RNA, protein sequences, protein structures, and ligands. These functions expect DNA, RNA, and Protein sequences in plain text or FASTA format, protein structures in the Protein Data Bank (PDB)[1] or Crystallographic Information File (CIF) format, and ligands as SMILES encoding or their files in the Spatial Data File (SDF) format. Feature descriptors module contains wide array of encoding methods for transforming supported data types into statistical vectors. It offers 173 encoding schemes across all types with 65 encoding schemes for protein sequences, 43 for DNA sequences, 33 for RNA sequences, 18 for ligands, and 14 for protein structures. Feature analysis module contains 10 clustering algorithms, 3 dimensionality reduction algorithms, and 2 feature normalization algorithms for

[1] https://www.rcsb.org/.

in-depth exploration of the generated statistical vectors. Data visualization module offers 9 different plot types for visualization, such as histograms, heatmaps, and specialized protein and ligand structure visualizations to perform comprehensive data interpretation and presentation. Furthermore, cross-platform implementation provides iFeatureOmega in three formats: a web server, a graphical user interface (GUI) application, and a command-line interface (CLI) to accommodate users with varying levels of computational expertise and different usage preferences. This framework does not support the development of end-to-end predictive pipelines for different data types by taking distinct combinations of algorithms from each module.

MathFeature [106] is an open-source Python package developed in 2021. It provides functionality in 5 distinct modules: compatible data types, feature descriptor, feature analysis, predictor construction, and performance evaluation. The data type module supports the reading of DNA sequences, RNA sequences, and protein sequences in plain text or FASTA format. The feature descriptor module contains a wide array of encoding methods, offering a total of 37 descriptors. These are categorized into 20 mathematical descriptors and 17 conventional descriptors. The mathematical descriptors are further organized into five groups: numerical mapping (7 descriptors), chaos game representation (2 descriptors), Fourier transform (7 descriptors), entropy-based (2 descriptors), and graph-based (2 descriptors) categories. The conventional descriptors cover composition-based features, physicochemical properties, and sequence order effects. For feature analysis, MathFeature incorporates feature selection methods to reduce dimensionality. The predictor construction module encompasses various machine learning algorithms such as Random Forest, Support Vector Machines, and CatBoost. The performance evaluation module contains distinct metrics including accuracy, F1-score, Matthews Correlation Coefficient, and area under the ROC curve. MathFeature is available as a Python package that can be run via a command-line interface. Additionally, it provides a graphical user interface (GUI)-based platform, that ensures accessibility for users with varying levels of computational expertise. The platform allows for hybrid feature sets combining mathematical and conventional descriptors. The key advantages of MathFeature include the implementation of unique mathematical descriptors not available in other packages, high predictive performance in various biological applications, and suitability for a wide range of biological sequence analysis problems. Table 3.9 provides a functional summary of all 5 frameworks in terms of different capabilities.

Table 3.9 A comprehensive functional scope analysis of existing frameworks

Category	iFeature [155]	iLearn [157]	BioSeq-Analysis2.0 [638]	*MathFeature* [109]	*iLearnPlus* [156]	*iFeatureOmega* [151]	*BioSeq-Analysis* [97]
Number of feature sets for DNA sequence	0	26	36	38	46	43	20
Number of feature sets for RNA sequence	0	18	27	38	35	33	14
Number of feature sets for protein sequence	53	53	53	25	66	65	22
Number of clustering algorithms	5	6	0	–	10	10	0
Number of feature selection algorithms	4	5	2	–	5	0	2
Number of feature normalization algorithms	0	2	0	–	2	2	0
Number of dimension reduction algorithms	3	3	0	–	3	3	0
Number of machine-learning classifiers	0	5	5	4	21	0	4
Number of machine-learning regressors	–	–	–	–	–	0	0
Number of machine-learning multilabel classifiers	–	–	–	–	–	0	0
Number of cross-validation methods	0	2	3	2	2	0	3
Number of evaluation metrics	0	8	5	4	8	0	5

Sequence interaction analysis	–	–	–	–	–	**–**	–
Can build machine-learning pipeline	No	Yes	No	Yes	Yes	**No**	Yes
Can perform evaluation of the feature sets/machine learning models in a batch manner	No	Yes	No	No	Yes	**No**	Yes
Case studies and species evaluation	Limited	Limited	Limited	Limited	Limited	**Limited**	Limited
Results visualization	Limited	Limited	Limited	Limited	Limited	**Comprehensive**	Limited

Chapter 4
Insights of Biological Databases

The recent surge in the volume of Genomics and Proteomics Sequence analysis data and its widespread availability has opened new avenues for gaining deeper understanding of biological systems through the application of statistical models and AI algorithms. Notably, Margaret Dayhoff, an American physical chemist, played a pivotal role in advancing this field by introducing the concept of biological databases. She collected all available protein sequence data and created very first protein sequence database named Protein Information Resource (PIR) [73]. She compiled database information into a book titled "Atlas of Protein Sequence and Structure" [238]. Moreover, she developed different utilities to facilitate efficient search and retrieval of sequences from PIR database. The same utilities are even part of many current biological databases [73]. Apart from PIR database, with the passage of time, several other databases have emerged such as European Molecular Biology Laboratory (EMBL) was created in 1980, DNA Data Bank of Japan (DDBJ) in 1987, and National Center for Biotechnology Information (NCBI) in 1988. Moreover, the refinement of DNA sequencing technologies gave rise to next generation sequencing (NGS) technologies, which have contributed to the creation of numerous databases. NGS technologies are capable of reading thousands and millions of DNA fragments at a time [1027]. These technologies have significantly accelerated the sequencing of larger and more complex genome (e.g humans) [1044]. Fast Genome sequencing has produced large amount of genomic and proteomic data that enabled deeper exploration of genetic code such as genes, regulatory elements, coding/non-coding RNAs and proteins.

As of the present day, around 1727 biological databases [463] have been developed. These databases are classified into three main categories: Primary, Secondary, and Specialized. The relationship between these categories is visually depicted in Fig. 4.1. In primary databases, data is acquired from biological laboratories where it is generated using wet-lab experimental methods. Secondary databases encompass information derived from the analysis and interpretation of primary data. Specialized databases acquired data from either primary or secondary databases,

M. Nabeel Asim et al., *Artificial Intelligence for Molecular Biology*,
https://doi.org/10.1007/978-3-031-90450-9_4

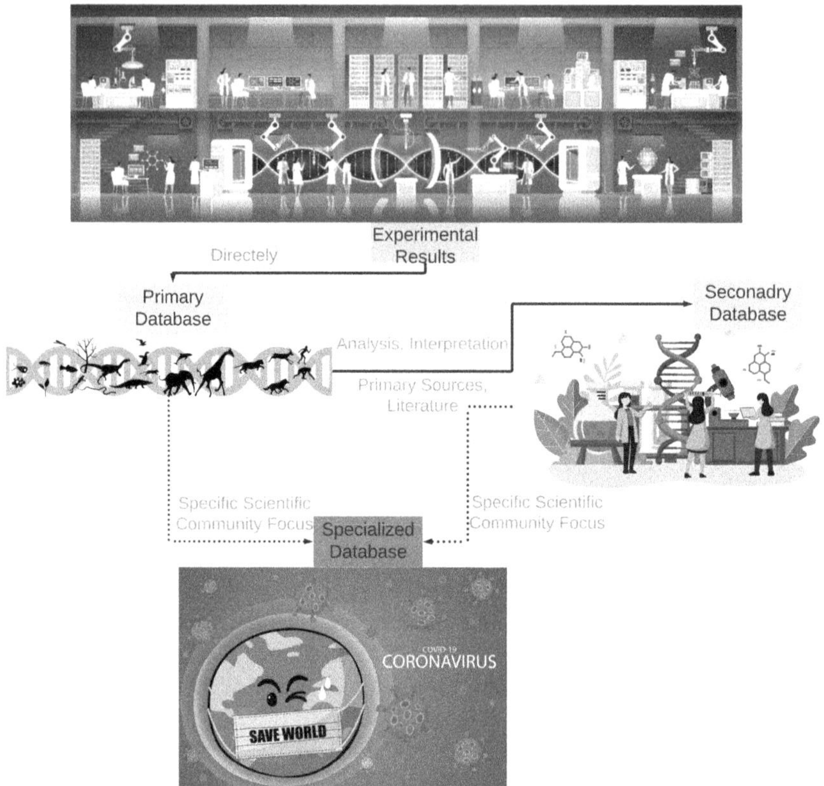

Fig. 4.1 Relationship between Primary, Secondary, and Specialized databases

focusing on a particular living organism, organ, disease, or biological analysis task. To better illustrate the key difference between these databases, let's say that a group of scientists performed an experiment to test the effects of a particular medicine on the human body. The raw data obtained during the experiment, such as drug dosage, experiment duration, and human physiological reactions, would be kept in a primary database namely National Center for Biotechnology Information's Gene Expression Omnibus (GEO). Afterwards raw data is subjected to both computational analysis and expert interpretations to produce novel insights. The analyzed or processed data, like the statistical importance of the drug's effect, would be saved in a secondary database namely Pharmacogenomics Knowledge base. On the other hand, the Alzheimer's Disease Neuroimaging Initiative (ADNI) database is a dedicated database that stores Alzheimer's disease-related data including brain imaging and cognitive information.

With an aim to accelerate and expedite research related to biological sequence analysis, Artificial Intelligence competence is being used for diverse types of tasks. For example, AI is being used to find duplicate data within biological databases.

Likewise, it is being used to retrieve biological data. Another objective of AI researchers is to replace expensive and time consuming wet-lab experiments with robust computational approaches. Additionally, AI researchers have developed robust methods to automatically identify enhancers in DNA sequences or modifications of DNA just using raw DNA sequences. Likewise, using available raw genomics and proteomics raw sequences, AI applications have been developed for histone occupancy prediction, RNA sequence family classification, protein-protein interaction prediction, protein-virus interaction prediction, and many other sequence analysis tasks related to different biomolecules [149]. Consequently, these AI applications significantly accelerates the completion of sequence analysis tasks that typically take several weeks when utilizing wet-lab experimental methods. The streamlined process enable these tasks to be accomplished within a few hours.

Although AI has been applied across a broad spectrum of genomics and proteomics sequence analysis tasks yet the field holds substantial potential for innovative applications. A comprehensive overview of biological databases is essential for AI researchers to leverage available resources. AI researchers can use this information to refine their search criteria and only explore databases relevant to specific sequence analysis task. Selection of appropriate sequences related data and understanding of data retrieval systems will enable them to train large language models in an unsupervised fashion. These models are capable of generating rich representations of biological sequences and can facilitate as foundational components for a variety of downstream biological sequence analysis tasks. Furthermore, a thorough understanding of various data formats and database categories will empower AI researchers to effortlessly access and utilize the necessary data, accelerating their research and development process. Besides this, partial content of this chapter is published in [57–60]. The focus of this chapter revolves around following key points.

- This chapter briefly describes 7 distinct data format namely, tabular format, structured formats, FASTA format, FASTQ format, GenBank Sequence flatfile format, EMBL-Bank sequence flatfile format, RefSeq sequence flatfile format. Although there exist more than 80 different data formats but most of the databases uses aforementioned data format so we only focused on these data formats.A comprehensive understanding of these data formats will assist AI researcher to acquire data from these databases and develop novel biological applications.
- This chapter comprehensively discusses 13 different data retrieval systems namely, Entrez/GQuery, Entrez Web Interface, Entrez E-utils, e-search, efetch, Entrez Direct, DBGET toolkit, bget command, bfind command, DBGET web interface, DBFetch Web Interface, DBFetch URL, DBFetch toolkit. These systems provide AI developers with versatile tools for accessing and utilizing biological data.
- This chapter illustrates details of 37 distinct databases that has been employed for development of datasets for DNA sequence analysis tasks namely: Descartes, PPD, EnhancerAtlas 2.0, DREAM Base, EmExplorer database, COSMIC, Dis-

GeNet, genomAD, ClinVar, HOCOMOCO Human v11 database, DeOri, BioLip, DeOri6.0, GWAS, Broad DepMap, CCLE, GENCODE, Consensus Coding Sequence Database, MSigDB, Gene Ontology, JASPAR, Database of Essential Genes, ENCODE, DataBase of Transcriptional Start Sites, MGC, GEO, Exon-Intron Database, Ensembl, RegulonDB, EPD, KEGG, NCBI, Eukaryotic Promoter Database, GenBank and OMIM.

- This chapter illustrates details of 11 distinct databases that has been employed for development of datasets for CRISPER systems namely: CRISPRCasDb, AcrHub, Anti-Crisprdb, AcrDb, AcrCatalog, UniProt Database, CasPDB, CasPedia, crisprSQL, Ensembl BioMart and CRISPOR.
- This chapter illustrates details of 64 distinct databases that has been employed for development of datasets for RNA sequence analysis tasks namely: SPENCER, m6A-Atlas v2, RNALocate v2.0, GENCODE Release 43, Circad, MNDR3.0, cantataDB 2.0, EVLncRNAs 2.0, piRBase, EuRBPDB, PanglaoDB, CSCD, RefSeq (version 90), GENCODE.v28, GENCODE.vM18, lncRNASNP2, LncRNADisease v2.0, CircRNADisease, CircBank, bpRNA, RNALocate, RMBase2.0, miRmine, CircInteractome, ATract, HMDAD, dbDEMC, NONCODEV5, RMBase, circRNADb, DisGeNET, NDB, LNCipedia, RefSeq (version 60), doRiNA, lncRNADisease, miRCancer, Encori, CircBase, EPDnew, PLncDB 2.0, ClinVar, GENCODE.v17, RNAcentral, miR2Disease, HMDD, dbGap, HGMD, TarBase, Gencode, NCBI, GtRNAdb, NPInter V4.0, miRBase, Rfam, CTD, ENCODE3, ENCODE, FANTOM5, GEO, ENSEMBL, KEGG, EMBL-EBI and OMIM.
- This chapter illustrates details of 68 distinct databases that has been employed for development of datasets for protein sequence analysis tasks namely: AlphaFoldDB, AmyPro, BindingDB, BRENDA, CARD, CATH, ChEMBL, ConSurf-DB, dbPTM, DIP, DisProt, DUD, DUD-E, GeneCards, GLASS, GOA, GPCRdb, HIPPIE, HPIDB, IEDB, IMGT, intAct, interPro, IPD-MHC, MalaCards, McPAS-TCR, MGnify, MINT database, MobiDB, MtSSPdb, Negatome database, OAS database, OGEE, PDBbind database, Phospho.ELM, PhosphoSitesPlus, PHROGs, PINA, PIRD, PPT-Ohmnet, PubChem, RCSB PDB, SAbDab, SCOP, SCOPe, STCRDab, TCDB, Therapeutic Targets Database, Uniclust30, VariBench, VDJdb, CCLE, KEGG, CTD, STRING, GEO, UniProtKB, DisGeNET, NCBI, BioGRID, OMIM, PDB, STITCH, EMBL-EBI, BioLiP, ClinVar, COSMIC and Prosite.
- This chapter illustrates details of 47 distinct databases that has been employed for development of datasets for peptide sequence analysis tasks namely: APD3, Peplab, B3Pdb, Biofilm resource, PlantPepDB, VARIDT, StarPep, DRAMP 2.0, AntiTbPdb, BioPepDB, dbAMP, DINeR, Flavor, THPdb, DRAMP, LAMP2, ADP2, BaAMP, CPPsite 2.0, INTEDE, NeuroPep, SATPdb, DBAASP, AHTPDB, Hemolytik, Hmrbase2, LAMP, Quorumpeps, CPPsite, DADP, PDB, CAMP, APD2, Brainpeps, IEDB, PepBank, DrugBank, SPdb5.1, PeptideAtlas, UniProtKB, AAIndex, Pubmed, SGD, Prosite, NCBI, SwissProt and UniProtKB.

4.1 Comprehensive Overview of Data Formats Utilized in Biological Databases

Biological databases utilize standardized formats to efficiently store, organize and access vast amount of biological data. Standardized formats ensures that sequence essential characteristics such as specie type, accession number and other relevant attributes are accurately represented and readily accessible. A well structured format enhances the efficiency and coherence of managing extensive biological data within databases. To accommodate diverse nature of biological data, biological databases utilize a range of data formats. Moreover, certain databases supports multiple formats simultaneously. The most widely used data formats in biological databases [337, 775, 987] include tabular formats (CSV and TSV), JSON, XML and specialized formats (FASTA, FASTQ [113, 310], Flat File format). The choice of format is determined by data's characteristics, intended use cases and compatibility with relevant systems [469] and tools.[1,2] Following subsections briefly describes details of 7 different data formats.

4.1.1 *Tabular Formats*

The most commonly employed tabular data representation schemes are Comma-Separated Values (CSV) files, Tab-Separated Values (TSV) files, and Binary format (XLSX) files. Each format organize data in rows and columns and offers distinct advantages for data storage and seamless exchange. CSV and TSV are text-based formats where values in rows are separated by commas or tabs, respectively. In both file formats, header row defines column name ("SeqID", "SeqName", "Type", "Function") and subsequent rows represent corresponding data. Ensembl[3] and ALEdb[4] databases provide data in CSV format, while HGNC,[5] InterPro,[6] UniParc,[7] RegulonDB,[8] and Ensembl[9] databases provide data in TSV format. Figure 4.2 illustrates TSV file format based data retrieved from RegulonDB database while Fig. 4.3 depicts the CSV file format based data retrieved from ALEdb database.

---

[1] https://www.ncbi.nlm.nih.gov/Banklt/oldbankit.html.

[2] https://www.ddbj.nig.ac.jp/sub/websub-e.html.

[3] http://www.ensembl.org/.

[4] https://aledb.org/.

[5] http://genenames.org/.

[6] https://www.ebi.ac.uk/interpro/.

[7] https://www.ebi.ac.uk/interpro/.

[8] https://regulondb.ccg.unam.mx/.

[9] http://www.ensembl.org/.

RISet - Notepad

File Edit Format View Help

```
1)riId  2)riType        3)regulatorId    4)regulatorName 5)cnfName    6)tfrsID        7)tfrsLeft        8)tfrsRight
RDBECOLIRIC00001        tf-promoter      RDBECOLITFC00023          GadW    GadW     RDBECOLIBSC01689          3653866 3653885
RDBECOLIRIC00002        tf-promoter      RDBECOLITFC00023          GadW    GadW     RDBECOLIBSC01690          3656835 3656854
RDBECOLIRIC00003        tf-promoter      RDBECOLITFC00088          GalR    GalR-D-galactose         RDBECOLIBSC01691
RDBECOLIRIC00004        tf-promoter      RDBECOLITFC00033          GalS    GalS-D-galactose         RDBECOLIBSC01692
RDBECOLIRIC00005        tf-promoter      RDBECOLITFC00088          GalR    GalR-D-galactose         RDBECOLIBSC01693
RDBECOLIRIC00006        tf-promoter      RDBECOLITFC00033          GalS    GalS-D-galactose         RDBECOLIBSC01694
RDBECOLIRIC00007        tf-promoter      RDBECOLITFC00088          GalR    GalR-D-galactose         RDBECOLIBSC01695
RDBECOLIRIC00008        tf-promoter      RDBECOLITFC00033          GalS    GalS-D-galactose         RDBECOLIBSC01696
RDBECOLIRIC00009        tf-promoter      RDBECOLITFC00088          GalR    GalR-D-galactose         RDBECOLIBSC01697
RDBECOLIRIC00010        tf-promoter      RDBECOLITFC00033          GalS    GalS-D-galactose         RDBECOLIBSC01698
```

Fig. 4.2 A sample TSV file format data downloaded from RegulonDB database

42C_42C_mut - Notepad

File Edit Format View Help

```
Reference Seq,Position,Mutation Type,Sequence Change,Gene (Scrollable),Function,Product,GO Process,
NC_000913,"702,352",DEL,Δ21 bp,nagA,(),(N-acetylglucosamine-6-phosphate deacetylase),(),(GO:0005737
NC_000913,"1,308,318",SNP,G→C,clsA,(enzyme; Macromolecule synthesis: Phospholipids),(cardiolipin sy
NC_000913,"2,173,364",DEL,Δ1 bp,"gatC, gatC",,,,,45881,intergenic (-2/+1),1.00/0.00,,,,,,,,,
NC_000913,"3,815,859",DEL,Δ82 bp,"[rph], [rph]",,,,,45309,,1.00/0.00,1.00/0.00,1.00/0.00,1.00/0.00,
NC_000913,"4,187,550",SNP,C→T,rpoC,(),"(RNA polymerase, beta prime subunit)",(GO:0006350 - transcri
NC_000913,"4,400,313",SNP,A→C,hfq,(factor; Phage-related functions and prophages),(),(GO:0009386 -
```

Fig. 4.3 A sample CSV file format data downloaded from ALEdb database

4.1.2 Structured Formats

Hierarchical or nested structured data formats organize data in tree based structures. This data format is well-suited for storing complex data with intricate relationships between distinct aspects of the data and multiple sub-records within the data. Nested structured data contains convoluted hierarchies of objects and arrays enabling complex data modelling. JavaScript Object Notation (JSON) and eXtensible Markup Language (XML) are most widely utilized formats to represent hierarchical or nested structured data. In JSON file format, nested-structured data is organized in key and value pairs enclosed in curly brackets {}. Contrarily, XML file format represents hierarchical data using tags. Each tag can contain attributes and nested tags which represents intricate relationships between data elements. AlphaFold DB[10] and bioDBnet[11] databases provide data in JSON format, while IPD,[12] InterPro,[13] IMGT/HLA[14] and UniParc[15] databases provide data in XML format.

Figure 4.4 illustrates toy genomic data obtained from the Non Redundant Patent Sequence (NRN) Database using the EMBL-EBI platform. The data is

[10] https://alphafold.com/.

[11] https://biodbnet-abcc.ncifcrf.gov/.

[12] https://www.ebi.ac.uk/ipd/kir/.

[13] https://www.ebi.ac.uk/interpro/.

[14] https://www.ebi.ac.uk/imgt/hla/.

[15] https://www.uniprot.org/.

```
<?xml version="1.0" encoding="UTF-8"?>
<seqXML seqXMLversion="0.4" xmlns:xsi="http://www.w3.org/2001/XML
    <entry id="NRN_DJ207917" source="NRNL1">
        <description>Method for identification of useful proteins
        <DNAseq>ATGAGTGATTTTGAGATAATTGTTGGAATTTCATCGTTGTTACAGGTTA
        <property name="sequence_length" value="291"/>
    </entry>
</seqXML>
```

Fig. 4.4 XML file format of genomic data from NRN database at EMBL-EBI

presented in XML format and information in Fig. 4.4 illustrates structure and content organization in this format.

As depicted in Fig. 4.4, each XML tag represents a particular attribute of sequence data i.e., information inside $<?xml\_\_\_\_...\_\_? >$ represents version of xml, subsequent line offers XML related directives such as "xsi:no Name space Schema Location" or "xmlns:xsi". Third line represents entry id and source where entry id encompass identifier information i.e. a unique sequence ID. This unique ID is often used to obtain its corresponding sequence (if not provided). Moreover, source refers to database from where the sequence is retrieved, In this particular example, database is NRNL1 which refers to non-redundant patent sequences databases and L1 represents nucleotide sequence cluster level 1. In case of other databases such as RefSeq or BioSample, it will be RS and PRJNA. The very next line contain sequence related information such as identification methods of useful proteins derived from yeast or gene's analysis methods of industrial yeast etc., enclosed in $< description >$ and $< /description >$ tags. Nucleotide sequence is given inside $< DNAseq >$ and $< /DNAseq >$ and finally next tag $< / >$ contains sequence property namely sequence length and its value which is 291 in this case.

4.1.3 Specialized Formats

Following subsections describes details of specialized formats including FASTA, FASTQ and flat file format.

4.1.3.1 FASTA Format

In the early 1980s, William Pearson introduced FASTA format that stands for "fast all". This format is widely accepted by most of the sequence analysis platforms and is considered fundamental data representation format for sequences. Databases

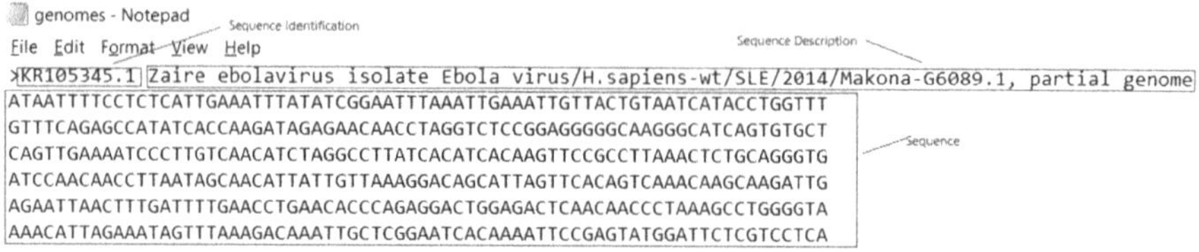

Fig. 4.5 FASTA file format of genomic sequence from NCBI

that utilizes FASTA formats are AlphaFold DB,[16] ChEMBL,[17] ENA sequence database,[18] Ensembl,[19] IMGT/LIGM-DB[20] and many more. Figure 4.5 depicts NCBI based toy genomic sequence retrieved from genomes database in FASTA format.

It can be seen in Fig. 4.5, each sequence entry starts with > symbol and everything after > symbol to first white space is considered as sequence identifier (ID) i.e., KR105345.1. Here KR refers to genome database and 105345.1 represents unique sequence identification number. This sequence identifier is usually written in different formats according to requirements of different databases. Table 4.1 depicts all possible formats that can be used as sequence identifier in FASTA header section. In Table 4.1, first 2 or 3 letters such as gb, lcl, emb, in format represents corresponding database of sequence whereas numerals, strings, accession number, locus, patent and application number separated by vertical line | represent other properties of that particular sequence. The information other than sequence identifier is sequence description.

Genome sequence is a long sequence primarily comprising of four base nucleotides (A, C, T, G). However, it can also contain other characters such as N, W, B, etc., which shows the uncertainty about presence of base nucleotides. For instance, nucleotide R in sequence indicates ambiguity in presence of A or G. All these types of uncertainties about presence of base nucleotides are sorted by the International Union of Pure and Applied Chemistry (IUPAC)[21] and different codes alphabets are given to each uncertainty. Table 4.2 depicts all IUPAC code alphabets for uncertainties in base nucleotide sequence.

[16] https://alphafold.ebi.ac.uk/.

[17] https://www.ebi.ac.uk/chembl/.

[18] https://www.ebi.ac.uk/ena/.

[19] http://www.ensembl.org/.

[20] https://www.ebi.ac.uk/imgt/hla/.

[21] https://www.bioinformatics.org/sms/iupac.html.

Table 4.1 FASTA header format of different databases

Database name	Abbreviation	Used format	Example
No database reference (Local)	lcl	(a) lcl|Integer-value, (b) lcl|String	(a) lcl|456, (b) lcl|hsa271
Genbank Information BackBone Sequence-ID	bbs	bbs|Integer-value	bbs|768
Genbank Information BackBone moltype	bbm	bbm|Integer-value	bbm|432
Genbank Information Import ID	gim	gim|Integer-value	gim|654
GenBank database	gb	gb|Accession Number|Locus	gb|MS22491|AGMA13TG
Protein Information Resource (PIR)	pir	pir|Accession Number|Name	pir|G46363
Swiss-Prot (SP)		sp|Accession Number|Name	sp|P31010|OVAXCHICK
Patent	pat	pat|Country Name|Patent|Sequence Value	pat|US|RE33188|1
Pre-Grant Patent	pgp	pgp|Country Number|Application Number|Sequence Value	pgp|EP|0238993|7
Reference Sequence	ref	ref|Accession Number|Accession Number.version	ref|NM010450.1|
Reference to a database that is not in this list (General Database Reference)	gnl	(a) gnl|Database Name|Integer Value, (b) gnl|Database Name|String	(a) gnl|taxon|7890, (b) gnl|PDB|e1632
Genbank Information Integrated Database	gi	gi|Integer Value	gi|27234723
DDBJ	dbj	dbj|Accession Number|Locus	dbj|REF85684.1|
PRF	prf	prf|Accession Number|Name	prf|0806162D
Protein Data Bank (PDB)	pdb	pdb|Entry|Chain	pdb|1I4L|C
Third-Party GenBank	tpg	tpg|Accession Number|Name	tpg|BK003321|
Third-Party EMBL	tpe	tpe|Accession Number|Name	tpe|BN001230|
Third-Party DDBJ	tpd	tpd|Accession Number|Name	tpd|FAA00170|
TrEMBL	tr	tr|Accession Number|Name	tr|Q90RT2|Q90TRT29HIV1

Table 4.2 IUPAC nucleotide code

IUPAC nucleotide code	Representing base	IUPAC nucleotide code	Representing base	IUPAC nucleotide code	Representing base	IUPAC nucleotide code	Representing base
A	Adenine	R	A or G	K	G or T	H	A or C or T
C	Cytosine	Y	C or T	M	A or C	V	A or C or G
G	Guanine	S	G or C	B	C or G or T	N	Any base
T (or U)	Thymine (or Uracil)	W	A or T	D	A or G or T	. or -	Gap

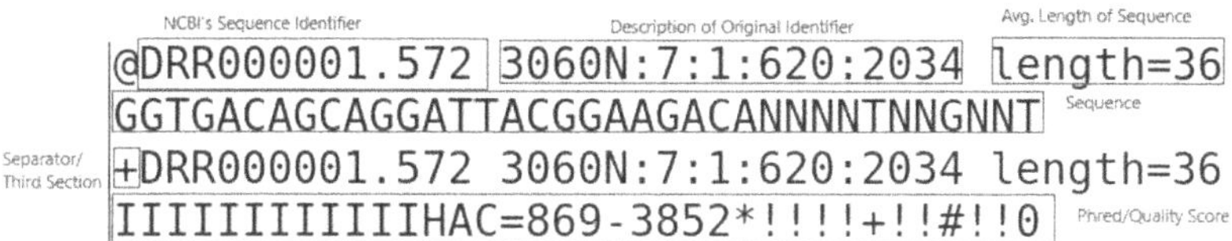

Fig. 4.6 FASTQ file

4.1.3.2 FASTQ Format

The FASTQ format is a widely used structured format in bioinformatics which efficiently captures both biological sequences and their associated quality scores from high-throughput sequencing. Moreover, reliability measure is associate with every base (A, C, G, T) and probability of being wrong is 1/1000. Major bioinformatics databases including NCBI's Sequence Read Archive (SRA)[22] and the European Nucleotide Archive (ENA) provide data in FASTQ format. Figure 4.6 depicts sequence reads archives, sourced from DDBJ[23] in FASTQ format.

It can be seen in Fig. 4.6 that FASTQ format comprises of four lines per entry: (a) header, (b) sequence, (c) separator, (d) quality scores. FASTQ-header looks similar to FASTA-header but it uses @ symbol instead of > symbol. FASTQ header consists of instrument name, run ID, flow cell ID, flow cell lane, if read is filtered then Y (otherwise N), tile number within Flow Cell Lane, x-coordinate of cluster within tile, y-Coordinate of cluster within tile, pair member (1 or 2 depending on Paired End or Mate Pair), control bits (on=0, otherwise an even number) and index sequence length. In Fig. 4.6, DRR000001 represents instrument name, run ID is 572, 3060 refers flow cell ID, read is not filtered therefore N, flow cell lane is 7, 1 indicates tile number within flow cell lane, 620 and 2034 depicts x and y-Coordinate of cluster within tile, respectively and 36 represents index sequence length.

Afterwards, second section contains measured sequence which may span multiple lines until the start of next section denoted by + sign. Third section starts with + sign and can have same sequence identifier and header information. Finally, last section encodes quality values for sequence and it must be of same length as sequence. The wired characters in this section are known as encoded numeral values. Each character $IIIIIIIIIIIIHAC = 869 - 3852\star$!!!!+!#!!0 represent numerical value which is known as Phred Score. Prime objective of computing Phred Score is to map two digit numbers to single character (likelihood error), so that length of quality string remains same as length of sequence. Equation 4.1 depicts a mathematical expression to compute probability of a base call being incorrect which is also known as Phred Score (Q).[24]

[22] https://www.ncbi.nlm.nih.gov/Traces/sra.

[23] https://www.ddbj.nig.ac.jp/dra/index-e.html.

[24] https://en.wikipedia.org/wiki/Phred_quality_score.

Fig. 4.7 Mapping of two digit numbers to special characters

```
!"#$%&'()*+,-./0123456789:;<=>?@ABCDEFGHI
|    |    |    |    |    |    |    |    |
0....5...10...15...20...25...30...35...40
|    |    |    |    |    |    |    |    |
worst................................best
```

$$P = 10^{\frac{-Q}{10}} \tag{4.1}$$

In the Eq. 4.1, P refers to probability measure associated with every base (A, C, G, T) while Q indicates Phred score. Moreover, there are two different formats to measure Phred score namely: (a) Sanger format, (b) (+ 64) format. Earlier, Sanger format, also known as (+ 33) format, was used which shifts ASCII code by 33. (+ 64) format has shifted the code values by 64 and ! is mapped to 0, and 'I' corresponds to 40. (+ 64) format is used as standardized format to represent confidence levels associated with every base (A, C, G, T) in sequencing data. Figure 4.7 illustrates (+ 64) format based associated Phred values of different wired characters.

4.1.3.3 Flat File Format

Flat file format typically refers to a simple, text-based file structure used to store information related to sequences, annotations or other biological details. Following section provides a detailed understanding of three different types of flat files formats.

GenBank Sequence Flatfile Format

GenBank Flatfile format contain meta data of sequence in terms of name, source, relevant annotations, open reading frame information, and potential translation product details. Information stored in GenBank flatfile format consists of three different parts namely: (a) Header section, (b) feature table section and (c) sequence section. Figures 4.8, 4.9, and 4.10 illustrate header section, feature table section and genomic sequence information retrieved from GenBank database using NCBI platform.

It can be seen in Fig. 4.8 that first line in header section known as locus line or locus field, contains locus name, length of sequence and a three-letter term designated by GenBank database based on different sequence types such as PRI to primate sequences, mammalian sequences are represented using MAM and so on. Table 4.3 describes all different sequence types along with their abbreviations. It also contains information related to accession number (unique identifier), keywords, organism, data source, author and submitter information.

Example in Fig. 4.8 is a plant, fungal and alagal sequence of Saccharomyces cerevisiae TCP1-beta gene with Axl2p (AXL2) and Rev7p (REV7) genes. It is

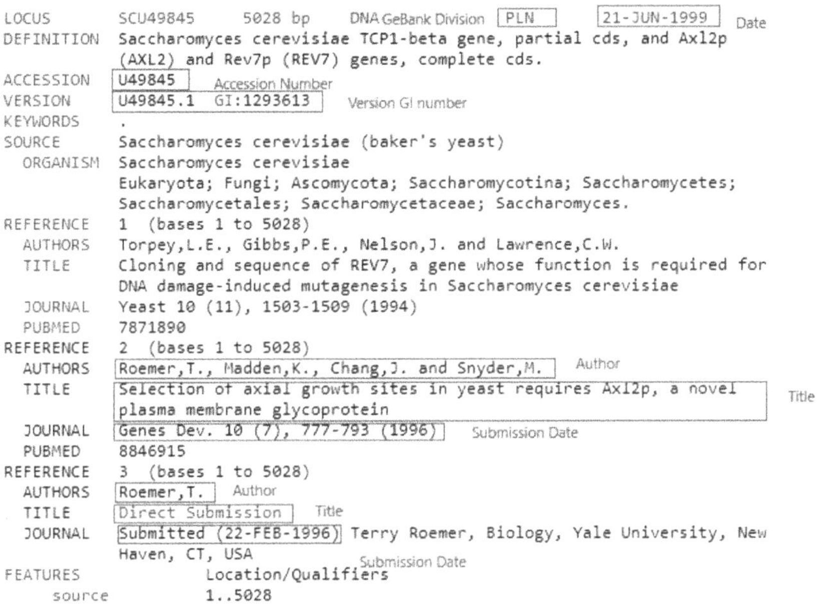

```
LOCUS       SCU49845     5028 bp    DNA GeBank Division  PLN     21-JUN-1999  Date
DEFINITION  Saccharomyces cerevisiae TCP1-beta gene, partial cds, and Axl2p
            (AXL2) and Rev7p (REV7) genes, complete cds.
ACCESSION   U49845    Accession Number
VERSION     U49845.1  GI:1293613    Version GI number
KEYWORDS    .
SOURCE      Saccharomyces cerevisiae (baker's yeast)
  ORGANISM  Saccharomyces cerevisiae
            Eukaryota; Fungi; Ascomycota; Saccharomycotina; Saccharomycetes;
            Saccharomycetales; Saccharomycetaceae; Saccharomyces.
REFERENCE   1  (bases 1 to 5028)
  AUTHORS   Torpey,L.E., Gibbs,P.E., Nelson,J. and Lawrence,C.W.
  TITLE     Cloning and sequence of REV7, a gene whose function is required for
            DNA damage-induced mutagenesis in Saccharomyces cerevisiae
  JOURNAL   Yeast 10 (11), 1503-1509 (1994)
  PUBMED    7871890
REFERENCE   2  (bases 1 to 5028)
  AUTHORS   Roemer,T., Madden,K., Chang,J. and Snyder,M.    Author
  TITLE     Selection of axial growth sites in yeast requires Axl2p, a novel     Title
            plasma membrane glycoprotein
  JOURNAL   Genes Dev. 10 (7), 777-793 (1996)   Submission Date
  PUBMED    8846915
REFERENCE   3  (bases 1 to 5028)
  AUTHORS   Roemer,T.   Author
  TITLE     Direct Submission   Title
  JOURNAL   Submitted (22-FEB-1996) Terry Roemer, Biology, Yale University, New
            Haven, CT, USA       Submission Date
FEATURES             Location/Qualifiers
     source          1..5028
```

Fig. 4.8 GenBank sequence Flatfile-header section

observed that saccharomyces cerevisiae is a yeast specie which is used in backing and brewing. Saccharomyces cerevisiae is a single cell fungus and a model organism in biological search and the TCP1-beta gene refers to its association with T-complex protein 1(TCP1), which is a sub-unit of the Chaperonin-containing TCP1(CCT) complex. Axl2p is protein encoded by the AXL2 gene and plays significant role in cell polarity and bud side selection during the yeast cell cycle which contributes in establishing axial budding pattern in yeast cells. Rev7p is a protein encoded by REV7 gene and is associated with DNA damage response and repair. Partial CDS refer to partial coding sequence of gene which indicates that only a portion of the gene's coding region has been reported while complete CDS indicates that entire coding sequence of a gene has been identified or reported. Version of sequence is indicated by version of accession number (accession number is U49845; first version is U49845.1). Moreover, last modification date refers to recent update in that file which may also be mentioned in comment section. Lastly, the information about publication and corresponding authors is mentioned in reference field.

Figure 4.9 depicts feature table information which includes information related to coding sequence (CDS) and genes information. It is observed that coding sequence (CDS) related to TCP1-beta gene spans from base 1 to 206 while CDS associated with the production of a plasma membrane glycoprotein known as Axl2p ranges from 687 to 3158 and CDS dealing with Rev7p gene spans from base 3300 to 4037.

```
                     /organism="Saccharomyces cerevisiae"
                     /db_xref="taxon:4932"
                     /chromosome="IX"
                     /map="9"
     CDS             <1..206
                     /codon_start=3
                     /product="TCP1-beta"
                     /protein_id="AAA98665.1"    Protein Ids related to TCP1-beta
                     /db_xref="GI:1293614"
                     /translation="SSIYNGISTSGLDLNNGTIADMRQLGIVESYKLKRAVVSSASEA
                     AEVLLRVDNIIRARPRTANRQHM"
     gene            687..3158
                     /gene="AXL2"
     CDS             687..3158
                     /gene="AXL2"
                     /note="plasma membrane glycoprotein"
                     /codon_start=1
                     /function="required for axial budding pattern of S.
                     cerevisiae"
                     /product="Axl2p"
                     /protein_id="AAA98666.1"    Protein Ids related to gene AXL2
                     /db_xref="GI:1293615"
                     /translation="MTQLQISLLLTATISLLHLVVATPYEAYPIGKQYPPVARVNESF
                     TFQISNDTYKSSVDKTAQITYNCFDLPSWLSFDSSSRTFSGEPSSDLLSDANTTLYFN
                     VILEGTDSADSTSLNNTYQFVVTNRPSISLSSDFNLLALLKNYGYTNGKNALKLDPNE
                     VFNVTFDRSMFTNEESIVSYYGRSQLYNAPLPNWLFFDSGELKFTGTAPVINSAIAPE
                     TSYSFVIIATDIEGFSAVEVEFELVIGAHQLTTSIQNSLIINVTDTGNVSYDLPLNYV
                     YLDDDPISSDKLGSINLLDAPDWVALDNATISGSVPDELLGKNSNPANFSVSIYDTYG
                     DVIYFNFEVVSTTDLFAISSLPNINATRGEWFSYYFLPSQFTDYVNTNVSLEFTNSSQ
                     DHDWVKFQSSNLTLAGEVPKNFDKLSLGLKANQGSQSQELYFNIIGMDSKITHSNHSA
                     NATSTRSSHHSTSTSSYTSSTYTAKISSTSAAATSSAPAALPAANKTSSHNKKAVAIA
                     CGVAIPLGVILVALICFLIFWRRRRENPDDENLPHAISGPDLNNPANKPNQENATPLN
                     NPFDDDASSYDDTSIARRLAALNTLKLDNHSATESDISSVDEKRDSLSGMNTYNDQFQ
                     SQSKEELLAKPPVQPPESPFFDPQNRSSSVYMDSEPAVNKSWRYTGNLSPVSDIVRDS
                     YGSQKTVDTEKLFDLEAPEKEKRTSRDVTMSSLDPWNSNISPSPVRKSVTPSPYNVTK
                     HRNRHLQNIQDSQSGKNGITPTTMSTSSSDDFVPVKDGENFCWVHSMEPDRRPSKKRL
                     VDFSNKSNVNVGQVKDIHGRIPEML"
     gene            complement(3300..4037)
                     /gene="REV7"
     CDS             complement(3300..4037)
                     /gene="REV7"
```

Fig. 4.9 GenBank sequence Flatfile-feature table section

Figure 4.10 depicts genomic sequence corresponding to that accession number which starts from origin and terminates with // symbol. Ensemble,[25] INSDC,[26] ENA[27] and DDBJ[28] use GenBank[29] sequence flatfile format.

[25] http://www.ensembl.org/.

[26] https://www.insdc.org/.

[27] https://www.ebi.ac.uk/ena/.

[28] https://www.ddbj.nig.ac.jp/index-e.html.

[29] https://www.ncbi.nlm.nih.gov/genbank/.

```
ORIGIN
        1 gatcctccat atacaacggt atctccacct caggtttaga tctcaacaac ggaaccattg
       61 ccgacatgag acagttaggt atcgtcgaga gttacaagct aaaacgagca gtagtcagct
      121 ctgcatctga agccgctgaa gttctactaa gggtggataa catcatccgt gcaagaccaa
      181 gaaccgccaa tagacaacat atgtaacata tttaggatat acctcgaaaa taataaaccg
      241 ccacactgtc attattataa ttagaaacag aacgcaaaaa ttatccacta tataattcaa
      301 agacgcgaaa aaaaaagaac aacgcgtcat agaacttttg gcaattcgcg tcacaaataa
      361 attttggcaa cttatgtttc ctcttcgagc agtactcgag ccctgtctca agaatgtaat
      421 aatacccatc gtaggtatgg ttaaagatag catctccaca acctcaaagc tccttgccga
      481 gagtcgccct cctttgtcga gtaattttca cttttcatat gagaacttat tttcttattc
      541 tttactctca catcctgtag tgattgacac tgcaacagcc accatcacta gaagaacaga
      601 acaattactt aatagaaaaa ttatatcttc ctcgaaacga tttcctgctt ccaacatcta
      661 cgtatatcaa gaagcattca cttaccatga cacagcttca gatttcatta ttgctgacag
      721 ctactatatc actactccat ctagtagtgg ccacgcccta tgaggcatat cctatcggaa
      781 aacaataccc cccagtggca agagtcaatg aatcgtttac atttcaaatt tccaatgata
      841 cctataaatc gtctgtagac aagacagctc aaataacata caattgcttc gacttaccga
      901 gctggctttc gtttgactct agttctagaa cgttctcagg tgaaccttct tctgacttac
      961 tatctgatgc gaacaccacg ttgtatttca atgtaatact cgagggtacg gactctgccg
     1021 acagcacgtc tttgaacaat acataccaat ttgttgttac aaaccgtcca tccatctcgc
     1081 tatcgtcaga tttcaatcta ttggcgttgt taaaaaacta tggttatact aacggcaaaa
     1141 acgctctgaa actagatcct aatgaagtct tcaacgtgac ttttgaccgt tcaatgttca
     1201 ctaacgaaga atccattgtg tcgtattacg gacgttctca gttgtataat gcgccgttac
     1261 ccaattggct gttcttcgat tctggcgagt tgaagtttac tgggacggca ccggtgataa
     1321 actcggcgat tgctccagaa acaagctaca gttttgtcat catcgctaca gacattgaag
     1381 gattttctgc cgttgaggta gaattcgaat tagtcatcgg ggctcaccag ttaactacct
```

Fig. 4.10 GenBank sequence Flatfile-sequence section

Table 4.3 Three-letter abbreviations of GenBank divisions

Sequence type	Abbreviation
Primate Sequences	PRI
Rodent Sequences	ROD
Mammalian Sequences	MAM
Vertebrate Sequences	VRT
Invertebrate Sequences	INV
Plant, Fungal and Alagal Sequences	PLN
Bacterial Sequences	BCT
Viral Sequences	VRL
Bacteriophage Sequences	PHG
Synthetic Sequences	SYN
Unannotated Sequences	UNA
Expressed Sequence Tag	EST
Patent Sequences	PAT
Sequence Tagged Sites Sequences	STS
Genome Survey Sequences	GSS
High Throughput Genomic Sequences	HTG
Unfinished High Throughput cDNA Sequences	HTC
Environmental Sampling Sequences	ENV

EMBL-Bank Sequence Flatfile Format

Another data format is EMBL-Bank flatfile format which uses two-letter abbreviations such as ID for unique identifier, AC indicating accession number, FT

Table 4.4 Two-letter abbreviations in EMBL-Bank flatfiles

Abbreviation	Description	Abbreviation	Description
ID	Identification	RX	Reference Cross-Reference
SV	Sequence Variation	RA	Reference Author
AC	Accession Number	RT	Reference Title
DT	Date	RL	Reference Location
DE	Description	DR	Database Cross-Reference
KW	Keyword	CC	Comments
OS	Organism Species	FH	Feature Table Header
OC	Organism Classification	FT	Feature Table Data
RN	Reference Number	SQ	Sequence Header
RP	Reference Position	XX	Spacer Line

representing feature table data, SQ sequence header. Table 4.4 describes all different abbreviations used in EMBL-Bank flatfile format with their details.

Various databases including IMGT/HLA,[30] IPD-KIR,[31] IPD-MHC,[32] JPO,[33] ENA,[34] Ensembl[35] databases provides data in EMBL-bank flatfile format. Similar to GenBank flatfile format, EMBL-Bank flatfile format consists of three sections such that Fig. 4.11 depicts header section, Fig. 4.12 refers to feature table section and Fig. 4.13 illustrates genomic sequence information retrieved from GenBank database using NCBI platform.

4.1.3.4 RefSeq Sequence Flatfile Format

A RefSeq sequence flatfile looks like regular GenBank flatfile except that it has a RefSeq accession number and a COMMENT section. RefSeq sequences have a different format of accession number as compared to other databases. Mainly, it has a two-letter prefix and numerals separated by an underscore sign. The two-letter prefix indicates sequence type. Table 4.5 describes all molecular types along with their abbreviation and associated project or database. RefSeq flat file format lists all the sources from where information about the sequence has been obtained and the COMMENT section cites the accession number(s) of the sequence record(s) used to derive the sequence.

Figures 4.14 and 4.15 depict differences of RefSeq flat file format with GenBank flat file format.

---

[30] https://www.ebi.ac.uk/imgt/hla/.

[31] https://www.ebi.ac.uk/ipd/kir/.

[32] https://www.ebi.ac.uk/ipd/mhc/.

[33] https://www.ebi.ac.uk/patentdata/proteins/.

[34] https://www.ebi.ac.uk/ena/.

[35] https://ftp.ensembl.org/pub/current_embl/.

```
ID   U49845; SV 1; linear; genomic DNA; STD; FUN; 5028 BP.
XX
AC   U49845;
XX
DT   07-MAY-1996 (Rel. 47, Created)
DT   09-NOV-2018 (Rel. 138, Last updated, Version 6)
XX
DE   Saccharomyces cerevisiae TCP1-beta gene, partial cds; and Axl2p (AXL2) and
DE   Rev7p (REV7) genes, complete cds.
XX
KW   .
XX
OS   Saccharomyces cerevisiae (baker's yeast)
OC   Eukaryota; Fungi; Dikarya; Ascomycota; Saccharomycotina; Saccharomycetes;
OC   Saccharomycetales; Saccharomycetaceae; Saccharomyces.
XX
RN   [1]
RP   1-5028
RX   PUBMED; 8846915.
RA   Roemer T., Madden K., Chang J., Snyder M.;
RT   "Selection of axial growth sites in yeast requires Axl2p, a novel plasma
RT   membrane glycoprotein";
RL   Genes Dev. 10(7):777-793(1996).
XX
RN   [2]
RP   1-5028
RA   Roemer T.;
RT   ;
RL   Submitted (22-FEB-1996) to the INSDC.
RL   Biology, Yale University, New Haven, CT 06520, USA
XX
DR   MD5; f152907ff924e11e159c909e145a77dd.
DR   EuropePMC; PMC1242257; 16143041.
DR   EuropePMC; PMC3667078; 23734176.
DR   EuropePMC; PMC4393579; 25888209.
XX
FH   Key             Location/Qualifiers
FH
FT   source          1..5028
FT                   /organism="Saccharomyces cerevisiae"
FT                   /chromosome="IX"
FT                   /mol_type="genomic DNA"
FT                   /db_xref="taxon:4932"
FT   mRNA            <1..>206
FT                   /product="TCP1-beta"
FT   CDS             <1..206
FT                   /codon_start=3
FT                   /product="TCP1-beta"
FT                   /db_xref="GOA:P39076"
FT                   /db_xref="InterPro:IPR002194"
FT                   /db_xref="InterPro:IPR002423"
FT                   /db_xref="InterPro:IPR012716"
FT                   /db_xref="InterPro:IPR017998"
```

Fig. 4.11 EMBL-bank sequence Flatfile-header section

```
FT                   /gene="AXL2"
FT   mRNA            <687..>3158
FT                   /gene="AXL2"
FT                   /product="Axl2p"
FT   CDS             687..3158
FT                   /codon_start=1
FT                   /gene="AXL2"
FT                   /product="Axl2p"
FT                   /note="plasma membrane glycoprotein"
FT                   /db_xref="GOA:P38928"
FT                   /db_xref="InterPro:IPR006644"
FT                   /db_xref="InterPro:IPR008009"
FT                   /db_xref="InterPro:IPR013783"
FT                   /db_xref="InterPro:IPR014805"
FT                   /db_xref="InterPro:IPR015919"
FT                   /db_xref="SGD:S000001402"
FT                   /db_xref="UniProtKB/Swiss-Prot:P38928"
FT                   /protein_id="AAA98666.1"
FT                   /translation="MTQLQISLLLTATISLLHLVVATPYEAYPIGKQYPPVARVNESFT
FT                   FQISNDTYKSSVDKTAQITYNCFDLPSWLSFDSSSRTFSGEPSSDLLSDANTTLYFNVI
FT                   LEGTDSADSTSLNNTYQFVVTNRPSISLSSDFNLLALLKNYGYTNGKNALKLDPNEVFN
FT                   VTFDRSMFTNEESIVSYYGRSQLYNAPLPNWLFFDSGELKFTGTAPVINSAIAPETSYS
FT                   FVIIATDIEGFSAVEVEFELVIGAHQLTTSIQNSLIINVTDTGNVSYDLPLNYVYLDDD
FT                   PISSDKLGSINLLDAPDWVALDNATISGSVPDELLGKNSNPANFSVSIYDTYGDVIYFN
FT                   FEVVSTTDLFAISSLPNINATRGEWFSYYFLPSQFTDYVNTNVSLEFTNSSQDHDWVKF
FT                   QSSNLTLAGEVPKNFDKLSLGLKANQGSQSQELYFNIIGMDSKITHSNHSANATSTRSS
FT                   HHSTSTSSYTSSTYTAKISSTSAAATSSAPAALPAANKTSSHNKKAVAIACGVAIPLGV
FT                   ILVALICFLIFWRRRRENPDDENLPHAISGPDLNNPANKPNQENATPLNNPFDDDASSY
FT                   DDTSIARRLAALNTLKLDNHSATESDISSVDEKRDSLSGMNTYNDQFQSQSKEELLAKP
FT                   PVQPPESPFFDPQNRSSSVYMDSEPAVNKSWRYTGNLSPVSDIVRDSYGSQKTVDTEKL
FT                   FDLEAPEKEKRTSRDVTMSSLDPWNSNISPSPVRKSVTPSPYNVTKHRNRHLQNIQDSQ
FT                   SGKNGITPTTMSTSSSDDFVPVKDGENFCWVHSMEPDRRPSKKRLVDFSNKSNVNVGQV
FT                   KDIHGRIPEML"
FT   gene            complement(<3300..>4037)
FT                   /gene="REV7"
FT   mRNA            complement(<3300..>4037)
FT                   /gene="REV7"
FT                   /product="Rev7p"
FT   CDS             complement(3300..4037)
FT                   /codon_start=1
FT                   /gene="REV7"
FT                   /product="Rev7p"
FT                   /db_xref="GOA:P38927"
FT                   /db_xref="InterPro:IPR003511"
FT                   /db_xref="InterPro:IPR036570"
FT                   /db_xref="SGD:S000001401"
FT                   /db_xref="UniProtKB/Swiss-Prot:P38927"
FT                   /protein_id="AAA98667.1"
FT                   /translation="MNRWVEKWLRVYLKCYINLILFYRNVYPPQSFDYTTYQSFNLPQF
FT                   VPINRHPALIDYIEELILDVLSKLTHVYRFSICIINKKNDLCIEKYVLDFSELQHVDKD
FT                   DQIITETEVFDEFRSSLNSLIMHLEKLPKVNDDTITFEAVINAIELELGHKLDRNRRVD
FT                   SLEEKAEIERDSNWVKCQEDENLPDNNGFQPPKIKLTSLVGSDVGPLIIHQFSEKLISG
FT                   DDKILNGVYSQYEEGESIFGSLF"
```

Fig. 4.12 EMBL-bank sequence Flatfile-feature table data

Figure 4.14 reveals the difference of accession number from aforementioned two formats, Let's if accession number is NC_027264, "NC" indicates that sequence is complete genomic molecule from reference assembly and numeral value represents unique identification number.

```
SQ   Sequence 5028 BP; 1510 A; 1074 C; 835 G; 1609 T; 0 other;
     gatcctccat atacaacggt atctccacct caggtttaga tctcaacaac ggaaccattg        60
     ccgacatgag acagttaggt atcgtcgaga gttacaagct aaaacgagca gtagtcagct       120
     ctgcatctga agccgctgaa gttctactaa gggtggataa catcatccgt gcaagaccaa       180
     gaaccgccaa tagacaacat atgtaacata tttaggatat acctcgaaaa taataaaccg       240
     ccacactgtc attattataa ttagaaacag aacgcaaaaa ttatccacta tataattcaa       300
     agacgcgaaa aaaaaagaac aacgcgtcat agaacttttg gcaattcgcg tcacaaataa       360
     attttggcaa cttatgtttc ctcttcgagc agtactcgag ccctgtctca agaatgtaat       420
     aatacccatc gtaggtatgg ttaaagatag catctccaca acctcaaagc tccttgccga       480
     gagtcgccct cctttgtcga gtaattttca cttttcatat gagaacttat tttcttattc       540
     tttactctca catcctgtag tgattgacac tgcaacagcc accatcacta gaagaacaga       600
     acaattactt aatagaaaaa ttatatcttc ctcgaaacga tttcctgctt ccaacatcta       660
     cgtatatcaa gaagcattca cttaccatga cacagcttca gatttcatta ttgctgacag       720
     ctactatatc actactccat ctagtagtgg ccacgcccta tgaggcatat cctatcggaa       780
     aacaataccc cccagtggca agagtcaatg aatcgtttac atttcaaatt tccaatgata       840
     cctataaatc gtctgtagac aagacagctc aaataacata caattgcttc gacttaccga       900
     gctggctttc gtttgactct agttctagaa cgttctcagg tgaaccttct tctgacttac       960
     tatctgatgc gaacaccacg ttgtatttca atgtaatact cgagggtacg gactctgccg      1020
     acagcacgtc tttgaacaat acataccaat ttgttgttac aaaccgtcca tccatctcgc      1080
     tatcgtcaga tttcaatcta ttggcgttgt taaaaaacta tggttatact aacggcaaaa      1140
     acgctctgaa actagatcct aatgaagtct tcaacgtgac ttttgaccgt tcaatgttca      1200
     ctaacgaaga atccattgtg tcgtattacg gacgttctca gttgtataat gcgccgttac      1260
     ccaattggct gttcttcgat tctggcgagt tgaagtttac tgggacggca ccggtgataa      1320
     actcggcgat tgctccagaa acaagctaca gttttgtcat catcgctaca gacattgaag      1380
     gattttctgc cgttgaggta gaattcgaat tagtcatcgg ggctcaccag ttaactacct      1440
     ctattcaaaa tagtttgata atcaacgtta ctgacacagg taacgtttca tatgacttac      1500
     ctctaaacta tgtttatctc gatgacgatc ctatttcttc tgataaattg ggttctataa      1560
     acttattgga tgctccagac tgggtggcat tagataatgc taccatttcc gggtctgtcc      1620
     cagatgaatt actcggtaag aactccaatc ctgccaattt ttctgtgtcc atttatgata      1680
     cttatggtga tgtgatttat ttcaacttcg aagttgtctc cacaacggat ttgtttgcca      1740
     ttagttctct tcccaatatt aacgctacaa ggggtgaatg gttctcctac tattttttgc      1800
     cttctcagtt tacagactac gtgaatacaa acgtttcatt agagtttact aattcaagcc      1860
     aagaccatga ctgggtgaaa ttccaatcat ctaatttaac attagctgga gaagtgccca      1920
     agaatttcga caagctttca ttaggtttga aagcgaacca aggttcacaa tctcaagagc      1980
```

Fig. 4.13 EMBL-bank sequence Flatfile-sequence information

4.2 Primary Sequences Databases—GENBANK, EMBL-BANK and DDBJ

Archival primary sequence databases store raw sequence data, often derived from experimental result and include some interpretation and explanation. However, these databases lack curation and suffer from redundancies as identical sequences may be submitted under varying names by various laboratories. The majority of protein sequences in these primary databases are the result of computational translation of open reading frames (ORFs) and lack extensive experimental verification. Currently, three main primary databases namely; GenBank, EMBL-Bank and DDBJ collectively encompass all publicly available sequence data. These three databases are interconnected,sharing the data to ensure efficient accessability.

Table 4.5 Two-letter abbreviations prefix of RefSeq accession number

Abbreviation	Molecule type	Description
AC_	Genomic	Complete genomic molecule, usually alternate assembly
NC_	Genomic	Complete genomic molecule, usually reference assembly
NG_	Genomic	Incomplete genomic region
NT_	Genomic	Contig or scaffold, clone-based or WGS[a]
NW_	Genomic	Contig or scaffold, primarily WGS
NW_	Genomic	Contig or scaffold, primarily WGS
NZ_	Genomic	Complete genomes and unfinished WGS data
NM_	mRNA	Protein-coding transcripts (usually curated)
NR_	RNA	Non-protein-coding transcripts
XM_	mRNA	Predicted model protein-coding transcript
XR	RNA	Predicted model non-protein-coding transcript
AP_	Protein	Annotated on AC_ alternate assembly
NP_	Protein	Associated with an NM_ or NC_ accession
YP_	Protein	Annotated on genomic molecules without an instantiated transcript record
XP_	Protein	Predicted model, associated with an XM_ accession
WP_	Protein	Non-redundant across multiple strains and species

[a] Whole genome shotgun sequence data

4.2.1 *GENBANK*

GenBank, also known as Los Alamos Sequence Database, was introduced in 1979 at Los Alamos National Laboratory. In 1982, it became a public database and during 1989 to 1992, it was merged with NCBI,[36] which is a consortium of multiple databases.The database is consistently growing, doubling in size every 18 months [89, 205].

4.2.1.1 Sequence Submission to NCBI/GenBank

During initial phase of database development, data was manually incorporated by extracting sequences from relevant publications. The laborious process evolved in 1993, when GenBank introduced direct submission of sequence information, bypassing requirement of prior publication. Nowadays, new inventions of gene or mRNA are first submitted to any primary databases and their accession number is required in scientific publications.

The GenBank database facilitates the submission of sequences through 2 web-based tool (Banklt, Sequin) and a command-line tool tbl2asn. In the past, Banklt[37] potential was limited as it offered only submission of one exon at a time. Each

[36] http://www.ncbi.nlm.nih.gov/.

[37] https://www.ncbi.nlm.nih.gov/Banklt/oldbankit.html.

```
NCBI Reference Sequence: NC_027264.1
FASTA   Graphics

Go to:
                                                      GenBank Division     Date
LOCUS       NC_027264             78917 bp    DNA     circular PLN 03-APR-2023
DEFINITION  Saccharomyces cerevisiae isolate NCYC3594 mitochondrion, complete
            genome.
ACCESSION   NC_027264   RefSeq Accession Number
VERSION     NC_027264.1   Version
DBLINK      BioProject: PRJNA927338
KEYWORDS    RefSeq.
SOURCE      mitochondrion Saccharomyces cerevisiae (brewer's yeast)
  ORGANISM  Saccharomyces cerevisiae
            Eukaryota; Fungi; Dikarya; Ascomycota; Saccharomycotina;
Published   Saccharomycetes; Saccharomycetales; Saccharomycetaceae;
Reference 1 Saccharomyces.
REFERENCE   1  (bases 1 to 78917)
  AUTHORS   Wolters,J.F., Chiu,K. and Fiumera,H.L.
  TITLE     Population structure of mitochondrial genomes in Saccharomyces
Published   cerevisiae
Reference 2
  JOURNAL   Unpublished
REFERENCE   2  (bases 1 to 78917)
  CONSRTM   NCBI Genome Project
  TITLE     Direct Submission
  JOURNAL   Submitted (18-JUN-2015) National Center for Biotechnology
Published
Reference 3 Information, NIH, Bethesda, MD 20894, USA
REFERENCE   3  (bases 1 to 78917)
  AUTHORS   Wolters,J.F., Chiu,K. and Fiumera,H.L.
  TITLE     Direct Submission
  JOURNAL   Submitted (23-APR-2015) Biological Sciences, Binghamton University,
            4400 Vestal Parkway East, Binghamton, NY 13902, USA
```

Fig. 4.14 RefSeq sequence flatfile format with accession number difference

```
COMMENT     PROVISIONAL REFSEQ: This record has not yet been subject to final
            NCBI review. The reference sequence is identical to KR260476.
                                          Information from where sequence obtaines
            ##Assembly-Data-START##
            Assembly Method       :: MIRA v. 3.4.1
            Coverage              :: 44
            Sequencing Technology :: PacBio
            ##Assembly-Data-END##
            COMPLETENESS: full length.
```

Fig. 4.15 RefSeq sequence flatfile format with comment section difference

submission underwent assignment of a unique accession number that represents a sequential and individualized process for incorporating genetic information into the database. Currently, a set of sequences can be submitted at the same time. That allows to submit entire sequence encompassing both exons and introns by assigning a proper identifier for each segment at submission time.

For complex submissions involving long sequences, multiple annotations, or studies in phylogenetics and population genetics, Sequin submission tool[38] is

[38] http://www.ncbi.nlm.nih.gov/Sequin/.

recommended. Each Sequin file should contain fewer than 10,000 sequences for optimal performance. Larger submissions are advised to be made using tbl2asn.[39] Unlike BankIt, which operates through the web, Sequin and tbl2asn are standalone submission tools compatible with Mac, PC, and UNIX platforms. Submitter can download Sequin or tbl2asn, work offline to prepare the submission in the required format, and then proceed with the final submission.

4.2.2 EMBL-BANK

In July 1974 nine European countries and Israel established, European Molecular Biology Laboratory (EMBL)[40] through an intergovernmental treaty. In 1918, transitioned to EMBL Data Library became central depository of nucleotide sequences. Later, in 1993, it evolved to EBI and became precursor of present EMBL-Bank also known as EMBL nucleotide sequence database or EMBL Database. European Nucleotide Archive(ENA) also collaborated with EMBL Database to widen the scope and now it contains three databases namely: (a) Sequence Read Archive(SRA), (b) Trace Archive, (c) EMBL-Bank. In a nutshell, EMBL-EBI serve as central hub of databases and tools which can be easily accessible through EMBL Services.[41]

4.2.2.1 Sequence Submission to ENA/EMBL-Bank

Earlier, ENA/EMBL-Bank was restricted to record sequence length having 350,000 base pairs. To accommodate longer genomes, sequences were splitted into sub-sequences and stored as multiple entries in database [505]. This restriction was lifted after few years and allowed submission of sequence with any length by year 2013. Currently, Webin, a web-based platform, is used for sequence submission at EMBL-Bank. It allows submission of single and multiple sequences as well as very large number of sequences(bulk submissions), links and directions.[42] Moreover, certain genomes that were previously split to meet 350,000-bp limit, have now been updated as a single entry in EMBL-Bank.

[39] http://www.ncbi.nlm.nih.gov/genbank/tbl2asn2/.

[40] https://www.ebi.ac.uk/.

[41] http://www.ebi.ac.uk/services.

[42] https://www.ebi.ac.uk/ena/about/embl_bank_submissions.

4.2.3 DDBJ

DDBJ is only nucleotide sequence data bank in Asia which is established in 1986. It is maintained at National Institutes of Genetics at Mishima, Japan. Recent improvements and added features of DDBJ are discussed in number of research studies [500, 531, 532, 775]. Recently, DDBJ's sequence submission webpage[43] had a complete makeover. Initially, Sakura had been used to submit the sequences for 17 years till November 2012 and it was replaced by online submission faculty termed as Nucleotide Sequence Submission System (NSSS).[44] To facilitate long sequences submission or for simultaneously submission of multiple sequence, DDBJ introduces Mass Submission System (MSS).[45] Moreover, it [532, 575] maintains a DDBJ Trace Archive(DTA) and Sequence Read Archive(SRA) housing permanent archive of base calls, quality estimates and DNA sequence chromatogram (traces) for single-pass reads generated from large-scale sequencing.

4.3 Redundancy and Contamination in Databases (GenBank, EMBL-Bank and DDBJ)

All three databases namely GenBank, EMBL-Bank and DDBJ go through nucleotides and amino acids sequence submission process. At the time of submission, an accession number is assigned to each sequence as per prior agreement with INSDC collaboration which is a unique identifier for a sequence record. It is the combination of letter(s) and numbers such as AF50301, AAG67541, AL672111 etc. The letters of an accession number indicates the database where the sequence information is originally submitted. The format of accession number for nucleotide and protein sequences are distinct as:

Nucleotide: 1 letters + 5 numerals (e.g. J00750) or 2 letter + 6 numerals (e.g.AF208545)

Protein: 3 letters + 5 numerals (e.g. AAG60350)

Submitted sequence may be contaminated reflecting inaccurate genetic information of the original organism or organelle, due to contamination from an external source. Various reasons of contamination are discussed on NCBI[46] web page. To help the submitter to check their cloned sequence for possible contamination with the vector

[43] https://www.ddbj.nig.ac.jp/faq/datasub-e.html.

[44] https://www.ddbj.nig.ac.jp/sub/websub-e.html.

[45] https://www.ddbj.nig.ac.jp/sub/mss_flow-e.html.

[46] http://www.ncbi.nlm.nih.gov/VecScreen/contam.html.

sequences, NCBI offers VecScreen[47] program, that compare submitted sequence with UniVec vector sequence database. It also detects contamination with many of the adapters, linkers and PCR primers commonly employed in the most popular cDNA cloning strategies.

The sequence information of a specific gene/mRNA can be submitted by multiple authors in the database (GenBank, EMBL-Bank and DDBJ) as different groups may end up cloning gene/mRNA resulting in redundant entries.

4.3.1 RefSeq Database

RefSeq database was established to resolve the problem of redundancy and other potential errors prevalent in primary databases. It contains rich, annotated and non-redundant nucleotide and amino acid sequences, providing a single record for each organism including viruses, bacteria and eukaryotes etc. Prime objective of RefSeq is to create one non-redundant sequence by integrating all relevant information of a single sequence with different accession number from different databases.

4.3.2 INSDC

The INSDC,[48] a collaborative consortium, was initiated between GenBank, EMBL (ENA) and DDBJ. Initially, INSDC has maintained the primary nucleotide sequence database [742], for 30 years. Moreover, it has a policy of providing free and unrestricted access of available data to researcher [119].

4.3.2.1 Sequence Submission to INSDC

The Sequence Read Archive (SRA), which is run by the INSDC with assistance from the NCBI, EMBL-EBI, and DDBJ, was created as a public repository for next-generation sequence data. You can reach the SRA from the NCBI,[49] from EBI,[50] and from DDBJ.[51]

[47] http://www.ncbi.nlm.nih.gov/VecScreen/VecScreen.html.

[48] http://www.insdc.org/.

[49] https://www.ncbi.nlm.nih.gov/Traces/sra.

[50] https://www.ebi.ac.uk/ena.

[51] https://trace.ddbj.nig.ac.jp.

4.4 Data Retrieval Systems

This section provides details of 5 different retrieval systems namely; Entrez/GQuery, Entrez Web Interface, Entrez E-utils, Entrez Direct and DBGET.

4.4.1 Entrez/GQuery

The NCBI devised user-friendly, resourceful, text-based search and data retrieval system known as Entrez. Entrez takes keywords as an input and subsequently retrieves data specific to the keywords from public databases associated with the NCBI consortium. To retrieve desired data, Entrez can be utilized in three different ways namely: (a) Entrez Web interface, (b) Entrez E-utilities (web API method), (c) Entrez Direct toolset. Following subsections provide a brief description of all three ways.

4.4.2 Entrez Web Interface

Figure 4.16 illustrates the homepage of Entrez web Interface (https://www.ncbi.nlm.nih.gov/Entrez/ or https://www.ncbi.nlm.nih.gov/search/). In search field of interface, keywords related to species, diseases and genes or accession numbers can be searched. Keywords based search will furnish all databases that contain relevant data. From retrieved databases users can get more detail by clicking on desired database.

Figure 4.17 depicts a toy example of usage of Entrez web interface method for keyword Mus musculus Slco1a6. It can be seen that results of Mus musculus Slco1a6 related to literature, genes, proteins and genomes are obtained from several databases on a single platform.

4.4.3 Entrez E-utils

Another method for utilizing the Entrez tool is to interact with it through a standardized URL syntax. It offers a URL that enables users to perform either a search or retrieve data using the "e-search" or "e-fetch" method.

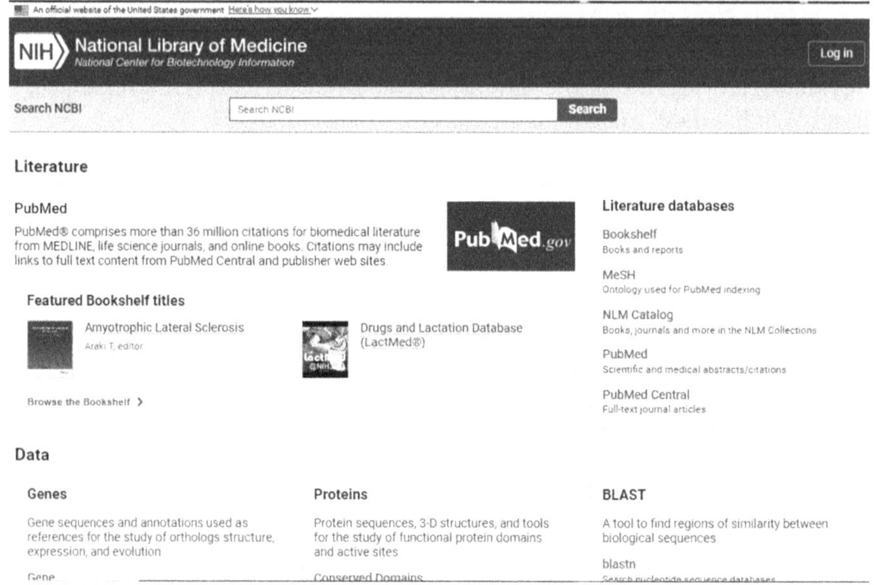

Fig. 4.16 Entrez homepage

4.4.3.1 e-Search

e-Search outputs a XML response corresponding to queries. Following URL is used to search data.
https://eutils.ncbi.nlm.nih.gov/entrez/eutils/esearch.fcgi?db=<databasename>/& term=<Accessionnumber>or<Diseasename>or<speciename>or<gene>

In above URL, https://eutils.ncbi.nlm.nih.gov/entrez/eutils/ is used to access Entrez e-utils, while "esearch.fcgi" represents the queries in only searched and not fetch yet. Moreover, different queries such as database name or terms can be searched. Following examples describe working of e-utils using esearch method with accession number and disease name. If database is "nucleotide" and accession number is "MW015936", search request to server can be written as follows. https://eutils.ncbi.nlm.nih.gov/entrez/eutils/esearch.fcgi?db=nucleotide& term=MW015936

The output obtained for aforementioned query is shown in Fig. 4.18.

4.4.3.2 efetch

efetch is a method used for data retrieval. By using this method, data can be downloaded in two formats namely "fasta" and "genbank". following URL, similar to esearch method is used to fetch data.
https://eutils.ncbi.nlm.nih.gov/entrez/eutils/efetch.fcgi?db=<database>&id=<uid_ list>&rettype=<retrieval_type>&retmode=<retrieval_mode>

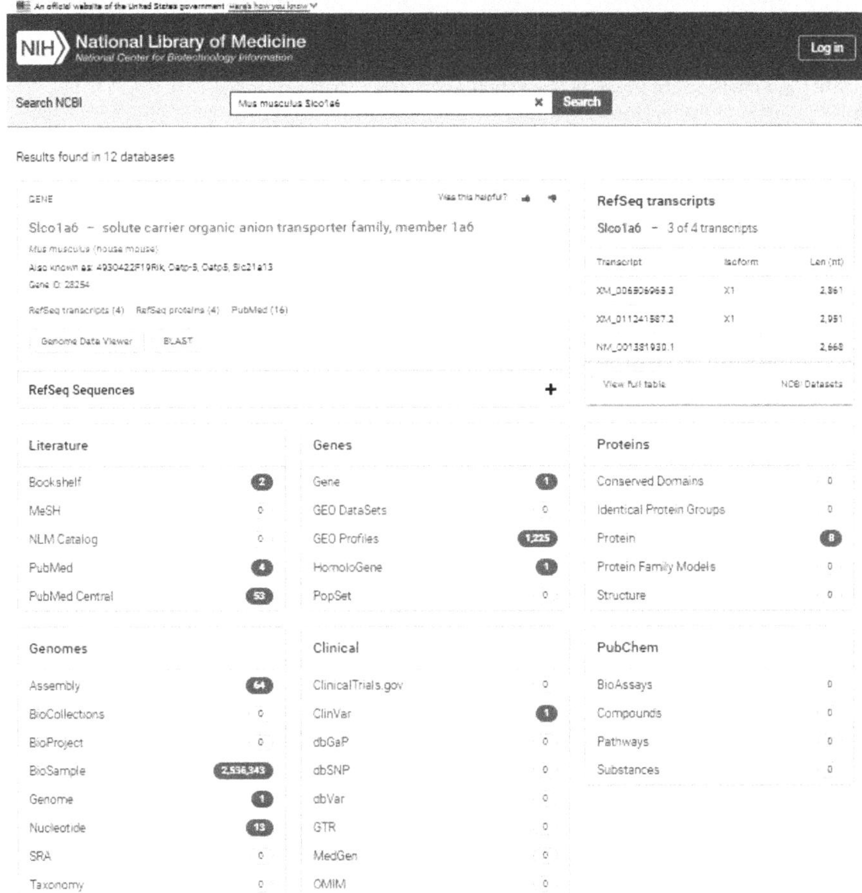

Fig. 4.17 Entrez search for “Mus musculus Slco1a6”

Fig. 4.18 e-Search results using E-utils with accession number MW015936

```
▼<eSearchResult>
   <Count>1</Count>
   <RetMax>1</RetMax>
   <RetStart>0</RetStart>
  ▼<IdList>
     <Id>1913350235</Id>
   </IdList>
   <TranslationSet/>
   <QueryTranslation/>
 </eSearchResult>
```

```
>MW015936.1 Zika virus isolate Zika virus/H.sapiens-tc/THA/2006/CVD_06-020, complete genome
AGTTGTTGATCTGTGTGAATCAGACTGCGACAGTTCGAGTTTGAAGCGAAAGCTAGCAACAGTATCAACA
GGTTTTATTTTGGATTTGGAAACGAGAGTTTCTGGTCATGAAAAACCCAAAGAAGAAATCCGGAGGATTC
CGGATTGTCAATATGCTAAAACGCGGAGTAGCCCGTGTGAGCCCCTTTGGGGGCTTGAAGAGGCTGCCAG
CCGGACTTCTGCTGGGTCATGGGCCCATCAGGATGGTCTTGGCGATTCTAGCCTTTTTGAGATTCACGGC
AATCAAGCCATCACTGGGTCTCATCAATAGATGGGGTTCAGTGGGGAAAAAAGAGGCTATGGAAATAATA
AAGAAGTTCAAGAAAGATCTGGCTGCCATGCTGAGAATAATCAATGCTAGGAAGGAGAAGAAGAGACGAG
GCACAGATACTAGTGTCGGAATTGTTGGCCTCCTGCTGACCACAGCCATGGCAGTGGAGGTCACTAGACG
TGGGAGTGCATACTATATGTACTTGGACAGAAGTGATGCTGGGGAGGCCATATCTTTTCCAACCACACTG
GGGATGAATAAGTGTTATATACAGATCATGGATCTTGGACACATGTGTGATGCCACCATGAGCTATGAAT
GCCCTATGCTGGATGAGGGGGTAGAACCAGATGACGTCGATTGTTGGTGCAACACGACGTCAACTTGGGT
TGTGTACGGAACCTGCCACCACAAAAAAGGTGAAGCACGGAGATCTAGAAGAGCTGTGACGCTCCCCTCC
```

Fig. 4.19 e-Fetch results using E-utils with keyword “zika virus”

In above URL, database name, accession number and format type is required to retrieve data and two different data formats namely fasta and genbank can be downloaded using this hyperlink. Figure 4.19 illustrates fasta sequence file downloaded from using aforementioned URL.

4.4.4 Entrez Direct

It is a window terminal method which executes commands and their corresponding arguments to retrieve data. Users need to install this toolkit locally on their operating systems. For UNIX operating system, Entrez Direct can be installed by following two steps as given below.

- Run any one of the following commands on terminal window
  sh -c “(curl -fsSL https://ftp.ncbi.nlm.nih.gov/entrez/entrezdirect/install-edirect.sh)” or
  sh -c “(wget -q https://ftp.ncbi.nlm.nih.gov/entrez/entrezdirect/install-edirect.sh)”
- Once it is done, configure the PATH for active terminal session by using following command
  `export PATH=HOME/edirect:PATH`

Genomic data is searched and retrieved by using esearch and efetch commands (as in E-utils) on terminal. Both commands requires database name, and any one from “name of specie/ disease” or “accession number”. Table 4.6 describes commands along with their defined tasks.

4.4.5 DBGET

DBGET, a command-line toolkit associated with GenomeNet using DDBJ platform, is used for data retrieval purposes. Users mainly use DBGET tool to retrieve specific

Table 4.6 Usage of Enterz Direct commands

Commands	Description
esearch -db **Database** -query **Accession Number**	Searches data by providing database name and accession number
esearch -db **Database** -query **Disease Name**	Searches data by providing database name and disease name
esearch -db **Database** -query **Specie Name**	Searches data by providing database name and specie name
esearch -db **Database** -query **Gene**	Searches data by providing database name and gene name
efetch -db **Database** -format **gb** -id **Accession Number** >filename.gb	Downloads filename.gb file in GenBank format by providing database name and accession number
efetch -db **Database** -format **fasta** -id **Accession Number** >filename.fa	Downloads filename.fa in fasta format by providing database name and accession number

entries or data by providing accession numbers or other relevant information about nucleotide sequences, genes or other biological data. DBGET also entails entries from KEGG database but subscription to KEGG is needed to retrieve data.

4.4.5.1 DBGET Toolkit

This toolkit is manually installed in operating systems and DBGET installation process on UNIX operating systems involves 5 different steps.

- First of all, download DBGET software by using https://www.genome.jp/ftp/tools/dbget/dbget.6.5.tar.gz link and execute `/wget` <URL> or `/curl` <URL> command on terminal for package installation as shown below.
  `./wget`https://www.genome.jp/ftp/tools/dbget/dbget.6.5.tar.gz
  or
  `./curl`https://www.genome.jp/ftp/tools/dbget/dbget.6.5.tar.gz
- Downloaded folder is a zip folder namely "`dbget.6.5.tar.gz`" which is then extracted by running `./tar -xf dbget.6.5.tar.gz` command on terminal
- Then DBGET package is explored by using `./cd dbget-6.5` command
- Afterwards, configure all installation files for REST framework by executing `./configure -rest` and then "`make`" commands on terminal
- Finally, call the bin directory where all commands are present by running `./cd bin/` command

There are three basic commands namely (a) `bfind`, (b) `bget`, (c) `blink` which are used for searching and retrieving database entries by DBGET.

Table 4.7 Utilization of bget commands

Commands	Descriptions
./bget **database:accession number**	Searches entities by specifying the name of database and accession number
./bget **-format** f **database:accession number**	Searches entities in fasta format by specifying the name of database and accession number
./bget **-format** f **-nucleotide** n n **database:accession number**	Searches entities nucleotide sequence in fasta format by specifying the name of database and accession number
./bget **-format** f **<filename>**	Downloads filename.fa in fasta format by specifying the name of database and accession number
./bget **-fasta** f **MOL/MDL** m compound/drug- **database:accession number**	Searches chemical structure information for MOL/MDL in fasta format by specifying compound/drug database and accession number

bget Command

bget is often used in context of retrieving sequences from public databases such as GenBank, EMBL, DDBJ, UniProt, RefSeq. Generic syntax of bget command is as follow.

```
./bget [options] [database] [accession_numbers or query]
```

Above syntax consists of three parts. First part is options which can be of three different types such that (a) Entries number `(-n)`, (b) Output file `(-o)`, (c) Log file `(-l)`. Second part includes public database name or its abbreviation depending on different types of commands while third part refers to accession number or biological information related data. Table 4.7 narrates some commands along with their specified tasks which are used to retrieve data from databases using different type of information.

bfind Command

Another command of DBGET is the **bfind** command that searches and retrieves database entries by using keywords. Different queries such as the properties of the species/disease can be found by integrating different keywords. Generic syntax of bget command is as follow.

```
./bfind [options] [database] [keyword]
```

Above syntax consists of three parts. First and second part is exactly similar to bget command. Third parts are keywords which depicts information related to genomic data. Table 4.8 narrates some commands along with their specified tasks which are used to retrieve data from databases using different type of information.

Table 4.8 Exertion of bfind commands

Comments	Descriptions
./bfind **database keyword**	Searches keyword related to species/disease from the specified database
./bfind **database** \(virus **.or.** virus \) **.and. species**	Searches information for the specified keywords using default words: (.and.),(.or.),(.not.)

Table 4.9 Usage of blink commands

Commands	Descriptions
./blink -u **database:accession number**	Searches the cross-reference databases of the specified database and the accession number
./blink -u **database:accession number** -t **tax**	Searches the taxonomic identity of the specified database and the accession number
./blink -u -t **tax <filename >**	Downloads the file containing taxonomic identifiers of the specified databases and the accession numbers

blink Command

The subsequent command is the **blink** command which searches related entries from both the specified database and all other databases. The blink demonstrates the cross-references of the databases for the specified database with the accession number. Table 4.9 illustrates the commands along with their prescribed tasks.

4.4.5.2 DBGET Web Interface

bget and bfind command can also be used retrieve data using web interface. Using bget command, data can be retrieved using following URL.

https://www.genome.jp/dbget-bin/www_bget?-format+Database+Accession number

Entries can be retrieved by providing specified database and accession number with desired format in above mentioned URL. Moreover, format defines the desired format in which data is required. For a better understanding, let's consider a toy example by providing database name and accession number along with desired format in above mentioned URL. For fasta format, RefSeq database and accession number to be "YP009227196", the search request to server can be written as follows.

https://www.genome.jp/dbget-bin/www_bget?-f+RefSeq+YP_009227196

Figure 4.20 depicts result obtained for this query.

It can be seen in Fig. 4.20, sequence corresponding to RefSeq accession number is obtained. In addition, a search engine appears on screen if bget is replaced with bfind in aforementioned URL. Figure 4.21 illustrates search engine obtained after passing following search request to server.

https://www.genome.jp/dbget-bin/www_bfind?-f+RefSeq+YP_009227196

GenomeNet

```
>rs:YP_009227196 [YP_009227196] capsid protein C [Zika virus].
mknpkeeirririvnmlkrgvarvnplgglkrlpaglllghgpirmvlailaflrftaik
pslglinrwgsvgkkeameiikkfkkdlaamlriinarkerkrr
```

Fig. 4.20 bget result using URL with accession number YP009227196

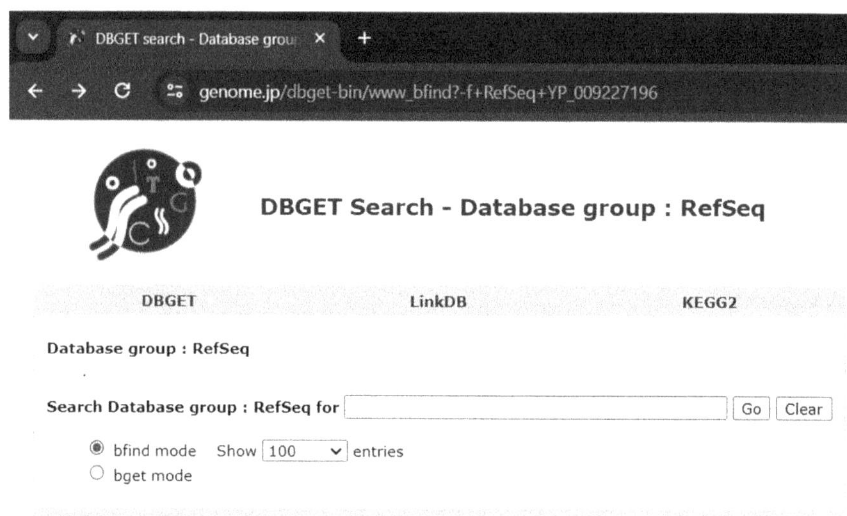

Fig. 4.21 bfind result using URL with accession number YP009227196

4.4.6 Sequence Retrieval System

Sequence retrieval systems (SRS) provide another way for data retrieval in databases. Earlier, many SRS tools were used such as EMB-NET SRS and DKFZ-SRS which are not publicly available now. One of the publicly available Sequence Retrieval System (SRS) server is "dbfetch" which can be accessed through EMBL-EBI platform (http://www.ebi.ac.uk/Tools/dbfetch/dbfetch/). User can retrieve desired sequences by providing information related to database and accession number/s. There are three different methods to utilize DBFetch for retrieving data namely: (a) DBFetch Web Interface, (b) DBFetch using URL, (c) DBFetch toolkit. Following subsection provides a brief description of all three methods.

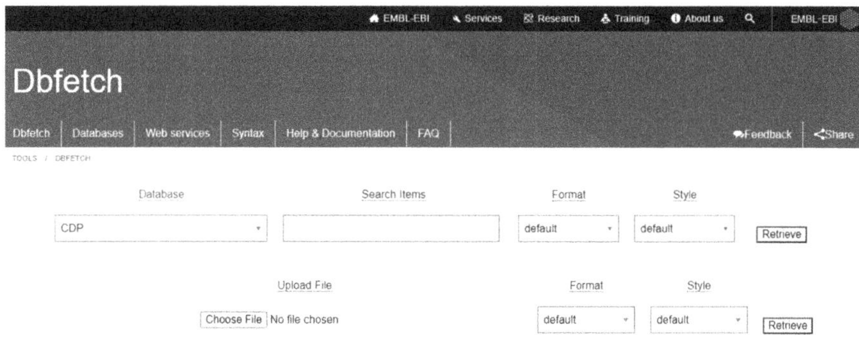

Fig. 4.22 DBFetch homepage

```
>CDP:ERR7457558 not applicable|Ireland:Europe / Ireland / Wicklow|2021-10-12|SARS-CoV-2 complete genome
NNNNNNNNNNNNNNNNNNNNNNNNNNNNNNNNNNNNNNNNNNNNNNNNNNNNNNNNNAAACGAACTTTAA
AATCTGTGTGGCTGTCACTCGGCTGCATGCTTAGTGCACTCACGCAGTATAATTAATAACTAATTACTGTCGTTGACAGG
ACACGAGTAACTCGTCTATCTTCTGCAGGCTGCTTACGGTTTCGTCCGTTTTGCAGCCGATCATCAGCACATCTAGGTTT
TGTCCGGGTGTGACCGAAAGGTAAGATGGAGAGCCTTGTCCCTGGTTTCAACGAGAAAACACACGTCCAACTCAGTTTGC
CTGTTTTACAGGTTCGCGACGTGCTCGTACGTGGCTTTGGAGACTCCGTGGAGGAGGTCTTATCAGAGGCACGTCAACAT
CTTAAAGATGGCACTTGTGGCTTAGTAGAAGTTGAAAAAGGCGTTTTGCCTCAACTTGAACAGCCCTATGTGTTCATCAA
ACGTTCGGATGCTCGAACTGCACCTCATGGTCATGTTATGGTTGAGCTGGTAGCAGAACTCGAAGGCATTCAGTACGGTC
GTAGTGGTGAGACACTTGGTGTCCTTGTCCCTCATGTGGGCGAAATACCAGTGGCTTACCGCAAGGTTCTTCTTCGTAAG
AACGGTAATAAAGGAGCTGGTGGCCATAGTTACGGCGCCGATCTAAAGTCATTTGACTTAGGCGACGAGCTTGGCACTGA
TCCTTATGAAGATTTTCAAGAAAACTGGAACACTAAACATAGCAGTGGTGTTACCCGTGAACTCATGCGTGAGCTTAACG
GAGGGGCATACACTCGCTATGTCGATAACAACTTCTGTGGCCCTGATGGCTACCCTCTTGAGTGCATTAAAGACCTTCTA
GCACGTGCTGGTAAAGCTTCATGCACTTTGTCCGAACAACTGGACTTTATTGACACTAAGAGGGGTGTATACTGCTGCCG
TGAACATGAGCATGAAATTGCTTGGTACACGGAACGTTCTGAAAAGAGCTATGAATTGCAGACACCTTTTGAAATTAAAT
TGGCAAAGAAATTTGACACCTTCAATGGGGAATGTCCAAATTTTGTATTTCCCTTAAATTCCATAATCAAGACTATTCAA
```

Fig. 4.23 DBFetch web interface output for "ERR7457558" accession number

4.4.6.1 DBFetch Web Interface

Figure 4.22 depicts homepage Dbfetch sequence retrieval system (http://www.ebi.ac.uk/Tools/dbfetch/). It can be seen in Figure that there are two different ways to retrieve data from Dbfetch Web interface: (1) One can retrieve data by providing database name, accession number/s, desired format and style (html or raw); (2) A file containing database number along with accession number is uploaded and desired format and style is selected.

Figure 4.23 illustrates a toy example of genomic data in fasta format which is retrieved by providing database name as CDP and accession number information as "ERR7457558" using DBFetch Web Interface.

4.4.6.2 DBFetch URL

Another method for utilizing the Entrez tool is to interact with it through a standardized URL syntax. It offers a URL that enables users to search or retrieve data using following URL

Table 4.10 Usage of DBFetch URLs

URLs	Descriptions
dbfetch?**id**	Retrieves data for specified accession number
dbfetch?db=**database&id=**id1,id2,id3,id4	Retrieves data for multiple accession numbers of specified database
dbfetch?db=**database&id=**id1,id2,id3,id4&**style=**raw	Retrieves data for multiple accession numbers of specified database in raw form
dbfetch?db=**database&id=**id1,id2,id3,id4&**format=**fasta	Retrieves data for multiple accession numbers of specified database in fasta format

https://www.ebi.ac.uk/Tools/dbfetch/dbfetch?db=DB_NAME&id=IDS&format =FORMATNAME&style=STYLE_NAME

In this, https://www.ebi.ac.uk/Tools/dbfetch/ is the base URL used to access dbfetch tool, where "dbfetch" indicates tool. Moreover, four parameters in URL specifies queries such that db refers to database name, id represents accession number/s, format provides format description and style specifies raw or html style. Table 4.10 illustrates different types of parameters that can be used in URL and its description of using dbfetch.

4.4.6.3 DBFetch Toolkit

DBFetch is a command-line toolkit for data retrieving which needs to be installed manually on operating system. DBFetch toolkit is installed using three computer languages namely python, perl and JAVA. Installation procedure of DBFetch on operating system using python language consists of 3 different steps.

- Clone the github repository of DBfetch on terminal window to download DBFetch toolkit using following commands
  1) `./git clone` https://github.com/ebi-wp/webservice-clients.git
  2) `./cd webservice-clients`
- In order to proceed, make sure python environment is actively working on system and navigate cloned file by executing following command
  `./cd python`
- Afterwards, install required python package using command given below
  `./pip install xmltramp2 requests`

After installing DBFetch toolkit on operating system, data from different databases is retrieved using following generic syntax on terminal window.

```
[language] [dbfetch.ext] [Database name] [Accession Number/s]
[format] [style]
```

Table 4.11 Various DBFetch commands to retrieve genomic data

Commands	Descriptions
python dbfetch.py **getSupportedDBs**	Displays list of the databases easily accessible by DBFetch
python dbfetch.py **getSupportedFormats**	Displays a list of databases along with their downloadable formats
python dbfetch.py fetchData **Database**:**SeqID format style**	Retrieves data for accession number of specified database in fasta format
python dbfetch.py fetchData **Database**:**SeqID1,SeqID2 format style**	Retrieves entries for multiple accession numbers of specified database in fasta format
python dbfetch.py fetchData **Database**:**SeqID1,SeqID2 format style >filename.ext**	Downloads raw fasta file of multiple accession numbers with the specified database

In above syntax, first two parameters are language specific; language specifies `language` which is used to install DBFetch toolkit such as JAVA, python or perl and `.ext` extension of file, if DBFetch is installed using python, it is `dbfetch.py` (`dbfetch.pl` for perl and same with java). Remaining parameters namely `Database name`, `Accession Number/s`, `format` and `style` are similar to DBFetch web interface. Table 4.11 illustrates widely used commands along with their task description for python based DBFetch toolkit.

4.5 Classification of Biological Databases

4.5.1 DNA Sequence Analysis Databases

This section provides a comprehensive overview of various databases employed to develop benchmark datasets for development of AI-based applications for 44 distinct DNA sequence analysis tasks. A total of 44 DNA sequence databases have been identified from 127 existing studies. Among these, 35 databases are publicly accessible, while the remaining 9 databases are either inaccessible or no longer exist. To ease the lives of researchers and practitioners, Table 4.12 summarizes accessible databases in terms of their release year, types of inherent genetic data (DNA, RNA, protein), details of species and organisms, statistics of raw sequences, and supported data formats.

A holistic view of the Table 4.12 reveals that, 12 databases provide RNA and Protein sequences as well in addition to providing DNA sequences. AS word embeddings methods and Large language models are trained in unsupervised fashion and when they are trained on large sequence data usually they produce better representations. To efficiently train word embedding methods and large language models, raw data can be acquired from these databases. To facilitate researchers, we have categorized 35 databases into three different categories on

Table 4.12 A summary of publicly accessible biological databases, their inherent data types, species diversity, and statistics of raw sequences related to different genomic and proteomic data

Database name	Release date	Types of data	Species	Organism	Sequences statistics	Data format
SPENCE	2022	ncRNAs	Homo sapiens	–	1700 patient samples, 6800 ncRNA transcripts, 29,526 ncRNA-encoded peptides from 15 cancer types, 8060 tumor-specific peptides, 4497 peptides with potential immunogenicity	.txt
m6A-Atlas v2	2022	mRNAs, lncRNAs, miRNAs	42 species	–	2813 samples, 16,868,200 m6A peaks, 797,091 m6A sites	.txt
RNALocate v2.0	2021	RNA	104 species	–	Number of entries: 213,260, Number of subcellular localization: 171	.txt
GENCODE Release 43	2021	ncRNAs	Animal, Homo sapiens, Mus musculus	–	63,086 genes, 19,411 protein-coding genes, 20,310 lncRNA genes, 7565 ncRNA genes, 14,716 pseudogenes, 254,070 transcripts, 89,581 Protein-coding transcripts, 21,774 Nonsense mediated decay transcripts, 59,927 Long non-coding RNA loci transcripts, 65,650 total No of distinct translations, 13,620 genes that have more than one distinct translations	–
Circad	2020	circRNAs	Homo sapien, Mus musculus, Rattus rattus	–	Number of disease related circRNA: 1388, Number of diseases: 150, No. of circRNAs in: Homo sapiens = 1270, Mus musculus = 66, Rattus rattus = 42	–

MNDR3.0	2020	lncRNAs, piRNAs, circRNAs, miRNAs, tRNAs, snoRNAs	117 species	–	Experimental data: 343,273 all RNA-disease entries, Predicted data: 237,329 entries miRNA-disease information, 348,176 entries lncRNA-disease information, 362,454 entries circRNA-disease information, 48,779 entries piRNA-disease information	.txt
cantataDB 2.0	2020	lncRNAs	39 species	–	239,631 lncRNAs	FASTA, .gtf
EVLncRNAs 2.0	2020	RNA	124 species	–	4010 lncRNAs, 1082 Diseases, 11,257 lncRNA-disease associations, 1665 Function Annotations (excluding interactions), 6244 Interactions, 37 Peptide-coding, 8 Structure, 33 Exosomal, 188 CircRNAs, 1079 Drug/chemoresistance/stress	.xlsx
piRBase	2019	piRNAs	44 species	–	181 million unique piRNA sequences	FASTA, .bed, .csv, .tsv, .json, .txt
EuRBPDB	2019	RBPs	162 species	–	315,222 RBPs	.txt, .fa
PanglaoDB	2019	RNA	Animal, Homo sapiens, Mus musculus	–	Mus musculus: 1063 samples, 184 tissues, 4,459,768 cells, 8651 clusters, Homo sapiens: 305 samples, 74 tissues, 1,126,580 cells, 1248 clusters	.tar

(continued)

Table 4.12 (continued)

Database name	Release date	Types of data	Species	Organism	Sequences statistics	Data format
CSCD	2018	circRNAs	Homo sapiens	–	Samples > 1000, including ~800 tissue samples and ~300 cell line samples, 1,013,461 cancer-specific circRNAs, 1,533,704 circRNAs normal samples and 354,422 circRNAs from both cancer and normal samples	.txt
RefSeq (version 90)	2018	DNA, RNA, Proteins	–	–	23,838,836 entries	.csv, .json
GENCODE.v28	2018	RNA, Proteins	Homo sapiens	–	58,381 Total No of Genes, 19,901 Protein-coding genes, 15,779 Long non-coding RNA genes, 7569 Small non-coding RNA genes, 147,723 Pseudogenes, 10,693—processed pseudogenes, 3519—unprocessed pseudogenes, 218—unitary pseudogenes, 38—polymorphic pseudogenes, 18—pseudogenes, 408 Immunoglobulin/T-cell receptor gene segments–protein coding segments, 237—pseudogenes, 203,835 Total No of Transcripts, 82,335 Protein-coding transcripts, 56,541—full length protein-coding, 25,794—partial length protein-coding, 14,889 Nonsense mediated decay transcripts, 28,468 Long non-coding RNA loci transcripts, 61,132 Total No of distinct translations, 13,641 Genes that have more than one distinct translations	.gtf, .gff, FASTA, .bed, .json, .tsv

GENCODE. vM18	2018	RNA, Proteins	_	Mouse	54,146 Total No of Genes, 21,978 Protein-coding genes, 12,726 Long non-coding RNA genes, 6108 Small non-coding RNA genes, 12,838 Pseudogenes, 9612—processed pseudogenes, 2842—unprocessed pseudogenes, 37—unitary pseudogenes, 79—polymorphic pseudogenes, 65—pseudogenes, 494 Immunoglobulin/T-cell receptor gene segments—protein coding segments, 203—pseudogenes, 136,535 Total No of Transcripts, 57,388 Protein-coding transcripts, 44,118—full length protein-coding, 13,270—partial length protein-coding, 6679 Nonsense mediated decay transcripts, 17,855 Long non-coding RNA loci transcripts, 44,166 Total No of distinct translations, 10,491 Genes that have more than one distinct translations	.gtf, .gff, FASTA, .bed, .json, .tsv
lncRNASNP2	2018	RNA	_	Human, Mouse	10,205,295 SNPs in 141,353 human lncRNA transcripts of 90,062 lncRNA genes, 859,534 Cosmic Noncoding Variations and 315,234 TCGA cancer mutations	.xlsx

(continued)

Table 4.12 (continued)

Database name	Release date	Types of data	Species	Organism	Sequences statistics	Data format
LncRNA Disease v2.0	2018	lncRNAs, circRNAs	Animal, Homo sapiens, Mus musculus, Rattus norvegicus, Gallus gallus	–	19,166 lncRNAs, 823 circRNAs, 529 diseases, 205,959 lncRNA-disease associations, 1004 circRNA-disease associations	.xlsx
CircRNA Disease	2018	circRNAs	12 species	Human, Chicken, Cow, Mouse, Rat	4246 circRNAs, 330 DO diseases, 6998 circRNA-diseases, 7,159,865 mutation-circRNAs	.txt, .xlsx
CircBank	2018	circRNAs, miRNAs	Plants	Human, Mouse, Fly, Worm, Yeast	More than 140,000 human annotated circRNAs, 1439 associations between 1135 circRNAs and 82 cancers	.bed, .txt, .xlsx
bpRNA	2018	RNA	–	–	708,144 hairpins, 517,672 bulges, 317,046 multi loops, 538,670 internal loops, 57,686 pseudoknots, 2,075,928 stems, 229,468 unpaired regions, 1,019,586 segments	FASTA, .pdf, .jpg
RNALocate	2017	RNA	65 species	–	42,190 Number of entries, 41 Number of subcellular localization, 23,100 RNAs	.txt, .xlsx, FASTA
RMBase2.0	2017	miRNAs	Homo sapiens, Mus musculus, Rhesus, Rattus, A.thaliana, S.cerevisiae, P.aeruginosa, Escherichia coli, S.pombe	Chimpanzee, Pig, Zebrafish, Fly	5411 m1A, 988 m5C, 1,373,355 m6A, 5096 2'-O-Me, 9570 pseudoU, 2824 others	.txt
miRmine	2016	miRNAs	Homo sapiens	–	2822 cell lines, 2822 tissues	excel, .csv, .pdf

Circ Interactome	2016	RNA	Homo sapiens	_	No of entries: 65,535	.xlsx
ATract	2016	RBPs	38 species	_	370 RBPs and 1583 RBP consensus binding motifs	.txt, .csv, .tsv
HMDAD	2016	DNA, RNA, Proteins	_	_	483 disease-microbe entries which include 39 diseases and 292 microbes	.txt
dbDEMC	2016	miRNAs	_	Human, Mouse, Rat	3268 miRNAs, 40 cancer types, 149 cancer subtypes, 403 datasets, 807 experiments, 46,388 samples	.txt
NONCODEV5	2016	lncRNAs	Arabidopsis, Caenorhabditis elegans	15 organisms	354,855 lncRNA genes, 548,640 lncRNA transcripts	FASTA
RMBase	2016	RNA	62 species	_	1,074,100 RNA modification, 73 types of RNA	.tar.gz
circRNADb	2015	circRNAs	Homo sapiens	_	32,914 annotated exonic circRNAs	FASTA, .tsv
DisGeNET	2015	DNA, RNA, Protein	Animals	Human	1,134,942 GDAs between 21,671 Genes, 30,170 diseases,and traits, 369,554 VDAs between 194,515 variants and 14,155 diseases and traits	.txt, RDF, SQL Dump
NDB	2014	RNA, DNA, Protein	Homo sapiens	_	17,894 3D structures containing nucleic acids	.csv, .json
LNCipedia	2013	lncRNAs	Homo sapiens	_	127,802 transcripts and 56,946 genes	.bed, FASTA, .gff, .gtf
RefSeq (version 60)	2013	DNA, RNA, Proteins	_	_	4,243,209 entries	.csv, .json

(continued)

Table 4.12 (continued)

doRiNA	2013	RNA	Homo sapiens, Mus Musculus, Caenorhabditis elegans, Drosophila melanogaster	–	–	.bed
lncRNA Disease	2013	lncRNAs, circRNAs	Animal, Homo sapiens, Mus musculus, Rattus norvegicus, Oryctolagus cuniculus	–	6066 lncRNAs, 10,732 circRNAs, 566 diseases, 13,191 lncRNA-disease associations, 12,249 circRNA-disease associations	.tsv, .xlsx
miRCancer	2013	miRNAs	–	34 organisms	57,984 miRNAs, 196 cancers, 9080 miR-Cancers	.txt
Encori	2013	mRNAs, miRNAs, ceRNAs, lncRNAs	23 species	Human, Mouse	2725 CLIP-seq datasets, 100 Degradome-seq datasets, 59 RNA-RNA interactome datasets, RNA-seq data: more than 10,800 samples from 32 cancer types, miRNA-seq data: 10,500 samples from 32 cancer types, Disease data: 1,800,000 mutations from 531 disease types, miRNA-ncRNA(CLIP): 460,000 interactions, miRNA-mRNA(CLIP): 1,200,000 interactions, RBP-mRNA:1,290,000 interactions, RBP-ncRNA: 1,600,000 interactions, RNA-RNA: gt;3,700,000 interactions, miRNA-ncRNA(degradome): 32,000 interactions, miRNA-mRNA(degradome): 459,000 interactions, ceRNA: 2,900,000 pairs, function annotation: gt;34,000 functional terms from 21 categories, Pan-Cancer: Differential Expression, Survival Analysis,CoExpression	.txt, .xlsx

CircBase	2013	circRNAs	Homo sapiens, Mus musculus, Caenorhabditis elegans, Latimeria chalumnae, Latimeria menadoensis	_	Human: 8483 circRNAs, Caenorhabditis elegans: 2399 circRNAs, Drosophila melanogaster: 5795 circRNAs	FASTA, .txt, .xlsx, .bed
EPDnew	2013	RNA	Animals, Plants, Fungi, Invertebrates	_	Animal: 13,1870 promoters, Plants: 39,784 promoters, Fungi: 9919 promoters, Invertebrates: 5597 promoters	.bed, .dat, .fps, .bb, .idx, FASTA
PLncDB 2.0	2013	lncRNAs	80 species	_	1,246,372 lncRNAs, 13,834 RNA-Seq datasets	.fa, .txt, .gff3
ClinVar	2013	DNA, RNA, Protein	Animals	Human	4,391,341 Records, 92,225 Total Genes	.xml, .tsv, VCF
GENCODE. v17	2012	RNA, Proteins	Homo sapiens	_	57,281 Total No of Genes, 20,330 Protein-coding genes, 13,333 Long non-coding RNA genes, 9078 Small non-coding RNA genes, 14,154 Pseudogenes, 29 polymorphic pseudogenes, 13,897 pseudogenes, Immunoglobulin/T-cell receptor gene segments; 386—protein coding segments, 228—pseudogenes, 194,871 Total No of Transcripts, 81,565 Protein-coding transcripts, 56,950 full length protein-coding, 24,615 partial length protein-coding, 12,913 Nonsense mediated decay transcripts, 22,631 Long non-coding RNA loci transcripts, 61,102 Total No of distinct translations, 13,569 Genes that have more than one distinct translations	.gtf, .gff, FASTA, .bed, .json, .tsv

(continued)

Table 4.12 (continued)

Database name	Release date	Types of data	Species	Organism	Sequences statistics	Data format
RNAcentral	2011	ncRNAs	–	–	96,670 sequences	.txt, FASTA, .json
miR2Disease	2009	miRNAs	Animal, Homo sapiens	–	349 miRNAs, 163 diseases, 3273 entries	.txt
HMDD	2008	miRNAs	–	Human	53,530 miRNA-disease association entries which include 1817 human miRNA genes, 79 virus-derived miRNAs, 2360 diseases from 37,090 papers	.txt, .xlsx
dbGap	2007	RNA	–	–	12,815 phenotype datasets, 430,727 datasets, 4.64 million samples	.xml, .csv
HGMD	2007	DNA, RNA, Protein	Animal, Homo sapiens	–	Mutation totals: (public release for academic/non-profits only): 291,339 or HGMD Professional release 2023.4: 504,008	.txt
TarBase	2006	miRNAs	24 species	–	5,878,998 interactions, 103 tissues, 3300 unique miRNAs, 57 cell types	.tsv.gz
Gencode	2006	DNA, RNA, Protein	Animals, Homo sapiens, Mus musculus	–	Homo sapiens: Total Genes = 63,086, Total Transcripts = 254,070, Total distinct Translations = 65,650, Mus musculus: Total Genes = 57,132, Total Transcripts = 149,138, Total distinct Translations = 44,819	.txt
NCBI	2005	DNA, RNA, Protein	Animals, Homo sapiens, Mus musculus	–	35,608 CCDS IDs that correspond to 19,107 Genes, with 48,062 Protein Sequences	FASTA
GtRNAdb	2005	tRNAs	740 species	–	Eukaryota: 599 Number of Genomes, 74,048 Number of tRNA Genes, Archaea: 220 Number of Genomes, 10,476 Number of tRNA Genes, Bacteria: 4038 Number of Genomes, 242,068 Number of tRNA Genes	.fa, .bed, .txt, .gtf, .tsv.gz

NPInter V4.0	2005	lncRNAs, miRNAs, circRNAs, snoRNAs, snRNAs	Homo sapiens, Mus musculus, Saccharomyces cerevisiae, Agrobacterium tumefaciens, Escherichia coli, Caenorhabditis elegans, Drosophila melanogaster, Kaposi sarcoma-associated herpesvirus	_	658,171 lncRNA interactions, 488,025 miRNA interactions, 61,700 snoRNA interactions, 12,789 snRNA interactions, 335 circRNA interactions, 488,315 RNA-Protein interactions	.txt, .xlsx, .tsv
miRBase	2004	miRNAs	_	271 organisms	38,589 hairpin precursors and 48,860 mature microRNAs	.gff3, .dat, FASTA
Rfam	2003	RNA	_	_	4170 families, 3,026,773 regions, ENA 133/134 Rfamseq	.txt, .fa, .tar.gz
CTD	2003	mRNAs	_	632 organisms	2,915,515 Chemical–gene interactions, 406,571 Phenotype–based interactions, 32,694,093 Gene–disease associations, 3,489,469 Chemical–disease associations, 6,577,078 Chemical–GO associations, 1,570,026 Chemical–pathway associations, 305,622 Disease–pathway associations, 1,358,371 Gene–gene interactions, 39,776,068 Gene–GO annotations, 135,792 Gene–pathway annotations, 3,133,281 GO–disease associations, 17,667 Chemicals with curated data, 7285 Diseases with curated data, 55,128 Genes with curated data	.csv, .tsv, .xml

(continued)

Table 4.12 (continued)

Database name	Release date	Types of data	Species	Organism	Sequences statistics	Data format
ENCODE3	2003	scRNAs, siRNAs, miRNAs, small RNAs	Homo sapiens, Mus Musculus, Caenorhabditis elegans, Drosophila melanogaster	–	9000 high-throughput sequencing libraries from assays	.txt, .hic, .fastq, .bed
ENCODE	2003	DNA, RNA, Protein	Animals, Homo sapiens, Mus musculus	–	17,238 sequences	FASTA, BAM, BigWig, .bed, VCF
FANTOM5	2002	lncRNAs, miRNAs, circRNAs, snoRNAs, snRNAs	–	Human, Mouse, Dog, Chicken, Rat, Rhesus Monkey	–	.bed, .txt, .xlsx
GEO	2000	DNA, RNA, Protein	21 species	–	Samples = 7,209,691	SOFT, MINiML, .txt
ENSEMBL	1999	DNA, RNA, Protein	Animals, Homo sapiens, Mus musculus, Danio rerio, Sus scrofa	–	44,048 Genomes, 1014 Ensembl Fungi Genomes, 78 Ensembl Metazoa Genomes for invertebrate species, 236 Genomes for vertebrate Species, 67 Ensembl Plants Genomes, 237 Ensembl Protists Genomes	FASTA, .gtf, .gff, MySQL Dump
KEGG	1995	DNA, RNA, Protein	Animals, Plants, Fungi, Protists, Bacteria, Archaea	14 organisms	Genes: 53,674,741, Addendum Proteins: 4181, Viral Genes: 688,823, Viral mature Peptides: 377	KGML, FASTA, .txt
EMBL-EBI	1994	DNA, RNA, Protein	–	–	–	.xml, FASTA, .txt, .tsv, .json
OMIM	1960	DNA, RNA, Protein	Animals	Homo sapiens	17,290 Gene descriptions, 18 Gene and Phenotypes combined, 6859 Phenotype description molecular basis known, 1502 Phenotype description molecular basis unknown, 1736 mainly Phenotypes with suspected mendelian basis	.txt

the basis of volume of raw sequences: low sequence facilitators, medium sequence facilitators, and high sequence facilitators. Specifically, 13 low sequence facilitators namely HOCOMOCO Human v11 database [544], Consensus Coding Sequence Database [820], MSigDB [621], Broad DepMap [915], JASPAR [311], Database of Essential Genes [1206], ENCODE [942], MGC [288], Exon-Intron Database [884], Ensembl [451], RegulonDB [867], EPD2 [803], OMIM [391], offer up to 100,000 DNA sequences each, while 9 medium sequence facilitators including PPD [956], DREAM [604], EmExplorer database [444], GenomAD [506], DeOri [329], BioLip [1154], DeOri6.0 [329], GWAS [120], Eukaryotic Promoter Database [803], provide up to 1 million DNA sequences. In contrast, 13 high sequence facilitators such as Descartes [127], EnhancerAtlas 2.0 [324], COSMIC [982], DisGeNet [809], ClinVar [555], CCLE [339], GENCODE [315], Gene Ontology [202], DataBase of Transcriptional Start Sites [967], GEO [198], KEGG [502], NCBI [337], and GenBank [87], offer more than 1 million DNA sequences each. These databases predominantly house DNA sequences from a diverse array of species, including humans, mice, plants, bacteria, and fungi. A comprehensive analysis reveals that approximately 22 databases namely Descartes [127], DREAM [604], EmExplorer database [444], COSMIC [982], DisGeNet [809], GenomAD [506], ClinVar [555], HOCOMOCO Human v11 database [544], DeOri [329], BioLip [1154], GWAS [120], Broad DepMap [915], CCLE [339], GENCODE [315], Consensus Coding Sequence Database [820], MSigDB [621], ENCODE [942], DataBase of Transcriptional Start Sites [967], MGC [288], GEO [198], Ensembl [451], and OMIM [391] focus on animal DNA sequences, 4 databases including PPD [956], Database of Essential Genes [1206], RegulonDB and [867] on bacterial sequences, and JASPAR [311] on plant DNA sequences. EnhancerAtlas 2.0 [324] is the only database that facilitate with both animal and bacteria DNA sequences, while 4 databases namely DeOri6.0 [329], Exon-Intron Database [884], EPD2 [803], and Eukaryotic Promoter Database [803] focus on animal and plants DNA sequences, whereas Gene Ontology [202], KEGG [502], and GenBank [87] provide DNA sequences for animal, plant and bacteria. Additionally, sequences from other organisms such as eukaryotes, invertebrates, fungi, and various microorganisms are also well-represented. Some databases encompass a broad spectrum of species. For instance, the EDP2 [803] database includes genomics data for 139 species, GenBank [87] houses sequences for 557,000 species, and PPD [956] has genomics data of 63 species.

A rigorous analysis of Table 4.12 reveals that out of 35 publicly accessible databases, several key categories of data emerge. Four databases namely Broad DepMap [915], genomAD, COSMIC, and MGC, provide data for DNA functional analysis tasks such as prediction of context-specific functional impact of genetic variants and conserved non-coding element classification. Seven databases namely BioLip, HOCOMOCO Human v11, GWAS, EnhancerAtlas 2.0, DataBase of Transcriptional Start Sites, Exon-Intron Database, and Eukaryotic Promoter Database offer data on gene expression regulation. Three databases namely PPD, CCLE, and EmExplorer focus on DNA modification data including methylcytosine and methyladenine modifications. Additionally, DeOri, Descartes, DeOri6.0, and JASPAR provide information on gene structure and stability, including chromatin accessibility prediction, YY1-mediated chromatin loop identification, and DNA replication origins identification. GENCODE, Consensus Coding Sequence Database, MSigDB, Gene Ontology, DisGeNet, Database of Essential Genes, KEGG, and NCBI offer comprehensive gene analysis data. Furthermore, eight other databases namely EPD, ENCODE, RegulonDB, GEO, Ensembl, ClinVar, GenBank, and OMIM provide a range of data on gene expression regulation, DNA modification prediction, genome structure and stability, DNA functional analysis, disease information, and gene analysis.

4.5.2 RNA Sequence Analysis Databases

This section highlights critical role of public databases in facilitating the development of AI-driven RNA-sequence analysis applications. Biological databases house a wealth of RNA information that serves as the foundation for development of benchmark datasets. A comprehensive understanding about contents of RNA molecule related databases may enable researchers to perform large scale AI-driven RNA sequence analysis. Deep understanding of public databases can empower researchers to develop different RNA sequence analysis tasks and distinct species related benchmark datasets. Distinct species datasets of a RNA sequence analysis task is important for conducting cross-species experiments using AI pipelines. This comparative analysis is essential for gaining a broader understanding of biological processes at a more fundamental level.

The ever-expanding nature of public databases facilitates researchers by providing access to increasingly larger data. As new sequences are added, researchers can use expanded data to benchmark the performance of existing AI-driven RNA sequence analysis pipelines. This benchmarking process offers valuable insights into how well current predictors perform with large data and helps researchers in identifying potential areas for improvement and development of more robust AI applications. Moreover, researchers can utilize these databases to acquire large volumes of RNA sequence data. This data can then be used to train word embedding methods and large language models in an unsupervised manner. The pre-trained models can be utilized to develop diverse types of RNA sequence analysis applications. Specifically, this section provides an extensive overview of databases that have been used to create benchmark datasets for 47 distinct RNA sequence analysis tasks. A comprehensive review of 172 research articles focused on AI-driven RNA sequence analysis tasks reveals that a total of 90 distinct databases have been utilized to develop 47 different RNA sequence analysis tasks related benchmark datasets.

From 90 databases, **64** databases are publicly accessible, while the remaining **26** are either inaccessible or no longer exist. To aid research community, Table 4.12 provides a detailed summary of accessible databases in terms of their release year, types of inherent RNA data, species and organisms details, raw sequence statistics, and supported data formats. A thorough analysis of Table 4.12 reveals that out of 64 accessible databases, 6 databases encompass data related to three different types of molecules namely DNA, RNA, and Proteins. Similarly, 2 databases contain data related to Proteins and RNA molecules. Among all accessible databases, 56 databases have dedicated information related to only RNA molecule. Specifically, miRNA sequences are available in 15 different databases namely m6A-Atlas v2 [617], MNDR3.0 [150], CircBank [658], RMBase2.0 [1138], miRmine [796], dbDEMC [1151], miRCancer [1117], Encori [586], miR2Disease [484], HMDD [588], TarBase [896], NPInter V4.0 [983], miRBase [364], ENCODE3 [233], FANTOM5 [763]. Furthermore, long non-coding RNA molecule related diverse types of information is available in 11 databases including m6A-Atlas v2 [617], MNDR3.0 [150], cantataDB 2.0 [971], LncRNADisease v2.0 [166], NONCODEV5 [299], LNCipedia [1022], lncRNADisease [166], Encori [586], PLncDB 2.0, NPInter V4.0 [983], and FANTOM5 [763]. Additionally, 11 databases namely Circad [852], MNDR3.0 [150], CSCD [1108], LncRNADisease v2.0 [166], CircRNADisease [960], CircBank [658], circRNADb [171], lncRNADisease [166], CircBase [348], NPInter V4.0 [983], and FANTOM5 [763] databases provide circular RNA sequences. Similarly, 6 databases (m6A-Atlas v2 [617], Encori [586], CTD [235], MNDR3.0 [150], NPInter V4.0 [983], FANTOM5 [763]) offer mRNA and snoRNA sequences. Also, 6 databases including NPInter V4.0 [983], FANTOM5 [763], MNDR3.0 [150], GtRNAdb [139], piRBase [1059], and ENCODE3 [233] contain information about four distinct RNA molecules namely snRNA, tRNA, piRNA, and siRNA.

Since word embedding methods and large language models are trained on large raw sequences data in an unsupervised manner to generate better representations, these databases can be utilized to efficiently train these language models. To assist researchers and practitioners, we categorized these databases based on the volume of raw sequences into three categories: (1) low sequence facilitators, (2) medium sequence facilitators, and (3) high sequence facilitators. Specifically, 38 low sequences facilitator databases provide 100,000 RNA sequences each and these database include SPENCER [672], m6A-Atlas v2 [617], RNALocate v2.0 [208], Lnc2Cancer v3.0 [327], GENCODE Release 43 [314], circR2Cancer [554], Circad [852], EVLncRNAs 2.0 [1228], PanglaoDB [317], GENCODE.v28 [315], GENCODE v18 [401], LncRNADisease v2.0 [166], CircRNADisease [960], RNALocate [208], miRmine [796], CircInteractome [279], ATtRACT [347], HMDAD [679], dbDEMC[1151], circRNADb [171], NDB [752], lncRNADisease [166], miRCancer [1117], Encori [586], CircBase [348], GENCODE v.17 [892], RNAcentral [79], miR2Disease [484], HMDD[588], TarBase [896], Gencode [400], NCBI [303], miRBase [364], ENCODE3 [233], ENCODE [203], ENSEMBL [1161], OMIM [391]. A total of 11 public databases fall into the "medium sequence facilitators" category and each database contain approximately 1 million sequences. Medium sequences facilitator databases are MNDR3.0 [150], cantataDB 2.0 [971], EuRBPDB [619], CSCD [1108], CircBank [658], NONCODEV5 [299], lncRNA2Target [483], LNCipedia [1022], EPDnew [271], HGMD [949], GtRNAdb

[139]. Whereas, a total of 18 high sequence facilitator databases are piRBase [1059], RefSeq [818], lncRNASNP2 [714], bpRNA [217], RMBase2.0 [1138], RMBase [961], DisGeNET [808], RefSeq (version 60) [776], PLncDB 2.0 [487], ClinVar [555], dbGap [688], NPInter V4.0 [983], Rfam [365], CTD [235], GEO [76], KEGG [502], EMBL-EBI [598], FANTOM5 [763], and doRiNA [38]. These databases predominantly house RNA sequences from a diverse array of species, including humans, mice, plants, bacteria, and fungi.

An extensive analysis of different databases reveals that about 9 databases, such as SPENCER [672], CSCD [1108], GENCODE.v28 [315], miRmine [796], CircInteractome [279], circRNADb [171], NDB [752], LNCipedia [1022], and GENCODE.v17 [892], focus on Homo sapiens RNA sequences, miR2Disease [484], and HGMD [949]. Whereas, OMIM [391] databases provide both homo sapiens and animal RNA sequences. Additionally, 6 databases namely GENCODE Release 43 [314], PanglaoDB [317], Gencode [400], NCBI [303], and ENCODE [203] facilitate Homo sapiens, animals and mus musculus RNA sequence. On the other hand, Circad [852] offers RNA sequences of Homo sapien, Mus musculus, and Rattus rattus. Sequences from other organisms, such as eukaryotes, invertebrates, fungi, and various microorganisms, are also well-represented in this database. Databases can be categorized into three distinct groups based on the variety of species they accommodate; (1) Broad coverage databases, (2) Moderate coverage databases, (3) Limited coverage databases. A total of 33 limited coverage databases facilitate RNA sequences of up to 20 different species including SPENCER [672], GENCODE Release 43 [314], Circad [852], PanglaoDB [317], CSCD [1108], GENCODE.v28 [315], LncRNADisease v2.0 [166], CircBank [658], RMBase2.0 [1138], miRmine [796], CircInteractome [279], NONCODEV5 [299], circRNADb [171], DisGeNET [808], NDB [752], LNCipedia [1022], doRiNA [38], lncRNADisease [166], CircBase [348], EPDnew [271], ClinVar [555], GENCODE.v17 [892], miR2Disease [484], HGMD [949], Gencode [400], NCBI [303], NPInter V4.0 [983], ENCODE3 [233], ENCODE [203], ENSEMBL [1161], KEGG [502], OMIM [391], and CircRNADisease [960].

A total of 9 moderate coverage databases encompass data related to 80 species. Moderate coverage databases are GEO [76], Encori [586], TarBase [896], ATract [347], cantataDB 2.0 [971], m6A-Atlas v2 [617], piRBase [1059], RMBase [961], RNALocate [208]. Whereas, a total of 22 broad coverage databases contain data of more than 80 different species. These databases are PLncDB 2.0 [487], RNALocate v2.0 [208], MNDR3.0 [150], EVLncRNAs 2.0 [1228], EuRBPDB [619], GtRNAdb [139], RefSeq (version 90) [749], GENCODE.vM18 [401], lncRNASNP2 [714], bpRNA [217], HMDAD [679], dbDEMC [1151], RefSeq (version 60) [776], miRCancer [1117], EMBL-EBI [598], RNAcentral [79], HMDD [588], dbGap [688], miRBase [364], Rfam [365], CTD [235], and FANTOM5 [763]. For example, pirbase [1059] offers RNA sequences of 44 species, EuRBPDB [619] houses sequences of 162 species, EVLncRNAs 2.0 [1228] has RNA sequence data of 124 species, RNALocate [208] contains RNA sequences of 104 species, m6A-Atlas v2 [617] houses RNA sequences of 42 species, and MNDR [150] has RNA sequence data of 117 species.

From 64 publicly available databases, RNA categorization and identification tasks related data is available in 13 different databases namely SPENCER [672], cantataDB 2.0 [971], piRBase [1059], EVLncRNAs 2.0 [1228], CSCD [1108], RefSeq (version 90) [749], LNCipedia [1022], RefSeq (version 60) [776], GtRNAdb [139], Rfam [365], circRNADb [171], EPDnew [271], PLncDB 2.0 [487]. Similarly, different RNA interaction and binding sites tasks including RNA-protein binding sites prediction [54], coding RNA–protein interaction prediction, and RNA-protein binding affinity prediction related data is available in 10 databases namely CircBank [658], ClinVar [555], GENCODE Release 43 [314], ENCODE3 [233], EuRBPDB [619], CircInteractome [279], ATtRACT [347], ENCODE [203], NDB [752], doRiNA [38]. In addition, RNA-disease association prediction task related data is available in 12 databases namely miR2Disease [484], HMDD [588], HMDAD [679], dbDEMC [1151], Circad [852], MNDR3.0 [150], lncRNADisease [166], NPInter V4.0 [983], CTD [235], miRCancer [1117], LncRNADisease v2.0 [166], and CircRNADisease [960]. RNA modification prediction tasks related data is available in RMBase [961], m6Atlas [617], and RMBase2.0 [1138]. Furthermore, GENCODE [400] provides RNA sequences for RNA categorization, identification and interaction tasks. RNA sequences data related to sub-cellular localization prediction, gene analysis, RNA single cell analysis, RNA special characteristics analysis, RNA categorization, association and interaction tasks are available in remaining databases namely NCBI [303], dbGap [688], RNAcentral [79], OMIM [391], ENSEMBL [1161], GEO [76], TarBase [896], HGMD [949], RNALocate [208], PanglaoDB [317], KEGG [502], EMBL-EBI [598], FANTOM5 [763].

Apart from aforementioned databases, RNAcentral database contains diverse types of non-coding RNAs information about multiple species and organisms. This database was established in 2014 and until November 2023, it's 23 different versions have been released. Before the establishment of RNAcentral database, non-coding RNA (ncRNA) sequence data was distributed across different databases such as 5S rRNAdb , Ensemble Plants and Ensemble Fungi. The primary objective behind development of RNAcentral database was to centralize scattered non-coding RNA sequence data from various databases into a single platform. Currently, RNAcentral database achieves this goal by integrating data from 53 different databases that are illustrates in Table 4.13.

Overall, RNAcentral database contains 34,179,431 unique non-coding RNA sequences that are systematically categorized into 28 distinct classes. Each sequence represents a unique non-coding RNA type including, transfer RNA (tRNA), long non-coding RNA (lncRNA), short nuclear RNA (snRNA), micro RNA (miRNA), piwi-interacting RNA (piRNA) and many others. Non-coding RNA type specific categorization of sequences enables researchers to delve into distinct biological functionalities associated with each particular non-coding RNA type. Furthermore, RNAcentral database is enriched with online similarity search tool that facilitates users to perform sequences similarity search across an extensive collection of non-coding RNAs.

Table 4.13 A comprehensive review of databases imported into RNAcentral database

Database name	Description	URL
5S rRNA	5SrRNA database refers to the nucleotide sequence information of 5S ribosomal RNAs and their genes, along with non-ribosomal RNAs	http://combio.pl/rrna/
CRW	Comparative RNA Web (CRW) is an open-source repository for information related to ribosomal, introns and other RNA sequences and structures	https://crw-site.chemistry.gatech.edu/
dictyBase	This database serves as a comprehensive repository that contains information related to the social amoeba Dictyostelium discoideum model organism	http://dictybase.org/
ENA	European Nucleotide Archive (ENA) serves as a globally comprehensive data repository for both DNA and RNA sequences	https://www.ebi.ac.uk/ena/browser/home
Ensemble	Ensemble is a genome browser specifically designed for vertebrate genomes that facilitate the sequence variation and transcriptional regulation	https://asia.ensembl.org/index.html
Ensembl Fungi	Ensembl Fungi is a repository specifically offering sequences for fungal genomes	https://fungi.ensembl.org/index.html
Ensembl Metazoa	This database contains sequences of Metazoa species	https://metazoa.ensembl.org/index.html
Ensembl Plants	The Ensembl Plants database provides genome-scale data exclusively for plant species	https://plants.ensembl.org/index.html
Ensembl Protists	Ensembl Protists is a repository that specifically designed for Protists organisms	https://protists.ensembl.org/index.html
GENCODE	GENCODE provides gene features for human and mouse genomes that are used for identification and mapping of all protein-coding genes	https://www.gencodegenes.org/#
EVLncRNAs	EVlncRNAs is a comprehensive database offering phenotypic information on long noncoding RNAs across multiple species	https://www.sdklab-biophysics-dzu.net/EVLncRNAs2/
Expression Atlas	Expression Atlas is a repository that provides information about genes and protein expression	https://www.ebi.ac.uk/gxa/about.html
FlyBase	FlyBase serves as a comprehensive repository, offering gene and genomes information specifically for Drosophila (Fruit Fly)	https://flybase.org/
GeneCards	A database offering information related to human genes that provides genomic, proteomic, functional, transcriptomic and genetic data	https://www.genecards.org/

(continued)

Table 4.13 (continued)

Database name	Description	URL
Greengenes	Greengenes is a comprehensive database of full-length 16S rRNA genes, offering a taxonomy derived from de novo tree	https://greengenes.secondgenome.com/
GtRNAdb	Genomic tRNA database offering tRNA gene information derived from 740 species	http://gtrnadb.ucsc.edu/
HGNC	The HGNC approves distinctive symbols and names for human loci, covering both protein-coding genes and ncRNA genes	https://www.genenames.org/
IntAct	The IntAct database is an open-source system and tool designed for managing molecular interaction data	https://www.ebi.ac.uk/intact/home
LncBase	This database serves as a collection of long non-coding RNAs	https://diana.e-ce.uth.gr/lncbasev3
LncBook	LncBook is a repository that provides a high-quality collection of human long non-coding RNAs	https://ngdc.cncb.ac.cn/lncbook/
LNCipedia	LNCipedia is a publicly accessible database that offers non-coding RNA sequences	https://lncipedia.org/
lncRNAdb	This database provides lncRNAs (long non-coding RNA sequences) derived from eukaryotes	http://lncrnadb.org/
MalaCards	MalaCards is specifically designed for human maladies disease that provide sequences related to human gene	https://www.malacards.org/#
MGI	MGI serves as a database that performs laboratory experiment on mice, facilitate the study of human health and disease	https://www.informatics.jax.org/
MGnify	This database offers comprehensive information on various microbiome data	https://www.ebi.ac.uk/metagenomics
miRBase	miRBase is a repository comprising sequences of all published mature miRNAs	https://mirbase.org/
MirGeneDB	MirGeneDB is a database containing validated and annotated microRNA genes that are manually curated	https://mirgenedb.org/
Modomics	Modomics is a database that specializes in RNA modifications, offering detailed information on the chemical structures of modified ribonucleosides, positioning of modified residues and enzymes for RNA modifications	https://genesilico.pl/modomics/
NONCODE	NONCODE serves as a comprehensive database specializing in the collection and annotation of long non-coding RNAs	http://www.noncode.org/index.php
PDBe	Protein Data Bank (PDB) is a repository specifically designed for three-dimensional structure data for protein and nucleic acid molecules	https://www.ebi.ac.uk/pdbe/

(continued)

Table 4.13 (continued)

Database name	Description	URL
piRBase	This database serves as a collection of piRNA sequences	http://bigdata.ibp.ac.cn/piRBase/index.php
PLncDB	The PLncDB provides a long non-coding RNA sequences (lncRNAs) derived from plants	https://www.tobaccodb.org/plncdb/
PomBase	PomBase is a repository offering precise and reliable structural and functional information related to yeast Schizosaccharomyces pombe	https://www.pombase.org/
PSICQUIC	PSICQUIC is a publicly accessible database that offers molecular-interaction data	http://www.ebi.ac.uk/Tools/webservices/psicquic/view/home.xhtml
RDP	The Ribosomal Database Project (RDP) is a repository offering ribosomal data derived from phylogenetic trees	https://www.glbrc.org/about
REDIportal	REDIportal is a tool for editing RNA that undergoes modification through the base substitutions, insertions or deletions in human or other organisms	http://srv00.recas.ba.infn.it/atlas/
RefSeq	RefSeq is a database specifically designed for nucleotide (DNA and RNA) sequences originating from viruses, bacteria eukaryotes and other organisms	https://www.ncbi.nlm.nih.gov/refseq/
Rfam	The Rfam database provides information related to non-coding RNA families and other structural RNA elements	https://rfam.org/
RGD	Rat Genome database (RGD) provides information related to rat genetic, genomic, physical attributes and physiology	https://rgd.mcw.edu/
Ribocentre	Ribocentre is a specialized repository designed for all-natural ribozymes that includes representative structures and chemical mechanisms	https://www.ribocentre.org/
RiboVision	RiboVision provides in-depth information and visual analysis of ribosomal proteins and their molecular interaction with rRNA	http://apollo.chemistry.gatech.edu/RiboVision2/
SGD	The Saccharomyces Genome Database (SGD) serves as a repository of biological information that accesses the functional relationship between sequences and gene products	https://yeastgenome.org/
SILVA	SILVA is a publicly accessible database offering gene sequences of ribosomal RNA (rRNA) from Bacteria, Archaea and Eukaryota	https://www.arb-silva.de/

(continued)

Table 4.13 (continued)

Database name	Description	URL
snoDB	snoDB is an interactive database focused on human small nucleolar RNAs (snoRNAs) that offer current details on snoRNA features, location of genomic, conservation, host gene snoRNA-RNA targets	https://bioinfo-scottgroup.med.usherbrooke.ca/snoDB/
snOPY	snOPY is a database specifically designed for studying RNA modification including snoRNAs, snoRNA gene loci and target RNAs	http://snoopy.med.miyazaki-u.ac.jp/
snoRNA Database	The snoRNA Database serves as a collection of archaeal snoRNAs that managed by the Lowe Lab	http://lowelab.ucsc.edu/snoRNAdb/index.html
SRPDB	Single Recognition Particle Database (SRPDB) offers organized and annotated phylogenetic sequences related to the structure and function of SRP	https://rth.dk/resources/rnp/SRPDB/
TAIR	The Arabidopsis Information Resource (TAIR) database is designed for molecular and genetic data originating from Arabidopsis thaliana plant	https://www.arabidopsis.org/
TarBase	TarBase database serves as a collection of miRNA gene-interaction	https://dianalab.e-ce.uth.gr/tarbasev9
tmRNA Website	The tmRNA website is designed for bacterial RNA molecules, specifically focusing on the properties of tRNA and mRNA that are described in tmRNA website	https://rth.dk/resources/rnp/tmRDB/
WormBase	WormBase is a specialized repository that contains information related to genomic and genetic data about nematodes	https://wormbase.org/#012-34-5
ZFIN	ZFIN (The Zebrafish Information Network) provides information related to genetic and genomic data that are originating from zebrafish organism	https://zfin.org/
ZWD	ZWD serves as a repository of non-coding RNA sequences	https://bitbucket.org/zashaw/zashaweinbergdata/src/master/

The RNAcentral database aims to provide comprehensive and freely accessible high-quality data regarding non-coding RNA sequences. Users can access data in two ways: through FTP files or using API. In the FTP files, users can choose to access data either from individual databases or combined database data files. Moreover, this database also facilitates users in accessing selective non-coding RNA data by using the browser option, which provides all data related to that specific non-coding RNA type.

4.5.3 Protein Sequence Analysis Databases

This section presents a comprehensive survey of protein databases that encompasses essential data for the development of AI-driven applications across 62 diverse protein sequence analysis tasks. It equips AI researchers with essential information required to identify appropriate databases for the development of high-quality benchmark datasets, which are the cornerstone for development of AI-driven protein sequence analysis applications.

In the realm of AI-driven protein sequence analysis, a detailed review of 295 research articles indicates that researchers have harnessed a remarkable diversity of 100 unique protein databases to develop 627 benchmark datasets for 62 protein sequence analysis tasks. To the best of our knowledge, 68 of these databases are currently publicly accessible, while the remaining 32 are either restricted or no longer available. Table 4.14 presents a valuable road-map for AI researchers to select optimal databases for development of high-quality benchmark datasets. It offers a concise yet informative overview of 68 publicly accessible databases by highlighting their diverse characteristics such as database name, release date, data types, related species and organisms, data statistics, and data formats.

A closer examination of the 'data type' feature in Table 4.1 reveals that, all databases contain protein data and out of the 68 databases, 13 also contain information related to DNA and RNA. These databases include DisGeNET [972], CARD [703], VariBench [741], ClinVar [464], BioLip [1218], CCLE [771], NCBI [477], MtSSPdb [110], GEO [283], KEGG [502], PINA [1102], EMBL-EBI [687] and OMIM [577]. Moreover, 13 databases contain different data types as follows, transcriptomics: MtSSPdb [110], immune repertoires: OAS database [540], genes, mutations and drugs: GeneCards [860], IMGT [695], and COSMIC [940], host proteins: HPIDB [34], gene and diseases: MalaCards [294], molecules, drugs, compounds, and drugs: ChEMBL [1184], DUD-E[737], and BindingDB [647], and chemicals: [973], and DUD [447]. Moreover, data related to TCR sequences, antigens, immunoglobulins (IGs), T cell epitopes, microbiome and antibodies is available in McPAS-TCR [991], VDJdb [69], PIRD [792], MGnify [848] and IEDB [1020] databases. In addition, Negatome database [102] contains domain pairs sequences, PubChem [520] provides compounds strings, genes, and cell lines, CTD [236] houses data related to chemical-gene interaction, chemical-disease interaction, and chemical-phenotype interactions, intAct [418], provides data related to interactions, interactors, and mutations, and enzymes data is available in BRENDA [142].

In Table 4.14, we performed a detailed analysis of 'Species' feature to categorize databases into 3 classes: (1) Fewer species coverage, (2) Moderate species coverage, (3) Large species coverage. In the category of fewer species coverage, we have included 35 databases housing data for 20 species or fewer. This category databases names are DisProt [825], PHROGs [984], MtSSPdb [110], PPT-Ohmnet [1249], COSMIC [940], HPIDB [34], McPAS-TCR [991], VDJdb [69], DisGeNET [972], HIPPIE [21], MalaCards [294], ClinVar [464], BioLip [1218], PDB [121], ConSurf-

Table 4.14 An overview of publicly available biological databases: data types, species diversity, and raw sequence statistics for genomic and proteomic information

Database name	Release date	Types of data	Species	Organism	Sequences statistics	Data format
Alpha FoldDB	2021	Protein	48 species	–	214,683,839 protein structures	.txt, .csv, .json, FASTA
DisProt	2021	Protein	Viruses, Archaea, Eukaryota	Bacteria	Disorder function: 558 proteins, 874 regions, Structural state: 3022 proteins, 6922 regions, Structural transition: 543 proteins, 894 regions, Cellular component: 29 proteins, 54 regions, Biological process: 248 proteins, 531 regions, Molecular function: 1203 proteins, 3889 regions	.json, .tsv, GAF, FASTA
PHROGs	2021	Protein	Viruses infecting bacteria or Archaea	–	Protein orthologous groups: 38,880, Proteins: 868,340, Prophages: 12,498	.tsv, .csv, .xlsx, .pdf, FASTA, MSA, HMM
MtSSPdb	2020	Protein, Genomics, Transcriptomics	Medicago truncatula, Panicum virgatum, Arabidopsis thaliana	Plant	Re-annotated genes: 70,094, Small Peptides genes: 4439, Known SSP gene families: 72	FASTA, .gff, .txt, HMM
OAS database	2018	Protein, Immune repertoires	–	Rabbit, Human, Mouse, Rhesus, Camel, Rat	Unpaired sequences: 2,428,016,345 unique sequences, Paired sequences: 2,038,528 filtered sequences	.csv
PPT-Ohmnet	2018	Protein	Homo sapiens	–	Nodes (human proteins): 4510, Edges (tissue specific interactions): 70,338, Nodes in largest SCC: 4488, Edges in largest SCC: 70,316, Number of triangles: 6,698,541	.txt, .edgelist

(continued)

Table 4.14 (continued)

Database name	Release date	Types of data	Species	Organism	Sequences statistics	Data format
COSMIC	2018	Protein, Genes, Mutations, Drugs	Homo sapiens	Animal	Total Genomic variants: 24,599,940, Genomic non-coding variants: 16,748,366,406, Genomic mutations within Exons: 768, Genomic mutations within Intronic and other intragenic regions: 9,217,664, Samples: 1,531,613, Fusions: 19,428, Gene expression variants: 9,215,470, Differentially Methylated CpGs: 7,930,489	FASTA, .tsv
AmyPro	2017	Protein	39 species	–	125 amyloid precursor proteins	.txt, .json, FASTA
HPIDB	2017	Protein, Host	11 species	1	9957 Influenza interactions, 8174 Herpes viruses interactions, 6862 Saccharomyces cerevisiae interactions, 6515 Papillomaviruses interactions, 4366 Human immunodeficiency virus interactions, 4026 Yersinia interactions, 3069 Bacillus interactions, 2617 Hepatitis C virus interactions, 1371 Francisella tularensis, 1030 Measles virus	FASTA
McPAS-TCR	2017	TCR sequences, Protein	Homo sapiens, Mus musculus	–	386 Human TCRα, 3887 Human TCRβ, 254 Mouse TCRα, 1194 Mouse TCRβ	.csv
MobiDB	2017	Protein	24 species	–	Total proteins: 219.7M, Total residues: 75.5B	.tsv, .json
STCRDab	2017	Protein	–	–	Number of PDB entries with a TCR structure: 618, Number of $\alpha\beta$ TCRs: 851, Number of $\gamma\delta$ TCRs: 18, Number of TCRs complexed to MHC/MHC-like molecules: 680	.csv, .txt

VDJdb	2017	Protein, TCRs Antigens	Homo sapiens, Macaca mulatta, Mus musculus	–	Homo sapiens Chain TRA: Records: 30,937, Paired records: 24,797, Unique epitopes: 943, Homo sapiens Chain TRB: Records: 43,806, Paired records: 25,722, Unique epitopes: 1131, Macaca mulatta Chain TRA: Records: 74, Paired records: 0, Unique epitopes: 1, Macaca mulatta Chain TRB: Records: 1290, Paired records: 0, Unique epitopes: 3, Mus musculus Chain TRA: Records: 1680, Paired records: 1620, Unique epitopes: 55, Mus musculus Chain TRB: Records: 2210, Paired records: 1626, Unique epitopes: 63	.tsv
PIRD	2016	Protein, IGs, TCRs	–	–	11.395 million sequences, and the phenotypes with the top 3 abundant sequences were 2.539 million in IgA nephropathy project, 1.924 million in minimal residual disease (MRD) project and 1.920 million in healthy samples	.irf
Uniclust30	2016	Protein	–	–	9.7 million clusters, 7 million singletons	.tsv, FASTA
IPD-MHC	2015	Protein	77 species	92 organisms	629 genes, 11,940 alleles	.dat, .txt, .xml, FASTA
DisGeNET	2015	DNA, RNA, Protein	Homo sapiens	Animal	1,134,942 GDAs between 21,671 Genes, 30,170 diseases, and traits, 369,554 VDAs between 194,515 variants and 14,155 diseases and traits	.txt, RDF, SQL Dump
GLASS	2014	Protein	–	–	562,871 unique GPCR-ligand entries, 1,046,026 experimentally data entries, 3056 GPCR entries, 825 human GPCR, 733 GPCRs that have experimental association data, 342,539 ligand entries, 241,243 Lipinski-druglike ligand	.tsv, .sdf

(continued)

Table 4.14 (continued)

Database name	Release date	Types of data	Species	Organism	Sequences statistics	Data format
MGnify	2014	Microbiome, Protein	–	–	Residues: Sequence: 577,410,242,951, Cluster: 131,163,572,133, Total Sequences: 2,973,257,435, Clusters: 729,215,663, Biome: 491	.tsv, FASTA
SAbDab	2014	Protein	–	–	Total number of antibody structures: 8634, Number of structures with at least one paired VH/VL: 6947, Number of FV regions: 17,150, Number of structures with antigen: 8205, Number of antibodies with affinity data: 739	.tsv, .pdb
SCOPe	2014	Protein	–	–	Class: All alpha proteins, Number of folds: 290, Number of superfamilies: 519, Number of families: 1089, Class: All beta proteins, Number of folds: 180, Number of superfamilies: 375, Number of families: 993, Class: Alpha and beta proteins (a/b), Number of folds: 148, Number of superfamilies: 247, Number of families: 1,003, Class: Alpha and beta proteins (a+b), Number of folds: 396, Number of superfamilies: 580, Number of families: 1387, Class: Multi-domain proteins (alpha and beta), Number of folds: 74, Number of superfamilies: 74, Number of families: 128, Class: Membrane and cell surface proteins and peptides, Number of folds: 69, Number of superfamilies: 131, Number of families: 204, Class: Small proteins, Number of folds: 100, Number of superfamilies: 141, Number of families: 280, Totals: Number of folds: 1257, Number of superfamilies: 2067, Number of families: 5084	FASTA
MINT database	2013	Protein	674 species	–	Interactions: 139,547, Interactors: 27,756	.mitab
BindingDB	2013	Protein, Compounds	–	–	2,903,069 binding data for 9319 proteins and over 1,253,918 drug-like molecules	.tsv

CARD	2013	Protein, RNA, DNA, compounds, molecules	40 species	–	377 pathogens, 21,079 chromosomes, 2662 genomic islands, 41,828 plasmids and 155,606 whole-genome shotgun assemblies, resulting in collation of 322,710 unique ARG allele sequences	.tsv, .json, .gz, .tar, .pdf, .txt, tab, FASTA, OBL, OWL
HIPPIE	2013	Protein	Homo sapiens	1	more than 270,000 confidence scored and annotated PPIs	.txt, .tsv, .json
MalaCards	2013	Protein, Genes, Disease	Homo sapiens	Human	22,960 entries, 15,278 with associated genes, Total disorders: 22,960, Gene-related Disorders: 15,278	–
VariBench	2013	Protein, RNA, DNA	–	–	19,335 Pathogenic tolerance affecting variations, 21,170 Neutral human nonsynonymous coding SNPs (neutral tolerance data), 17,525 Clustered pathogenic tolerance affecting variations, 15,745 Clustered neutral tolerance affecting variations, 14,610 Pathogenic tolerance affecting variations, 17,393 Neutral human nonsynonymous coding SNPs (neutral tolerance data), 13,096 Clustered pathogenic tolerance affecting variations, 13,107 Clustered neutral tolerance affecting variations, 1760 Functional and nonfunctional variants extracted from the Protein Mutant Database (PMD), 1592 Clustered variants from the Protein Mutant database, 2156 Variations from ProTherm, 1784 Missense variations from 80 proteins, 964: 339 Variants in 9 proteins and 625 variants from ProTherm database, 19 MLH1 and MSH2 gene variants	.xlsx
ClinVar	2013	DNA, RNA, Protein	Homo sapiens	Animal	4,391,341 records, 92,225 genes	.xml, .tsv, .vcf
BioLip	2012	DNA, RNA, Protein	Homo sapiens	Animal	873,925 Entries, 448,816 regular ligands, 191,485 mental ligands, 37,492 Peptide ligands, 43,448 DNA ligands, 152,684 RNA ligands, 873,925 binding affinity data, 451,485 Protein receptors	FASTA

(continued)

Table 4.14 (continued)

Database name	Release date	Types of data	Species	Organism	Sequences statistics	Data format
OGEE	2011	Protein, Genes	91 species		Human cell lines: 931, Human tissues: 27, Human essential genes more than 57,878, Genes: 213,608, Conditional essential genes: 15,440	.txt
PDB	2011	Protein	Homo sapiens, Mus musculus, Arabidopsis thaliana, Saccharomyces cerevisiae	–	~150,000 entries	FASTA
Negatome database	2010	Protein, Domain pairs	–	–	Number of pairs: 30,756	.txt
ChEMBL	2009	Protein, Molecules, Compounds, Drugs	–	–	15,598 targets, 2,431,025 distinct compounds, 20,772,701 activities, 89,892 publications, 262 deposited datasets	.sdf, FASTA
ConSurf-DB	2009	Protein	Homo sapiens, Mus musculus	–	473,197 PDB chains, 108,958 non-redundant PDB chains	FASTA
dbPTM	2009	Protein	Homo sapiens	–	2,235,664 experimental sites, 542,107 putative sites, 2,777,771 sites, 82,444 literatures	FASTA
DUD-E	2009	Protein, Compounds	–	–	22,886 active compounds, 102 targets, 224 ligands	–
CCLE	2008	DNA, RNA, Protein	Homo sapiens	Animal	1019 RNA cell lines, 954 microRNA expression profiles, 899 Protein lines, 897 Genome-wide histone modifications, 843 DNA methylation, 329 whole Genome Sequencing, 326 whole exome Sequencing	.csv

STITCH	2007	Protein, Chemical	Eukaryote, Prokaryote	2031 organisms	more than 9,600,000 proteins, 340,000 to 430,000 compounds	.tsv.gz
DUD	2006	Protein, Compounds	–	–	2950 active compounds, 40 targets	.mol2, .pdb, .sdf
PINA	2006	mRNA, Protein	–	–	Homo sapiens: Binary Interactions: 439,714, Complexes: 15,252, Saccharomyces cerevisiae: Binary Interactions: 128,319, Complexes: 6302, Caenorhabditis elegans: Binary Interactions: 22,305, Complexes: 105, Drosophila melanogaster: Binary Interactions: 57,578, Complexes: 810, Mus musculus: Binary Interactions: 57,669, Complexes: 1304, Rattus norvegicus: Binary Interactions: 5796, Complexes: 307, Arabidopsis thaliana: Binary Interactions: 56,282, Complexes: 431, mRNA expression: Number of patients: 9870, Number of genes: 608,188, Protein expression: Number of patients: 936, Number of proteins: 73,330	.csv, .excel
TCDB	2005	Protein	–	–	Protein sequences: 23,572, Transporter families: 1929	FASTA
NCBI	2005	DNA, RNA, Protein	Homo sapiens, Mus musculus	Animal	35,608 CCDS IDs that correspond to 19,107 Genes, with 48,062 Protein Sequences	FASTA
PDBbind database	2004	Protein	–	–	Biomolecular complexes: 23,496, Protein-ligand: 19,443, Protein-protein: 2852, Protein-nucleic acid: 1052, Nucleic acid-ligand complexes: 149	.mol2, .sdf

(continued)

Table 4.14 (continued)

Database name	Release date	Types of data	Species	Organism	Sequences statistics	Data format
PubChem	2004	Compounds, Genes, Protein, Cell lines	–	–	Compounds: 118,372,533, Substances: 319,659,057, BioAssays: 1,671,253, Bioactivities: 295,155,009, Genes: 113,242, Proteins: 247,869, Taxonomy: 108,194, Pathways: 241,163, Cell Lines: 2005	.csv, .json, .xml, .sdf, .asnt
GOA	2003	Protein	–	–	68 million GO annotations to almost 54 million proteins in more than 480,000 taxonomic groups	GPAD, GPI
IEDB	2003	T Cell Epitopes, Antibodies, Protein	–	4505 organisms	Peptidic Epitopes 1,619,619, Non-Peptidic Epitopes 3188, T Cell Assays 536,844, B Cell Assays 1,405,550, MHC Ligand Assays 4,879,690, Restricting MHC Alleles 1010, References 24,908	.xlsx, .tsv, .json, .csv
Phospho SitesPlus	2003	Protein	–	Human, Mouse, Rat	Proteins: Non-redundant: 20,205, Total: 59,514, PTMs, all types: Non-redundant: 485,813, Total: 600,912, PTMs, low-throughput (LTP) methods: Non-redundant: 25,499, Total: 31,609, PTMs, high-throughput (HTP) MS/MS: Non-redundant: 478,249, Total: 588,707, MS peptides: Non-redundant: 640,925, Total: 2,631,035	.txt, .xlsx, FASTA, OWL

CTD	2003	Protein, Chemical, Genes, Phenotypes, Diseases, Chemical–Gene/Protein Interactions, Gene–Disease Associations, Chemical–Disease Associations, Chemical–Phenotype Interactions, Gene–Gene Interactions, Pathways	–	632 organisms	2,915,515 Chemical–gene interactions, 406,571 Phenotype–based interactions, 32,694,093 Gene–disease associations, 3,489,469 Chemical–disease associations, 6,577,078 Chemical–GO associations, 1,570,026 Chemical–pathway associations, 305,622 Disease–pathway associations, 1,358,371 Gene–gene interactions, 39,776,068 Gene–GO annotations, 135,792 Gene–pathway annotations, 3,133,281 GO–disease associations, 17,667 Chemicals with curated data, 7285 Diseases with curated data, 55,128 Genes with curated data	.csv, .tsv, .xml
STRING	2003	Protein	–	12,535 organisms	59.3 million proteins, 20 billion interactions	.txt, .sql
BioGRID	2003	Protein	74 species	–	2,694,446 protein and genetic interactions, 31,144 chemical interactions, 1,128,339 post translational modifications, non-redundant interactions to 2,091,895, raw interactions to 2,694,446, non-redundant chemical associations to 13,719, raw chemical associations to 31,144, Non-Redundant PTM Sites to 563,757 and Un-Assigned PTMs to 57,396	.mitab, psi, psi25, tab, tab2, tab3
intAct	2002	Protein, Molecules	16 species	3671 organisms	Binary Interactions 1,572,071, Interactions 844,973, Interactors 143,194, Proteins 124,275, Mutation Features 79,805, Experiments 75,229, Publications 23,417, Nucleic Acids 12,142, Controlled Vocabulary Terms 4058, Genes 1289, Interaction Detection Methods 246	.xml, tab, .json, xgmml

(continued)

Table 4.14 (continued)

Database name	Release date	Types of data	Species	Organism	Sequences statistics	Data format
interPro	2002	Protein	–	12 organisms	3510 homologous superfamily, 25,772 family, 14,524 domain, 379 repeat, 133 active sites, 75 binding sites, 741 conserved sites, 17 PTM	.tsv, .json, .txt
Therapeutic Targets Database	2001	Protein, Disease, Pathways, Drugs	Homo sapiens	–	Targets: 3730, Drugs: 39,863	.xlsx, .txt
GEO	2000	DNA, RNA, Protein	21 species	–	7,209,691 samples	SOFT, MINiML, .txt
DIP	1999	Protein	834 species	–	28,850 proteins, 81,923 interactions	FASTA
Phospho. ELM	1999	Protein	Caenor_ habditis, Drosophila, Vertebrate	–	8718 substrate proteins covering 3370 tyrosine, 31,754 serine and 7449 threonine instances	.dump
RCSB PDB	1998	Protein	–	–	Structures from the PDB: 222,036, Computed Structure Models (CSM): 1,068,577	.txt, FASTA, .pdb, .xml, .sdf, .mol2, .cif, API
GeneCards	1997	Genes, Protein, RNA	Homo sapiens	Human	43,839 HGNC approved, 21,601 Protein coding, 291,492 RNA genes including 130,365 lncRNAs, 111,811 piRNAs, and 49,316 other ncRNAs	–

IMGT	1995	Genes, Protein	IMGT/ LIGM-DB: 369 species, IMGT/ PRIMER-DB: 11 species, IMGT/ GENE-DB: 38 species	–	IMGT/LIGM-DB: Nucleotide sequences of IG and TR from 369 species (251,528 entries), IMGT/PRIMER-DB: Oligonucleotides (primers) of IG and TR from 11 species (1864 entries), IMGT/GENE-DB: International nomenclature for IG and TR genes from 38 species (11,391 genes, 15,659 alleles), IMGT/3Dstructure-DB and IMGT/2Dstructure-DB: 3D structures (IMGT Colliers de Perles) of IG antibodies, TR, MH and RPI (8751 entries), IMGT/mAb-DB: Monoclonal antibodies (IG, mAb), fusion proteins for immune applications (FPIA), composite proteins for clinical applications (CPCA), and related proteins (RPI) of therapeutic interest (1489 entries)	FASTA
KEGG	1995	DNA, RNA, Protein	6 species	14 organisms	53,674,741 Genes, 4181 Addendum Proteins, 6,88,823 Viral Genes, 377 Viral mature Peptides	KGML, FASTA, .txt
SCOP	1994	Protein	–	–	Number of folds: 1562, Number of IUPR: 24, Number of hyperfamilies: 22, Number of superfamilies: 2816, Number of families: 5936, Number of inter-relationships: 60, Non-redundant domains: 72,544, Protein structures: 861,631	.txt, FASTA
EMBL-EBI	1994	DNA, RNA, Protein	–	–	~130 million sequences	.xml, FASTA, .txt, .tsv, .json

(continued)

Table 4.14 (continued)

Database name	Release date	Types of data	Species	Organism	Sequences statistics	Data format
GPCRdb	1993	Protein, Drugs	_	_	424 Human proteins, 40,450 Species orthologs, 69,580 Genetic variants, 968 Drugs, 175 Drug targets, 405 Disease indications, 217,578 Ligands, 527 Endogenous ligands, 481,718 Ligand bioactivities, 35,606 Ligand site mutations, 48,039 Ligand interactions, 1160 GPCRs structures, 842 GPCRs structure models, 2922 Generic residues, 504 Refined structures	.json
CATH	1990	Protein	_	_	41 architectures, 1390 topology, 6631 homologous superfamily, 32,388 S35 superfamily, 45,835 S60 family, 62,915 S95 family, 122,727 S100 family, 500,238 domains	.txt, .gz, FASTA
Prosite	1989	Protein	Mammals	_	1559 documentation entries, 1308 patterns, 863 profiles and 869 ProRules	.dat, .doc, .txt
BRENDA	1987	Protein, Enzyme	_	16,018,959 organisms	38,623 active compounds, 32,832,265 sequences	.json, .txt
UniProtKB	1986	Protein	Archaea, Eukaryotes, Viruses	Bacteria	11,206 Peptides	FASTA, .xml, .dat
OMIM	1960	DNA, RNA, Protein	Homo sapiens	Animal	17,290 Gene descriptions, 18 Gene and Phenotypes combined, 6859 Phenotype description molecular basis known, 1502 Phenotype description molecular basis unknown, 1736 mainly Phenotypes with suspected mendelian basis	.txt

DB [85], dbPTM [585], CCLE [771], STITCH [973], NCBI [477], intAct [418], Therapeutic Targets Databases [1041], Phospho.ELM [256], GeneCards [860], KEGG [502], Prosite [923], UniProtKB [111], OMIM [577], OAS database [540], SAbDab [280], Negatome database [102], DUD-E [737], DUD [447], PDBbind database [1050], PhosphoSitesPlus [436] and interPro [103]. On the other hand, in the moderate species coverage category, we included 8 databases encompassing data for a range of 21 to 80 species. These databases include AlphaFoldDB [1013], AmyPro [1014], MobiDB [812], IPD-MHC [683], CARD [703], BioGRID [789], GEO [283] and ChEMBL [1184]. In large species coverage category, we included 25 databases encompassing data more than 80 species. This category related databases are MINT database [622], OGEE [382], DIP [1107], IMGT [695], STCRDab [573], PIRD [792], Uniclust30 [720], GLASS [140], MGnify [848], SCOPe [312], BindingDB [647], VariBench [741], PINA [1102], TCDB [866], PubChem [520], GOA [456], IEDB [1020], CTD [236], STRING [972], RCSB PDB [90], SCOP [36], EMBL-EBI [687], GPCRdb [794], CATH [466] and BRENDA [142].

Based on an in-depth analysis of 'organism' feature in Table 4.14, we have categorized these databases into 2 different classes: (1) Narrow-organisms range, (2) Wide-organisms range databases. In narrow-organisms range databases, 20 or fewer organisms are present and we have included 12 databases to this category. This names of these databases are DisProt [825], MtSSPdb [110], OAS database [540], COSMIC [940], HPIDB [34], DisGeNET [972], HIPPIE [21], MalaCards [294], ClinVar [464], BioLip [1218], CCLE [771] and NCBI [477]. In contrast, remaining databases, also known as wide-organism range, contain more than 20 databases such as BRENDA [142], intAct [418], STRING [972], and CTD [236] etc.

Since word embeddings and large language models based predictive pipelines require large amount of raw data for training in an unsupervised fashion, these databases act as facilitators for development of these predictive pipelines. For this, we have categorized these databases based on the volume of data into three different categories: (1) Low sequence facilitator, (2) Medium sequence facilitator, (3) High sequence facilitator. Low sequence facilitator databases provide with up to 100,000 sequence. A total of 26 databases are low sequence facilitator databases which include AmyPro [1014], BindingDB [647], ChEMBL [1184], DisProt [825], DUD [447], DUD-E [737], HPIDB [34], interPro [103], IPD-MHC [683], MalaCards [294], MtSSPdb [110], Negatome database [102], PDBbind database [1050], Phospho.ELM [256], PPT-Ohmnet [1249], SAbDab [280], SCOPe [312], STCRDab [573], TCDB [866], Therapeutic Targets Database [1041], VDJdb [69], CCLE [771], UniProtKB [111], NCBI [477], OMIM [577] and Prosite [923]. Similarly, medium sequence facilitator databases contain data sequences within a range of 100,000 to 1 million. There are 17 medium sequence facilitators databases namely CARD [703], CATH [466], ConSurf-DB [85], DIP [1107], GeneCards [860], GLASS [140], GPCRdb [794], IMGT [695], MINT database [622], OGEE [382], PhosphoSitesPlus [436], PHROGs [984], PINA [1102], RCSB [90], PDB [90], SCOP [36], GEO [283] and BioLip [1218]. In this study, 25 databases are identified as high sequence facilitator databases including AlphaFoldDB [1013], BRENDA [142], dbPTM [585], GOA [456], IEDB [1020], intAct [418], MGnify [848], MobiDB [812], OAS database [540], PubChem [520], KEGG [502], CTD [236], STRING [972], DisGeNET [809], BioGRID [789], STITCH [973], ClinVar [464], COSMIC [940], HIPPIE [21], McPAS-TCR [991], PIRD [792], Uniclust30 [720], VariBench [741], PDB [121] and EMBL-EBI [687].

Furthermore, data related to protein function prediction is available in 7 databases namely CARD [703], AlphaFoldDB [1013], DisProt [825], GOA [456], MobiDB [812], SCOPe [312] and STCRDab [573]. Similarly, data related to structure prediction, bitter peptides identification, domain boundary prediction, variant effects prediction, protein complexes identification, intrinsically disorder protein prediction, G-Protein coupled receptors identification and virus-host protein interaction prediction task is available in 7 databases including AlphaFoldDB [1013], AmyPro [1014], BindingDB [647], CATH [466], ConSurf-DB [85], DIP [1107] and DisProt [825], respectively. In addition, data for drug-target interaction and drug-protein interaction prediction is present in 6 databases namely DUD-E [737], BindingDB [647], ChEMBL [1184], BRENDA [142], PubChem [520] and DUD [447] databases. Moreover, data for multiple interaction types prediction and compound-protein binding affinity prediction utilize tasks is extracted from ChEMBL [1184] database, MINT database [622], intAct [418], and Therapeutic Targets Database [1041]. Additionally, data related to virus-host interaction prediction, protein-protein interaction prediction [55], gene functions prediction, secreted peptides prediction, antibody sequence infilling, phage-host interaction prediction, TRP channels classification and mutation effects prediction is available at 11 databases including GPCRdb [794], HIPPIE [21], HPIDB [34], intAct [418], MGnify [848], MtSSPdb [110], OAS database [540], PHROGs [984], RCSB PDB [90], TCDB [866] and VariBench [741] databases. Similarly, data regarding post-translational modification prediction is sourced from 3 different databases namely dbPTM [585], Phospho.ELM [256] and PhosphoSitesPlus [436]. Moreover, 2 databases namely DUD [447] and PDBbind database [1050] houses data for commercially available inhibitors prediction against SARS-CoV-2. Moreover, DUD [447] database also contains data related to drug-target binding affinity prediction, whereas GLASS [140], BindingDB [647], ChEMBL [1184] facilitate with data related to compound-protein interaction prediction. IEDB [1020] database is specific for providing sequences for anti-inflammatory peptides identification, Protein Binding Sites Prediction and MHC–peptide class II interaction prediction. Moreover, data related to enzyme substrate prediction and protein function identification is also available at interPro [103]. In contrast, data related to disease genes identification is available at MalaCards [294], ChEMBL [1184], intAct [418], MINT database [622] and GeneCards [860] database. Specifically, data related to protein-protein interaction prediction is provided by 7 databases including DIP [1107], HIPPIE [21], intAct [418], PINA [1102], PPT-Ohmnet [1249], PPT-Ohmnet [1249], and MINT database [622]. Furthermore, data related to essential genes identification is available at OGEE [382] and DIP [1107] but OGEE [382] also facilitates with data related to essential gene identification. Data related to vascular calcification, protein properties prediction, remote homology detection, solubility, fold prediction and subcellular location identification tasks is available in 5 databases namely PDBbind database [1050], BindingDB [647], PubChem [520], SCOP [36] and SCOPe [312] databases. Additionally, data related to nucleic acid binding protein prediction, secondary structure prediction, and binding affinity prediction task is present at 10 databases including Uniclust30 [720], MGnify [848], VDJdb [69], SCOPe [312], PIRD [792], DisProt [825], SCOP [36], BindingDB [647], ChEMBL [1184] and PubChem [520] database.

4.5.4 Peptides Databases

This section provides an overview of the various databases that are used to develop peptide classification datasets related to 22 distinct types of peptides. A thorough analysis of 204 peptide classification articles reveals, approximately 90 biological databases contain sequences and different types of peptides information. Out of these, 47 databases are accessible, while the remaining are either inaccessible or no longer exist. The characteristics of all accessible databases, including release date, types of peptide data, data format, data statistics, and the species and organisms to which the data belongs, are presented in Table 4.15. A high level analysis of Table 4.15 reveals that most of the databases contain therapeutic peptides sequences. Specifically:

- Antimicrobial peptides sequences are present in 24 distinct databases including APD3 [1054], Peplab [985], PlantPepDB [230], DRAMP2.0 [503], BioPepDB [599], dbAMP [480], UniProt [204], DRAMP [296], BaAMP [254], ADP2 [1105], SATPdb [930], DBAASP [350], AHTPDB [546], Hemolytik [333], LAMP [1224], DADP [768], APD2 [1051], DrugBank [1097], UniProtKB [204], PubMed [129], SGD [182], Prosite [454], NCBI [337], and SwissProt [204].
- Anti-fungal peptides sequences are present in 16 different databases namely Peplab [985], PlantPepDB [230], StarPep [10], DRAMP 2.0 [503], UniProt [204], DRAMP [296], LAMP2 [1162], ADP2 [1105], SATPdb [930], AHTPDB [546], CAMP [988], APD2 [1051], UniProtKB [204], PubMed [129], Prosite [454], and SwissProt [204].
- Anticancer peptides sequences are present in 13 different databases namely Peplab [985], PlantPepDB [230], StarPep [10], DRAMP 2.0 [503], BioPepDB [599], DRAMP [296], ADP2 [1105], SATPdb [930], AHTPDB [546], APD2 [1051], AAIndex [511], PubMed [129], and SGD [182].
- Antibacterial peptides sequences are present in 12 databases including Peplab [985], PlantPepDB [230], StarPep [10], DRAMP 2.0 [503], dbAMP [480], DRAMP [296], LAMP2 [1162], ADP2 [1105], SATPdb [930], CAMP [988], APD2 [1051], and Pubmed [129].
- Antiviral peptides sequences are available in 11 different databases including Peplab [985], PlantPepDB [230], StarPep [10], DRAMP 2.0 [503], LAMP2

Table 4.15 A summary of publicly accessible biological databases, their inherent peptides data types, species diversity, and statistics of raw sequences related to different peptides

Database name	Release date	Types of peptides	Organism name	Species name	Sequences statistics	Data format
APD3	2024	Anti-Microbial	Bacteria, Archaea, Protists, Fungi, Plants, Animals	–	3146 Anti-Microbial Peptides, 190 predicted, 314 synthetic Anti-Microbial Peptides	FASTA
Peplab	2022	Anti-Inflammatory, Anti-Bacterial, Anti-Cancer, Anti-Fungal, Anti-Microbial, Anti-Viral, DPP-IV inhibitor, Neuropeptide	Animals, Marine, Micro-Organisms, Plants	Human	2784 Bioactive Peptides	FASTA
B3Pdb	2021	Blood Brain Barrier	Bovine, T5 Bacteriophage derived, Bacteria	Human, Mouse	1225 Blood Brain Barrier Peptides	SMILES
Biofilm resource	2020	Biofilm Inhibitory Peptides	Bacteria	–	501 Protein structures	.csv
PlantPepDB	2020	Anti-Cancer, Anti-Hypertensive, Anti-Microbial, Anti-Bacterial, Anti-Fungal, Anti-Viral, Insecticidal	Plants	–	3848 Peptides	.txt
VARIDT	2020	Anti-Cancer	–	124 Microbe Species	585 approved, 301 clinical trial Drug for treating 572 disease (2072 MBIs, 10 610 TSRs, 46,748 EGRs, 12,209 EGMs and 10,255 PTMs)	.txt

StarPep	2019	Anti-Bacterial, Anti-Viral, Anti-Cancer, Anti-Parasitic, Anti-Fungal	Archaea, Bacteria, Animals etc.	–	45,120 Peptides	Graph Structure
DRAMP 2.0	2019	Anti-Microbial, Anti-Bacterial, Anti-Fungal, Anti-Viral, Anti-Cancer	Animals, Plants, Bacteria, Fungi, Archaea, Protozoal	–	19,899 entries, 5084 general entries, 14,739 patent entries, and 76 clinical entries.	.txt, FASTA, .xlxs
AntiTbPdb	2018	Anti-Tubercular	Mycobacterium	–	542 Peptides	FASTA
BioPepDB	2018	Anti-Microbial, Anti-Hypertensive, Anti-Cancer, Therapeutic Peptides	Cereal, Legume, Milk, Bovine, Porcine, Shrimp, Fish, Amphibian, Fungi	Chicken, Egg, Potato	4000 Peptides	FASTA
dbAMP	2018	Anti-Microbial, Anti-Bacterial	Amphibia, Mammals, Arthropods, Viridiplantae	–	18,345 validated Anti-Microbial Peptides, 10,364 non-validated Anti-Microbial Peptides	FASTA
DINeR	2017	Neuro Peptides	–	–	4700 Peptides	FASTA
Flavor	2017	Bitter	–	–	179 Peptides	.mol2, .json, 2Dimage, .sdf
THPdb	2017	Therapeutic	–	Human	69 Drug, contains 239 Peptides/Proteins, 380 variants	.txt, .csv, .xlsx
UniProt	2004	Anti-Microbial, Anti-Fungal, Signal	Bacteria, Eukaryotes, Archaea, Proteomes	Viruses	UniProtKB/SwissProt (11206 Peptides), UniProtKB/TrEMBL (550565 Peptides), UniRef100 (7286 Peptides), UniRef90 (3586 Peptides), UniRef50 (1085 Peptides), UniParc (10,399 Peptides)	FASTA, .xml, .dat

(continued)

Table 4.15 (continued)

Database name	Release date	Types of peptides	Organism name	Species name	Sequences statistics	Data format
DRAMP	2016	Anti-Microbial, Anti-Bacterial, Anti-Fungal, Anti-Viral, Anti-Cancer	Plants	–	17,349 Anti-Microbial Peptides, including 4571, 12,704 patented Sequences, 74 Peptides in Drug development	.txt, FASTA, .xlxs
LAMP2	2016	Anti-Bacterial, Anti-Viral, Anti-Fungal, Anti-Tumor	Bacteria, Plants	In-Vertebrates, Vertebrates	23,253 Anti-Microbial Peptides, 7824 Natural, 15,429 Synthetic Anti-Microbial Peptides	FASTA
BaAMP	2015	Anti-Microbial, Anti-Biofilm	Amphibian, Murine	Human	221 Peptides	.csv
ADP2	2015	Anti-Microbial, Anti-Bacterial, Anti-Viral, Anti-Fungal, Anti-Cancer	Bacteria, Archaea, Fungi, Protists, Plants, Animals	–	3940 Peptides	FASTA
CPPsite 2.0	2015	Cell Penetrating Peptides	–	Human, Mouse, Chinese hamster	1699 Unique Peptides, 1753 Linear Peptides, 102 Cyclic Peptides, 714 Cationic Peptides, 391 Amphipathic Peptides, 1558 Peptides	.tgz, .txt
INTEDE	2015	Drug-metabolizing enzymes	Bacteria	Homo sapiens	1047 Unique Drug-metabolizing enzymes (448 host and 599 Microbial)	FASTA, .txt
NeuroPep	2015	Neuropeptides	–	Vertebrates, In-Vertebrates	5949 Peptides	FASTA

SATPdb	2015	Anti-Cancer, Anti-Viral, Anti-Fungal, Anti-Bacterial, Anti-Microbial, Anti-Hypertensive	Eukaryotes, Archaea, Bacteria	Mouse, Rat, Dog, Human, Anuran, Viruses	19,192 Peptides	.pdb, .txt
DBAASP	2014	Anti-Microbial	Animals, Plants, Protista, Fungi, Bacteria, Viruses, Archaea	–	21,848 Peptides	.csv, FASTA
AHTPDB	2014	Anti-Microbial, Anti-Cancer, Anti-Fungal, Tumor-homing	Natural food, Fungi, Algae, Micro-Organism, Insects, Snake venom	–	5978 Peptides	.txt
Hemolytik	2013	Anti-Microbial	Bovine, Porcine, Ovine, Fish	Human, Sheep, Rat, Rabbit, Mouse, Guinea-Pig, Horse, Pig, Dog, Chicken, Trout, HagFish, Goat	2970 Hemolytic, 1750 Unique Hemolytic, 295 non-Hemolytic Sequences	FASTA
Hmrbase2	2013	Hormone Peptides	Animals, Plants	–	12,056 entries, including 7406 Peptides Hormones, 753 non-Hormones Peptides, 3897 Hormones receptors	FASTA
LAMP	2013	Anti-Microbial	Bacteria, Plants	In-Vertebrates, Vertebrates	5547 Anti-Microbial Peptides including 3904 Natural, 1643 Synthetic Anti-Microbial Peptides	FASTA

(continued)

Table 4.15 (continued)

Database name	Release date	Types of peptides	Organism name	Species name	Sequences statistics	Data format
Quorumpeps	2013	Quorum sensing	–	Bacillus anthracis, Bacillus cereus, Bacillus halodurans, Bacillus mojavensis, Bacillus pumilus, Bacillus stearothermophilus, Bacillus subtilis, Bacillus thuringiensis, Carnobacterium maltaromaticum, Carnobacterium piscicola, Clostridium acetobutylicum, Clostridium botulinum, Clostridium perfringens, Clostridium sporogenes, Clostridium thermocellum, Cryptococcus neoformans, Eikenella corodens, Enterococcus faecalis, Enterococcus faecium, Escherichia coli, Lactobacillus plantarum, Lactobacillus sakei, Lactococcus lactis, Listeria monocytogenes, Phage-produced, Pseudomonas aeruginosa, Staphylococcus arlettae, Staphylococcus aureus, Staphylococcus auricularis, Staphylococcus capitis, Staphylococcus caprae, Staphylococcus carnosus, Staphylococcus cochnii subsp. cochnii, Staphylococcus cochnii subsp. urealyticum, Staphylococcus epidermidis, Staphylococcus gallinarum, Staphylococcus intermedius, Staphylococcus lugdunensis, Staphylococcus simulans, Staphylococcus warneri, Staphylococcus xylosus, Streptococcus agalactiae, Streptococcus anginosus, Streptococcus constellatus, Streptococcus crista, Streptococcus dysgalactiae, Streptococcus gordonii, Streptococcus milleri, Streptococcus mitis, Streptococcus mutans, Streptococcus oralis, Streptococcus pneumoniae, Streptococcus pyogenes, Streptococcus sanguis, Streptococcus thermophilus, Streptomyces, Synthetic, Thermotoga maritima	92 Peptides	.hin

CPPsite	2012	Cell Penetrating Peptides	–	Human, Mouse, Chinese hamster	741 Cell Penetrating Peptides Sequences with 85 different cell lines	.tgz, .txt
DADP	2012	Anti-Microbial	–	Anuran Species from 12 families	1766 Signal Peptides, 805 Anti-Microbial Peptides	FASTA, .txt, .xlxs
PDB	2011	Cell Penetrating Peptides	–	Homo sapiens, Mus musculus, Arabidopsis thaliana, Saccharomyces cerevisiae	–	FASTA
CAMP	2010	Anti-Bacterial, Anti-Fungal, Anti-Viral	Prokaryotes, Eukaryotes	–	10,247 Anti-Microbial, 2915 Anti-Bacterial, 1144 Anti-Fungal, 117 Anti-Viral Peptides	.txt
APD2	2009	Anti-Bacterial, Anti-Cancer, Anti-Fungal, Anti-Hypertensive, Anti-Microbial, Anti-Viral	Bacteria, Insects, Plants	Frog	65 Anti-Cancer, 76 Anti-Viral (53 Anti-HIV), 327 Anti-Fungal, 944 Anti-Bacteria Peptides	FASTA
Brainpeps	2008	Blood Brain Barrier	–	Mouse, Rat, Dog, Human	1734 Peptides	FASTA
IEDB	2008	Peptidic Epitopes, Non-Peptidic Epitopes	Archeobacterium, Bacteria, Eukaryote	Viruses	1,612,732 Peptidic Epitopes, 3188 non-Peptidic Epitopes, 512,749 T Cell Assays, 1,402,194 B Cell Assays, 4,803,127 MHC Ligand Assays	.txt
PepBank	2007	Blood Brain Barrier	–	–	21,691 Peptides	–
DrugBank	2006	Anti-Microbial	–	–	>100 Peptides	.json, .csv, .sql, .xml
SPdb5.1	2005	Signal	Archaea, Bacteria, Eukaryotes	Viruses	18,146 Peptides	FASTA

(continued)

Table 4.15 (continued)

Database name	Release date	Types of peptides	Organism name	Species name	Sequences statistics	Data format
PeptideAtlas	2004	Cell Penetrating Peptides	Eukaryotes	–	35,391 Peptides	FASTA
UniProtKB	1986	Anti-Microbial, Anti-Fungal, Signal	Archaea, Bacteria, Eukaryotes	Viruses	561,771 Peptides	FASTA, .xml, .dat
AAIndex	1996	Anti-cancer Peptides	–	–	AAIndex1: 437 amino acids indices, AAIndex2: 71 amino acid mutation matrices	.txt, .jp
Pubmed	1996	Therapeutic Peptide, Anti-Fungal, Anti-Microbial, Cell Penetrating Peptides, Bitter, Anti-Viral, Signal, Neuro, Anti-Hypertensive, Anti-Tubercular, Anti-Bacterial, Anti-MRSA, Blood Brain Barrier, Tumor-homing, Anti-Angiogenic, Insecticidal, DPP-IV inhibitor, Anti-Sepsis, Biofilm inhibitor, Anti-Biofilm, Anti-Cancer, Anti-Inflammatory, Hormone	Animals, Plants, Micro-Organisms, others	Human	3,116,945 Peptides	.xml

SGD	1996	Signal, Anti-Microbial, Anti-Cancer, Hormone	Yeast, Fungi	Viruses	2360 Peptides	.txt, FASTA
Prosite	1989	Anti-Microbial, Anti-Fungal, Signal	Mammals	–	1559 documentation entries, 1308 patterns, 863 profiles and 869 ProRules	.dat, .doc, .txt
NCBI	1988	Anti-Microbial Peptides	–	–	–	.db, .txt, .csv, .bq, seq.gz etc.
SwissProt	1986	Anti-Microbial, Anti-Fungal, Signal	Archaea, Bacteria, Eukaryotes	Viruses	11,206 Peptides	FASTA, .xml, .dat

[1162], DRAMP [296], ADP2 [1105], SATPdb [930], CAMP [988], APD2 [1051], and Pubmed [129].

Several databases, such as Prosite [454], PubMed [129], SGD [182], SPdb5.1 [186], and UniProt (UniProtKB and SwissProt) [204], contain signal peptide sequences. Others, like APD2 [1051], BioPepDB [599], PlantPepDB [230], PubMed [129], and SATPdb [930], specialize in cataloging sequences of antihypertensive peptides. There are also four databases dedicated to insecticidal peptides and three databases for hormone peptides. Only a couple of databases encompass sequences related to anti-inflammatory, antibiofilm, biofilm inhibitory, dipeptidyl peptidase-IV, antitubercular, tumor-homing, and neuropeptides. A deep analysis of Table 4.15 reveals that, non of public database contain anti-MRSA peptides sequences data.

A comprehensive analysis of databases concerning various species and organisms reveals that nearly ten databases offer peptide sequences for archaea and bacteria. Additionally, data related to peptides present in plants is available in 11 databases namely ADP3, APD, DBAASP, DRAMP, DRAMP 2.0, Hmrbase2, LAMP, LAMP 2.0, PepLab, PlantPepDB, and PubMed. These databases collectively encompass data from over 35 species, including frogs, humans, chickens, rats, amphibians, and vertebrates. Moreover, these databases cover a broad spectrum of living organisms, exceeding 100 organisms, including bacteria, bovines, eukaryotes, plants, mammals, and other microorganisms.

4.5.5 CRISPR Databases

This section provides an overview of databases that can be used to develop novel CRISPR-related benchmark datasets. Additionally, it entails the types and quantities of data available in 17 different databases which can help researchers to identify valuable resources for compiling comprehensive and diverse datasets necessary for effective CRISPR research and applications.

The rapid advancements in CRISPR technology have generated a vast amount of data, leading to the creation of numerous databases. Table 4.16 summarizes the list of public databases categorized based on different CRISPR areas. These databases encompass data related to various aspects of CRISPR systems, such as CRISPR arrays, acr proteins, operons, Cas proteins, and on/off-target activities. This abundance of data presents a significant opportunity for the development of novel benchmark datasets, which can enhance the performance and accuracy of AI tools designed for CRISPR research.

Databases such as CRISPRBank [98], and CRISPRCasDb [814] provide a wealth of sequences and annotations for CRISPR arrays. These databases include data on hundreds of thousands of CRISPR arrays and spacers from a vast number of bacterial and archaeal genomes. This creates opportunities for training AI models to accurately identify and annotate CRISPR arrays in newly sequenced genomes.

Table 4.16 A pool of CRISPR related databases

	Database	Data type	URL	Description
CRISPR Arrays	CRISPRBank [98]	Arrays, repeats, and spacers in FASTA	Link	CRISPRBank contains analysis of genome from RefSeq 95 July, 2019. Particularly, CRISPRDetect 2.4 was employed to comprehensively analyze all 151,845 bacterial genomes and 855 archaeal genomes. In total, there are 132,379 CRISPR arrays and 1,992,510 spacers
	CRISPRCasDb [814]	FASTA	Link	CRISPRCasdb contains comprehensive data on CRISPR-Cas systems including information on 2086 CRISPR arrays and 130,293 spacers, along with details on 19,232 Cas proteins and 7125 associated Cas proteins
Acr	AcrHub [1063]	XLSX, FASTA	Link	AcrHub offers extensive annotations and functional data for anti-CRISPR associated proteins. It features information on 1800 proteins and their interactions, spanning various species within the bacterial and archaeal domains
	Anti-Crisprdb [450]	XLSX, CSV, JSON	Link	Anti-CRISPRdb catalogs a wide array of anti-CRISPR proteins with detailed annotations, encompassing sequences and structural data for 1200 proteins. The database covers anti-CRISPR proteins found in numerous bacterial species, providing insights into their diversity and functionalities
	AcrDb			AcrDB is a comprehensive database providing sequences and structural information of anti-CRISPR proteins. It includes data on 2500 anti-CRISPR proteins across diverse bacterial and archaeal species
	AcrCatalog [385]	FASTA	Link	AcrCatalog is a specialized database that catalogs anti-CRISPR proteins and their interactions across various CRISPR-Cas systems. It contains sequences, structural information, and functional annotations for approximately 16,919 putative acr proteins. These proteins are associated with specific CRISPR-Cas systems, including Cas-IA to IE, Cas-IIA to IIC, Cas-IIIA to IIID, Cas-IVA, Cas-VA, and Cas-VIA to VIC

(continued)

Table 4.16 (continued)

	Database	Data type	URL	Description
Aca	UniProt Database [204]	TXT, FASTA, XML, JSON	Link	Universal Protein Resource database is a resource for protein sequence and annotation data
	AcrCatalog [385]	TXT	Link	Anti-CRISPR proteins predicted with ML [384]
	AcrHub [1063]	XLSX, FASTA	Link	AcrHub predicts Anti-CRISPR proteins
Cas	CasPDB [978]	FASTA	Link	CasPDB is an integrated database housing 287 reviewed Cas proteins, 257,745 putative Cas proteins, and 3593 Cas operons from 32,023 bacterial species and 1802 archaeal species. The database comprehensively contains all 3593 putative Cas operons, including 328 operons associated with the type II CRISPR-Cas system
	CRISPRCasdb [814]	SQL, FASTA	Link	CRISPRCasdb contains CRISPR arrays andcas genes from complete genome sequences
	UniProt Database [204]	TXT, FASTA, XML, JSON	Link	Universal Protein Resource database is a resource for protein sequence and annotation data
	CasPedia [6]	FASTA	Link	CasPedia is an annotated database for Cas proteins from bacteria and archaea, featuring 287 reviewed Cas proteins, 257,745 putative Cas proteins, and 3593 Cas operons from 32,023 bacterial and 1802 archaeal species. It offers free access, a user-friendly interface, and details on all operons, including 328 from the type II CRISPR-Cas system
On/Off target activity	cripsrSQL	CSV	Link	crisprSQL is a SQL-based database forCRISPR/Cas9 off-target cleavage assays and epigenetically annotated, base-pair resolved cleavage frequency distributions
	Ensembl BioMart	FASTA, GTF,GFF, SQL	Link	Ensembl is a genome browser for vertebrate genomes that supports research in comparative genomics, evolution, sequence variation and transcriptional regulation,BioMart is a data mining tool
	crisprSQL	CSV	Link	crisprSQL is a SQL-based database forCRISPR/Cas9 off-target cleavage assays and epigenetically annotated, base-pair resolved cleavage frequency distributions
	CRISPOR	TSV	Link	Data from CRISPOR paper [387]

Databases such as AcrHub [1063], Anti-Crisprdb [267], AcrDb [450], and AcrCatalog [385] offer extensive data on anti-CRISPR proteins, including their sequences, structures, and functional annotations. These datasets span a wide variety of species which provides a comprehensive view of acr proteins diversity. The richness of this data can be utilized to create benchmark datasets for training AI models that predict anti-CRISPR proteins based on sequence data. Moreover, these datasets can be used to develop models to map and predict interactions between anti-CRISPR proteins and CRISPR-Cas systems, as well as benchmark tools that annotate the function and efficacy of anti-CRISPR proteins.

The CasPDB [978], CRISPRCasdb [814], UniProt [204], and CasPedia [6] contain extensive data on Cas proteins, including reviewed and putative proteins, as well as comprehensive operon information. This presents opportunities for using these datasets to benchmark AI tools that predict the three-dimensional structures of Cas proteins. Additionally, benchmark datasets can be developed for annotating the function of Cas proteins in various CRISPR-Cas systems, and tools can be created to study the evolutionary relationships between different Cas proteins and operons.

Databases such as crisprSQL [951], Ensembl BioMart [1161] provide valuable data on the on/off-target activity of CRISPR-Cas systems. These datasets include detailed information on off-target cleavage assays and epigenetically annotated cleavage frequency distributions. They offer opportunities to benchmark AI models that predict off-target effects of CRISPR-Cas editing and develop datasets that help optimize CRISPR tools for higher specificity and reduced off-target activity. Additionally, these datasets can be used to create benchmarks for assessing the safety and efficacy of CRISPR-based gene editing in various organisms.

In conclusion, the diverse and extensive CRISPR-related databases provide a rich source of data for the development of novel benchmark datasets. These datasets can significantly advance AI tools in CRISPR research, improving the understanding and application of CRISPR technology. By leveraging these opportunities, researchers can enhance the accuracy, functionality, and safety of CRISPR-based applications, driving forward the fields of genetics and biotechnology.

Chapter 5
DNA and RNA Sequence Representation Learning Methods

DNA sequence is made up of repetitive patterns of 4 fundamental nucleotides namely Thymine (T), Cytosine(C), Adenine (A), and Guanine (G) [865]. DNA is transcribed into RNA sequence where Thymine (T) nucleotide is substituted with uracil (U) and other nucleotides remain unchanged. RNA sequence undergoes a translation process to form a protein sequence, which is composed of repeating patterns of 20 distinct amino acids. These amino acids are named as Alanine, Arginine, Asparagine, Aspartic Acid, Cysteine, Glutamic acid, Glutamine, Glycine, Histidine, Isoleucine, Leucine, Lysine, Methionine, Phenylalanine, Proline, Serine, Threonine, Tryptophan, Tyrosine, Valine, Selenocysteine, and Pyrrolysine.

Within the realm of DNA, RNA, and protein sequence analysis, a majority of the tasks can be categorized into three main groups: clustering, classification, and regression. Across all these tasks, prime objective is to examine the arrangement of nucleotides or amino acids within the sequences. Then, leverage these distribution patterns to make clusters of similar sequences or assign pre-defined class labels or float values to the sequences. More specifically, in classification task, distribution patterns are utilized to identify whether a DNA sequence is a promoter or a coding sequence, or whether a protein sequence is an enzyme or a structural protein. In case of regression, these patterns are utilized to predict stability of a protein or the binding affinity of a ligand to a protein.

While developing Artificial Intelligence supported biological sequence classification applications, it is considered that sequences that belong to same class have similar distribution of nucleotides or amino acids. Similarly, sequences that belong to different classes have dissimilar distributions of nucleotides or amino acids. However, machine and deep learning predictors cannot directly operate on raw DNA, RNA, and Protein sequences due to their inherent dependency over statistical values. This illustrates that to perform AI-supported DNA, RNA or protein sequence analysis, an important task is the conversion of DNA, RNA, and protein sequences into statistical vectors.

M. Nabeel Asim et al., *Artificial Intelligence for Molecular Biology*,
https://doi.org/10.1007/978-3-031-90450-9_5

The performance of machine and deep learning predictors completely relies upon the quality of generated statistical vectors. This is mainly because machine and deep learning predictors learn discriminative patterns from statistical vectors of different classes. Overall, it can be concluded, a simple classifier can produce better performance when it is fed with statistical vectors having discriminative patterns of different classes. On the other hand, a sophisticated classifier may not produce better performance when it is fed with statistical vectors having less discriminative patterns. While transforming raw sequences into statistical vectors, discriminative patterns come from nucleotides or amino acids distribution patterns that exist in raw DNA, RNA and protein sequences. It means for transforming raw sequences into statistical vectors, we need robust and precise encoding methods that capture long-range dependencies and discriminative distribution patterns of nucleic or amino acids and encode such information into statistical vectors. The primary goal of this chapter is to provide a brief description of various encoding methods that have been proposed for the conversion of raw DNA or RNA sequences into statistical vectors.

5.1 Nucleotides Distribution Based Encoding Methods

DNA and RNA sequence analysis tasks are similar to Natural Language Processing tasks. DNA and RNA sequences can be considered equivalent to documents or sentences which contain words as features. In DNA and RNA sequences, we can generate features in the form of kmers that are sub-sequences of length k. The kmers are generated by sliding a fixed-size window over the whole sequence. Figures 5.1 and 5.2 illustrate the process of overlapped and non-overlapped kmer generation, where it can be seen for window size or kmer size k = 1, monomers are generated. In case of window size k = 2, bimers are generated and for window size k = 3, trimers are generated. Similarly, with the increase in number k, the length of generated kmers will increase. Here, it is important to mention that window size

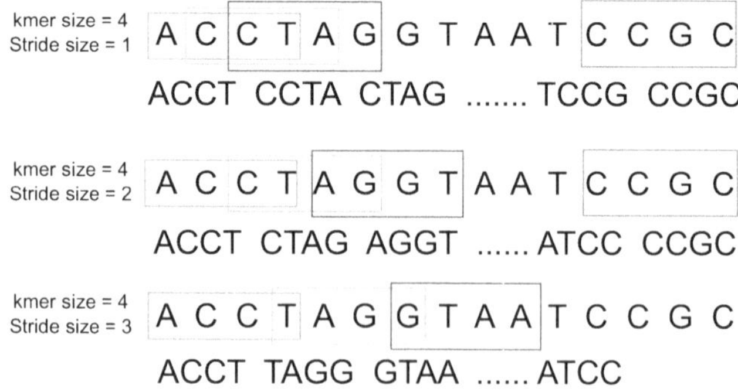

Fig. 5.1 A high level overview of generation of over-lapped kmers

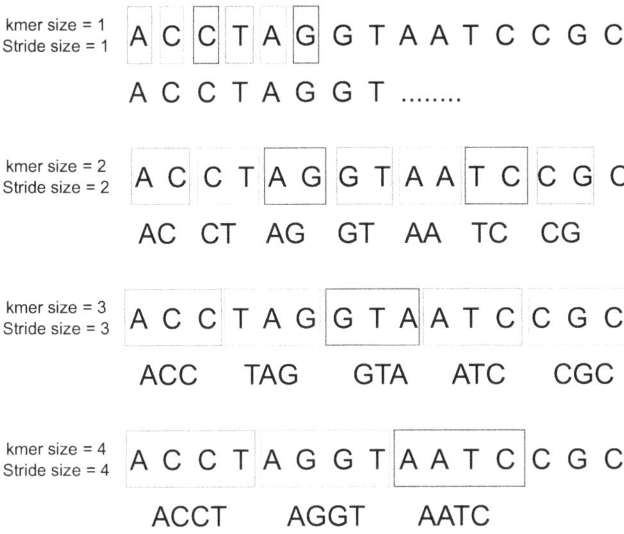

Fig. 5.2 A high level overview of generation of non-overlapped kmers

not only controls the length of kmers but also controls the unique vocabulary of corpus. As DNA and RNA sequences are made up from repetitive patterns of 4 basic nucleotides so unique vocabulary of corpus can be computed using following expression 4^k. Here k denotes the size of window. In case of k = 1 unique vocabulary will be $4^1 = 4$, and for k = 2 unique vocabulary will be $4^2 = 16$.

Furthermore, while generating kmers, apart from window size, selection of appropriate stride size of window is also important. Stride size refers to the step size or the amount by which the sliding window moves along the sequence when generating k-mers. When stride size is equal to window size then non-overlapped kmers are generated and when stride size is less than window size overlapped kmers are generated. Non-overlapped kmers offer more discriminative patterns that are important to discriminate sequences into different classes. While over-lapped kmers contain both discriminative and semantic information about distribution of nucleotides in the sequences. In both overlapped and non-overlapped kmers, window and stride size generate same length kmers as well as vocabulary of kmers. Only difference is discriminative and semantic potential of nucleotides in generated kmers. Figures 5.1 and 5.2 illustrate the process of overlapped and non-overlapped kmer generation.

After generating kmers, next step is to compute distribution of kmers and transform kmers based sequence samples into statistical vectors. Following sub-sections briefly describe working paradigm of kmer encoder variants including nucleic acid composition (NAC), Di-nucleotide composition, tri-nucleotide composition, and higher order nucleotide composition.

5.1.1 Nucleic Acid Composition (NAC)

Nucleic Acid composition (NAC) encoder makes use of 1 window and 1 stride size to generate kmers (monomers or unimers) of sequences and computes unique unimers occurrence frequencies in each sequence. Furthermore, it normalizes occurrence frequencies with sequence length. Equation 5.1 illustrates mathematical expression for computing normalized frequencies of unimers.

$$Encoding(unimer_i) = \frac{\text{Occurrence frequency of } unimer_i}{\text{Sequence length}} \quad \text{where } unimer_i \; \varepsilon \; \{A, C, G, T\} \tag{5.1}$$

In the above expression, $unimer_i$ represents ith unimer which is a member of a set comprising of four distinct unimers. Figure 5.3a illustrates complete workflow for transforming raw sequence into statistical vector using NAC encoder. It can be seen in Fig. 5.3 a toy DNA sequence AAGTATC is segregated into unimers A A G T A T C. Furthermore, occurrence frequencies of 4 unique unimers are computed. As depicted in Fig. 5.3a, Unimer A occurred thrice times while unimers C and G occurred once and unimer T occurred twice. Therefore, occurrence frequencies of A, C, G and T are 3, 1, 1 and 2 respectively. The length of sequence is 7, so according to Eq. 5.1, after normalizing unique unimers frequencies with 7, Final statistical vector of sequence 'AAGTATC' is {0.42, 0.14, 0.14, 0.28}. NAC encoder transforms raw sequences into 4 dimensional statistical vectors.

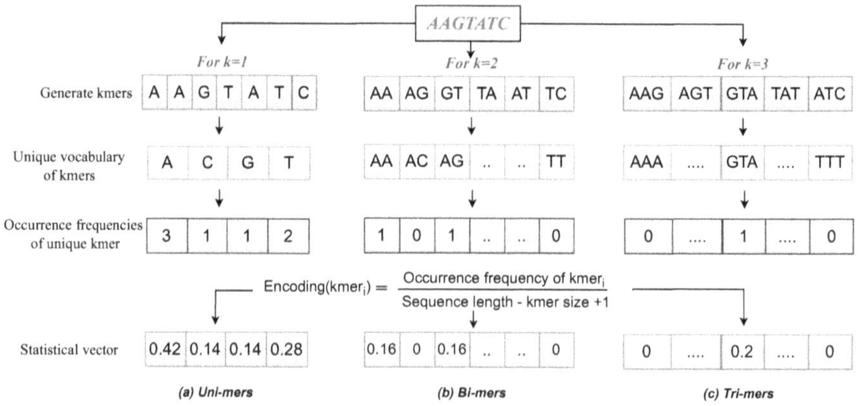

Fig. 5.3 A comprehensive graphical illustration depicting the workflow of three different encoding methods: (**a**) Nucleic acid composition. (**b**) Di-nucleotide composition (DNC). (**c**) Tri-nucleotide composition

5.1.2 Di-nucleotide Composition (DNC)

Di-nucleotide composition encoder can be used for transforming raw sequences into statistical vectors by computing occurrence frequencies of both overlapped and non-overlapped bimers. To generate non-overlapped bimers, it utilizes window and stride size 2. On the other hand, to generate overlapped kmers, it utilizes window size 2 and stride size 1. After generating bimers, it computes occurrence frequencies of 16 unique bimers. Next, it normalizes occurrence frequencies with sequence length − kmer size + 1. Equation 5.2 illustrates mathematical expression for computing distribution of unique bimers.

$$Encoding(bimer_i) = \frac{\text{Occurrence frequency of } bimer_i}{\text{Sequence length} - 1}$$

$$\text{where } bimer_i \; \varepsilon \; \{AA, AC, AG, AT, CA, CC, CG, CT, GA, GC, GG, GT, TA, TC, TG, TT\} \tag{5.2}$$

In the above expression, $bimer_i$ is the ith bimer from 16 unique bimers dictionary. Figure 5.3b illustrates a toy sequence AAGTATC and transformation into statistical vector using DNC encoder. This encoder transforms raw sequences into 16 dimensional statistical vectors. It is rarely possible that a sequence may not contain all 16 unique bimers. However, if any bimer does not occur in a sequence then it is represented with 0 value.

5.1.3 Tri-nucleotide Composition (TNC) and Higher-Order

Similar to DNC, Tri-nucleotide Composition (TNC) encoder also generates overlapped and non-overlapped trimers. In the process of non-overlapped trimers generation, a 3 size window with 3 stride size is slided over raw sequences. To generate overlapped trimers, there is an option to choose either stride size 1 or 2, with both stride sizes and window size 3, overlapped trimers are generated. While developing a computational predictor, a comprehensive experimentation is required to choose an appropriate value of stride size. After generating trimers, it computes unique trimers occurrence frequencies and normalizes them using *sequence length − kmer size + 1*. Equation 5.3 illustrates mathematical expression to compute distribution of a trimer.

$$Encoding(Trimer_i) = \frac{\text{Occurrence frequency of } trimer_i}{\text{Sequence length} \; - \; 2} \tag{5.3}$$

where $trimer_i$ ε { *AAA, AAC, AAG, AAT, ACA, ACC, ACG, ACT, AGA, AGC, AGG, AGT, ATA, ATC, ATG, ATT, CAA, CAC, CAG, CAT, CCA, CCC, CCG, CCT, CGA,*

CGC, CGG, CGT, CTA, CTC, CTG, CTT, GAA, GAC, GAG, GAT, GCA, GCC, GCG, GCT, GGA, GGC, GGG, GGT, GTA, GTC, GTG, GTT, TAA, TAC, TAG, TAT, TCA, TCC, TCG, TCT, TGA, TGC, TGG, TGT, TTA, TTC, TTG, TTT}.

It can be seen from Fig. 5.3c, using window size 3 and stride size 1, overlapped trimers AAG, AGT, GTA, TAT, ATC are generated. After, computing occurrence frequencies of unique trimers and normalizing them final statistical vector of sequence is computed. This encoder transforms raw sequences into 64 dimensional statistical vectors.

Similarly, for higher values of k ($k \geq 4$), this encoder generates higher length kmers and sequences are transformed into 4^k dimensional statistical vectors.

5.1.4 *One Hot Vector Encoding Method*

One-hot vector encoding method transforms raw DNA and RNA sequences into statistical vectors by capturing categorical relationships of nucleotides or groups of nucleotides known as k-mers. This method is widely used in various DNA and RNA sequence analysis tasks such as prediction of N6-methyladenine sites [674], DNA/RNA sequence binding specificities [48, 999], arginine methylation sites [1225], RBP binding sites in LNCRNAs [1210]. This encoder first segregates raw sequences into k-mers by sliding a k size window wit a particular stride size. Once the k-mers are generated, it identifies the unique k-mers and assigns a vector to each of these k-mers. In each k-mer vector, only one entry is set to 1, while the rest of the entries are set to 0. This encoder generates statistical vectors of dimensions unique k-mers × sequence length. To gain a basic understanding of how the One-hot vector encoding method operates, consider a toy sequence, "ACGUCGUAGU." When we use a window size of 1, the sequence individual units, or unimers, are created. Equation 5.4 illustrates the one-hot vectors for each of the four distinct unimers.

$$A = \begin{pmatrix} 1 \\ 0 \\ 0 \\ 0 \end{pmatrix} G = \begin{pmatrix} 0 \\ 1 \\ 0 \\ 0 \end{pmatrix} U = \begin{pmatrix} 0 \\ 0 \\ 1 \\ 0 \end{pmatrix} C = \begin{pmatrix} 0 \\ 0 \\ 0 \\ 1 \end{pmatrix} \tag{5.4}$$

Using one hot vectors of 4 unique unimers, toy sequence is transformed into statistical vector as "1000000101000010000101000010100001000010". However, one-hot vector representation is not suitable for higher order k-mers such as 4-mers, 5-mers. In this case, the resulting vector is high-dimensional, however almost all parts of the vector are zero and contain no relevant information, which forces a high computational complexity on the following deep learning methodology without any merit. Furthermore, the location of the value which is set to one is arbitrarily set and therefore no semantic connection exists between two different k-mers.

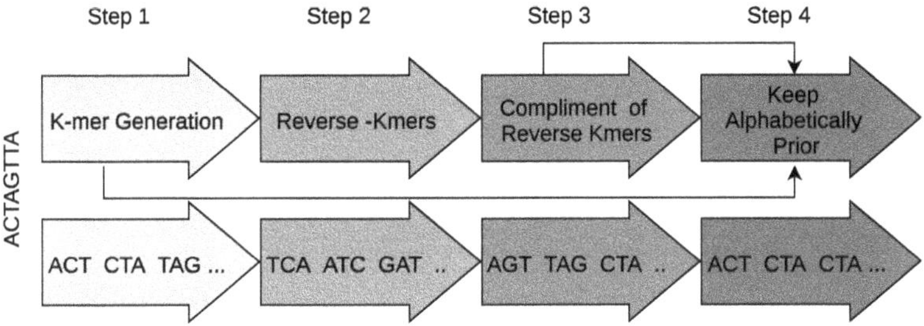

Fig. 5.4 Flow diagram of RC-Kmers encoding

5.1.5 Reverse Compliment-Kmer (RCKmer)

This encoder has been utilized in different types of DNA and RNA sequence analysis tasks such as Promoters identification [1215], enhancers identification [593], DNA N4-methylcytosine sites prediction [1124]. To transform DNA and RNA sequences into statistical feature space, working paradigm of this encoder can be summarized into 4 different steps shown in Fig. 5.4. First, it generates kmers by sliding k-size window over whole sequence. Then, it reverses the order of nucleotides in the generated kmers and computes compliments of reversed kmers by exchanging A with T/T with A and C with G/G with C. Let's if the generated kmers are CTT and ACA, their reverse compliments can be formulated as AAG and TGT. In the next step, it compares complimented kmers with original kmers and keeps kmers that is alphabetically prior. Furthermore, it computes occurrence frequencies of selected kmers to map on dictionary.

Following subsections elaborate RCkmer based encoding process for unimers, also known as nucleic acid composition, for bimers known as di-nucleotide composition, and tri-nucleotide composition for trimers.

5.1.5.1 Nucleic Acid Composition (NAC)

NAC encoder segregates sequences into unimers by sliding a 1 size window. In this particular case as each kmer contain only one nucleotide so there is no need to reverse the order of nucleotide. Let's consider a toy sequence ATCGGACCA. Its segregated sequence is processed to generate complimented kmers as illustrated in Fig. 5.5. Then, it computes their count and normalizes it with sequence length − kmer size + 1 by using Eq. 5.5.

$$Encoding(unimer_i) = \frac{\text{Occurrence frequency of } unimer_i}{\text{Sequence length}} \tag{5.5}$$

$$\text{where } unimer_i \; \varepsilon \; \{A, C, G, T\}$$

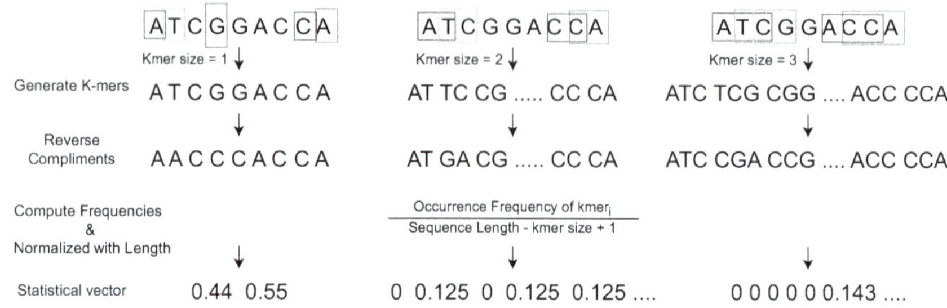

Fig. 5.5 Overall encoding process of Reverse Compliment-Kmer (RCKmer) encoder

Table 5.1 Reverse compliment of bimers

Kmer	AA	AC	AG	AT	CA	CC	CG	CT	GA	GC	GG	GT	TA	TC	TG	TT
Reverse compliment	AA	AC	AG	AT	CA	CC	CG	AG	GA	GC	CC	AC	TA	GA	CA	AA

In the reverse complement method, only the reverse complements of modified k-mers are highlighted in red to indicate that reverse compliment of others remain unchanged

In the above expression, $unimer_i$ is the generated reverse complimented unimers that belongs to $\{A, C\}$. The statistical representation of the given sequence in Fig. 5.5 is {0.44, 0.55}.

5.1.5.2 Di-nucleotide Composition (DNC)

DNC encoder first generates unique bimers with window and stride size 2 and then reverses their order of nucleotides. In the next step, it computes compliments and removes repeated kmers to generate vocabulary of unique bimers. Then, it computes frequency of each bimer and normalize it with sequence length − kmer size + 1. This particular encoder generates a 10 dimension unique vector representation of bimers (Table 5.1).

5.1.5.3 Tri-nucleotide Composition (TNC) and Higher Order

TNC encoder generates trimers of raw sequence and computes their reverse compliments. Reverse complement of all 64 trimers are shown in Table 5.2. After segregating unique trimers, it counts their occurrence frequencies, normalizes using Eq. 5.5, and generates a feature vector. In this particular case, 32-dimensional feature space is generated for raw sequence. For higher values of k ($k \geq 4$), it generates kmers and computes their reverse compliments. After removing repeated kmers to generate feature space of unique kmers, it computes frequency for higher length subsequences/kmers.

Table 5.2 Reverse compliment of trimers

kmer	Reverse compliment	kmer	Reverse compliment	kmer	Reverse compliment	kmer	Reverse compliment
AAA	AAA	CAA	CAA	GAA	GAA	TAA	TAA
AAC	AAC	CAC	CAC	GAC	GAC	TAC	GTA
AAG	AAG	CAG	CAG	GAG	CTC	TAG	CTA
AAT	AAT	CAT	ATG	GAT	ATC	TAT	ATA
ACA	ACA	CCA	CCA	GCA	GCA	TCA	TCA
ACC	ACC	CCC	CCC	GCC	GCC	TCC	GGA
ACG	ACG	CCG	CCG	GCG	CGC	TCG	CGA
ACT	ACT	CCT	AGG	GCT	AGC	TCT	AGA
AGA	AGA	CGA	CGA	GGA	GGA	TGA	TCA
AGC	AGC	CGC	CGC	GGC	GCC	TGC	GCA
AGG	AGG	CGG	CCG	GGG	CCC	TGG	CCA
AGT	ACT	CGT	ACG	GGT	ACC	TGT	ACA
ATA	ATA	CTA	CTA	GTA	GTA	TTA	TAA
ATC	ATC	CTC	CTC	GTC	GAC	TTC	GAA
ATG	ATG	CTG	CAG	GTG	CAG	TTG	CAA
ATT	AAT	CTT	AAG	GTT	AAC	TTT	AAA

In reverse complement method, only reverse complements of modified k-mers are highlighted in red to indicate that reverse compliment of others remain unchanged

5.1.6 *Cumulative Skew*

This encoder has been utilized in different types of DNA and RNA sequence analysis tasks such as genome analysis [366], genomic compositional asymmetry and replication selection degree prediction [41, 452, 1080]. Commutative Skew encoder transforms raw DNA or RNA sequences into 2-dimensional statistical vectors. Figure 5.6 provides a visual representation of the process involved in converting raw sequences into statistical vectors. First, it calculates occurrence frequencies of four unique nucleotides (A, C, G, and T/U). Next, for the computation of AT/U skewness, it computes difference between occurrence frequencies of Adenine (A) and Thymine/Uracil (T/U) nucleotides. This score is normalized with combined frequencies of Adenine and Thymine/Uracil nucleotides. Similarly in case of GCskew, it computes difference between G and C nucleotides frequencies and normalizes this score with the combined frequencies of G and C nucleotides. Finally, it creates a two-dimensional statistical vector by combining scores of both AT/U skew and GC skew. Equations 5.6, 5.7 and 5.8 illustrate mathematical expressions for computing scores of AT/U, GC skew and final statistical vector respectively. In these equations, count(C), count(A), count(T) and count(G), are the occurrence frequencies of cytosine(C), adenine(A), thymine(T) and guanine(G) nucleotides respectively.

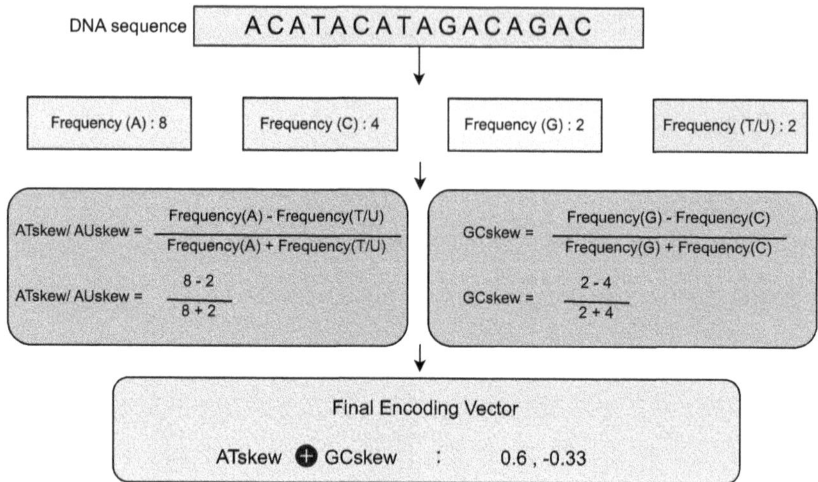

Fig. 5.6 Overall workflow of Cumulative Skew encoder method

$$AT/Uskew = \frac{count(A) - count(T/U)}{count(A) + count(T/U)} \tag{5.6}$$

$$GCskew = \frac{count(G) - count(C)}{count(G) + count(C)} \tag{5.7}$$

$$\text{Statistical Vector} = AT/Uskew \oplus GCskew \tag{5.8}$$

5.1.7 ATGC Ratio

This encoder has been utilized in different types of DNA and RNA sequence analysis tasks such as Promoters identification [37] and DNA modification prediction [738]. This encoder captures and encodes combined distribution patterns of A and T nucleotides with respect to combined distribution patterns of G and C nucleotides. Equation 5.9 illustrates mathematical expression for computing ATGC Ratio. In Eq. 5.9 count(C), count(A), count(T) and count(G), refers to occurrence frequencies of cytosine(C), adenine(A), thymine(T) and guanine (G) nucleotides. Figure 5.7 describes encoding process of ATGC Ratio encoder with a toy sequence example.

$$ATGC\ Ratio = \frac{count(A) + count(T/U)}{count(G) + count(C)} \tag{5.9}$$

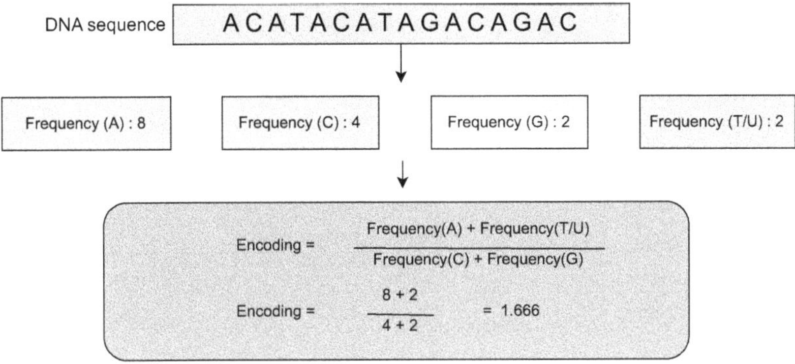

Fig. 5.7 Overall Workflow of ATGC ratio encoding method

5.1.8 GC Content

This encoder has been utilized in different types of DNA and RNA sequence analysis tasks such as growth temperature prediction in prokaryotes [446], Porphyridium mitogenomes analysis [526], shaping amino acids during Bacterial Evolution Process [272] and Saccharomyces cerevisiae genome analysis [596]. This encoder captures and encodes combine distribution of two nucleotides namely G and C. Equation 5.10 illustrates mathematical expression to compute GC content of a sequence where count(C), count(A), count(T) and count(G) refers to occurrence frequencies of cytosine(C), adenine(A), thymine(T) and guanine(G) nucleotides respectively.

$$GC\ content = \frac{count(G) + count(C)}{count(A) + count(T/U) + count(G) + count(C)} \tag{5.10}$$

This encoder transforms DNA or RNA sequences into single dimensional statistical vectors. It computes values in the range of 0 to 1. In a particular sequence, distribution of G and C nucleotides is higher than distribution of A and T nucleotides, if range of computed score is between 0 to 0.5. Conversely, in comparison to A and T nucleotides the distribution of G and C nucleotides will be higher if the computed score lies in the range of 0.5 to 1. Overall, this encoder captures whether distribution of GC is higher or lower than the distribution of AT nucleotides. Figure 5.8 provides a comprehensive illustration of the encoding process employed by the GC Content encoder using a sample toy sequence. The Fig. 5.8 showcases the step-by-step transformation of the sequence, highlighting the specific mechanisms and calculations involved in determining the GC content.

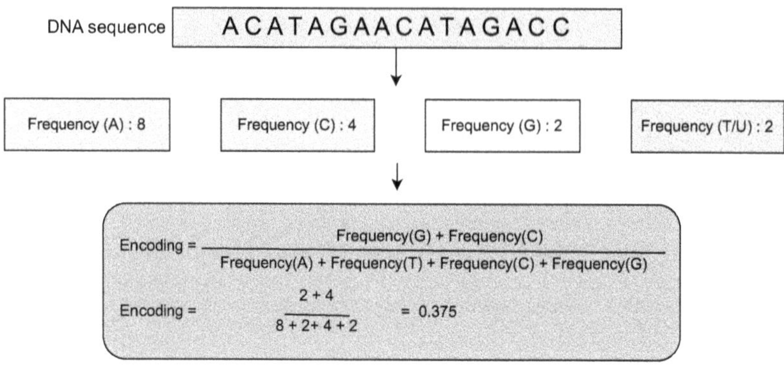

Fig. 5.8 Workflow of GCcontent encoding method

5.1.9 Spectrum

This encoder has been utilized in different types of DNA and RNA sequence analysis tasks such as Promoter identification [37] and DNA modification prediction [738]. Spectrum encoder computes occurrence frequencies of four unique nucleotides (A, T, C, G). Furthermore, it also captures special distribution of two nucleotides (T/U, A) and encodes this information into two special variables namely: X and Y. Both variables represent boolean distribution either 0 or 1. X is formulated as 1 if the first letter of the sequence is T or U, otherwise zero. Similarly, if 10th letter of sequence is A, Y represents to 1 otherwise, it is 0. Finally, it concatenates occurrence frequencies of four unique nucleotides and spatial distribution of two variables X and Y as $count(A) \oplus count(C) \oplus count(G) \oplus count(T/U) \oplus X \oplus Y$. Spectrum encoder transforms raw sequences into 6 dimensional statistical vectors. Figure 5.9 depicts a detailed illustration of the encoding process employed by spectrum encoder. The Figure showcases the step-by-step sequence transformation process and highlights specific mechanisms and calculations involved in determining spectrum of a toy sequence.

5.1.10 Z-curve

This encoder has been utilized in different types of DNA and RNA sequence analysis tasks such as DNA element identification [266], phosphorylation site identification [660] and genome analysis [1205]. Z-Curve encoder computes discriminative distribution of four unique nucleotides in 3 different scenarios. First, it computes distribution difference between A, G and C,T nucleotides. In second scenario, it computes distribution difference between A, C and G, T nucleotides. In third scenario, it computes distribution difference between A, T and C, G nucleotides. In all three scenarios it computes discriminative potential of four

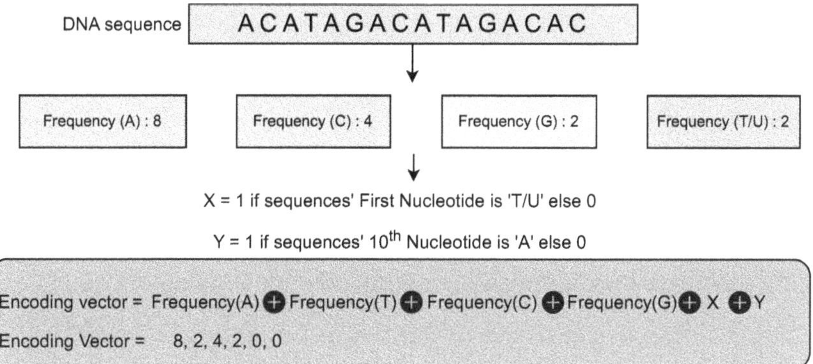

Fig. 5.9 Overall workflow of Spectrum encoding method

unique nucleotides in the form of pairs. Equations 5.11–5.13 represent mathematical expressions for computing discriminative distribution of nucleotides in all three scenarios.

$$X = (count(A) + count(G)) - (count(C) + count(T/U)) \tag{5.11}$$

$$Y = (count(A) + count(C)) - (count(G) + count(T/U)) \tag{5.12}$$

$$Z = (count(A) + count(T/U)) - (count(C) + count(G)) \tag{5.13}$$

In Eqs. 5.11–5.13, count(C), count(A), count(T) and count(G) refers to occurrence frequencies of cytosine(C), adenine(A), thymine(T) and guanine(G) nucleotides respectively. Finally, it concatenates all 3 values and generates a 3-dimensional feature vector as $X \oplus Y \oplus Z$.

Figure 5.10 showcases the step-by-step sequence transformation process and highlights specific mechanisms and calculations involved in determining Z-Curve of a toy sequence.

5.1.11 Accumulated Nucleotide Frequency

This encoder has been utilized in different types of DNA or RNA sequence analysis tasks such as prediction of N6-methyladenosine [170] and N2-methylguanosine Sites [174]. This encoder transforms raw DNA and RNA sequences into statistical vectors by computing accumulated distribution of nucleotides. Working paradigm of this encoder can be summarized in 3 different steps. It generates kmers and computes occurrence frequencies of kmers in an accumulated fashion. Specifically, at a particular position of a kmer, it computes occurrence frequencies of the kmer in

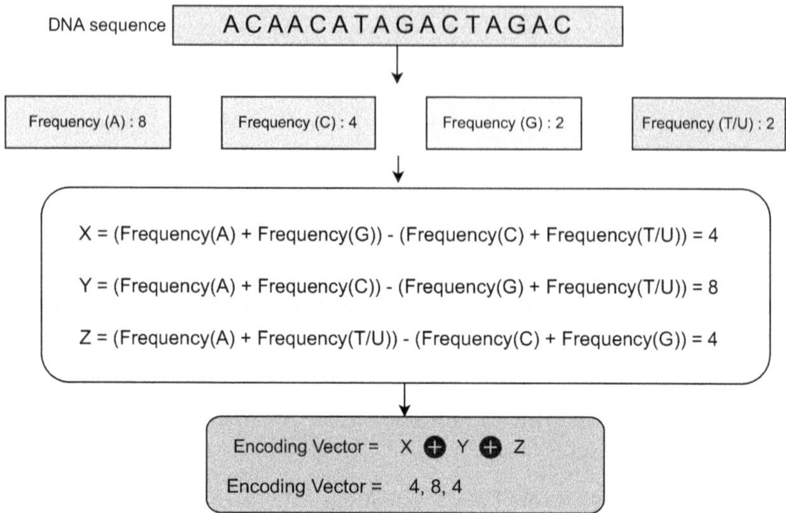

Fig. 5.10 Overall workflow of Zcurve encoding method

all previous positions and normalizes computed score with sequence length until that particular kmer position. It generates a feature vector of dimension sequence length − kmer size +1. Equation 5.14 illustrates mathematical expression to compute accumulated frequencies of kmers.

$$Encoding(kmer_i) = \frac{Occurrence\ frequency\ of\ kmer_i\ upto\ current\ position\ of\ kmer_i}{Sequence\ length\ upto\ current\ position\ of\ kmer_i} \tag{5.14}$$

To more briefly explain encoding process, consider a toy sequence ACATAGACTGCA and generated unimers such as $\{A, C, A, T, A, G, A, C, T, G, C, A\}$. It calculates the cumulative count of unimers up to the given position and then normalizes this count by the length of the kmer up to that particular position. This can be illustrated as follows: $\left\{1, \frac{1}{2}, \frac{2}{3}, \frac{1}{4}, \frac{3}{5}, \frac{1}{6}, \frac{4}{7}, \frac{2}{8}, \frac{2}{9}, \frac{2}{10}, \frac{3}{11}, \frac{5}{12}\right\}$. Following similar process, this encoder can be employed to convert DNA sequences into statistical vectors for k-mers of any desired size. Such as for kmer size 2, it segregates toy sequence into bimers as $\{AC, CA, AT, TA, AG, GA, AC, CT, TG, GC, CA\}$. It generates statistical vector as: $\left\{1, \frac{1}{2}, \frac{1}{3}, \frac{1}{4}, \frac{1}{5}, \frac{1}{6}, \frac{2}{7}, \frac{1}{8}, \frac{1}{9}, \frac{2}{10}, \frac{2}{11}, \right\}$. Figure 5.11, depicts a detailed illustration of the encoding process employed by ANF encoder.

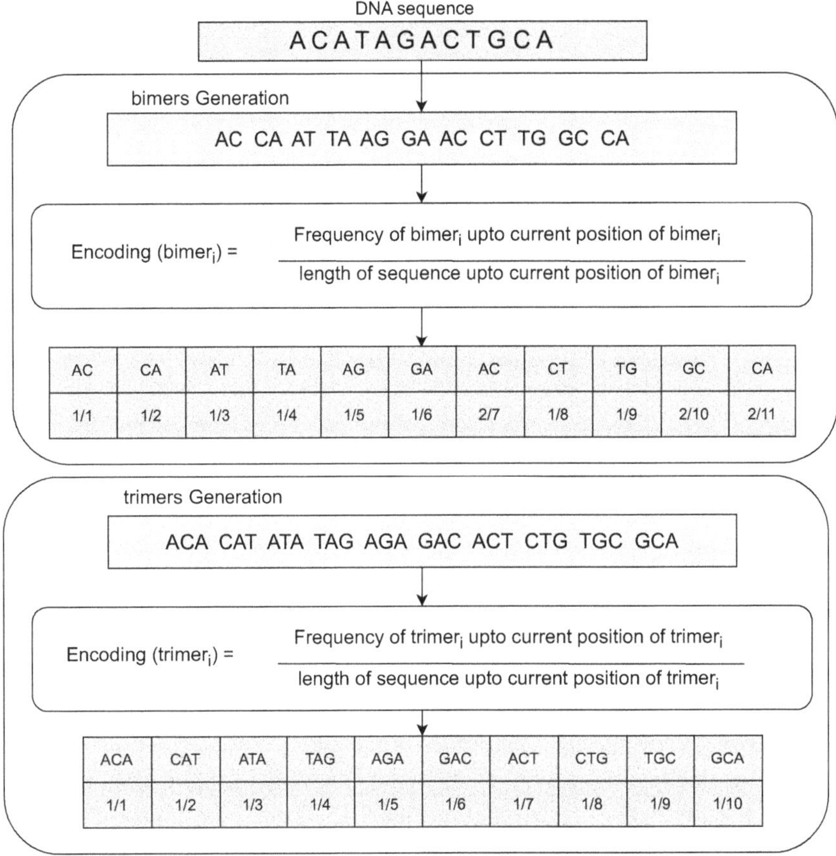

Fig. 5.11 Overall workflow of Accumulated Nucleotide Frequency (ANF) method

5.1.12 *Frequency Chaos*

This encoder has been utilized in different types of DNA and RNA sequence analysis tasks such as similarity comparison [428], and sequence alignment [664]. Working paradigm of frequency chaos encoder can be summarized in 3 different steps. First, it segregates sequences into kmers. Secondly, it computes occurrence frequencies of unique kmers and normalizes computed frequencies with sequence length − kmer size + 1. Furthermore, at every position of unique kmers it substitutes their computed normalized frequencies. It generates a final statistical vector of dimension sequence length − kmer size − unique kmers having zero frequencies + 1. Figure 5.12 shows a comprehensive visual representation of the encoding process employed by Frequency Chaos encoder using a sample toy sequence.

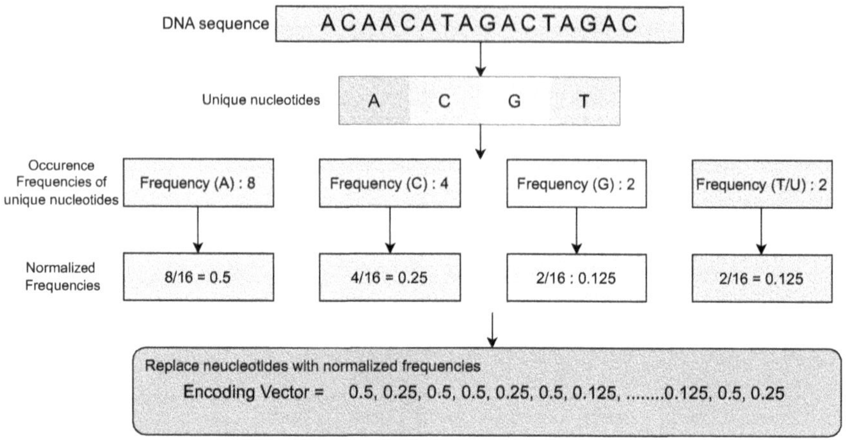

Fig. 5.12 Working of frequency chaos encoder

5.1.12.1 Fickitt Score

This encoder has been utilized in different types of DNA and RNA sequence analysis tasks such as long noncoding RNA Identification [392], RNA sequence coding potential prediction [996], RNA-associated interactions [1030] and plant lncRNA identification [1197]. Working paradigm of Fickett Score encoder can be summarized into 7 different steps. First, it computes occurrence frequencies of four basic nucleotides namely A, C, G, and T. Secondly, it normalizes computed frequencies with length of sequence. Equation 5.15 illustrates mathematical expression for computing normalized occurrence frequencies of nucleotides and $content_i$ represents to normalized frequency of each nucleotide.

$$Content_i = \frac{Occurrence\ Frequency\ of\ Nucleotide_i}{Sequence\ Length},\ Nucleotide_i \in \{A, C, G, T\} \tag{5.15}$$

Thirdly, it acquires precomputed occurrence probabilities of nucleotides by comparing each nucleotide content value with a pre-computed content parameter value shown in Table 5.3. Through aforementioned comparison, it acquires nucleotide pre-computed probability value corresponding to the range of precomputed content parameter value in which nucleotide content value lies. Furthermore, it also acquires nucleotide weight value from Table 5.4 and make use of mathematical expression of Eq. 5.16 for computing nucleotides content probability.

Table 5.3 Characteristic parameters of nucleotides for Fickitt Score 1

Position parameter	Probability of nucleotides				Content parameter	Probability of nucleotides			
	A	C	G	T		A	C	G	T
0.1 to 1.1	0.22	0.23	0.08	0.09	0.00 to 0.17	0.21	0.31	0.29	0.58
1.1 to 1.2	0.20	0.30	0.08	0.09	0.17 to 0.19	0.81	0.39	0.33	0.51
1.2 to 1.3	0.34	0.33	0.16	0.20	0.19 to 0.21	0.65	0.44	0.41	0.69
1.3 to 1.4	0.45	0.51	0.27	0.54	0.21 to 0.23	0.67	0.43	0.41	0.56
1.4 to 1.5	0.68	0.48	0.48	0.44	0.23 to 0.25	0.49	0.59	0.73	0.75
1.5 to 1.6	0.58	0.66	0.53	0.69	0.25 to 0.27	0.62	0.59	0.64	0.55
1.6 to 1.7	0.93	0.81	0.64	0.68	0.27 to 0.29	0.55	0.64	0.64	0.40
1.7 to 1.8	0.84	0.70	0.74	0.91	0.29 to 0.31	0.44	0.51	0.47	0.39
1.8 to 1.9	0.68	0.70	0.88	0.97	0.31 to 0.33	0.49	0.64	0.54	0.24
1.9 to 2.0+	0.94	0.80	0.90	0.97	0.33 to 0.99	0.28	0.82	0.40	0.28

Table 5.4 Weights to be given to nucleotides for Fickitt Score 1

Nucleotides	Position weight	Content weight
A	0.26	0.11
C	0.18	0.12
G	0.31	0.15
T	0.33	0.14

$$P(Content) = \sum_{i=1}^{n} content_i\ weight \times content_i\ probability \tag{5.16}$$

In the above expression, "$content_i\ weight$" represents the content weight of ith nucleotide, "$content_i\ probability$" is the probability of ith nucleotide and n indicates all four nucleotides (A, C, G, T). In fourth step, a mode operation is performed on each nucleotide position of raw sequence with a divisor of 3, and each nucleotide is categorized into one of three subsequences based on the remainder value. Equation 5.17 presents the mathematical expression used to categorize a nucleotide into one of three subsequences, labeled as subsequence 1, subsequence 2, and subsequence 3.

$$\begin{cases} i\%3 = 0, & Subsequence\,1\,(SS_1) \\ i\%3 = 1, & Subsequence\,2\,(SS_2) \\ i\%3 = 2, & Subsequence\,3\,(SS_3) \end{cases} \tag{5.17}$$

In the fifth step, it determines the position of a nucleotide relative to its occurrence within three distinct subsequences, considering both maximum and minimum occurrences. Equation 5.18 depicts mathematical expression for computing position of each nucleotide.

$$Position\ N_i = \frac{max(Freq(N_i)\ in\ SS_1, Freq(N_i)\ in\ SS_2, Freq(N_i)\ in\ SS_3)}{min((Freq(N_i)\ in\ SS_1, Freq(N_i)\ in\ SS_2, Freq(N_i)\ in\ SS_3) + 1},$$
$$N_i \in \{A, C, G, T\} \tag{5.18}$$

In Eq. 5.18, $Freq(N_i)$ in SS_1 represents to occurrence frequency of ith nucleotide in sub sequence 1 and max function refers to maximum occurrence count of a nucleotide in all 3 subsequences. Min function represents to minimum occurrence count of a nucleotide in all thee subsequences. Similar to $content_i$, it make use of Table 5.3 and Position value of ith nucleotide to acquire nucleotide pre-computed probability value corresponding to the range of precomputed Position Parameter value in which nucleotide computed position value lies. By using this position probability and position weight of nucleotide, it computes position probability of nucleotides as depicted in Eq. 5.19.

$$P(Position) = \sum_{i=1}^{n} position_i\ weight \times position_i\ probability \tag{5.19}$$

Sixth step computes Fickett score by taking sum of both position and content probabilities shown in Eq. 5.20.

$$Fickett\ Score\_1 = P(Content_i)_1 + P(Position\ N_i)_1 \tag{5.20}$$

Furthermore, it computes Fickket score 2 (Eq. 5.21) by reusing $Content_i$ and Position N_i computed through Eqs. 5.15 and 5.18. For both $Content_i$ and Position N_i values, it utilizes different weight values using Table 5.6 and acquires different precomputed values of nucleotides probabilities from Table 5.5.

$$Fickett\ Score\_2 = P(Content_i)_2 + P(Position\ N_i)_2 \tag{5.21}$$

Seventh step concatenates both values and output a 2 dimension feature vector as illustrated in Eq. 5.22.

$$Feature\ Vector = Fickett\ Score\_1 \oplus Fickett\ Score\_2 \tag{5.22}$$

To briefly understand working paradigm of this encoder, consider a toy sequence “ACGTAGTCTAGCTAGTA”. In this sequence occurrence frequencies of 4 nucleotides is as A:5, C:3, G:4, T:5. The length of sequence is 16 so normalized frequencies also known as content of nucleotides are $A = \frac{5}{17} = 0.294$, $C = \frac{3}{17} = 0.176$, $G = \frac{4}{17} = 0.235$, $T = \frac{5}{17} = 0.294$. In next step while comparing nucleotide A content value (0.294) with content parameter column of Table 5.3, A nucleotide content value lies in the range of 0.29 to 0.33. Its corresponding probability value 0.44 is acquired. Similarly, by comparing content values of other 3 nucleotides C, G and T the acquired probability values are 0.51,

Table 5.5 Characteristic parameters of nucleotides for Fickitt Score 2

	Probability of nucleotides					Probability of nucleotides			
Position parameter	A	C	G	T	Content parameter	A	C	G	T
0.1 to 1.1	0.36	0.38	0.17	0.24	0.00 to 0.17	0.19	0.33	0.23	0.51
1.1 to 1.2	0.50	0.51	0.31	0.39	0.17 to 0.19	0.38	0.59	0.44	0.60
1.2 to 1.3	0.54	0.56	0.41	0.48	0.19 to 0.21	0.45	0.56	0.52	0.57
1.3 to 1.4	0.57	0.60	0.52	0.58	0.21 to 0.23	0.45	0.47	0.57	0.52
1.4 to 1.5	0.58	0.60	0.62	0.67	0.23 to 0.25	0.48	0.45	0.56	0.48
1.5 to 1.6	0.48	0.52	0.61	0.62	0.25 to 0.27	0.52	0.46	0.52	0.48
1.6 to 1.7	0.52	0.49	0.65	0.64	0.27 to 0.29	0.58	0.50	0.50	0.53
1.7 to 1.8	0.57	0.55	0.74	0.69	0.29 to 0.31	0.58	0.59	0.47	0.56
1.8 to 1.9	0.55	0.44	0.67	0.60	0.31 to 0.33	0.55	0.63	0.40	0.49
1.9 to 2.0+	0.51	0.29	0.62	0.51	0.33 to 0.99	0.40	0.50	0.21	0.30

0.47 and 0.39, respectively. Furthermore, form Table 5.4, four nucleotides weights are acquired as A: 0.11, C: 0.12, G: 0.15 and T: 0.14. These weights are multiplied with nucleotides content probabilities and after summation final probability 0.2487 is computed by using Eq. 5.16. In the next step, it segregates whole sequence into three subsequence by taking sequence nucleotide position value mode with 3 as illustrated in Eq. 5.17. For given toy sequence, nucleotide at zeroth position is A, at first is C and at second position is G. It computes mode of position with 3 and based on remainder value, it categorizes the nucleotide to subsequences 1, 2 and 3. For above three nucleotide, it takes 0%3 which outputs 0 and based on remainder value, it categorizes A in subsequence 1. Similarly, 1%3 is equal to 1, therefore C is categorized in subsequence 2 and for G, 2%3 = 2, so G belongs to subsequence 3. It repeats the same procedure for all nucleotides in sequence and generates three subsequences as S_1: ATTATT, S_2: CACGAA and S_3: GGTCG. Then, it computes occurrence frequency of each nucleotides in all three subsequences such as S_1: (A= 2, C= 0, G= 0, T= 4); S_2: (A= 3, C= 2, G= 1, T= 0); S_3: (A= 0, C= 1, G= 3, T= 1). Equation 5.18 utilizes these occurrence frequencies and extracts positional information of each nucleotide in sequence which are $Position\ N_A = 3$, $Position\ N_C = 2$, $Position\ N_G = 3$ and $Position\ N_T = 4$, respectively.

It follows the same procedure for calculating position probabilities as it does for computing content probability. By comparing values with position parameter in Table 5.3, it can be seen that these position values lies in the range 1.9 to 2.0+. Therefore, position probabilities of nucleotide A, C, G and T are 0.94, 0.80, 0.90 and 0.97, respectively. In the next step, it acquires position weights of all four nucleotides as A: 0.26, C: 0.18, G: 0.31, T: 0.33. Equation 5.19 makes use of position weights and probabilities and computes probability with respect to position as P(Position) = 0.9875. Then, it calculates sum of content and position probability which is 1.2362, also known as $Fickket\ Score\_1$. It repeats the same process by utilizing probabilities of nucleotides with respect to computed content and position from Table 5.5. In this iteration, it utilizes precomputed weights of nucleotides from

Table 5.6 Weights to be given to nucleotides for Fickitt Score 2

Nucleotides	Position weight	Content weight
A	0.062	0.084
C	0.093	0.076
G	0.205	0.081
T	0.154	0.055

Table 5.6. The final score calculated for this iteration is known as $Fickket\ Score\_2$, which is equal to 0.4266. The final statistical vector generated by using Fickket score approach [1.1362 0.4266]

5.1.13 Enhanced Nucleic Acid Composition (ENAC)

This encoder has been utilized in different types of DNA and RNA sequence analysis tasks such as Promoters identification [1215], enhancers identification [593] and DNA N4-methylcytosine sites prediction [1124]. ENAC encoder transforms DNA and RNA sequences into statistical vectors by capturing distribution of nucleotides from different regions of sequences independently. First, it splits each sequence into subsequences by sliding n size window. Here, window size n is a hyperparameter that needs to be set according to nature and complexity of the corpus. In case of small window size each sequence is segregated into multiple subsequences and distribution of nucleotides is computed from small regions of sequence. While in case of large window size each sequence is segregated into few subsequences and distribution of nucleotides is computed from long regions of subsequences. After generating subsequences, it computes occurrence frequencies of unique nucleotides in each subsequence and normalizes computed frequencies with number of unique nucleotides of subsequence. Equation 5.23 illustrates mathematical expression for computing distribution of nucleotides in subsequences.

$$Encoding(Subsequence_k) = \frac{Occurrence\ frequency\ of\ unique\ neucleotide_i\ in\ subsequence_k}{Unique\ nucleotides\ in\ subsequence_k},$$
$$\text{where unique } nucleotide_i \in \{A, C, G, T\} \tag{5.23}$$

It generates a final statistical vector of dimension number of subsequences × 4.

Let's if raw sequence is ACATATAGAC, and window size is set as 5, this encoder will slide a 5-sized window over the sequence and generates sub-sequences as ACATA, and TAGAC. BY using Eq. 5.23 it calculates the normalized count of 4 nucleotides for each sequence as $\{1, 0.3, 0, 0.3\}$, and $\{0.5, 0.25, 0.25, 0.25\}$. The last step concatenates all the vectors and generates a (4×4) dimension feature

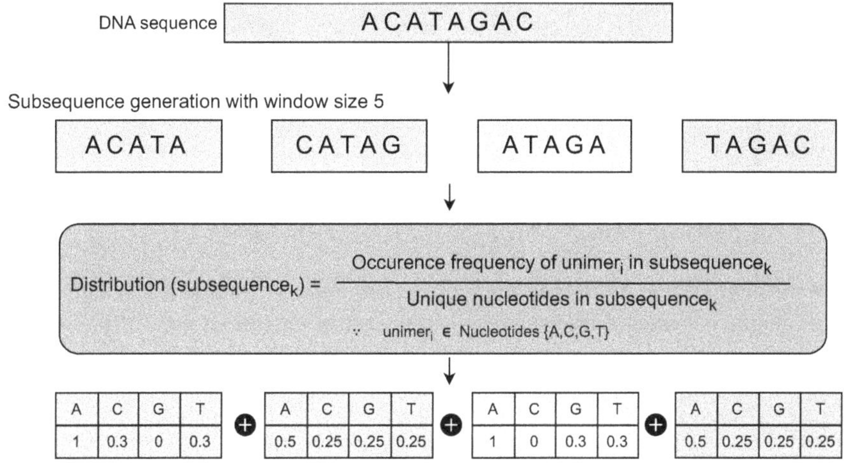

Fig. 5.13 Workflow of enhanced nucleic acid composition (ENAC) encoder

vector. Figure 5.13 depicts a detailed illustration of the encoding process employed by Enhanced Nucleic Acid Composition encoder using a sample toy sequence.

5.1.14 Composition of k-Spaced Nucleic Acid Pairs (CKSNAP)

This encoder has been utilized in different types of DNA and RNA sequence analysis tasks such as miRNA-disease associations identification [612], enhancers identification [593] and prediction of species-specific yeast DNA replication origin [692]. CKSNAP encoder transforms DNA and RNA sequences into statistical vectors by computing gap-based distribution of nucleotides. First, it segregates sequences into simple bimers and bimers with different gap values. Here, gap value is a hyperparameter that needs to be set according to the nature and complexity of task. If the gap value is defined as 3, the encoder will segregate sequences into 4 different types of bimers sequences: one type of bimers sequence with no gap value and 3 types of bimers sequences with gap values 1, 2 and 3. To understand different types of bimers sequences generation process, consider a generic sequence $A_0A_1A_2A_3A_4......A_l$ where A_i represents 4 basic nucleotides A, C, G, T, if gap value is defined as n, the encoder segregates generic sequence into n + 1 different types of bimers sequences that are briefly described in mathematical expression 5.24.

$$\begin{cases} A_0A_1, A_1A_2, A_2A_3, ..., A_{l-1}A_l & nogap \\ A_0A_2, A_1A_3, A_2A_4, ..., A_{l-2}A_l & gap = 1 \\ A_0A_3, A_1A_4, A_2A_5, ..., A_{l-3}A_l & gap = 2 \\ . & . \\ . & . \\ . & . \\ A_0A_{n+1}, A_1A_{n+2}, A_2A_{n+3}, ..., A_{l-(n+1)}A_l & gap = n \end{cases} \tag{5.24}$$

In the next step, it transforms each bimer sequence into statistical feature space by computing occurrence frequencies of 16 unique bimers and normalizes computed frequencies with sequence length − 1. Equation 5.25 illustrates mathematical expression for computing normalized occurrence frequency of each bimer.

$$Encoding(bimer_i) = \frac{\text{Occurrence frequency of } bimer_i}{\text{Sequence length}} - 1,$$
$$\text{where } bimer_i \ \varepsilon \ \{\text{AA, AC, AG, AT, CA, CC, CG, CT, GA, GC, GG, GT, TA, TC, TG, TT}\} \tag{5.25}$$

Equation 5.25 transforms each bimers sequence into 16 dimensional statistical vector. To generate final statistical vector of a sequence, simple bimers and gap based bimers sequence versions 16 dimensional vectors are concatenated. Overall, CKSNAP generates statistical vectors of dimension 16 × gap value + 16.

Figure 5.14 illustrates working paradigm of CKSNAP encoder with a toy sequence sample. In this example gap value is defined 2, so encoder segregates raw sequence into 3 different bimers sequence. Specifically, a bimers sequence without any gap, a bimers sequence with a gap value of 1 and a bimers sequence with a gap value of 2. Additionally, for each of these bimers sequences, the occurrence frequencies of unique bimers are calculated and subsequently normalized with respect to sequence length. The ultimate outcome is a statistical vector formed by the concatenation of the statistical vectors of all three bimers sequences.

5.2 Physio-Chemical Properties Based Encoding Methods

To transform raw DNA or RNA sequences into statistical vectors, researchers have experimentally computed diverse types of physio-chemical values of nucleotides. Raw sequences are transformed into statistical vectors by replacing standalone nucleotides or groups of neucleotides with these pre-computed physio-chemical values. This section briefly describes diverse types of encoding methods that transform DNA and RNA sequences into statistical vector by mapping nucleotides with their pre-computed physio-chemical values.

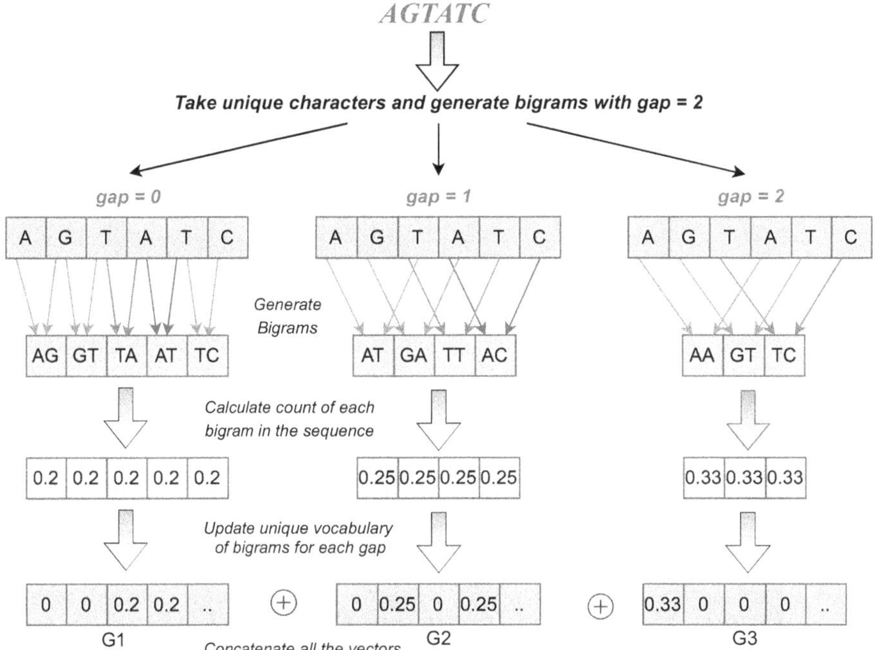

Fig. 5.14 Workflow of Composition of k-spaced Nucleic Acid Pairs (CKSNAP) encoder

5.2.1 *Ionization Constant*

The ionization constant of a chemical equation is the ratio of products and reactants raised to appropriate stoichiometric powers. It depends on the equilibrium between ions and molecules when they are not ionized in a solution. Pirogova et al., [813] experimentally calculated values of ionization constants of amino acids are illustrated in Table 6.20. Corresponding amino acid sequence of the protein sequence is transformed into the numerical vector by using these values. This mapping method has been utilized in exon location identification [231] task.

5.2.2 *Electron-Ion Interaction Potential (EIIP)*

This encoder has been utilized in different types of DNA and RNA sequence analysis tasks such as DNA replication prediction [692], exon location identification [740], cancer classification [739] and miRNA subcellular location prediction [46]. Electron-Ion Interaction Potential encoder transforms raw sequences into statistical vectors by mapping electron affinity values of nucleotides. A graphical illustration of EIIP encoder with a toy sequence is described in

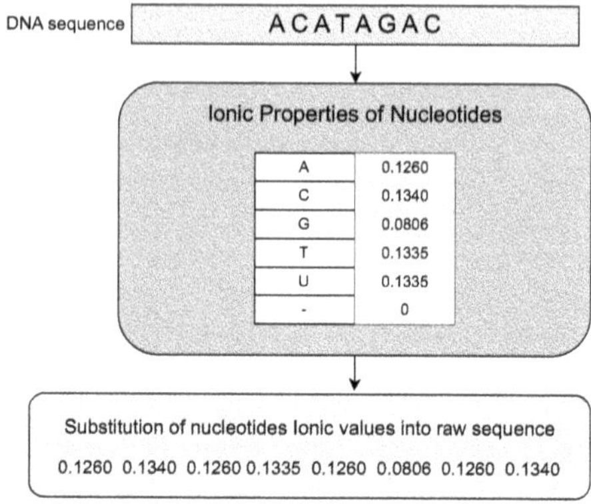

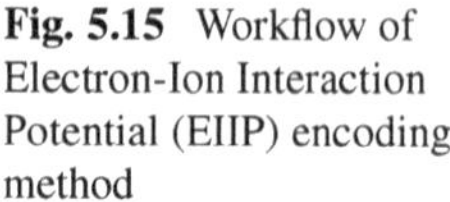
Fig. 5.15 Workflow of Electron-Ion Interaction Potential (EIIP) encoding method

Fig. 5.15. It can be seen in Fig. 5.15, a toy sequence ACATAGAC by substituting nucleotides electron affinity values is represented into a statistical vector as {0.12600.13400.12600.13350.12600.08060.12600.1340}.

5.2.3 *Pseudo Electron-Ion Interaction Potential (EIIP)*

This encoder has been utilized in diverse types of sequence analysis applications including exon location identification [290], detection of CpG Islands [331] and cancer classification [739]. Rather than transforming raw DNA or RNA sequences into statistical vectors by mapping nucleotides with their respective Ionic Properties, Pseudo EIIP encoder first generates bimers of sequences and transforms bimers sequences into statistical vectors by mapping nucleotides bimers with their respective Ionic Properties. Figure 5.16 provides a comprehensive overview of specific mechanisms and calculations involved in transforming raw sequences into statistical vectors by utilizing Pseudo EIIP encoder. PseudoEIIP encoder generates statistical vectors of raw sequences in 3 steps. In first step, it generates bimers of sequence. In next step, it calculates occurrence frequencies of bimers and normalizes them with sequence length $-$ 1. In the last step, feature vector is multiplied by corresponding EIIP values of kmers. Let's consider a toy sequence ATGACA, for k = 2, generated bimers are AT, TG, GA, AC, and CA. By incorporating values of bimers and their corresponding features, the final statistical vector of the raw sequence is {0, 0.26, 0, 0.2595, 0.26, 0, 0, 0, 0.2066, 0, 0, 0, 0, 0, 0.2141, 0} (Table 5.7).

5.2.4 *Nucleotide Chemical Property (NCP)*

This encoder has been utilized in the identification of DNA N4-methylcytosine sites [172], RNA pseudouridine sites identification [754], DNA modification sites predic-

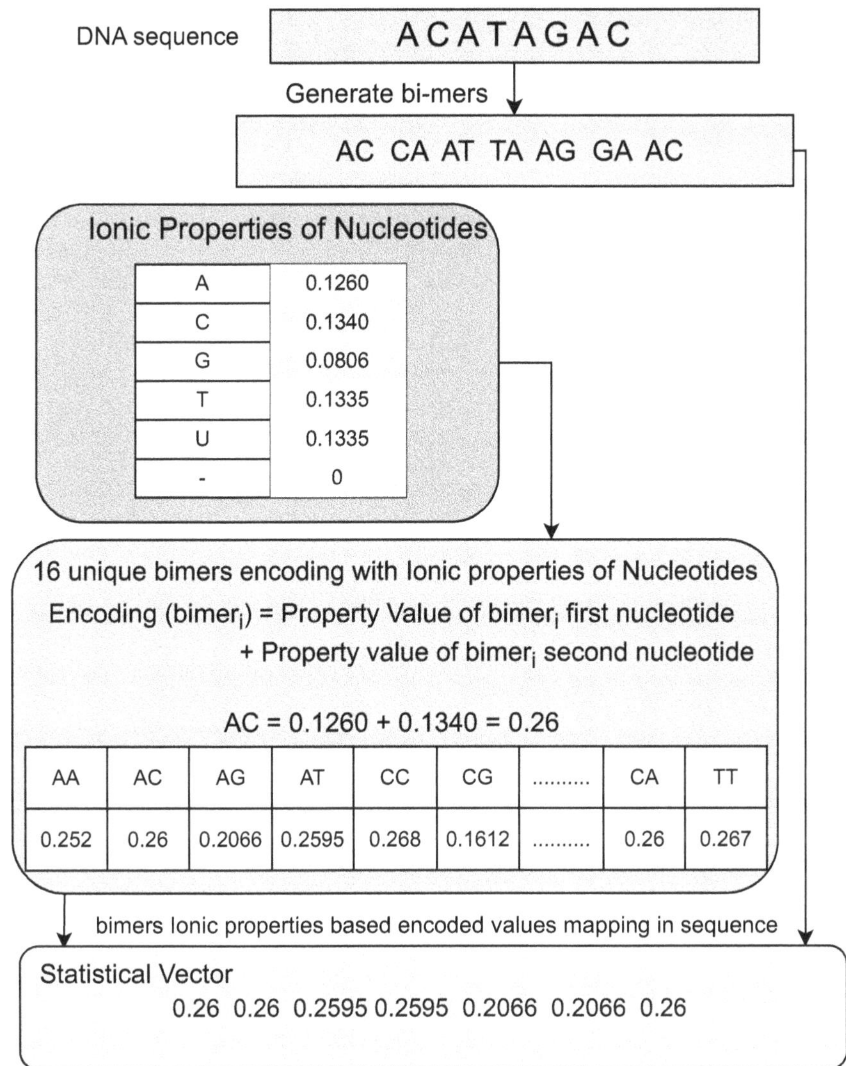

Fig. 5.16 Overall workflow of Pseudo Electron-Ion Interaction Potential (PseudoEIIP) encoding method

tion [1244] and N6-methyladenosine prediction [608]. NCP encoder transforms raw DNA or RNA sequences into statistical vectors by mapping 4 unique nucleotides with their precomputed chemical properties including A = 111, C = 010, G = 100, T/U = 001 and − = 000. It generates statistical vector of dimension $3 \times length$ of sequence. Figure 5.17 illustrates complete working paradigm of NCP encoder.

Table 5.7 Nucleotides mapping

Nucleotides	Symbol	Molecular mass	Atomic number	EIIP	Dipole moment	Trigonometric mapping	Complex mapping
Adenine	A	134	70	0.1260	0.4629	$-\cos\theta - j^*\sin\theta$	1+j
Guanine	G	150	78	0.0806	6.488	$-\cos\theta + j^*\sin\theta$	−1+j
Cytosine	C	110	58	0.1340	3.943	$\cos\theta - j^*\sin\theta$	− 1− j
Thymine	T	125	66	0.1335	1.052	$\cos\theta + j^*\sin\theta$	1 − j

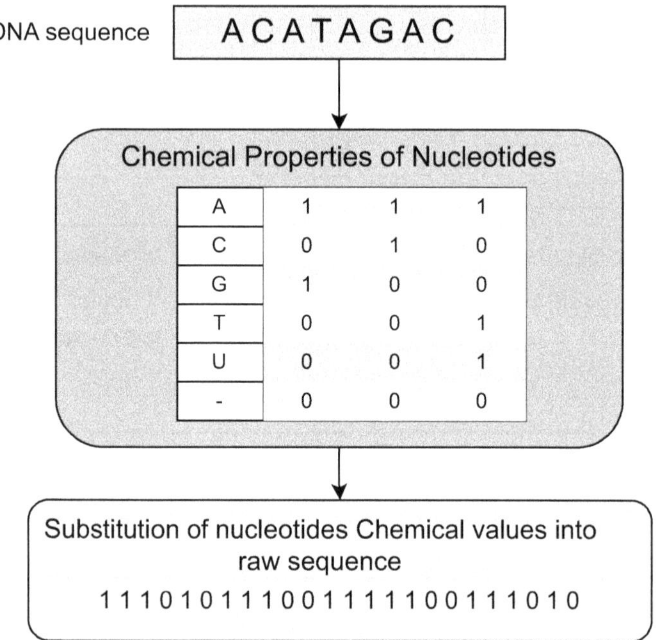

Fig. 5.17 Overall workflow of Nucleotide Chemical Property (NCP) encoder

5.2.5 *Atomic Number Mapping*

This encoder has been utilized in metabolic reactions prediction [817] and genetic code degeneracy reduction [934]. A DNA sequence is transformed into statistical vector by mapping 4 unique nucleotides with their atomic number values (as shown in Table 5.7) which represent total number of protons experimentally computed for each nucleotide [431]. Figure 5.18 describes transformation of a toy sequence into a statistical vector by mapping nucleotides with their corresponding atomic number values. The Figure provides a clear demonstration of how nucleotides atomic number mapping process converts raw DNA sequence into numerical representation.

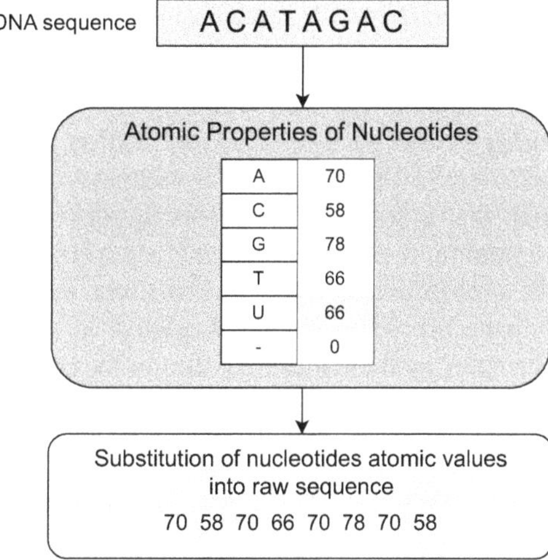

Fig. 5.18 Overall workflow of atomic number mapping method

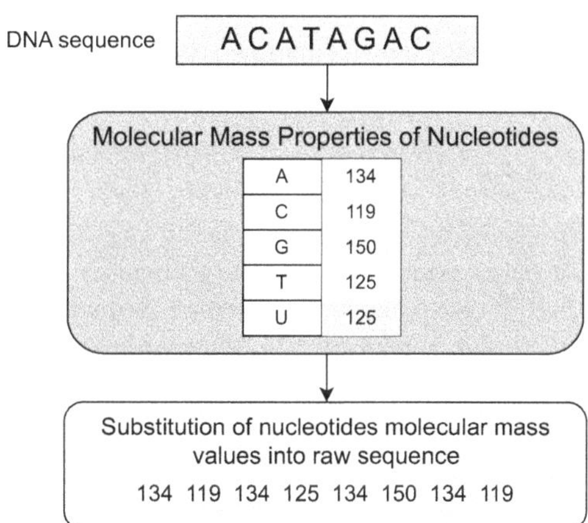

Fig. 5.19 Workflow of Molecular Mass Mapping method

5.2.6 *Molecular Mass Mapping*

This encoder has been utilized in different types of DNA and RNA sequence analysis tasks such as metabolic reactions prediction [817]. Molecular mass mapping is similar to atomic number encoding method but it uses a different set of nucleotide values. Experimentally computed molecular mass values of all four nucleotides are shown in Table 5.7. Figure 5.19 briefly describes sequence encoding process by employing nucleotides Molecular Mass values, Fig. 5.19 illustrates transformation of sequence 'ACATAGAC'. After replacing nucleotides with their Molecular Mass values a statistical vector is generated as {134, 119, 134, 125, 134, 150, 134, 119}.

5.2.7 *Minimum Entropy Mapping*

This encoder has been utilized in different types of DNA and RNA sequence analysis tasks such as DNA replication prediction [692], and exon location identification [740]. This encoder estimates randomness of nucleotides in a sequence such that a sequence with less entropy has less randomness. Working paradigm of minimum entropy mapping encoder comprises of 2 different steps. First of all, probabilities of the four distinct nucleotides are calculated by normalizing their occurrence frequencies relative to the length of the sequence. Equation 5.26 illustrates mathematical expression for computing probabilities of nucleotides.

$$P(Nucleotide_i) = \frac{Occurrence\ Frequency\ of\ Nucleotide_i}{Sequence\ Length}, Nucleotide_i \in \{A, C, GT\} \tag{5.26}$$

In next step, it utilizes these probabilities for calculating entropy H of sequence. This encoder transforms raw sequences into one dimensional statistical vectors.

$$H = \sum_{i=1}^{n} P(Nucleotide_i) \times log_2(P(Nucleotide_i)), Nucleotide_i \in \{A, C, G, T\} \tag{5.27}$$

For better understanding, let us consider a toy sequence "ACGTACGTACG". First of all, it computes the occurrence frequencies of four nucleotides A, C, G and T as 3, 3, 3 and 2, respectively. Then, it uses Eq. 5.26 and compute probabilities as $\frac{3}{11}$, $\frac{3}{11}$, $\frac{3}{11}$ and $\frac{2}{11}$. Equation 5.27 makes use of these probabilities and computes entropy as -0.599.

5.2.8 *Dipole Moments*

This encoder has been utilized in different types of DNA and RNA sequence analysis tasks such as DNA replication prediction [692, 1001], exon location identification [740] and DNA-binding proteins prediction [11]. Dipole moment is another physio-chemical property of amino acid that reflects Vanderwal forces between polar and non-polar molecules. It arises from interaction between polar molecules with permanent dipoles and non-polar molecules. The dipole moment is computed by adding up all the partial atomic charges within a molecule. Experimentally calculated dipole moments associated with nucleotides are provided in Fig. 5.20. Figure 5.20 illustrates the process of transforming a toy DNA sequence into statistical vector by employing pre-computed dipole moment values of nucleotides.

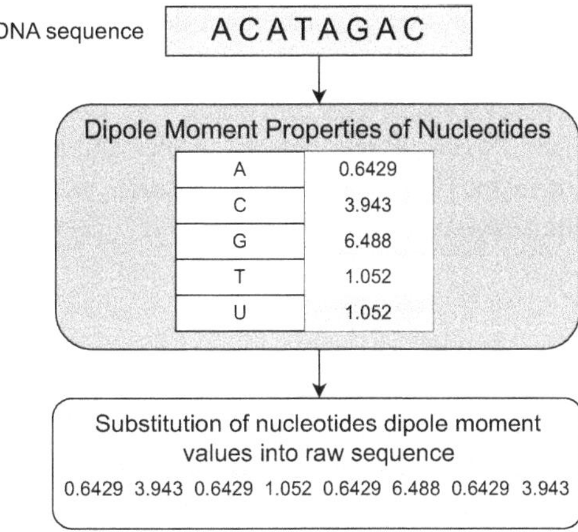

Fig. 5.20 Workflow of Dipole Moment encoder

5.2.9 *Trigonometric Mapping or Z-curve Representation*

This encoder has been utilized in different types of DNA and RNA sequence analysis tasks such as DNA element identification [266], genome analysis [1205], exon locations identification [228] and cancerous genes prediction [229]. Trigonometric Mapping or Z-Curve Representation approach aids to translate characteristics of nucleotides based on their trigonometric values. This encoder make use of following 4 trignometric values:

$$A = -cos\theta + j * sin\theta$$
$$C = -cos\theta - j * sin\theta$$
$$G = cos\theta + j * sin\theta$$
$$T = cos\theta - j * sin\theta$$

In above Equation utilization of $\theta = \frac{\pi}{3}$, makes nucleotides trigonometric values imaginary parts similar but real parts opposite [226]. The opposite signs of real parts represents the characteristics of DNA molecule. Two complementary strands of DNA molecule have two base pairs A-T and G-C which has opposite signs so that their sum is always equal to zero. Working paradigm of this encoder can be categorised into two different phases. In first phase, it maps trigonometric values of nucleotides and generates complex value vector of sequence. In second phase, it employs discrete fourier transform (DFT) strategy for the extraction of useful patterns from complex value vector. Following mathematical expression is used to generate DFT power spectrum.

$$x(k) = DFT\{x(n)\} = \sum_{n=1}^{N} x(n)e^{\frac{-j2\pi kn}{N}} \tag{5.28}$$

In Eq. 5.28, x(k) represents frequency domain representation of input sequence while x(n) refers to input complex value vector. This expression involves the used of complex exponential function $e^{\frac{-j2\pi kn}{N}}$, where j represents complex index, n is time index, N is total number of samples and k refers to frequency bin index. Frequency bin index (k) corresponds to discrete frequencies, ranging from 0 to N, at which input sequence is analyzed.

5.3 Correlation Based Encoding Methods

Researchers have provided precomputed physio chemical values of nucleotides bimers and trimers. Tables 5.8 and 5.9 represents different types of physio chemical values that correlation based encoding methods utilize for transformation of raw DNA sequence into statistical vectors.

5.3.1 Pseudo Dinucleotide Composition (PseDNC)

This encoder has been utilized in different types of DNA and RNA sequence analysis tasks such as recombination spots identification [168], enhancers identification [1104] and phage promoters prediction [919]. PseDNC encoder utilizes distributional information of nucleotides and their experimentally computed physio-chemical values for transforming raw DNA sequences into statistical vectors. Figure 5.21 provides a graphical illustration of the working paradigm of the PseDNC encoder. First, it generates all possible bimers (4^2) of 4 unique nucleotides and computes their occurrence frequencies in the raw sequence. Furthermore, it normalizes computed frequencies with length of sequence. Equation 5.29 illustrates mathematical expression for computing normalized frequencies of 16 unique bimers.

$$Distributed\ Frequency\,(bimer_i) = \frac{\text{Occurrence frequency of } bimer_i}{Sequence\ length\ \ -1},$$
$$\text{where} \quad bimer_i\ \varepsilon\ \{\text{AA, AC, AG, AT, CA, CC, CG, CT, GA, GC, GG, GT, TA, TC, TG, TT}\} \tag{5.29}$$

In Eq. 5.29, occurrence frequencies are normalized with sequence length − 1. The subtraction of 1 from sequence length is necessary because when sequence is segmented into bimers, its length decreases by one. Next, it computes correlations of sequence bimers by using six physio-chemical properties namely; rise, roll, twist, tilt, shift, and slide. With each physio-chemical property, it iteratively computes

Table 5.8 Neucleotides bimers physio chemical properties

Physicochemical properties	GG	GA	GC	GT	AG	AA	AC	AT	CG	CA	CC	CT	TG	TA	TC	TT
Base stacking [782]	−0.07	−0.65	−2.46	−0.92	0.49	1.02	−0.92	0.57	−0.58	0.57	−0.07	0.49	0.57	1.60	−0.65	1.02
Protein induced deformability [778]	0.36	−0.14	−0.30	−0.83	−0.89	−0.64	−0.83	−1.05	2.23	1.51	0.36	−0.89	1.51	0.42	−0.14	−0.64
B-DNA twist [358]	−0.06	1.11	0.79	−0.65	−1.33	0.00	−0.65	2.09	−1.14	0.60	−0.06	−1.33	0.60	−1.06	1.11	0.00
Dinucleotide GC Content [1021]	1.37	0.00	1.37	0.00	0.00	−1.37	0.00	−1.37	1.37	0.00	1.37	0.00	0.00	−1.37	0.00	−1.37
A-philicity [470]	−1.19	1.02	0.32	−1.36	−0.80	0.99	−1.36	−0.10	−0.27	1.19	−1.19	−0.80	1.19	0.32	1.02	0.99
Propeller twist [289]	1.40	−0.27	0.47	−0.16	−0.44	−1.89	−0.16	−0.75	0.80	0.98	1.40	−0.44	0.98	0.23	−0.27	−1.89
Duplex stabil-ity:(freeenergy) [958]	−1.23	0.27	−1.23	0.27	0.27	0.83	0.27	1.40	−2.17	−0.11	−1.23	0.27	−0.11	1.40	0.27	0.83
Duplex tabil-ity(disruptenergy) [117]	1.40	−0.50	1.40	−0.88	−0.50	−0.12	−0.88	−1.39	2.04	−0.12	1.40	−0.50	−0.12	−0.63	−0.50	−0.12
DNA denaturation [99]	0.67	−0.22	2.35	1.11	0.02	−0.84	1.11	−0.58	0.19	−0.92	0.67	0.02	−0.92	−1.60	−0.22	−0.84
Bending stiffness [932]	2.08	−0.12	0.67	−0.12	−0.12	−0.90	−0.12	−1.37	0.67	−0.12	2.08	−0.12	−0.12	−1.37	−0.12	−0.90
Protein DNA twist [778]	−0.50	0.83	−0.22	−1.04	−0.88	0.36	−1.04	−1.90	0.75	1.22	−0.50	−0.88	1.22	1.41	0.83	0.36

(continued)

Table 5.8 (continued)

Physicochemical properties	GG	GA	GC	GT	AG	AA	AC	AT	CG	CA	CC	CT	TG	TA	TC	TT
Stabilising energy of Z-DNA [427]	−0.58	0.15	0.59	1.03	0.15	0.51	1.03	1.97	−1.82	−1.38	−0.58	0.15	−1.38	−0.51	0.15	0.51
Aida_BA_transition [13]	−0.50	−0.19	1.52	0.26	−0.99	−0.93	0.26	1.03	2.36	0.45	−0.50	−0.99	0.45	−1.11	−0.19	−0.93
Breslauer_dG [117]	1.40	−0.50	1.40	−0.88	−0.50	−0.12	−0.88	−0.63	2.04	−0.12	1.40	−0.50	−0.12	−1.39	−0.50	−0.12
Breslauer_dH [117]	1.31	−1.10	1.35	−0.70	−0.12	0.46	−0.70	0.23	1.71	−1.02	1.31	−0.12	−1.02	−0.93	−1.10	0.46
Breslauer_dS [117]	1.14	−1.25	1.16	−0.56	0.08	0.67	−0.56	0.65	1.36	−1.36	1.14	0.08	−1.36	−0.63	−1.25	0.67
Electron_interaction [1021]	1.37	0.00	1.37	0.00	0.00	−1.37	0.00	−1.37	1.37	0.00	1.37	0.00	0.00	−1.37	0.00	−1.37
Hartman_trans_free_energy [403]	−0.49	−0.41	−1.71	−1.28	0.23	0.59	−1.28	−0.49	0.66	1.10	−0.49	0.23	1.10	2.03	−0.41	0.59
Helix-Coil_transition [138]	1.41	−0.49	1.41	−0.89	−0.49	−0.13	−0.89	−0.62	2.04	−0.13	1.41	−0.49	−0.13	−1.37	−0.49	−0.13
Ivanov_BA_transition [470]	−1.19	1.02	0.32	−1.36	−0.80	0.99	−1.36	−0.10	−0.27	1.19	−1.19	−0.80	1.19	0.32	1.02	0.99
Lisser_BZ_transition [635]	−0.49	−0.40	−1.71	−1.28	0.23	0.59	−1.28	−0.49	0.66	1.09	−0.49	0.23	1.09	2.04	−0.40	0.59
Polar_interaction [368]	1.37	0.00	1.37	0.00	0.00	−1.37	0.00	−1.37	1.37	0.00	1.37	0.00	0.00	−1.37	0.00	−1.37
SantaLucia_dG [873]	0.86	0.17	2.00	0.10	−0.50	−0.81	0.10	−1.46	1.57	−0.01	0.86	−0.50	−0.01	−1.75	0.17	−0.81
SantaLucia_dH [873]	−0.73	−0.03	2.35	0.60	−1.15	0.46	0.60	−0.87	1.65	−0.24	−0.73	−1.15	−0.24	−1.01	−0.03	0.46

SantaLucia_dS [873]	−1.28	−0.04	2.10	0.68	−1.14	0.83	0.68	−0.43	1.34	−0.30	−1.28	−1.14	−0.30	−0.51	−0.04	0.83
Sarai_flexibility [877]	−0.79	−0.43	0.08	0.85	−0.57	−0.70	0.85	3.16	−0.41	0.03	−0.79	−0.57	0.03	0.39	−0.43	−0.70
Stability [359]	0.50	0.52	2.52	0.97	−0.61	−0.77	0.97	−0.67	−0.76	−0.04	−0.76	0.50	−0.61	−1.49	0.52	−0.77
Stacking_energy [782]	0.07	0.65	2.45	0.91	−0.49	−1.02	0.91	−0.57	0.61	−0.57	0.07	−0.49	−0.57	−1.60	0.65	−1.02
Sugimoto_dG [873]	0.96	−0.24	1.35	−0.24	−0.24	−0.83	−0.24	−1.43	2.35	0.16	0.96	−0.24	0.16	−1.43	−0.24	−0.83
Sugimoto_dH [873]	1.26	0.09	1.04	0.42	−1.14	−0.36	0.42	−1.71	1.77	−0.25	1.26	−1.14	−0.25	−1.14	0.09	−0.36
Sugimoto_dS [873]	1.29	0.19	0.84	0.64	−1.41	−0.17	0.64	−1.68	1.43	−0.37	1.29	−1.41	−0.37	−0.96	0.19	−0.17
Watson-Crick_interaction [578]	1.37	0.00	1.37	0.00	0.00	−1.37	0.00	−1.37	1.37	0.00	1.37	0.00	0.00	−1.37	0.00	−1.37
Twist [354]	0.06	−0.08	−0.08	1.50	0.78	0.06	1.50	1.07	−1.66	−1.38	0.06	0.78	−1.38	−1.23	−0.08	0.06
Tilt [354]	1.08	0.50	0.22	0.50	0.36	0.50	0.50	0.22	−1.22	−1.36	1.08	0.36	−1.36	−2.37	0.50	0.50
Roll [354]	0.09	0.27	1.33	0.80	0.09	0.27	0.80	0.62	−0.44	−0.27	0.09	0.09	−0.27	−0.44	0.27	−3.28
Shift [354]	0.56	0.13	−0.35	0.13	0.68	1.59	0.13	−1.02	−0.82	−0.86	0.56	0.68	−0.86	−2.24	0.13	1.59
Slide [354]	−0.82	−0.39	0.65	1.29	−0.24	0.11	1.29	2.51	−0.29	−0.62	−0.82	−0.24	−0.62	−1.51	−0.39	0.11
Rise [354]	0.24	0.71	1.59	1.04	−0.62	−0.11	1.04	1.17	−1.39	−1.25	0.24	−0.62	−1.25	−1.39	0.71	−0.11

Table 5.9 Nucleotides trimers physio chemical properties

Trimers	Physicochemical properties											
	pendability (DNAse) [734]	Bendability (consensus) [734]	Trinucleotide GC content [1021]	Nucleosome positioning [356]	Consensus roll [368, 1021]	Consensus rigid [368, 1021]	Dnase I [118]	Dnase I rigid [118]	MW (Daltons) [1021]	MW (kg) [1021]	[879] Nucleosome	Nucleosome rigid [879]
AAA	−2.087	−2.745	−1.732	−2.349	−2.744	−2.744	2.274	2.118	−1	−1	−2.342	2.386
AAC	−1.509	−1.354	−0.577	−0.561	−1.363	−1.363	1.105	1.516	−1	−1	−0.555	0.548
AAG	−0.506	−0.257	−0.577	0.155	−0.26	−0.26	0.193	0.493	−1	−1	0.169	−0.179
AAT	−2.126	−2.585	−1.732	−1.991	−2.591	−2.591	2.141	2.158	−1	−1	−2.004	2.032
ACA	0.111	0.171	−0.577	0.155	0.164	0.164	−0.153	−0.123	1	1	0.169	−0.179
ACC	−0.121	0.064	0.577	0.274	0.071	0.071	−0.078	0.107	1	1	0.266	−0.275
ACG	−0.121	0.064	0.577	0.274	0.065	0.065	−0.074	0.107	1	1	0.266	−0.275
ACT	−1.354	−0.685	−0.577	0.453	−0.676	−0.676	0.536	1.357	1	1	0.459	−0.466
AGA	0.381	−0.15	−0.577	−0.74	−0.158	−0.158	0.109	−0.389	1	1	−0.748	0.743
AGC	0.304	0.92	0.577	1.287	0.911	0.911	−0.753	−0.313	1	1	1.28	−1.272
AGG	−0.313	−0.07	0.577	0.274	−0.07	−0.07	0.039	0.3	1	1	0.266	−0.275
AGT	−1.354	−0.685	−0.577	0.453	−0.676	−0.676	0.536	1.357	1	1	0.459	−0.466
ATA	1.615	0.572	−1.732	−0.978	0.584	0.584	−0.491	−1.585	−1	−1	−0.99	0.988
ATC	−0.737	−0.391	−0.577	0.214	−0.397	−0.397	0.307	0.727	−1	−1	0.217	−0.227
ATG	1.229	1.348	−0.577	0.87	1.358	1.358	−1.112	−1.215	−1	−1	0.893	−0.894
ATT	−2.126	−2.585	−1.732	−1.991	−2.591	−2.591	2.141	2.158	−1	−1	−2.004	2.032
CAA	0.265	−0.231	−0.577	−0.74	−0.226	−0.226	0.166	−0.275	−1	−1	−0.748	0.743
CAC	0.496	0.786	0.577	0.81	0.773	0.773	−0.646	−0.503	−1	−1	0.797	−0.8
CAG	1.576	0.92	0.577	−0.322	0.92	0.92	−0.762	−1.549	−1	−1	−0.314	0.304
CAT	1.229	1.348	−0.577	0.87	1.358	1.358	−1.112	−1.215	−1	−1	0.893	−0.894

CCA	−1.856	−1.14	0.577	0.274	−1.139	−1.139	0.917	1.876	1	1	0.266	−0.275
CCC	0.072	0.358	1.732	0.572	0.345	0.345	−0.3	−0.084	1	1	0.555	−0.562
CCG	−0.969	−0.712	1.732	−0.084	−0.705	−0.705	0.558	0.962	1	1	−0.072	0.062
CCT	−0.313	−0.07	0.577	0.274	−0.07	−0.07	0.039	0.3	1	1	0.266	−0.275
CGA	0.111	1	0.577	1.645	1.012	1.012	−0.834	−0.123	1	1	1.666	−1.646
CGC	−0.468	0.385	1.732	1.287	0.379	0.379	−0.326	0.455	1	1	1.28	−1.272
CGG	−0.969	−0.712	1.732	−0.084	−0.705	−0.705	0.558	0.962	1	1	−0.072	0.062
CGT	−0.121	0.064	0.577	0.274	0.065	0.065	−0.074	0.107	1	1	0.266	−0.275
CTA	0.882	−0.097	−0.577	−1.276	−0.097	−0.097	0.062	−0.88	−1	−1	−1.28	1.285
CTC	0.419	0.438	0.577	0.274	0.427	0.427	−0.365	−0.427	−1	−1	0.266	−0.275
CTG	1.576	0.92	0.577	−0.322	0.92	0.92	−0.762	−1.549	−1	−1	−0.314	0.304
CTT	−0.506	−0.257	−0.577	0.155	−0.26	−0.26	0.193	0.493	−1	−1	0.169	−0.179
GAA	−0.159	−0.605	−0.577	−0.918	−0.6	−0.6	0.474	0.146	−1	−1	−0.893	0.89
GAC	0.034	0.171	0.577	0.274	0.178	0.178	−0.165	−0.046	−1	−1	0.266	−0.275
GAG	0.419	0.438	0.577	0.274	0.427	0.427	−0.365	−0.427	−1	−1	0.266	−0.275
GAT	−0.737	−0.391	−0.577	0.214	−0.397	−0.397	0.307	0.727	−1	−1	0.217	−0.227
GCA	0.766	0.839	0.577	0.572	0.842	0.842	−0.702	−0.767	1	1	0.555	−0.562
GCC	1.036	2.097	1.732	2.479	2.089	2.089	−1.687	−1.029	1	1	2.487	−2.433
GCG	−0.468	0.385	1.732	1.287	0.379	0.379	−0.326	0.455	1	1	1.28	−1.272
GCT	0.304	0.92	0.577	1.287	0.911	0.911	−0.753	−0.313	1	1	1.28	−1.272
GGA	0.265	−0.097	0.577	−0.501	−0.103	−0.103	0.066	−0.275	1	1	−0.507	0.499
GGC	1.036	2.097	1.732	2.479	2.089	2.089	−1.687	−1.029	1	1	2.487	−2.433
GGG	0.072	0.358	1.732	0.572	0.345	0.345	−0.3	−0.084	1	1	0.555	−0.562
GGT	−0.121	0.064	0.577	0.274	0.071	0.071	−0.078	0.107	1	1	0.266	−0.275
GTA	0.342	−0.07	−0.577	−0.561	−0.062	−0.062	0.031	−0.351	−1	−1	−0.555	0.548

(continued)

Table 5.9 (continued)

Trimers	Physicochemical properties											
	Bendability (DNAse) [734]	Bendability (consensus) [734]	Trinucleotide GC content [1021]	Nucleosome positioning [356]	Consensus roll [368, 1021]	Consensus rigid [368, 1021]	Dnase I [118]	Dnase I rigid [118]	MW (Daltons) [1021]	MW (kg) [1021]	Nucleosome [879]	Nucleosome rigid [879]
GTC	0.034	0.171	0.577	0.274	0.178	0.178	−0.165	−0.046	−1	−1	0.266	−0.275
GTG	0.496	0.786	0.577	0.81	0.773	0.773	−0.646	−0.503	−1	−1	0.797	−0.8
GTT	−1.509	−1.354	−0.577	−0.561	−1.363	−1.363	1.105	1.516	−1	−1	−0.555	0.548
TAA	0.689	−0.284	−1.732	−1.395	−0.275	−0.275	0.206	−0.692	−1	−1	−1.376	1.384
TAC	0.342	−0.07	−0.577	−0.561	−0.062	−0.062	0.031	−0.351	−1	−1	−0.555	0.548
TAG	0.882	−0.097	−0.577	−1.276	−0.097	−0.097	0.062	−0.88	−1	−1	−1.28	1.285
TAT	1.615	0.572	−1.732	−0.978	0.584	0.584	−0.491	−1.585	−1	−1	−0.99	0.988
TCA	1.73	1.348	−0.577	0.274	1.348	1.348	−1.103	−1.696	1	1	0.266	−0.275
TCC	0.265	−0.097	0.577	−0.501	−0.103	−0.103	0.066	−0.275	1	1	−0.507	0.499
TCG	0.111	1	0.577	1.645	1.012	1.012	−0.834	−0.123	1	1	1.666	−1.646
TCT	0.381	−0.15	−0.577	−0.74	−0.158	−0.158	0.109	−0.389	1	1	−0.748	0.743
TGA	1.73	1.348	−0.577	0.274	1.348	1.348	4.522	−1.696	1	1	0.266	−0.275
TGC	0.766	0.839	0.577	0.572	0.842	0.842	−0.702	−0.767	1	1	0.555	−0.562
TGG	−1.856	−1.14	0.577	0.274	−1.139	−1.139	0.917	1.876	1	1	0.266	−0.275
TGT	0.111	0.171	−0.577	0.155	0.164	0.164	−0.153	−0.123	1	1	0.169	−0.179
TTA	0.689	−0.284	−1.732	−1.395	−0.275	−0.275	0.206	−0.692	−1	−1	−1.376	1.384
TTC	−0.159	−0.605	−0.577	−0.918	−0.6	−0.6	0.474	0.146	−1	−1	−0.893	0.89
TTG	0.265	−0.231	−0.577	−0.74	−0.226	−0.226	0.166	−0.275	−1	−1	−0.748	0.743
TTT	−2.087	−2.745	−1.732	−2.349	−2.744	−2.744	−2.615	2.118	−1	−1	−2.342	2.386

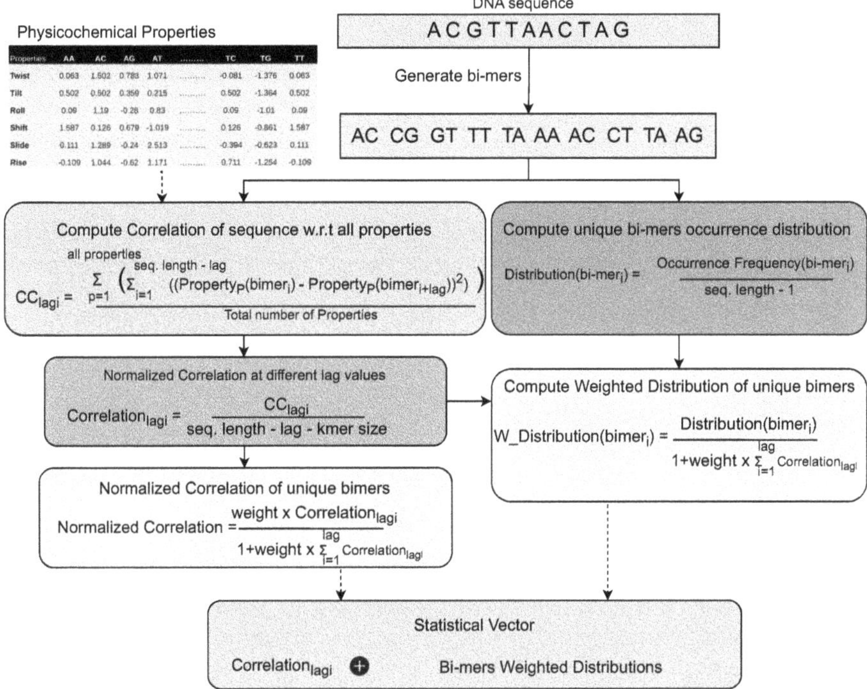

Fig. 5.21 Overall workflow of Pseudo dinucleotide composition (PseDNC) encoder

correlation values of sequence bimers for different lag values. Here, lag value is a hyper-parameter that needs to be set according to nature and complexity of task. Primarily, lag value decides range in which correlation of bimers is computed. The feasible value of the lag is usually determined through experimentation using a subset of the data.

To understand the process of lag-based correlation extraction, consider a toy sequence $R_0 R_1 R_2 ... R_L$ and lag value is set as 3. In the toy sequence R_i can be any nucleotide from (A,C,G,T) at the sequence position $i \epsilon \{1, 2, 3, 4, ... L\}$ and L is the length of the sequence. By sliding a 2 size window with stride size 1, sequence is segregated into bimers as $R_0 R_1$, $R_1 R_2$, $R_2 R_3$, $R_3 R_4$, ...$R_{L-1} R_L$. In Fig. 5.22, the graphical representation showcases the correlations between sequence bimers at three distinct lag values. Specifically, sub-figure (a) represents correlation of bimers at lag 1, sub-figure (b) represents correlation of bimers at lag 2 and sub-figure (c) represents correlation of bimers at lag 3. It can be seen in Fig. 5.22a at lag 1 correlation is computed between consecutive bimers while at lag 2 (Fig. 5.22b) correlation is computed by skipping one bimer from the sequence of bimers among different bimers. Overall, it can be concluded lag value decides range in which correlation of bimers is computed.

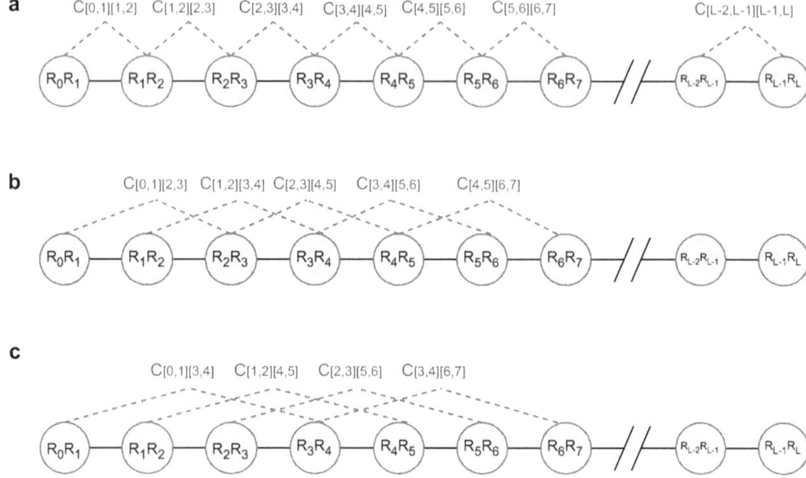

Fig. 5.22 Graphical illustration of Lag based correlation among bimers

Equation 5.30 illustrates mathematical expression for computing correlation of bimers.

$$cc_{lag_n} = \frac{\sum_{p=1}^{6}\left[\sum_{i=1}^{seq.length-lag-1}\left[Property_p(bimer_i) - Property_p(bimer_{i+lag})\right]^2\right]}{6} \tag{5.30}$$

In Eq. 5.30, P = 1 to 6 is a loop on 6 physio-chemical properties, internal loop k = 1 to seq. length-lag-1 represents all bimers of the sequence. $Property_p(bimer_i)$ denotes pth property value of ith bimer of the sequence and $Property_p(bimer_{i+lag})$ denotes pth property value of $(i + lag)$th bimer of the sequence.

It is important to note that encoder computes correlation of sequence bimers at all lag values. Lets if lag value is set as 3, encoder iteratively computes sequence bimers correlation at lag values 1, 2 and 3.

Next, it normalizes computed correlation values using Eq. 5.31.

$$Theta_{lag_n} = \frac{cc_{lag_n}}{Sequence\ length\ -\ lag\ value\ -2} \tag{5.31}$$

Equation 5.32 illustrates normalized correlation values of bimers at different lag values.

$$Theta\ Array = \left\{Theta_{lag_1} \oplus Theta_{lag_2} \oplus Theta_{lag_3} \oplus \ldots. \oplus Theta_{lag_n}\right\} \tag{5.32}$$

Furthermore, using normalized correlation values and distributed frequency of bimers, it computes weighted distribution of bimers. Equation 5.33 represents mathematical expression used to calculate the weighted distribution of bimers.

$$Bi\text{-}mers\ weighted\ distribution = \frac{\text{Distributed frequency of } bimer_i}{1 + weight \times sum(Theta\ Array)} \tag{5.33}$$

Equation 5.34 illustrates mathematical expression for computing weighted correlation values. In Eqs. 5.33 and 5.34 weight is a hyper-parameter that is usually set as [0.25, 1].

$$Weighted\ correlation\ value = \frac{weight \times ThetaArray[i]}{1 + weight \times sum(Theta\ Array)} \tag{5.34}$$

Finally, encoder concatenates outputs of Eqs. 5.33 and 5.34 and generates a $4^2 + lag$ dimensional feature vector.

Let's consider a toy sequence ATTCTA with lag value 2 and all possible bimers can be formulated as AT, TT, TC, CT and TA. First, the encoder computes their occurrence frequencies and normalizes them to generate statistical vector as: $\{0, 0, 0, 0.2, 0, 0, 0, 0.2, 0, 0, 0, 0, 0.2, 0.2, 0, 0.2\}$.

Then it calculates correlation between bimers for three different physio-chemical properties (PCP) namely: rise, tilt and roll. The encoder calculates correlation using Eq. 5.30 and PCP values from Table 5.8. It produces resulting value 5.9095 with lag 1 and 6.67583 value with lag 2. Then it generates theta array by computing theta values as 1.9698 with lag 1 and 3.3379 with lag 2 using Eq. 5.32. Theta array can be written as $[1.9698, 3.3379]$. To compute weighted distribution of bimers and bimers weighted correlation values, it normalizes bimer feature vector and theta array with $(sum[ThetaArray] * weight + 1)$ as illustrated in Eqs. 5.33 and 5.34, respectively. The final feature vector obtained by concatenating weighted values of bimers frequencies and bimers correlation values can be formulated as: $\{0, 0, 0, 0.08586, 0, 0, 0, 0.08586, 0, 0, 0, 0, 0.08586, 0.08586, 0, 0.08586\} \oplus \{0.1691, 0.4015\}$.

5.3.2 *PseTNC and PcpseKNC*

This encoder has been utilized in different types of DNA and RNA sequence analysis tasks such as phage promoters prediction [919], gene identification [564] and identification of translation initiation sites [169]. Similar to PseDNC, this encoder segregates raw sequences into trimers and computes occurrence frequencies of trimers. Furthermore, computed frequencies are normalized with length of sequence − 2. Equation 5.35 illustrates mathematical expression for computing normalized frequencies of trimes.

$$Distributed\ Frequency\ (trimer_i) = \frac{\text{Occurrence frequency of } trimer_i}{Sequence\ length\ -\ 2} \tag{5.35}$$

where $trimer_i\ \varepsilon$ { AAA, AAC, AAG, AAT,ACA, ACC, ACG, ACT, AGA, AGC, AGG, AGT, ATA, ATC, ATG, ATT, CAA, CAC, CAG, CAT, CCA, CCC, CCG, CCT, CGA, CGC, CGG, CGT, CTA, CTC, CTG, CTT, GAA, GAC, GAG, GAT, GCA, GCC, GCG, GCT, GGA, GGC, GGG, GGT, GTA, GTC, GTG, GTT, TAA, TAC, TAG, TAT, TCA, TCC, TCG, TCT, TGA, TGC, TGG, TGT, TTA, TTC, TTG, TTT }.

Furthermore, it makes use of Physicochemical Properties to compute correlation between sequence trimers at different lag values. A comprehensive motivation behind finding correlation among bimers at different lag values is summarized in Sect. 5.3.1. Table 5.9 depicts 64 trimers physcio chemical property values that are utilized for correlation computation. Equation 5.36 illustrates mathematical expression for computing correlation between trimers at different lag values.

$$cc_{lag_n} = \frac{\sum_{p=1}^{All Propoerties}\left[\sum_{i=1}^{seq.length-lag-1}\left[Property_p(trimer_i) - Property_p(trimer_{i+lag})\right]^2\right]}{Total\ number\ of\ properties} \tag{5.36}$$

In Eq. 5.36, $trimer_i$ is ith trimer and $trimer_{i+gap}$ is $i + lag$th trimer in trimers sequence sample. Furthermore, $Property_p(trimer_i)$ and $Property_p$ $(trimer_{i+gapth})$ are calculated values of two physio-chemical properties of trimers. At n different lag values, cc_{lag_n} can be explained as: $\{Theta_{lag_1} \oplus Theta_{lag_2} \oplus Theta_{lag_3} \oplus .... \oplus\ Theta_{lag_n}\}$.

As shown in Eq. 5.37, cc_{lag_n} values are normalized to compute theta array.

$$Theta_{lag_n} = \frac{cc_{lag_n}}{length\ of\ sequence\ -\ lag_n\ -\ 3} \tag{5.37}$$

Furthermore, it normalizes theta array and Distributed frequencies of trimers with $(sum[ThetaArray] * weight + 1)$ as illustrated in Eqs. 5.38 and 5.39, respectively.

$$Trimers\ Weighted\ Distribution(trimer_i) = \frac{\text{Distributed frequency of } trimer_i}{1 + weight \times sum(Theta\ Array)} \tag{5.38}$$

$$Weighted\ Theta\ Array = \frac{weight \times Theta\ Array[i]}{1 + weight \times sum(Theta\ Array)} \tag{5.39}$$

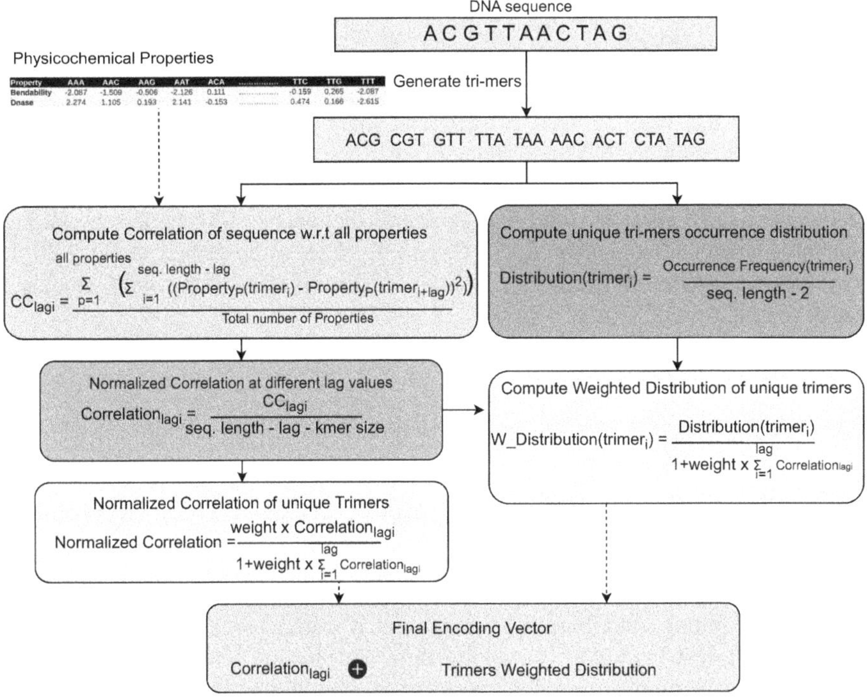

Fig. 5.23 Overall workflow of Pseudo trinucleotide composition (PseTNC) encoder

Finally, it concatenates both vector spaces (Trimers Weighted Distribution $(trimer_i)$ and Weighted Theta Array) and generates a $(4^3 + lag)$ dimensional statistical representation. Figure 5.23 summarises different steps that Pseudo trinucleotide composition (PseTNC) encoder employs for transforming raw DNA sequence into statistical vector.

For PseKNC, this encoder generates kmers for higher values of k $(k \geq 4)$ and generates 4^k dimensional feature vector comprising normalized frequencies of kmers. After generating feature space, it follows above steps to iteratively compute correlation between kmers based on physio-chemical properties of kmers with all lag values. Using extracted correlations, it computes theta with n different lag and correlation values to generate a theta array. Then it normalizes theta array and k-mers distributed frequencies with $(sum[ThetaArray] * weight + 1)$ and concatenates both vectors to generate a $(4^k + lag)$ dimensional statistical representation of sequence.

5.3.3 Series Correlation Pseudo Dinucleotide Composition (ScPseDNC)

This encoder has been utilized in different types of DNA and RNA sequence analysis tasks such as N7-Methylguanosine Sites Prediction [96], yeast DNA replication prediction [692] and RNA m5C sites prediction [661]. ScPSeDNC is another variant of kmer encoder that generates dictionary of 16 bimers by taking four basic nucleotides namely A, C, G, and T. Next, as described in Eq. 5.40, it computes occurrence frequencies of bimers and normalizes them with sequence length − 1.

$$Distribution\ frequency\ of\ (bimer_i) = \frac{\text{Occurrence frequency of } bimer_i}{\text{Sequence length} - 1}$$
$$\text{where } bimer_i\ \varepsilon\ \{\text{AA, AC, AG, AT, CA, CC, CG, CT, GA, Gc, GG, GT, TA, TC, TG, TT}\} \tag{5.40}$$

After computing distribution frequencies, it computes correlation between bimers at different lag values by taking different physio-chemical properties. To briefly understand motivation behind correlation computation at different lag values, consider a toy sequence $R_0R_1R_2R_3R_4.........R_L$ where R_i can be any nucleotide from (A, C, G, T) at the sequence position $i \in \{1, 2, 3, 4, ...L\}$ and L is the length of the sequence. It segregates sequence into bimers. Here lag is a hyper-parameter that need to be set according to statistics of sequences and corpus. Usually value of lag is selected by performing experimentation using a subset of data. Figure 5.22 illustrates 3 different lag based bimers where sub-figure (a) represents correlation of bimers at lag 1, sub-figure (b) represents correlation of bimers at lag 2 and sub-figure (c) represents correlation of bimers at lag 3. It can be seen in Fig. 5.22a at lag 1 correlation is computed between consecutive bimers while at lag 2 (Fig. 5.22b) correlation is computed by skipping one bimer from the sequence of bimers among different bimers. Overall, it can be concluded lag value decides range in which correlation of bimers is computed. Then it computes correlation for each property based on n different lag values using Eq. 5.41.

$$cc_{pnlag_n} = \sum_{p=1}^{AllProperties} \left[\sum_{i=1}^{seq.length-lag-1} [Property_p(bimer_i) \times Property_p(bimer_{i+lag})] \right] \tag{5.41}$$

$Property_p(bimer_i)$ and $Property_p(bimer_{i+lag})$ in above Equation represent pre-computed physio-chemical property values of ith and $i + lag$th bimers.

In next step, it makes use of cc_{pnlag_n} for computing normalized correlation values using mathematical expression shown in Eq. 5.42.

$$Theta_{pnlag_n} = \frac{cc_{pnlag_n}}{length\ of\ sequence\ -\ lag_n\ -\ 1} \tag{5.42}$$

Equation 5.43 refers to output of $Theta_{pnlag_n}$ at different lag values.

$$Theta\ Array = \{Theta_{lag_1} \oplus Theta_{lag_2} \oplus Theta_{lag_3} \oplus .... \oplus Theta_{lag_n}\} \tag{5.43}$$

Furthermore, using Eq. 5.44, it computes bimers weighted distribution.

$$Bi\text{-}mers\ weighted\ distribution = \frac{\text{Distributed frequency of } bimer_i}{1 + weight \times sum(Theta\ Array)} \tag{5.44}$$

Moreover, it make use of Eq. 5.45 mathematical expression for computing weighted theta array.

$$Weighted\ Theta\ Array = \frac{weight \times ThetaArray[i]}{1 + weight \times sum(Theta\ Array)} \tag{5.45}$$

Finally, encoder concatenates outputs of Eqs. 5.44 and 5.45 and generates a $4^2 + lag$ dimensional feature vector. Figure 5.24 graphically represents different steps that ScPseDNC encoder performs for transforming raw DNA sequences into statistical vectors.

5.3.4 ScPseTNC and ScPseKNC

This encoder has been utilized in different types of DNA and RNA sequence analysis tasks such as DNA replication prediction [692], N7-Methylguanosine Sites Prediction [96]. ScPSeTNC is another variant of correlation encoder that generates dictionary of 64 possible trimers by taking four basic nucleotides (A, C, G, T). Next, as shown in Eq. 5.46, it computes occurrence frequencies of trimers and normalizes them with sequence length − 2.

$$Distribution\ frequency\,(Trimer_i) = \frac{\text{Occurrence frequency of } trimer_i}{\text{Sequence length } - 2} \tag{5.46}$$

where $trimer_i\ \varepsilon$ { *AAA, AAC, AAG, AAT, ACA, ACC, ACG, ACT, AGA, AGC, AGG, AGT, ATA, ATC, ATG, ATT, CAA, CAC, CAG, CAT, CCA, CCC, CCG, CCT, CGA, CGC, CGG, CGT, CTA, CTC, CTG, CTT, GAA, GAC, GAG, GAT, GCA, GCC, GCG, GCT, GGA, GGC, GGG, GGT, GTA, GTC, GTG, GTT, TAA, TAC, TAG, TAT, TCA, TCC, TCG, TCT, TGA, TGC, TGG, TGT, TTA, TTC, TTG, TTT* }.

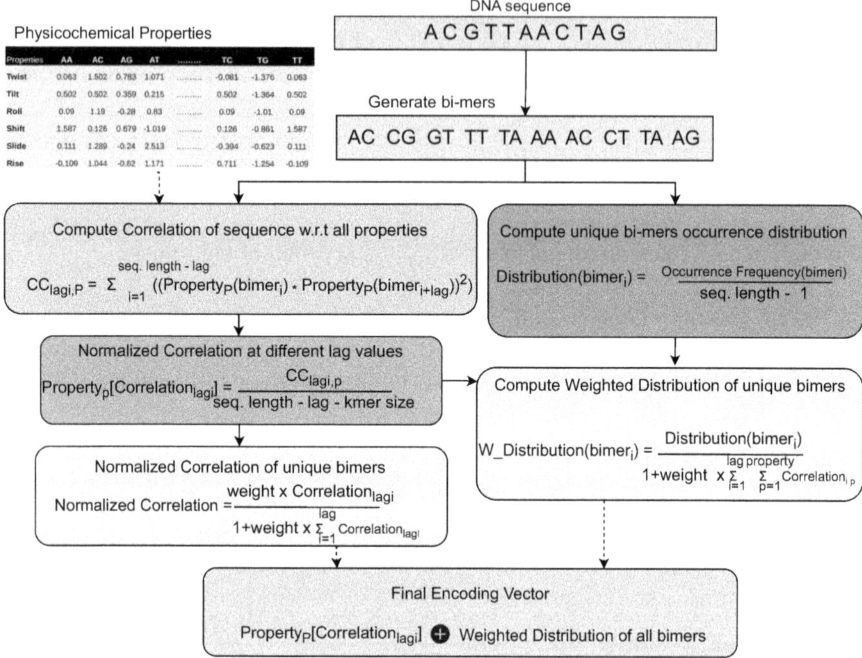

Fig. 5.24 Overall workflow of series correlation pseudo di-nucleotide composition (ScPseDNC) encoder

After computing distribution frequency, it computes sequence trimers correlation for n different lag values using different physio-chemical properties. To accomplish this, it computes correlation for each property based on n different lag values using Eq. 5.47.

$$cc_{pnlag_n} = \sum_{p=1}^{6} \left[\sum_{i=1}^{seq.length-lag-1} \left[Property_p(trimer_i) \times Property_p(trimer_{i+lag}) \right] \right] \tag{5.47}$$

In above Equation, $Property_p(trimer_i)$ and $Property_p(trimer_{i+lag})$ represent pre-computed physio-chemical property values of ith and $i + lag$th trimers. Furthermore, based on extracted correlation, it computes theta by using Eq. 5.48 and generates theta array for all calculated theta values. Theta array contains theta values corresponding to each combination of lag and property value resulting in $(theta \times properties)$ dimensional feature vector.

$$Theta_{pnlag_n} = \frac{cc_{pnlag_n}}{Sequence\ length\ - lag_n\ - 2} \tag{5.48}$$

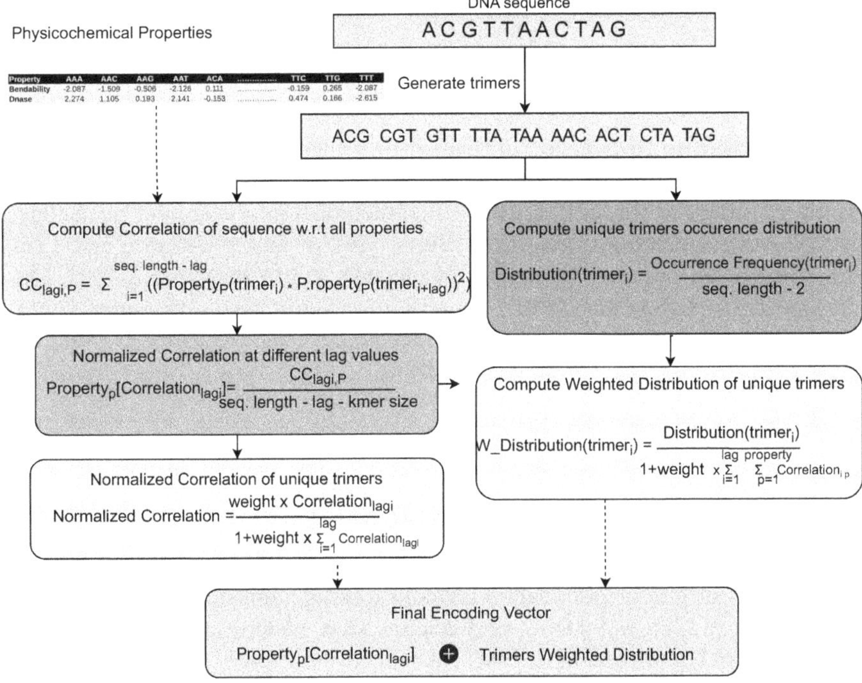

Fig. 5.25 Overall workflow of series correlation pseudo tri-nucleotide composition (ScPseTNC)

Furthermore, using Eq. 5.49, it computes bi-mers weighted distribution and employs Eq. 5.50 for computing weighted theta array.

$$Bi\text{-}mers\ weighted\ distribution = \frac{\text{Distributed frequency of } bimer_i}{1 + weight \times sum(Theta\ Array)} \tag{5.49}$$

$$Weighted\ Theta\ Array = \frac{weight \times ThetaArray[i]}{1 + weight \times sum(Theta\ Array)} \tag{5.50}$$

Finally, encoder concatenates bi-mers weighted distribution and weighted theta array and generates a $4^3 + lag$ dimensional feature vector. Figure 5.25 graphically represents different steps that ScPseTNC encoder performs for transforming raw DNA sequences into statistical vectors.

5.4 Covariance Based Encoding Methods

Researchers have proposed diverse types of encoders that are competent in transforming variable length DNA and RNA sequences into fixed length statistical vectors by utilizing covariance information of nucleotides. These encoders make use of pre-computed physio-chemical properties and auto covariance and cross-covariance strategies for the extraction of correlational information of nucleotides. The following subsections describe 6 divers types of auto covariance and cross-covariance based encoders namely: Dinucleotide Auto Covariance [1221], Dinucleotide Cross Covariance [919], Dinucleotide Auto-Cross Covariance [1221], Trinucleotide Auto Covariance [594], Trinucleotide Cross Covariance [594] and Trinucleotide Auto-Cross Covariance [594].

5.4.1 *Dinucleotide Auto Covariance (DAC)*

This encoder has been utilized in different types of DNA and RNA sequence analysis tasks such as recombination hot/cold spots identification [652], enhancers identification [1221] and DNA modification sites identification [1176]. Prime objective behind development of DAC encoder was to capture correlation between dinucleotides that are separated by a specific lag distance within the sequence, specifically focusing on their shared physicochemical index. Figure 5.26 provides a comprehensive overview of specific mechanisms and calculations involved in transforming raw sequences into statistical vectors by utilizing DAC encoder. To briefly understand encoding process of DAC encoder, consider a generic sequence $A_0A_1A_2A_3A_4......A_L$ where A_i can be any nucleotide from (A, C, G, T) at the sequence position $i \epsilon \{1, 2, 3, 4, ...L\}$ and L is the length of the sequence. Raw sequence is segregated into bimers as A_0A_1, A_1A_2, A_2A_3, ..., $A_{L-1}A_L$.

In the next step, generated bimers and their physio-chemical property values are utilized for computing mean value of sequence. Equation 5.51 illustrates mathematical expression for computing sequence mean value.

$$Sequence\ mean\ of\ property_p = \frac{\sum_{i=1}^{seq.length-1} Property_p(bimer_i)}{Sequence\ length - 1},$$
$$where\ \ 1 < p < \text{number of properties} \tag{5.51}$$

In above expression $property_p$ represents 6 different physio-chemical properties which means sequence mean value is computed for each physio-chemical property.

After computing sequence mean values, it calculates auto covariance of sequence bimers at different lag values. Here lag is a hyper-parameter that defines distance between sequence bimers. Primarily, this distance helps in computing covariance

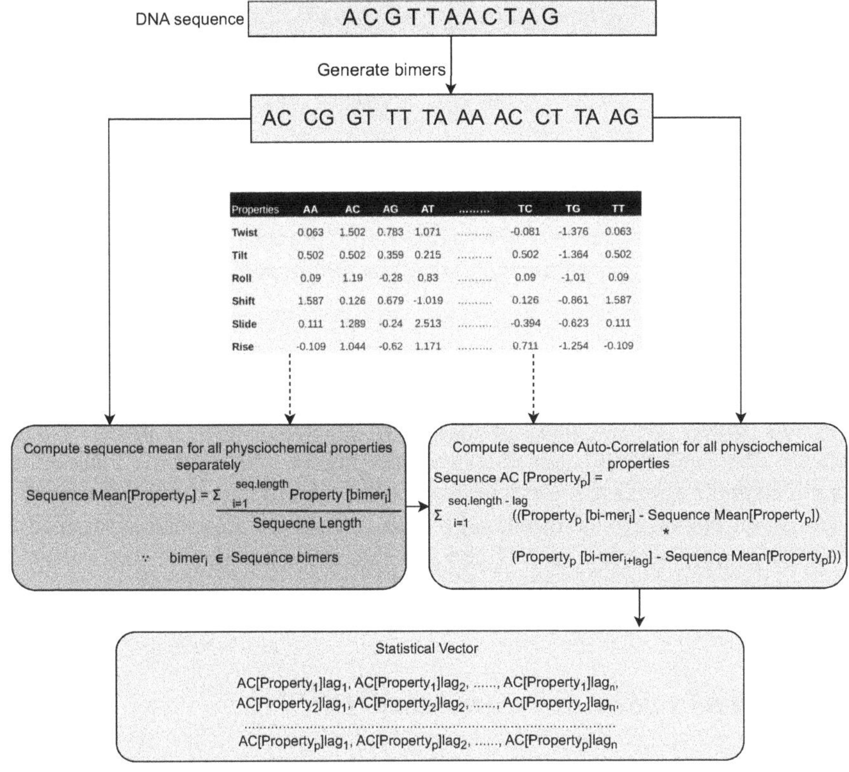

Fig. 5.26 Workflow of Dinucleotide Auto Covariance (DAC) method

between bimers at different scales. Let's if lag value is set as n, the encoder computes bimers covariance values using n different lag values separately. Mathematical expression 5.52 illustrates sequence bimers at different values of lag.

$$
\begin{cases}
A_0A_1,\ A_1A_2,\ A_2A_3,\ A_3A_4,\ A_4A_5...,\ A_{L-1}A_L & bimers\ \ with\ \ lag = 0 \\
A_0A_2,\ A_1A_3,\ A_2A_4,\ A_3A_5,\ A_4A_6...,\ A_{L-2}A_L & bimers\ \ with\ \ lag = 1 \\
A_0A_3,\ A_1A_4,\ A_2A_5,\ ...,\ A_{L-3}A_L & bimers\ \ with\ \ lag = 2 \\
. & . \\
. & . \\
. & . \\
A_0A_{n+1},\ A_1A_{n+2},\ ...,\ ...,\ A_{L-(n+1)}A_L & bimers\ \ with\ \ lag = n
\end{cases}
\tag{5.52}
$$

Usually value of lag is selected by performing experimentation using a subset of data. Equation 5.53 illustrates mathematical expression for computing auto covariance of sequence bimers.

$$Sequence\ auto\ correlation\ of\ property_{plag_k} = \sum_{i=1}^{Seq.length-lag-1} (Property_p(bimer_i) - Sequence\ mean\ property_p) \times (Property_p(bimer_{i+lag}) - Sequence\ mean\ property_p) \tag{5.53}$$

In the above expression, $Property_p(bimer_i)$ represents pth property value of ith bimer and $Property_p(bimer_{i+lag})$ denotes pth property value of $(i + lag)$th bimer of sequence.

This encoder transforms raw DNA sequences into fixed length statistical vectors of dimension number of physicochemical $properties \times lag$ value. Let's consider a toy sequence ACGTTAAC. Lag value is set as 2 and three physicochemical properties namely rise, tilt, and roll are used to generate statistical vector of given sequence. DAC encoder first generates sequence bimers with lag value 1 and utilize them to compute sequence mean with respect to each physicochemical property. It makes use of both sequence mean and bimers to compute sequence covariance. It repeat all process with lag value 2 and generate a 6 dimension feature vector of sequence.

5.4.2 Dinucleotide Cross Covariance (DCC)

This encoder has been utilized in different types of DNA and RNA sequence analysis tasks such as phage promoters prediction [919], Recombination hot/cold spots identification [377] and enhancers identification [1221] . Rather than computing sequence bimers covariance with separate physcio chemical properties, DCC encoder simultaneously reaps the benefit of two different properties. Figure 5.27 illustrates high-level overview about working paradigm of DCC encoder. This encoder makes possible pairs of properties that are considered for sequence covariance computation. Let's if 6 physio-chemical properties are selected for covariance computation than 36 possible pairs are generated as [['Rise', 'Roll'], ['Roll', 'Rise'], ['Rise', 'Shift'], ['Shift', 'Rise'], ['Rise', 'Slide'], ['Slide', 'Rise'], ['Rise', 'Tilt'], ['Tilt', 'Rise'], ['Rise', 'Twist'], ['Twist', 'Rise'], ['Roll', 'Shift'], ['Shift', 'Roll'], ['Roll', 'Slide'], ['Slide', 'Roll'], ['Roll', 'Tilt'], ['Tilt', 'Roll'], ['Roll', 'Twist'], ['Twist', 'Roll'], ['Shift', 'Slide'], ['Slide', 'Shift'], ['Shift', 'Tilt'], ['Tilt', 'Shift'], ['Shift', 'Twist'], ['Twist', 'Shift'], ['Slide', 'Tilt'], ['Tilt', 'Slide'], ['Slide', 'Twist'], ['Twist', 'Slide'], ['Tilt', 'Twist'], ['Twist', 'Tilt']]. After generating all possible pairs, it computes sequence mean for each pair of property by using Eqs. 5.54 and 5.55.

$$Sequence\ \ mean\ \ property_{P1} = \frac{\sum_{i=1}^{seq.length-lag-1} Property_{P1}(bimer_i)}{Sequence\ \ length\ \ -1} \tag{5.54}$$

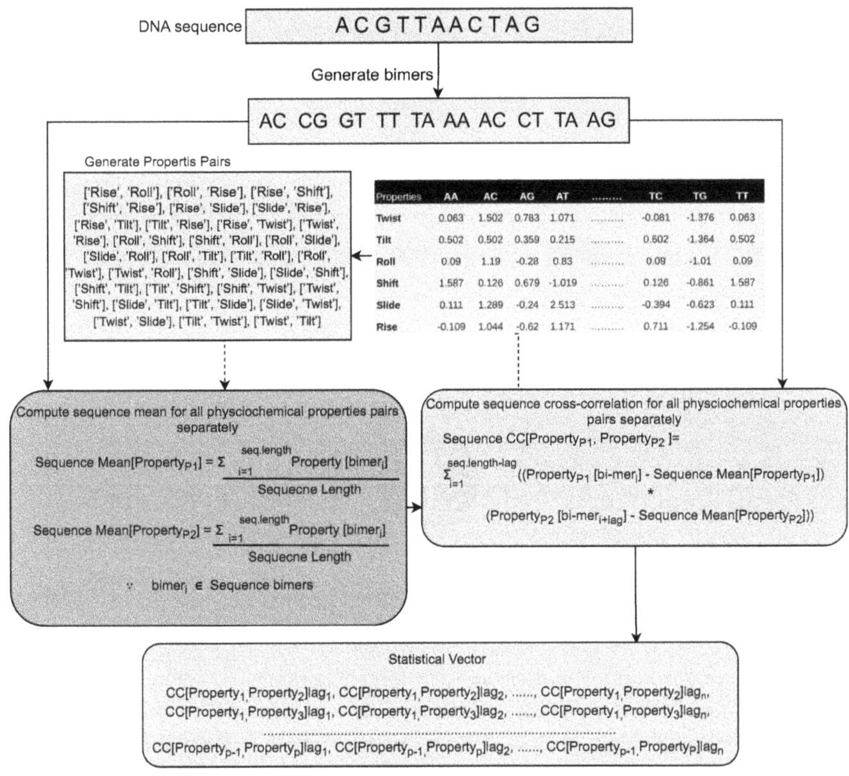

Fig. 5.27 Workflow of Dinucleotide-based Cross Covariance (DCC) encoding method

$$Sequence\ \ mean\ \ property_{P2} = \frac{\sum_{i=1}^{seq.length-lag-1} Property_{P2}(bimer_i)}{Sequence\ \ length\ \ -1} \tag{5.55}$$

Furthermore, it utilizes property pairs based computed sequence mean values for computing cross covariance of sequence bimers at different lag values using Eq. 5.56. Here, lag value represents distance between sequence bimers at which cross covariance is computed. A brief motivation about computing variance at different lag values is described in Sect. 5.4.1.

$$\begin{aligned} &Sequence\ covariance(property_{P1}, property_{P2}) = \\ &\sum_{i=1}^{Seq.length\ -\ lag\ -1} (Property_{P1}(bimer_i) - Sequence\ Mean(property_{P1})) \\ &\times (Property_{P2}(bimer_{i+lag}) - Sequence\ Mean(property_{P2})) \end{aligned} \tag{5.56}$$

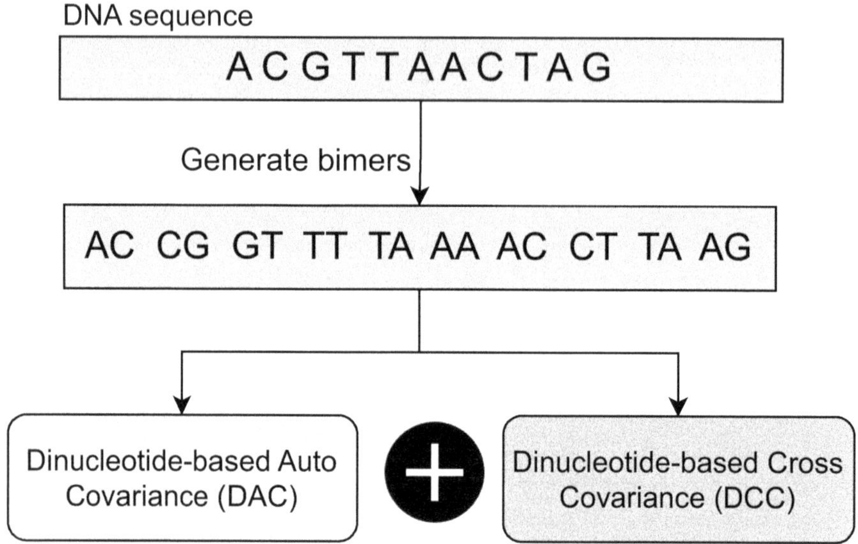

Fig. 5.28 Workflow of Dinucleotide-based Auto-Cross Covariance (DACC) encoding method

In the above equation, $Property_{P1}(bimer_i)$ represents to sequence ith bimer physcio chemical value for P_1th property and $Sequence\ Mean(property_{P1})$ represents sequence mean value for same P_1th property. Similarly, $Property_{P2}(bimer_{i+lag})$ represents to sequence $i + lag$th bimer physico chemical value for P_2th property and $Sequence\ Mean(property_p|_2)$ represents sequence mean value for same P_2th property. In the same way, it computes sequence bimers covariance for each pair of properties and generates statistical vector of dimension number of pairs of physio chemical properties $\times$ number of lag values.

5.4.3 *Dinucleotide-Based Auto-Cross Covariance (DACC)*

This encoder has been utilized in different types of DNA and RNA sequence analysis tasks such as recombination hot/cold spots identification [652] and enhancers identification [1221]. A high-level overview about Workflow of Dinucleotide-based Auto-Cross Covariance (DACC) encoding method is illustrated in Fig. 5.28. This encoder transforms raw DNA sequences into statistical vectors by utilizing sequence encoding criteria of two different sequence encoders namely; Dinucleotide Auto Covariance (DAC) and Dinucleotide-based Cross Covariance (DCC). DAC encoder computes sequence bimers covariance by utilizing individual physicochemical properties, while DCC encoder computes sequence bimers cross-covariance by using pairs of properties. A detailed description of both encoders is available in Sects. 5.4.1 and 5.4.2.

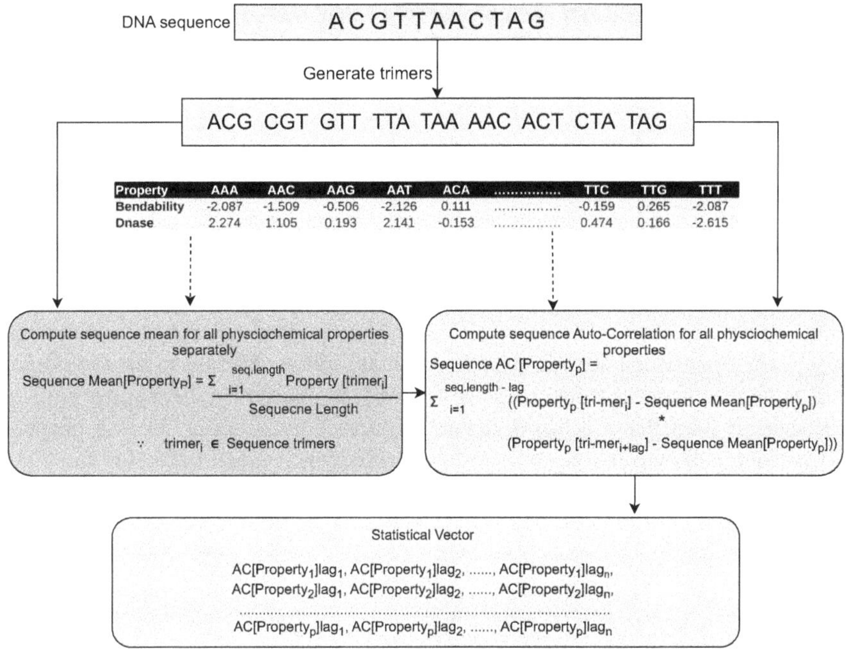

Fig. 5.29 Workflow of Trinucleotide Auto Covariance (TAC) method

5.4.4 Trinucleotide Auto Covariance (TAC)

This encoder has been utilized in different types of DNA and RNA sequence analysis tasks such as subcellular localization prediction [594], prediction of nucleosome positioning [396] and non-coding RNA prediction [1122]. Figure 5.29 provides a comprehensive overview of specific mechanisms and calculations involved in transforming raw sequences into statistical vectors by utilizing DAC encoder. Trinucleotide auto covariance encoder follows 5 different steps to generate a statistical vector of raw sequence. First, this encoder takes lag value and generates trimers with defined lag value. For example, if the given sequence is ACGTTAAC and lag value is defined as 2. It generates trimers with lag 1 and trimers with lag 2. In the next step, for each lag value it iteratively computes sequence trimers mean for each physio-chemical property by using Eq. 5.57

$$Sequence\ mean\ Property_p = \frac{\sum_{i=1}^{seq.length-lag-1} Property_p(trimer_i)}{Sequence\ length\ -\ lag\ -\ 1} \tag{5.57}$$

After computing mean of trimers for each property, it computes trimers auto covariance for n different lag values by using following mathematical expression 5.58.

Let's if the lag value is 2, for each property, it calculates covariance with lag 1 and lag 2 separately by using Eq. 5.52.

$$Sequence\ auto\ correlation\ property_p lag_k$$
$$= \sum_{i=1}^{Seq.length-lag-1} (Property_p(trimer_i) - Sequence\ mean\ Property_p)$$
$$\times (Property_p(trimer_{i+lag}) - Sequence\ mean\ property_p) \tag{5.58}$$

In the above equation, $Property_p(trimer_i)$ represents to sequence ith trimer physcio chemical value for pth property and $Sequence\ Mean(property_p)$ represents sequence mean value for same pth property. Similarly, $Property_p(trimer_{i+lag})$ represents to sequence $i + lag$th trimer physicochemical value for pth property and $Sequence\ Mean(property_p)$ represents sequence mean value for same pth property. This encoder transforms raw DNA sequence into statistical vector of dimension number of physio chemical properties × number of lag values.

5.4.5 Trinucleotide Cross Covariance (TCC)

This encoder has been utilized in different types of DNA and RNA sequence analysis tasks such as promoters identification [650], subcellular localization prediction [594]. Figure 5.30 provides a comprehensive overview of specific mechanisms and calculations involved in transforming raw sequences into statistical vectors by utilizing Trinucleotide Cross Covariance encoder. Similar to TAC encoder, this encoder first generates trimers. After generating trimers, next step is to make possible pairs of properties that are considered for computation. After generating pairs, for each pair, it computes sequence trimers mean by using Eqs. 5.59 and 5.60.

$$Sequence\ \ mean\ \ property_{P1} = \frac{\sum_{i=1}^{seq.length-lag-1} Property_{P1}(trimer_i)}{Sequence\ \ length\ \ -1} \tag{5.59}$$

$$Sequence\ \ mean\ \ property_{P2} = \frac{\sum_{i=1}^{seq.length-lag-1} Property_{P2}(trimer_i)}{Sequence\ \ length\ \ -1} \tag{5.60}$$

After computing mean, it computes the cross-covariance between trimers at k and k+lag position in the sequence of trimers by using Eq. 5.61.

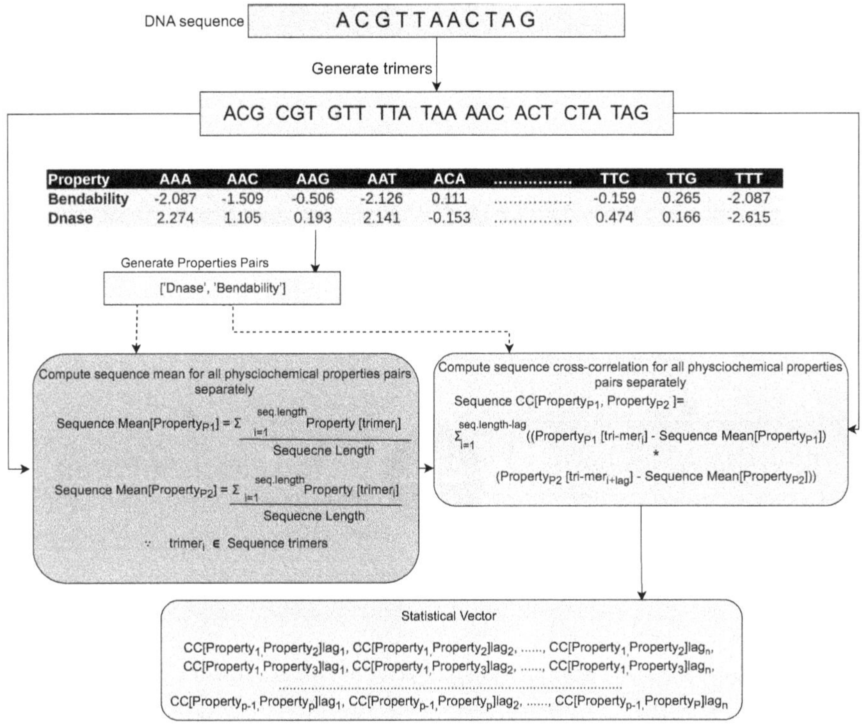

Fig. 5.30 Workflow of Trinucleotide-based Cross Covariance (TCC) method

$$Sequence\ covariance(property_{P1}, property_{P2}) = \sum_{i=1}^{Seq.length\ -\ lag\ -1} (Property_{P1}(trimer_i) - Sequence\ Mean(property_{P1})) \times (Property_{P2}(trimer_{i+lag}) - Sequence\ Mean(property_{P2})) \tag{5.61}$$

In the above equation, $Property_{P1}(trimer_i - SequenceMean(property_{P1})$ is the first term and it uses first property of pair for $Property_{P1}(trimer_i)$ and $SequenceMean(property_{P1})$. Similarly, $Property_{P2}(trimer_{i+lag}) - SequenceMean(property_{P2})$, second term uses second term of pair. In the same way, this encoder computes covariance for each pair of properties and in the last step generates possible pair's dimensions feature vector for final statistical representation.

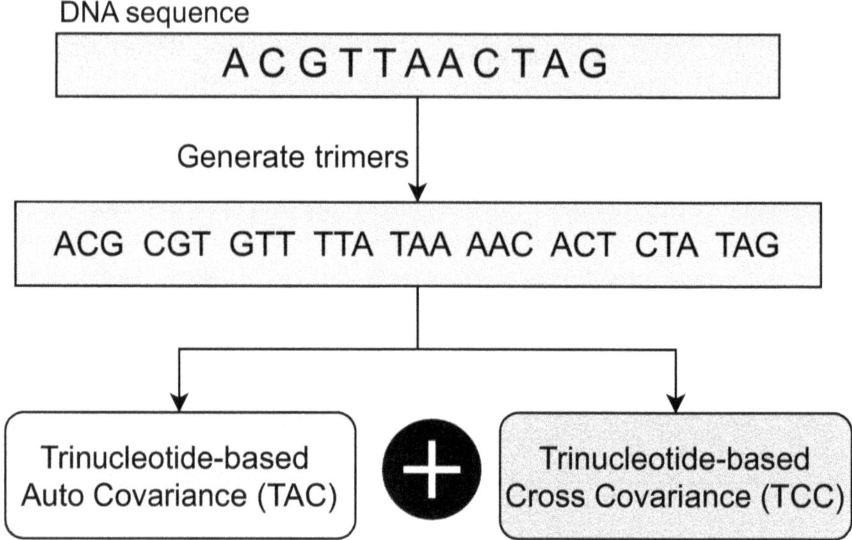

Fig. 5.31 Workflow of Trinucleotide-based Auto-Cross Covariance (TACC) method

5.4.6 Trinucleotide-Based Auto-Cross Covariance (TACC)

This encoder has been utilized in different types of DNA and RNA sequence analysis tasks such as promoters identification [650], subcellular localization prediction [594], generation of different modes of DNA, RNA and proteins [650] and DNA replication [651]. TACC encoder transforms raw DNA sequences into statistical vectors by reaping the benefits of two individual encoding methods namely; TAC and TCC. TAC computes the covariance between trimer for each property, while TCC calculates cross-covariance between trimers by using pairs of properties to reap the benefits of their combined effect on trimers. A detailed description of both encoders is available in Sects. 5.4.4 and 5.4.5. A graphical illustration about Workflow of Trinucleotide-based Auto-Cross Covariance (TACC) encoder is shown in Fig. 5.31.

5.5 Label Based Sequence Encoders

Label based Sequence encoding methods transform DNA or RNA sequences into statistical vectors by capturing position aware discriminative distributional information of nucleotides. These methods make use of both sequence and its class label to generate statistical vector. Position-specific trinucleotide propensity based on single stranded or double stranded characteristic of DNA (PSTNPSS/PSTNPDS) encoding methods are almost similar and have been widely used to perform diverse types of

DNA and RNA classification tasks such as identification of RNA Pseudouridine Sites [1199], promoters prediction [610], identification of N4-methylcytosine sites [413].

5.5.1 Position-Specific Trinucleotide Propensity Based on Single-Strand (PSTNPss)

The prime assumption behind the development of Position-specific trinucleotide propensity based on single-stranded characteristic (PSTNPss) encoder [210, 413] was to generate statistical representations of DNA sequences by generating 3-mers and assigning higher scores to those 3-mers which had more discriminative class densities.

To understand working paradigm of PSTNPSS Encoder, consider a corpus $C = S_1, S_2, ...S_M$, where X number of sequences belong to training set and Y number of sequences belong to test set. In the corpus C, each S_i represents a DNA sequence that is comprised of four repeated letters including A, C, G, T. After generating 3-mers of corpus sequences, each sequence S_i can be represented as $S_i = 3\text{-}mer_1, 3\text{-}mer_2, ...3\text{-}mer_n$. In each sequence, 3-mers positions can be represented as $P_i = P_1, P_2, ...P_z$. PSTNPSS Encoder computes the vocabulary $V = v_1, v_2, ...v_k$ which contains unique 3-mers of the corpus sequences.

Using positive and negative classes sequences of only training set X, PSTNPSS Encoder generates two matrices Z^{pos} and Z^{neg} where each matrix dimension consist of Vocabulary $(V) \times$ Positions (P). Mathematical expressions of both matrices are shown in Eqs. 5.62 and 5.63.

In positive class matrix Z^{pos}, each entry Z_{ij} represents the occurrence frequency of ith 3-mer at jth position across all sequences of positive class.

$$Z^{\text{pos}} = \begin{bmatrix} Z_{1,1} & Z_{1,2} & \cdots & Z_{1,z} \\ \vdots & \vdots & \ddots & \vdots \\ Z_{n,1} & Z_{n,2} & \cdots & Z_{n,z} \end{bmatrix} \tag{5.62}$$

Likewise, in negative class matrix Z^{neg}, each entry Z_{ij} represents the occurrence frequency of ith 3-mer at jth position across all sequences of negative class.

$$Z^{\text{neg}} = \begin{bmatrix} Z_{1,1} & Z_{1,2} & \cdots & Z_{1,z} \\ \vdots & \vdots & \ddots & \vdots \\ Z_{n,1} & Z_{n,2} & \cdots & Z_{n,z} \end{bmatrix} \tag{5.63}$$

In both matrices, position specific 3-mers occurrence frequencies Z_{ij} that assist to capture inherent 3-mer distribution patterns are computed using function described in Fig. 5.32

Input: $Z^{pos/neg}$: A Zero-Valued Matrix of K-mers Vocabulary × K-mers Positions for Positive/Negative Class Sequences
Output: $Z^{pos/neg}$: A K-mer Position Aware Distribution Matrix of K-mers Vocabulary × K-mers Positions for Positive/Negative Class Sequences

```
function OccurenceComputer(Matrix Z^{pos/neg}, Positive/Negative Class Sequences Corpus (C))
    n ← rowlength(Z^{pos/neg})
    z ← columnlength(Z^{pos/neg})
    NS^{pos/neg} ← length(C)
    for i ← 1 to n do
        for j ← 1 to z do
            occurrence ← 0
            for x ← 1 to NS^{pos/neg} do
                if K − mer_{i}{j} in C_x then
                    occurrence ← occurrence + 1

            Z_ij^{pos/neg} ← occurrence


    return Z^{pos/neg}

```

Fig. 5.32 Function for populating zero-valued matrix $Z^{pos/neg}$ of size K-mers vocabulary × positions with K-mer occurrence frequencies

It is evident in Fig. 5.32, from top to bottom, first loop index i is an iterator on vocabulary V of 3-mers, second loop index j is an iterator on all possible positions P, and last loop index x is an iterator on all sequences of positive class from training set NS^{pos} to compute the count of positive sequences in which ith 3-mer occurs at jth position.

Similarly, Z^{neg} is populated using the same function (Fig. 5.32) but using the negative class sequences of training set NS^{neg}, where we compute the count of negative class sequences in which ith 3-mer appears at jth position.

Afterward, PSTNPSS encoder computes 3-mer position specific density values in positive and negative classes. The positive class density of ith 3-mer at jth position can be computed by normalizing the Z_{ij} value with total number of positive class sequences NS^{pos}.

$$Z^{posden} = \frac{Z^{pos}}{NS^{pos}}, \quad 0 \leq Z^{posden} \leq 1 \tag{5.64}$$

Similarly, negative class density of ith 3-mer at jth position is computed by normalizing the Z_{ij} value with total number of negative class sequences NS^{neg}. Equations 5.64 and 5.65 represent mathematical expressions to compute positive and negative class densities, where each entry of Z^{pos} and Z^{neg} matrices is normalized with total number of sequences in positive and negative classes respectively.

$$Z^{negden} = \frac{Z^{neg}}{NS^{neg}}, \quad 0 \leq Z^{negden} \leq 1 \tag{5.65}$$

PSTNPss [210, 413] encoder score for ith 3-mer present at jth position denoted as $PSTNPss$ can be computed using class densities difference. Equation 5.66 denotes generic mathematical expression to compute PSTNPss score of 3-mers.

$$PSTNPss = Z^{posden} - Z^{negden} \tag{5.66}$$

5.5.2 *Position-Specific Trinucleotide Propensity Based on Double-Strand (PSTNPds)*

PSTNPds is a variant of PSTNPss encoder and it also generates 3-mers of DNA sequence, but before generating 3-mers, PSTNPDS first transforms 4 nucleotides (A,C,G,T) sequence to 2 neucleotides (A, G) sequence by replacing T nucleotide with A and G nucleotide with C. Lets if we have a toy sequence "ACGTCG-TATGCAAGCGTA". Its transformed sequence is "ACCACCAAACCAACCCAA". The transformed sequence is segregated into 3-mers and remaining whole working paradigm is similar to PSTNPss encoding method. Primarily, PSTNPds encoder is computationally less expensive because PSTNPss encoder generates 64 unique 3-mers and PSTNPds generates 16 unique 3-mers. Among two different classes sequences, at all possible positions it is easy to capture discriminative potential of 16 3-mers as compare to discriminative potential of 64 3-mers. How ever it is possible different discriminative patterns that exist with 64 3-mers may no remain similar discriminative for 16 3-mers. In this particular scenario PSTNPds encoder remains fail in encoding discriminative patterns to statistical vectors of two different classes.

5.5.3 *POCD-ND Encoder*

PSTNPss [210] and PSTNPDS encoders transform raw DNA sequences into statistical vectors by utilizing only 3-mers, however, different size k-mers may generate different types of discriminative patterns. With an aim to transform raw sequences into statistical vectors by extracting and encoding more discriminative patterns of 4 neucleotides, Nabeel et al. [738] proposed a more generalized version of PSTNPss [210, 413] encoder named POCD-ND that can be utilized for any k-mer.

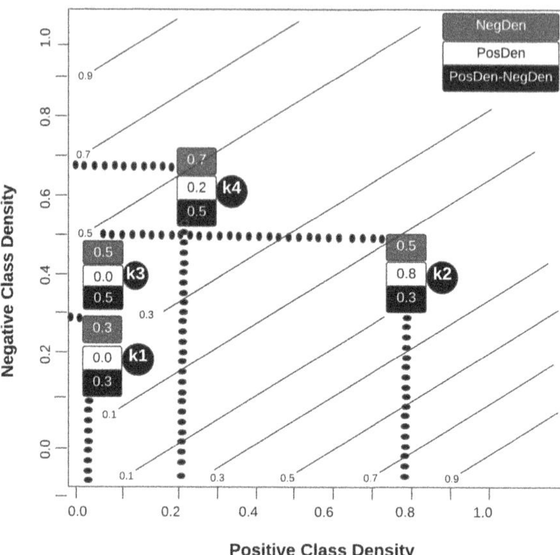

Fig. 5.33 Contour plots for PSTNPss encoder where contour lines going parallel to diagonal assigning equal scores to those K-mers having same positive to negative class density differences. Here K-mers are represented as K_i, their positive class densities in white colour, negative class densities in gray colour and difference of positive to negative class densities in black colour

$$PSTNPss = Z^{posden} - Z^{negden} \tag{5.67}$$

Another major drawback of PSTNPss [210, 413] and PSTNPDS encoders is that they assign same scores to k-mers having different level of discriminative potential. To more briefly understand, lets discuss this drawback using a contour plot [309]. Figure 5.33 illustrates contour lines with respect to positive and negative class densities. Figure 5.33 illustrates two different k-mers along the contour lines having positive to negative class density difference value equal to 0.3. Similarly, another pair of k-mers including k3 and k4 are shown where positive to negative class density difference value is equal to 0.5. Analysis of positive and negative class densities along with differences for all four k-mers (Fig. 5.33) indicates that from first pair, k1 (posden = 0, negden = 0.3, PSTNPss = 0.3) located near to y-axis is of utmost importance on the contour line. Similarly, from second pair, k3 (posden = 0, negden = 0.5, PSTNPss = 0.5) near to y-axis has more importance on the contour line. Across both pairs, as we move along the contour line away from the origin towards the top right-corner, posden and negden values are increasing. In first pair, k2 (posden = 0.2, negden = 0.7, PSTNPss = 0.5) and in second pair, k4 (posden = 0.8, negden = 0.5, PSTNPss = 0.3) are less important than k-mer k1 and k3 respectively. This is because k1 and k3 are present in only negative class and absent in positive class whereas k2 and k4 are present in both classes. PSTNPss [210, 413] encoder assigns equal score 0.3 to first pair of k-mers, and 0.5 to second pair of k-mers, indicating that it assigns equal score to k-mers regardless of their occurrences in positive and negative classes, which shall not be the case.

To overcome these drawbacks (POCD-ND) [738] encoder generate more comprehensive statistical representations of DNA sequences based on three main assumptions. (I) Like PSTNPss assumption [210, 413], those k-mers are discriminative which have large position aware occurrence based positive to negative class density difference, (II) Those k-mers are more discriminative whose position aware occurrence based density is high in only one particular class and close to zero in other class, (III) If two k-mers have equal $Z^{posden} - Z^{negden}$ difference, then the k-mers having lower $min(Z^{posden}, Z^{negden})$ value shall be assigned higher scores. Here min denotes the minimum function which returns the minimum value by comparing modification and non-modification class densities. A comprehensive details of these assumptions is provided in motivating example Sect. 5.5.3.1.

To generate statistical representations of DNA sequences, the first step is to generate k-mers by sliding a fixed-size window with a particular stride size. In this process, DNA sequences are segregated into sub-sequences where each sub-sequence called k-mer represents a group of nucleotides. In the generated sub-sequences, size of k-mer or sub-sequence depends upon the size of a window which is slided to generate them.

Suppose we have a corpus $C = S_1, S_2, ...S_M$, where X number of sequences belong to training set and Y number of sequences belong to test set. In the corpus C, each S_i represents a DNA sequence that is comprised of four repeated letters including A, C, G, T. After generating k-mers of corpus sequences, each sequence S_i can be represented as $S_i = k\text{-}mer_1, k\text{-}mer_2, ...k\text{-}mer_n$. In each sequence, k-mers positions can be represented as $P_i = P_1, P_2, ...P_z$. POCD-ND encoder computes vocabulary $V = v_1, v_2, ...v_k$ which contains unique k-mers of the corpus sequences and their positions in the sequences. The size of the vocabulary depends on the size of k-mer and can be computed using 4^k where k represents the size of k-mer. For example, in the case of 1-mer, vocabulary size will be $4^1 = 4$, for 2-mers, vocabulary size will be $4^2 = 16$, and so on. Whereas, number of positions entirely depend on the sequence length. With the increase in size of k-mers, the size of vocabulary also increases and with the increase of sequences length, number of positions also increase.

Using positive and negative classes sequences of only training set X, POCD-ND generates two matrices Z^{pos} and Z^{neg} where each matrix has the dimension of Vocabulary $(V) \times$ Positions (P). Snippets of both matrices are shown in Eqs. 5.68 and 5.69.

In positive class matrix Z^{pos}, each entry Z_{ij} represents the occurrence frequency of ith k-mer at jth position across all sequences of positive class.

$$Z^{\text{pos}} = \begin{bmatrix} Z_{1,1} & Z_{1,2} & \cdots & Z_{1,z} \\ \vdots & \vdots & \ddots & \vdots \\ Z_{n,1} & Z_{n,2} & \cdots & Z_{n,z} \end{bmatrix} \tag{5.68}$$

Likewise, in negative class matrix Z^{neg}, each entry Z_{ij} represents the occurrence frequency of ith k-mer at jth position across all sequences of negative class.

$$Z^{\text{neg}} = \begin{bmatrix} Z_{1,1} & Z_{1,2} & \cdots & Z_{1,z} \\ \vdots & \vdots & \ddots & \vdots \\ Z_{n,1} & Z_{n,2} & \cdots & Z_{n,z} \end{bmatrix} \tag{5.69}$$

In both matrices, position specific k-mers occurrence frequencies Z_{ij} that assist to capture inherent k-mer distribution patterns are computed using function described in Fig. 5.32

It is evident in Fig. 5.32, from top to bottom, first loop index i is an iterator on vocabulary V of k-mers, second loop index j is an iterator on all possible positions P, and last loop index x is an iterator on all sequences of positive class from training set NS^{pos} to compute the count of positive sequences in which ith k-mer occurs at jth position.

Similarly, Z^{neg} is populated using the same function (Fig. 5.32) but using the negative class sequences of training set NS^{neg}, where we compute the count of negative class sequences in which ith k-mer appears at jth position.

Afterward, POCD-ND encoder computes k-mer position specific density values in positive and negative classes. The positive class density of ith k-mer at jth position can be computed by normalizing the Z_{ij} value with total number of positive class sequences NS^{pos}.

$$Z^{posden} = \frac{Z^{pos}}{NS^{pos}}, \quad 0 \leq Z^{posden} \leq 1 \tag{5.70}$$

Similarly, negative class density of ith k-mer at jth position is computed by normalizing the Z_{ij} value with total number of negative class sequences NS^{neg}. Equations 5.70 and 5.71 represent mathematical expressions to compute positive and negative class densities, where each entry of Z^{pos} and Z^{neg} matrices is normalized with total number of sequences in positive and negative classes respectively.

$$Z^{negden} = \frac{Z^{neg}}{NS^{neg}}, \quad 0 \leq Z^{negden} \leq 1 \tag{5.71}$$

To generate statistical representations of DNA sequences based on three main assumptions, POCD-ND encoder makes use of powerful expression to assign score to ith k-mer present at jth position based on its discriminative potential. Equation 5.72 illustrates generic mathematical expression to compute POCD-ND scores of k-mers.

$$POCD - ND = \frac{PSTNPss}{min(Z^{posden},\ Z^{negden})}, \quad min(Z^{posden},\ Z^{negden}) > 0 \tag{5.72}$$

It is important to mention that, in Eq. 5.72, while computing k-mers scores, if positive or negative class density of a particular k-mer gets zero than denominator becomes zero which leads to unsigned infinity. To handle such scenario most effectively, we propose to select any arbitrary value grater than zero to perform

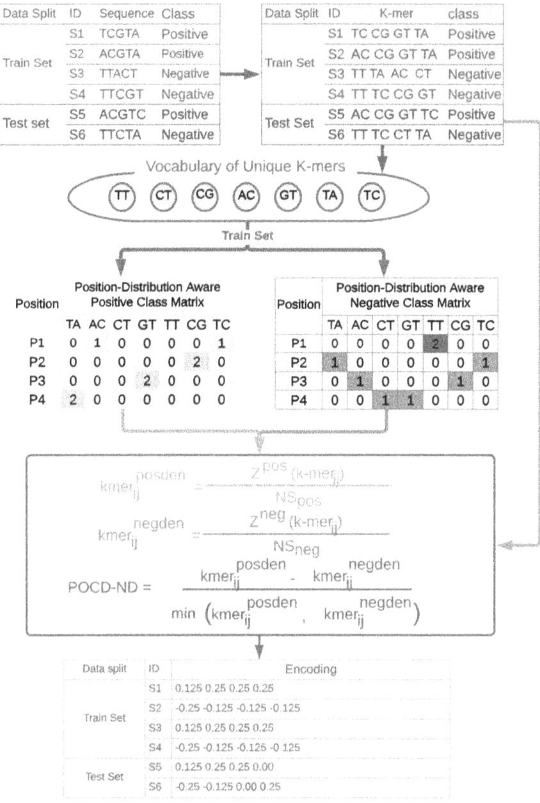

Fig. 5.34 Working paradigm of POCD-ND sequence encoding method. The genomic sequences are converted into k-mers, where k = 1 $\cdots k$. Position aware distribution matrices are generated for unique k-mers with respect to modification and non-modification classes represented as (Z^{pos}, Z^{neg}), containing k-mer occurrence information at each unique possible position in modification and non-modification sequences represented as (NS^{pos}, NS^{neg}). k-mer modification and non-modification class densities represented as ($k - mer_{ij}^{posden}$, $k - mer_{ij}^{negden}$), their differences and normalized values are computed to generate optimal statistical weights for k-mers. Concatenation of k-mers statistical weights generate statistical vectors of DNA sequences which are used to train machine learning classifier for the prediction of three different DNA modifications. Here different colour schemes are segregating various steps involved in the working of POCD-ND sequence encoding method

division. In our experimentation, we change this value from 0.1 to 0.9 with a step size of 0.1. Through performance analysis, we find that small value 0.1 is most appropriate, hence we have reported performance figures using small 0.1 value when denominator becomes zero.

A complete workflow of POCD-ND encoder using a hypothetical corpus of 6 sequences is illustrated in Fig. 5.34. POCD-ND encoder segregates the sequences into k-mers and divides the sequences into training and test k-mer sequences sets. It computes the vocabulary of unique k-mers, and utilizes vocabulary and only training sequences to precisely generate statistical representations of corpus sequences in three steps: (1) Generate k-mer$_{ij}$ position aware distribution matrices for modification and non-modification classes, (2) Compute k-mer$_{ij}$ densities in modification and non-modification classes, (3) Compute k-mer position aware distribution based modification and non-modification class densities normalized difference. We assume that distribution of k-mers in test sequences is close to the distribution in training sequences. Hence, we utilize the k-mer position aware distribution matrices constructed using training sequences in order to generate statistical weights of k-mer test sequences by following the aforementioned step 2 and step 3.

5.5.3.1 A Motivating Example

This section briefly describes the effect of division by minimum k-mer position specific class density using a hypothetical example. Table 5.10 indicates a hypothetical dataset containing 15 sequences related to two classes c1 and c2. In each sequence, k-mer occurrence at 10 different positions is provided. The dataset is un-balanced because only 5 sequences belong to c1 class and remaining 10 sequences belong to c2 class.

Table 5.11 shows k-mer positive and negative class densities for the sample dataset and the scores given by PSTNPss and POCD-ND encoders. Furthermore, we show the locations of k-mer ten different positions in the Fig. 5.35 where x-axis represents the k-mer density in c1 class and y-axis represents the k-mer density in c2 class.

Table 5.10 A hypothetical dataset of 15 sequences related to c_1 and c_2 classes. Occurrence frequency of a particular k-mer at 10 different positions is shown

Sequences	Class	P_1	P_2	P_3	P_4	P_5	P_6	P_7	P_8	P_9	P_{10}
1	c_1	1	0	1	1	0	0	0	1	1	1
2	c_1	0	0	1	1	1	0	0	0	0	1
3	c_1	1	0	1	0	0	0	0	0	0	0
4	c_1	1	1	1	1	0	0	0	0	1	1
5	c_1	1	0	1	0	0	0	0	0	0	0
6	c_2	0	1	0	0	0	1	0	0	0	1
7	c_2	0	1	1	0	0	1	1	0	0	1
8	c_2	1	1	0	0	0	1	0	0	0	1
9	c_2	1	1	0	0	0	0	0	0	0	1
10	c_2	0	0	0	1	0	0	0	0	0	0
11	c_2	1	0	0	1	1	0	0	0	0	0
12	c_2	1	0	0	0	0	1	0	0	0	0
13	c_2	0	1	0	0	0	1	1	0	0	1
14	c_2	0	1	1	0	0	0	0	0	0	1
15	c_2	1	1	1	0	1	0	1	0	0	1

Table 5.11 K-mers densities in C1 class ($k\text{-}mer_{ij}^{posden}$), C2 class ($k\text{-}mer_{ij}^{negden}$), scores and ranks assigned by PSTNPss [210, 413] and POCD-ND encoders to a particular k-mer present in 10 different positions in a hypothetical dataset

k-mer	P_1	P_2	P_3	P_4	P_5	P_6	P_7	P_8	P_9	P_{10}
$k\text{-}mer_{ij}^{negden}$	0.5	0.7	0.3	0.2	0.2	0.5	0.3	0	0	0.7
$k\text{-}mer_{ij}^{posden}$	0.8	0.2	1	0.6	0.2	0	0	0.2	0.4	0.6
PSTNPss [210, 413] Score	0.3	0.5	0.7	0.4	0	0.5	0.3	0.2	0.4	0.1
POCD-ND Score	0.6	2.5	2.23	2	0	5	3	2	4	0.17
PSTNPss [210, 413] Rank	7	2	1	4	10	3	6	8	5	9
POCD-ND Rank	8	4	5	6	10	1	3	7	2	9

In Fig. 5.35, k-mers positions located in the top left and bottom right corners are most discriminative. K-mers located along the diagonal are the least discriminative as their occurrences in both classes are equal. The k-mers located on axes, except the k-mer closer to the origin are the most discriminative because they occur in only one class. A good encoding method shall assign higher scores to the k-mers located in top left and bottom right corners. We discuss why POCD-ND scores and rankings for most k-mer positions are different than PSTNPss [210, 413] encoder scores and rankings.

- The k-mer at first and seventh positions has same score of 0.3 for PSTNPss [210, 413] encoder. POCD-ND encoder assigns higher score to k-mer at seventh position and lower score to k-mer at first position. It is evident in the Fig. 5.35 that k-mer at seventh position lies on y-axis as compared to k-mer at first position which lies on slight distance to diagonal at middle of lower and upper right corner. So intuitively, k-mer at seventh position is more discriminative and shall be assigned a higher score, as done by the POCD-ND encoder.
- The k-mer at fourth and ninth positions has equal PSTNPss [210, 413] score. We can see in the Fig. 5.35 that k-mer at ninth position is far more important than k-mer at fourth position as it is very close to x-axis, hence POCD-ND encoder assigns higher score to k-mer at ninth position and lower score to k-mer at fourth position.
- The k-mer at third position has the highest weight and rank for PSTNPss [210, 413] encoder among the ten k-mer positions. POCD-ND places k-mer at third position at fifth rank because of normalization with 0.3 that lowers its score as compared to k-mer at second, sixth, seventh, and ninth positions. It is clear from the Fig. 5.35 that k-mer at sixth position lies on y-axis. Furthermore. this k-mer position is the nearest to top left or bottom right corner, hence this k-mer position is the most discriminative among all k-mer positions.
- The k-mer at fifth position is assigned the lowest score and rank by both PSTNPss [210, 413] and POCD-ND as it lies on diagonal and has equal positive and negative class densities. Both encoders assign this k-mer position zero score.

As a whole, POCD-ND encoder assigns better scores and ranks to k-mers at different positions by correctly quantifying their discriminative potential.

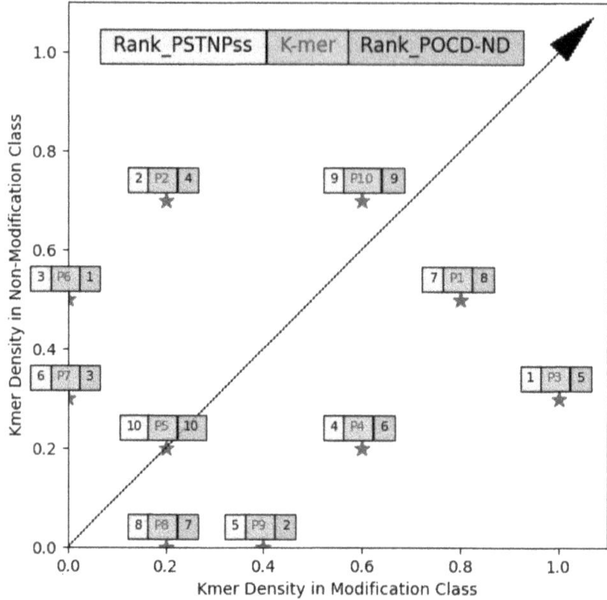

Fig. 5.35 Graphical representation of a particular k-mer at ten different positions in hypothetical dataset with ranks assigned by PSTNPss [210, 413] and POCD-ND encoders. Here K-mers at particular positions are shown in red colour, ranks assigned by existing PSTNPss sequence encoder in yellow colour, and ranks assigned by POCD-ND sequence encoder in pink colour

5.6 Fourier Transformation Based Encoding

Fourier Transformation based encoding comprises of two different stages. At first stage Raw DNA or RNA sequences are transformed into statistical vectors by utilizing any traditional encoding method including kmer distribution based encoding, auto covariance, cross covariance based encoding, or physcio chemical properties based encoding. At second stage generated statistical vectors are passed to Fourier Transformation method that is competent in extracting two different types of spectrum patterns of features. Equations 5.73 and 5.74 illustrate mathematical expressions for extracting two different types of spectrums.

$$\text{spectrum A} = (absolute(FX[i]))^2 \tag{5.73}$$

$$\text{spectrum B} = absolute(FX[i]) \tag{5.74}$$

Furthermore, it utilizes both spectrums to extract 19 different features namely average, median, maximum, minimum, standard deviation, population standard deviation, interquartile, semi-interquartile, coefficient of variation, skewness, kurtosis, peaks of both spectrums, percentile15, percentile25, percentile50, and percentile75. Equation 5.75 illustrates mathematical expression to compute average spectrum.

$$Average = \frac{sum(spectrum)}{len(spectrum)} \tag{5.75}$$

Then, Eq. 5.76 is used to compute the median of spectrum. After computing average and median of spectrum, it computes maximum and minimum values of spectrum using Eqs. 5.77 and 5.78

$$Median = median(spectrum) \tag{5.76}$$

$$Maximum = max(spectrum) \tag{5.77}$$

$$Minimum = min(spectrum) \tag{5.78}$$

In the next step, it computes peak values of both spectrums by using average and lengths of spectrums as illustrated in following mathematical expressions 5.79 and 5.80

$$Peak = \frac{1}{average(spectrum)} \times \frac{len(spectrum)}{3} \tag{5.79}$$

$$Peak2 = \frac{1}{average(spectrum2)} \times \frac{len(spectrum2)}{3} \tag{5.80}$$

It then computes standard deviation and population standard deviation of spectrum. Standard deviation measures the spread of distribution while population standard deviation is a parameter that is computed from every individual in the population. Equations 5.81 and 5.82 is mathematical expression used to compute standard deviation and population standard deviation.

$$Standard\ Deviation = \sqrt{\frac{\sum_{i=1}^{n}(X_i - \bar{X})}{n-1}} \tag{5.81}$$

$$Population\ Standard\ Deviation = \sqrt{\frac{\sum_{i=1}^{n}(X_i - \bar{X})}{N}} \tag{5.82}$$

After generating 8 informative features, it computes percentiles of the spectrum as 15, 25, 50, 75, and 90 by using Eq. 5.83. It computes variance of the spectrum as follows:

$$Percentile_n = \frac{\sum_{i=1}^{n} Spectrum[i]}{100} \times n \tag{5.83}$$

In the above expression $spectrum[i]$ denotes ith value of spectrum and n represents the percentile value which can be 15, 25, 50, 75, and 90. By using above-mentioned computed quantities, it determines interquartile, semi-interquartile, coefficient of variation, skewness, and kurtosis. Following Eqs. 5.84–5.88 are used to computes these values.

$$interquartile = percentile(spectrum, 75) - percentile(spectrum, 25) \tag{5.84}$$

$$semi\text{-}interquartile = \frac{percentile(spectrum, 75) - percentile(spectrum, 25)}{2} \tag{5.85}$$

$$Coefficient\ of\ variation = \frac{Standard\ Deviation}{Ave(Spectrum)} \tag{5.86}$$

$$Skewness = 3 \times \frac{Ave(Spectrum) - Median}{StandardDeviation} \tag{5.87}$$

$$kurtosis = \frac{percentile(spectrum, 75) - percentile(spectrum, 25)}{2 \times (percentile(spectrum, 90) - percentile(spectrum, 10))} \tag{5.88}$$

Interquartile represents the middle half of the data or contains the middle values of data. Semi-interquartile is the half of interquartile range. Skewness measures the asymmetry of distribution, which can be of three types; positive, negative, and zero skews. Kurtosis measures how often outliers occur in a distribution, which is also known as tailedness of distribution. By employing FFT on a feature vector, 19 discriminative features are extracted and generate a feature space of 19 dimensions.

Chapter 6
Protein Sequence Representation Learning Methods

This chapter briefly describes different encoding methods that transforms raw protein sequences into statistical vectors.

6.1 Amino Acids Distribution Based Encoding Methods

Protein sequence analysis is similar to DNA and RNA sequence analysis tasks. To transform protein sequences into feature vectors, we can generate kmers that are subsequences of length k. The kmers can be generated by sliding a fixed-size window over the whole sequence. Figure 6.3 illustrates the process of kmer generation, where monomers are generated for window size k = 1. In case of window size 2, bimers are generated and for window size k = 3, trimers are generated. Similarly, with the increase in number k, the length of generated kmers will increase. Here, it is important to mention that window size not only controls the length of kmers but also controls the unique vocabulary of corpus. As protein sequences are made from repetitive patterns of 20 different amino acids, so unique vocabulary of corpus can be computed using following expression 20^k. In case of k = 1 unique vocabulary will be $20^1 = 20$, and for k = 2 unique vocabulary will be $20^2 = 400$.

Furthermore, while generating kmers, apart from window size, selection of appropriate stride size of window is also important. When stride size is equal to window size then non-overlapped kmers are generated and when stride size is less than window size overlapped kmers are generated. Non-overlapped kmers have more discriminative potential that is important to categorise sequences into different classes. On the other hand, over-lapped kmers have both discriminative and semantic information about amino acids distribution in the sequences. In both overlapped and non-overlapped kmers, window and stride size generate same length kmers as well as vocabulary of kmers. Figures 6.1 and 6.2 illustrate the process of overlapped and non-overlapped kmer generation.

M. Nabeel Asim et al., *Artificial Intelligence for Molecular Biology*,
https://doi.org/10.1007/978-3-031-90450-9_6

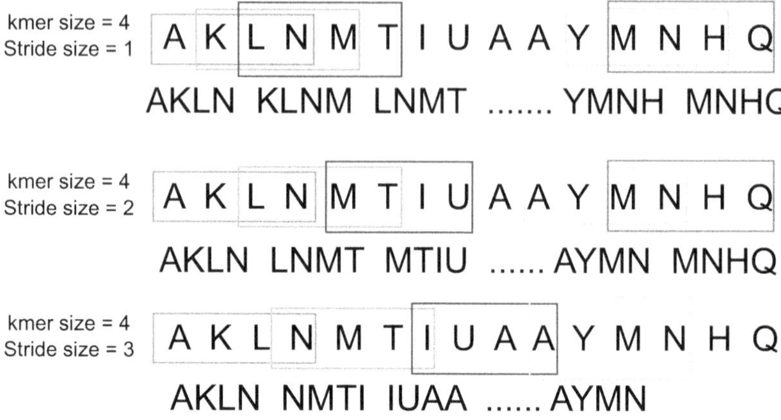

Fig. 6.1 Overlapped kmers generation

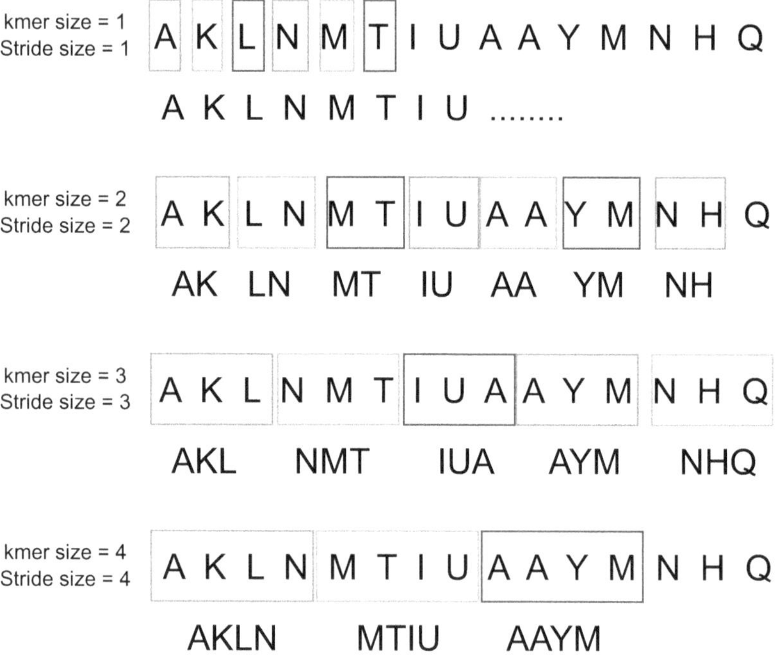

Fig. 6.2 Non-overlapped generation of kmers

Next step is to transform kmer based sequence samples into statistical vectors. Following sub-sections briefly describe working paradigm of kmer encoder variants including amino acid composition (AAC), di-amino acid composition, tri-amino acid composition, and higher order amino acid composition (Fig. 6.3).

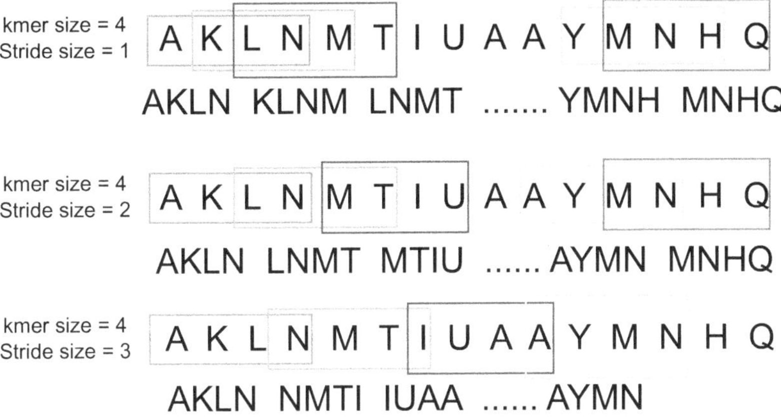

Fig. 6.3 Generation of overlapped kmers

6.1.1 Amino Acid Composition (AAC)

This encoder has been utilized in different types of protein sequence analysis tasks such as protein-protein interaction prediction [855], plant pentatricopeptide repeat coding gene/protein identification [824] and acetylation prediction [849]. AAC encoder makes use of window and stride sizes of 1 to generate monomers or unimers of the sequences and computes 20 unique unimers occurrence frequencies in each sequence. Furthermore, it normalizes occurrence frequencies with sequence length. Equation 6.1 illustrates mathematical expression for computing normalized frequencies of unimers.

$$Encoding(Unimer_i) = \frac{\text{Occurrence frequency of } unimer_i}{\text{Sequence length}},$$
$$\text{where } unimer_i \varepsilon \ \{A, C, D, E, F, G, H, I, K, L, M, N, P, Q, R, S, T, V, W, Y\} \tag{6.1}$$

In the above expression, $unimer_i$ is the ith unimer from 20 unique amino acids.

6.1.2 Di-amino Acid Composition (DAC)

This encoder has been utilized in different types of protein sequence analysis tasks such as protein-protein interaction prediction [855] and plant pentatricopeptide repeat coding gene/protein identification [824]. Di-amino acid composition (DAC) encoder can be used to transform raw sequences into statistical vectors by computing both overlapped and non-overlapped kmers or bimers distribution. In case of non-overlapped kmers, it utilizes window and stride size of 2 to generate bimers and counts their occurrences. Next, it normalizes count with sequence length − kmer size +1. Equation 6.2 is used to compute normalized count values for final statistical representation.

$$Encoding(bimer_i) = \frac{\text{Occurrence frequency of } bimer_i}{\text{Sequence length } - 1}, \quad \text{where } bimer_i \; \varepsilon \; \{AC, DE, FG, ...ST, VW\} \tag{6.2}$$

In the above expression, $bimer_i$ is the ith bimer that belongs to 20^4 unique bimers.

6.1.3 Tri-amino Acid Composition (TAC) and Higher

TAC encoder has been utilized in different types of protein sequence analysis tasks including protein-protein interaction prediction [855] and plant pentatricopeptide repeat coding gene/protein identification [824]. Similar to DAC, TAC encoder also generates overlapped and non-overlapped trimers. In the process of non-overlapped trimers generation, window and stride sizes of 3 need to rotate over raw sequences. After generating non-overlapped trimers, it computes unique trimers occurrence frequencies and normalizes them using Eq. 6.3. Overlapped trimers are generated with window size 3 and stride size can be 1 or 2. This encoder generates a 20^3 dimensional feature vector. Similarly, for higher values of k ($k \geq 4$), this encoder generates higher length kmers and 4^k dimensional feature space can be generated by computing normalized occurrence frequencies of unique k-mers.

$$Encoding(trimer_i) = \frac{\text{Occurrence frequency of } trimer_i}{\text{Sequence length } - 2}, \quad \text{where } trimer_i \; \varepsilon \; \{ACD, DEF, FGH, ...STV, VWY\} \tag{6.3}$$

6.2 Enhanced Amino Acid Composition (EAAC)

This encoder has been utilized in different types of protein sequence analysis tasks such as structural class prediction [372], subcellular localization prediction [594] and protein malonylation sites prediction [1062]. Working paradigm of EAAC encoder can be summarized into two different steps. First, it generates subsequences by sliding a k-size window over whole sequence. In the next step, it computes the count of each unique amino acid and normalizes it with unique amino acids in each subsequence as illustrated in Eq. 6.4.

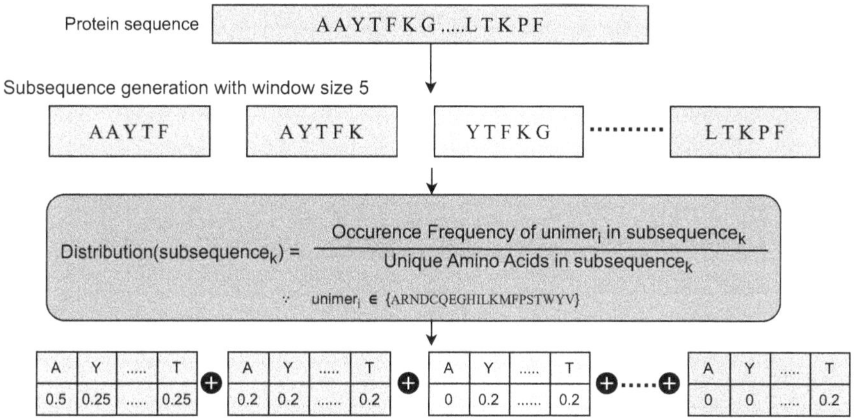

Fig. 6.4 Graphical illustration of enhanced amino acid composition (EAAC) encoder

$$
\begin{aligned}
&Encoding(Subsequence_k) \\
&\quad = \frac{Occurrence\ frequency\ of\ unique\ amino\ acid_i\ in\ subsequence_k}{Unique\ amino\ acid\ in\ subsequence_k}, \\
&\quad \text{where unique}\ amino\ acid_i \in \{A, C, D, E, F, G, H, I, K, L, M, N, P, Q, \\
&\qquad R, S, T, V, W, Y\}
\end{aligned} \tag{6.4}
$$

In the above expression, $unique\ amino\ acid_i$ represents ith amino acids and $subsequence_k$ refers to kth subsequence of sequence. In this way, for each subsequence a 20 dimensional feature vector is generated. Finally, it concatenates all subsequences feature vectors and generates a statistical vector of dimension (number of subsequences $\times$ 20). Figure 6.4 depicts a detailed illustration of the encoding process employed by EAAC encoder using a sample toy sequence and window size 5. The Figure showcases the step-by-step sequence transformation process and highlights specific mechanisms and calculations involved in determining distribution of amino acids in subsequences.

6.3 Accumulated Amino Acid Frequency (AAF)

This encoder has been utilized in different types of protein sequence analysis tasks such as protein domain identification [924], and structural class prediction [372]. It transforms raw protein sequences into statistical vectors by computing accumulated distribution of amino acids. Figure 6.5 graphically illustrates that working paradigm of this encoder can be summarized in 3 different steps. First, it generates unimers

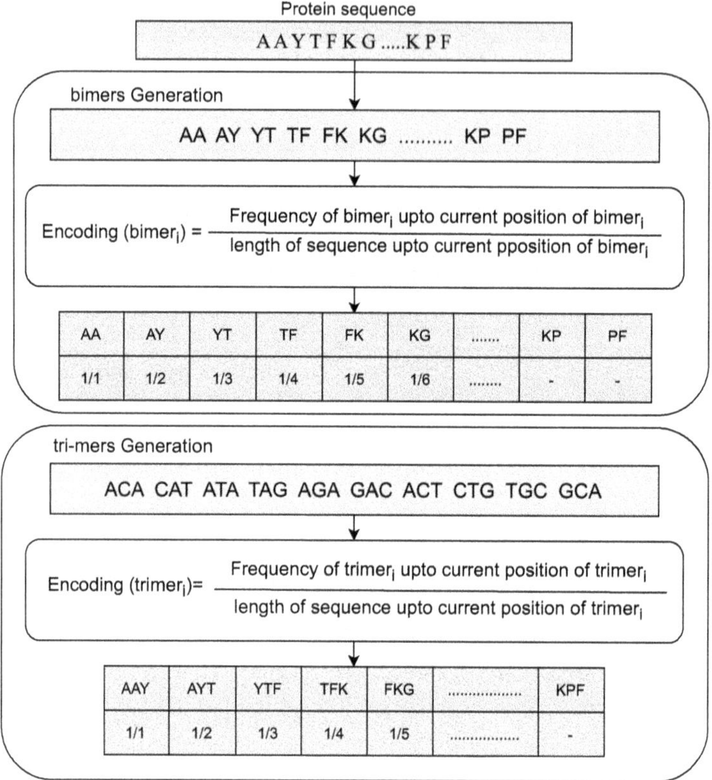

Fig. 6.5 Workflow of Accumulated Amino Acid Frequency (AAF) encoder

and computes occurrence frequencies of unimers in an accumulated fashion. Specifically, at a particular position of a unimer, it computes occurrence frequency of $unimer_i$ in all previous positions and normalizes occurrence frequency with sequence length until that particular unimer position. Equation 6.5 illustrates mathematical expression for computing occurrence frequencies of unimers in an accumulated fashion.

$$Encoding(unimer_i) = \frac{Occurrence\ frequency\ of\ unimer_i\ upto\ current\ position\ of\ unimer_i}{Sequence\ length\ upto\ current\ position\ of\ unimer_i}$$
$$\text{where unique } unimer_i \in \{A, C, D, E, F, G, H, I, K, L, M, N, P, Q, R, S, T, V, W, Y\} \quad (6.5)$$

6.4 Dipeptide Deviation from Expected Mean (DDE)

This encoder has been utilized in different types of protein sequence analysis tasks such as ion binding protein identification [650], drug-target interaction prediction [518], T-cell epitope prediction [248], protein domain identification [924], anticancer peptide prediction [343], discrimination of linear B-cell epitopes from non-epitopes [878] and identification of DNA-binding proteins [78]. DDE encoder measures difference between observed frequency and expected frequency of dipeptide within protein sequence. A positive computed value indicates that dipeptide occurs more frequently than expected, while a negative value accounts for reverse, indicating that the dipeptide occurs less frequently than expected. This measurement captures biases in presence of particular dipeptides within a protein sequence. It can be helpful for locating dipeptides that are overabundant or sparse and might be important structurally or functionally.

Figure 6.6 illustrates high level overview of 5 different steps of DDE encoding method. First, it generates 400 bimers of 20 unique amino acids. In the second step, it calculates theoretical mean of each bimer by using unique codon values corresponding to each amino acid, as shown in Table 6.1. Equation 6.6 illustrates mathematical expression to compute theoretical mean of bimers.

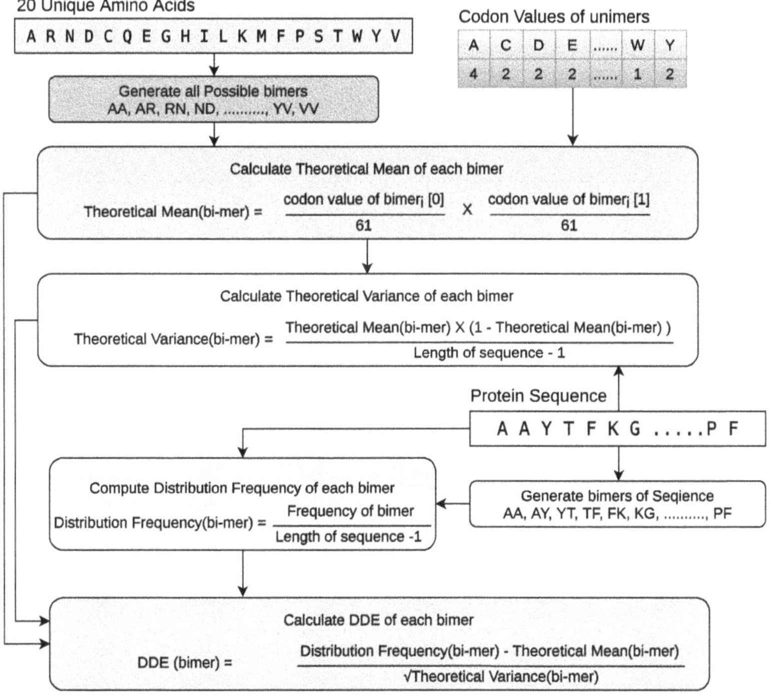

Fig. 6.6 Workflow of Dipeptide Deviation from expected Mean (DDE) encoding method

Table 6.1 Codon values of amino acids

A	C	D	E	F	G	H	I	K	L	M	N	P	Q	R	S	T	V	W	Y
4	2	2	2	2	4	2	3	2	6	1	2	4	2	6	6	4	4	1	2

$$Theoretical\ mean\ bimer_i = \frac{bimer_i\ first\ amino\ acid\ codon\ value}{61} \times \frac{bimer_i\ second\ amino\ acid\ codon\ value}{61},$$
$$\text{where } bimer_i\ \varepsilon\ \{AC, DE, FG, ...ST, VW\} \tag{6.6}$$

After computing theoretical mean in third step, it determines distribution frequency of each bimer. Equation 6.7 illustrates mathematical expression for computing bimers distribution frequencies.

$$Distribution\ Frequency(bimer_i) = \frac{\text{Occurrence frequency of } bimer_i}{\text{Sequence length } - 1},$$
$$\text{where } bimer_i\ \varepsilon\ \{AC, DE, FG, ...ST, VW\} \tag{6.7}$$

In Eq. 6.7, occurrence frequency of $bimer_i$ represents count of ith bimer and sequence length – 1 refers to total number of generated bimers. In fourth step, it computes theoretical variance which is mathematically illustrated in Eq. 6.8.

$$Theoretical\ variance\ bimer_i = \frac{Theoretical\ mean\ bimer_i(1 - Theoretical\ mean\ bimer_i)}{N} \tag{6.8}$$

Finally, theoretical mean and variance and distribution frequency of $bimer_i$ are used to compute $bimer_i$ dipeptide deviation from expected mean. Equation 6.9 illustrates mathematical expression to compute DDE encoding of raw protein sequence.

$$DDE[bimer_i] = \frac{Distribution\ frequency\ of\ bimer_i - Theoretical\ mean\ of\ bimer_i}{\sqrt{Theoretical\ variance\ of\ bimer_i}} \tag{6.9}$$

DDE encoder transforms raw protein sequence into 20^2 dimensional statistical vector as follows: $\{DDE[bimer_1], DDE[bimer_2], ..., DDE[bimer_400]\}$.

6.5 Amino Acids Position Aware Encoding Methods

This section briefly describes different encoding methods that transforms raw protein sequences into statistical feature space by capturing amino acids positioning information.

6.5.1 *Position Relative Incidence Matrix (PRIM)*

This encoder has been utilized in different types of protein sequence analysis tasks such as N-linked glycosylation sites prediction [19], DNA Replication Proteins identification [33], anti-inflammatory peptide prediction [29], DNA-Binding proteins prediction [77], dihydrouridine sites prediction [959], lysine SUMOylation sites prediction [1243]. This encoder governs the physical attributes of proteins and determines how amino acids are relatively placed in long protein sequences. To transform raw protein sequences into statistical vectors, it extracts relative positioning information of amino acids in polypeptide chains.

Relative positioning information of amino acids is preserved into a Position relative incidence matrix. Each entry of matrix refers to sum of relative position of jth amino acid with respect to first occurrence of ith amino acid. Position relative incidence matrix formulation process comprises of 2 different steps. First of all, it extracts unique amino acids from raw sequence and acquires first occurrence position of each amino acid within the sequence. Furthermore, it iteratively takes first occurrence position of each amino acid and computes its relative position with respect to all other amino acids positions. Equation 6.10 depicts mathematical expression for computing relative positions of amino acids.

$$\begin{aligned} P_{i,j} = \sum_{\forall J} & Position\, of\; j\text{th}\; Amino\, acid \\ & - Position\, of\; first\; occurrence\; of\; i\text{th}\; Amino\, acid \end{aligned} \tag{6.10}$$

In Eq. 6.10 i and j refers to 20 unique amino acids (A, C, D, E, F, G, H, I, K, L, M, N, P, Q, R, S, T, V, W, Y) and P_{ij} represents relative position of jth amino acid with respect to ith amino acid. As protein sequence is made up from repetitive patterns of 20 unique amino acids so this encoder generates a 20×20 dimensional matrix shown in Eq. 6.11. This matrix contains 400 relative positions between amino acids. If extracted amino acids from given sequence are less than 20 then corresponding to missing amino acid matrix entire row is filled with 0s.

$$P = \begin{bmatrix} P_{A,A} & P_{A,C} & .. & P_{A,j} & .. & P_{A,Y} \\ P_{C,A} & P_{C,C} & .. & P_{C,j} & .. & P_{C,Y} \\ .. & .. & .. & .. & .. & .. \\ P_{i,A} & P_{i,C} & .. & P_{i,j} & .. & .. \\ .. & .. & .. & .. & .. & .. \\ P_{Y,A} & P_{Y,C} & .. & .. & .. & P_{Y,Y} \end{bmatrix} \tag{6.11}$$

To briefly understand working paradigm of this encoder, consider a toy sequence "AJCTLAPRIIM". The sequence is made up of nine unique amino acids namely A, J, C, T, L, P, R, I, M. Each unique amino acid first occurrence position is acquired as A:1, J:2, C:3, T:4, L:5, P:7, R:8, I:9, M:11. Moreover, some of these amino acid occurs more than once, so all positions of nine unique amino acids are A:1, 6; J:2; C:3; T:4; L:5; P:7; R:8; I:9, 10; M:11. Afterwards, it utilizes Eq. 6.10 to compute relative positions of these unique amino acids. It takes first amino acid A and iteratively computes its relative position with itself and with all other amino acids. Since, first occurrence position of A is 1, therefore, first relative position can be computed as $1 - 1 = 0$. Second occurrence position of A is 6 and its relative position is $6 - 1 = 5$. Using Eq. 6.10, $P_{A,A}$ is computed by adding both values such that $P_{A,A} = 5 + 0 = 5$.

Afterwards, it takes second amino acid J and computes its relative occurrence with A which is $2 - 1 = 1$. Since, J occurs for only one time in sequence, so $P_{A,J} = 1$. Similarly, for all other amino acids, it computes $P_{A,C} = 2$, $P_{A,T} = 3$, $P_{A,L} = 4$, $P_{A,P} = 6$, $P_{A,R} = 7$, $P_{A,I} = 17$, $P_{A,M} = 10$. Equation 6.12 depicts a 20×20 position relative incidence matrix that is formed by taking toy sequence amino acids positional information.

$$P = \begin{bmatrix} 5 & 2 & .. & 4 & .. & 0 \\ -2 & 0 & .. & -1 & .. & 0 \\ .. & .. & .. & .. & .. & .. \\ -1 & 1 & .. & 5 & .. & 0 \\ .. & .. & .. & .. & .. & .. \\ 0 & 0 & .. & .. & .. & 0 \end{bmatrix} \tag{6.12}$$

Furthermore, to extract more useful information from each sequence matrix, researchers have utilized three different types moments extraction methods: (1) raw moments, (2) central moments, (3) Hahn moments. These moments are statistical measures which describes various properties of data. Following subsections briefly describes these moments.

Raw Moments

Raw moments provide information about central tendency and skewness of data. Raw moments can be computed by following 3 different steps. First of all, it defines order up to which moments are computed. Then, it uses Eq. 6.13 for computing raw moments up to defined order.

$$M_{ij} = \sum_{p=1}^{n} \sum_{q=1}^{n} p^i q^j (P_{pq}) \tag{6.13}$$

In the above expression, M_{ij} represents the order of moments. Therefore, moment up to order n means sum of i and j should always be less than or equal to n ($i + j \leq n$). p and q refer to rows and columns of matrix. It begins by taking i = 0 and $0 \leq j \leq n$ which means, it first computes moments such as M_{00}, M_{01}, M_{02}, M_{03},..., M_{0n}. Then, it takes i = 1 and $0 \leq j \leq n$ and computes moments such as M_{10}, M_{11}, M_{12}, M_{13},..., $M_{(1)(n)}$. Similarly, it generates moments for $i \geq 2$. Lastly, it concatenates all moments starting from i = 0 to n.

To understand raw moments extraction process for order 3, reconsider matrix P (Eq. 6.12) that is derived from toy sequence "AJCTLAPRIIM". For order 3, range of i and j is $0 \leq i \leq 3$ and $0 \leq j \leq 3$. Using Eq. 6.13, it begins with setting i = 0 and $0 \leq j \leq 3$. For each entry (P_{pq}) of matrix P , it first extract p and q which represents corresponding row and column of entry P_{pq}. In subsequent step, it computes product of p raised to power i (p^i) and q raised to power j (q^j). Afterwards, it multiplies each entry P_{pq} with corresponding computed product $p^i q^j$ and sums all computed values. Since i and j is equal to zero for M_{00}, therefore, product $p^i q^j$ is equal to 1 ($p^0 q^0 = 1$) for all ps and qs. Specifically for M_{00}, computed sum is simple summation of all entries of matrix P which is $M_{00} = \sum_{p=1}^{n} \sum_{q=1}^{n} (P_{pq}) = 32$. Similarly, i = 0 and j = 1 for M_{01} therefore, product $p^i q^j$ is equal to q ($p^0 q^1 = q$). In case of M_{01}, it multiplies each entry (P_{pq}) with corresponding column value (q) and lastly, computes the sum of all computes values which is $M_{01} = \sum_{p=1}^{n} \sum_{q=1}^{n} q(P_{pq}) = 792$. Afterwards, it computes product $p^0 q^2$ which is q^2 to determine M_{02}. To compute M_{02}, it multiples each entry (P_{pq}) with square of corresponding column value (q^2) and calculates sum of all computed values such that $M_{02} = \sum_{p=1}^{n} \sum_{q=1}^{n} q(P_{pq}) = 7816$. It computes all raw moments by following same procedure as $M_{03} = 66{,}648$, $M_{10} = -362$, $M_{11} = 1861$, $M_{12} = 5683$, $M_{20} = 3318$, $M_{21} = 34{,}321$ and $M_{30} = -9482$. Lastly, it concatenates all raw moments computed up to order 3 for matrix P. Equation 6.14 depicts vector generated by computing raw moments of matrix P up to order 3.

$$M = \{32, 792, 7816, 66648, -362, 1861, 5683, -3318, 34321, -9482\} \tag{6.14}$$

Central Moments

To compute central moment, this approach determines centroids ($\bar{x}$ and $\bar{y}$) which are data point where data is eventually distributed in all directions. This methods first computed raw moments to determine centroids. Equation 6.15 depicts mathematical expression for computing centroids.

$$\bar{x} = \frac{M_{10}}{M_{00}}, \bar{y} = \frac{M_{01}}{M_{00}} \tag{6.15}$$

In the above expression, M_{10}, M_{01} and M_{00} refers to raw moments. After calculating centroids, central moments are calculated by using following mathematical expression 6.16.

$$\eta_{ij} = \sum_{p=1}^{n} \sum_{q=1}^{n} (p - \bar{x})^i (q - \bar{y})^j (P_{pq}) \tag{6.16}$$

Reconsidering above-mentioned example to understand central moment computation process for order n, it first utilizes computed raw moments and determine the value of $\bar{x}$ and $\bar{y}$. Since, $M_{10} = -362$, $M_{01} = 792$ and $M_{00} = 32$, therefore, $\bar{x} = \frac{M_{10}}{M_{00}} = \frac{-362}{33} = 11.31$ and $\bar{y} = \frac{M_{01}}{M_{00}} = \frac{-792}{33} = 24.75$. For order 3, range of i and j is $0 \leq i \leq 3$ and $0 \leq j \leq 3$. In the next step, similar to raw moments, it takes i = 0 and $0 \leq j \leq 3$ for first computing η_{00}, η_{01}, η_{02} and η_{03}. To compute η_{00}, it first extracts p and q for each entry (P_{pq}) of matrix P and then subtracts $\bar{x}$ from p and $\bar{y}$ from q. Afterwards, it multiplies $(p - \bar{x})^i$ with $(q - \bar{y})^j$ and similar to raw moments product $(p - \bar{x})^i(q - \bar{y})^j$ is always equal to 1 for all ps and qs because values of i and j are 0s. η_{00} is computed by multiplying each entry P_{pq} with corresponding $(p - \bar{x})^i(q - \bar{y})^j$ value. Therefore, here η_{00} is simply the sum of all entries of matrix P which is 32. Similarly, for η_{01}, product $(p - \bar{x})^i(q - \bar{y})^j$ becomes $(q - \bar{y})$ because $(p - \bar{x})^0 = 1$ and $(q - \bar{y})^1 = (q - \bar{y})$. Primarily, η_{01} multiplies each entry of matrix with corresponding computed $(q - \bar{y})$ value and finally computes their sum as 0. By following the same process, it computes $\eta_{02} = -11{,}786$, $\eta_{03} = 456{,}609$, $\eta_{10} = 0$, $\eta_{11} = 10{,}820.5$, $M\eta_{12} = -441{,}513.25$, $\eta_{20} = -7413.12$, $\eta_{21} = 361{,}255.31$ and $\eta_{30} = -214{,}738.82$. Finally, all computed moments are concatenated and following vector represents central moments computed for matrix P up to order 3.

$$M = \{32, 0, -11786, 456609, 0, 10820.5, -441513.25, -7413.12, \\ 361255.31, -214738.82\} \tag{6.17}$$

Hahn Moments

Hahn moments are orthogonal moments that reconstruct the original data by using inverse functions of discrete Hahn moments. Equation 6.18 illustrates mathematical expression for computing Hahn moments.

$$H_{ij} = \sum_{p=1}^{N-1} \sum_{q=1}^{N-1} h_i^{\tilde{u,v}}(q, N) h_j^{\tilde{u,v}}(p, N) P_{pq} \tag{6.18}$$

In the above expression, $h_i^{\tilde{u,v}}(q, N)$ and $h_j^{\tilde{u,v}}(p, N)$ represent the raw values of Hahn moments with two different variables as p and q. Here, p refers to rows and q denotes column of matrix P. Equation 6.19 depicts mathematical expression for computing raw Hahn moments using a weighting function and square norm.

$$h_i^{\tilde{u,v}}(r, N) = h_i^{u,v}(r, N)\sqrt{\frac{p(r)}{d_n^2}} \tag{6.19}$$

Here, $h_i^{u,v}(r, N)$ represents Hahn polynomial order of "i" with variable r can be formulated as:

$$\begin{aligned} h_i^{u,v}(r, N) &= (N + V - 1)_n(N - 1)_n \\ &\times \sum_{k=1}^{n}(-1)^k \frac{(-n)_k(-r)_k(2N + u + v - n - 1)_k}{(N + v - 1)_k(N - 1)_k} \frac{1}{k!} \end{aligned} \tag{6.20}$$

In above expression, $(N+V-1)_n$, $(N-1)_n$, $(-n)_k$ and $(-r)_k$ utilizes Pochhammer operator. Pochhammer operator (a_k) is mathematical expression of product of consecutive descending positive integers, starting with 'a' and decreasing by 1 for k terms. Equation 6.21 illustrates mathematical expression of Pochhammer series with variable a.

$$(a)_k = a.(a + 1).....(a + k - 1) \tag{6.21}$$

Above expression can be simplifies by using Gamma operator as follow.

$$(a)_k = \frac{\Gamma(a + k)}{\Gamma(a)} \tag{6.22}$$

Here, Gamma operator $\Gamma(a)$ is a mathematical function to generalize factorial function of complex numbers and real numbers, excluding negative integers. Equation 6.23 depicts mathematical expression for generalizing function using Γ operator.

$$\Gamma(a) = \int_0^\infty t^{a-t} e^{-t} dt \tag{6.23}$$

Moreover, p(r) in Eq. 6.19 represents is another polynomial series which is also simplified using Gamma operator. Equation 6.24 depicts mathematical expression of this series.

$$p(r) = \frac{\Gamma(u + r + v)(v + r + 1)(u + v + r + 1)_N}{(u + v + 2r + 1)n!(N - r - 1)!} \tag{6.24}$$

In the above expression, u and v are variables represent positional and compositional information of a primary sequence.

6.5.2 Reverse Position Relative Incidence Matrix (RPRIM)

This encoder has been utilized in different types of protein sequence analysis tasks such as N-linked glycosylation sites prediction [19], DNA-Binding proteins prediction [77], peptide prediction [29], dihydrouridine sites prediction [959], lysine SUMOylation sites prediction [1243]. RPRIM encoder works the same way as PRIM but on reverse primary sequences. This encoder first reverses the order of amino acids in raw sequence and remaining encoding process is similar to PRIM encoder.

6.5.3 Accumulative Absolute Position Incidence Vector (AAPIV)

This encoder has been utilized in different types of protein sequence analysis tasks such as lysine crotonylation sites identification [689], Beta-Lactamases prediction [43], N-linked glycosylation sites prediction [19], dihydrouridine sites prediction [959], lysine SUMOylation sites prediction [1243]. Accumulative absolute position incidence vector (AAPIV) encoder extracts composition of polypeptide chain by computing exact position of amino acids in protein sequence. Accumulative absolute position incidence vector transforms raw protein sequence into statistical feature space in two different phases. First, it extracts unique amino acids from raw sequence and determine their position values. Afterwards, it iteratively computes the accumulative sum of different positions of each amino acid. Equation 6.25 illustrates mathematical expression for computing accumulative absolute positions.

$$X_i = \sum_{k=1}^{n} p_k \tag{6.25}$$

Here, $i \in \{A, C, D, E, F, ..., W, Y\}$ and p_k refers to different positions of kth amino acid in given sequence. This encoder transforms raw protein sequences into 20 dimensional statistical vector. If extracted amino acids are less than 20 then this encoder places 0 at positions corresponding to missing amino acids. Equation 6.26 depicts the vector generated by using AAPIV.

$$X = \{X_A, X_C, X_D, X_E, X_F, ..., X_V, X_W, X_Y\} \tag{6.26}$$

To more briefly understand working paradigm of this encoder, let's consider a toy sequence "AACTIVPTR". The sequence is made up of 6 unique amino acids namely A, C, T, I, V, P, R. Each amino acid position is acquired as A:1, 2; C:3; T:4, 8; I:5; V:6; P:7; R:9. In the next, it iteratively computes the sum of positions of amino acids. For A, computed sum is $1+2=3$ which mean $X_A = 3$ while

C exists for only one time in the sequence, therefore, its position is $X_C = 3$. Similarly, it computes accumulative sum of other amino acids as $X_T = 4 + 8 = 12$, $X_I = 5$, $X_V = 6$, $X_P = 7$, $X_R = 9$. Lastly, it populates accumulative absolute position incidence vector by using these computed value. Equation 6.27 depicts 20 dimensional statistical vector of toy sequence.

$$X = \{3, 3, 0, 0, 0, ..., 6, 0, 0\} \tag{6.27}$$

6.5.4 Reverse Accumulative Absolute Position Incidence Vector

This encoder has been utilized in different types of protein sequence analysis tasks such as lysine crotonylation sites identification [689], Beta-Lactamases prediction [43], N-linked glycosylation sites prediction [19], dihydrouridine sites prediction [959], lysine SUMOylation sites prediction [1243]. RAAPIV encoder works exactly similar to AAPIV but first reverses the order of raw sequence. Afterwards, it follows same procedure as in AAPIV and generates a 20 dimensional feature vector.

6.6 z-Scale

The z-scale descriptor encodes peptides of equal length where any sequence variation can heavily impact biological activity due to amino acids chemical structures changes. This variation occurs due to 3 factors: (1) correlation among activity and similarity of amino acid pairs, (2) total number of amino acids substitution and (3) unaccounted relationships. To predict this change in activity, Quantitative Sequence Modeling (QSAM) is used, which is a major part of Quantitative Structure Activity Relationship (QSAR). QSAR represents sequences activity as a function of structural variation. The quantitative descriptor variables for 20 amino acids have been derived from qualitative data [415, 936].

The changes in chemical structure of each amino acid are quantified using various physicochemical properties. Feature space of multitude data values is reduced through principal component analysis (PCA) to produce z-scales (z_1, z_2, z_3) also called 'Principal properties' of amino acids. Each of 20 amino acids are represented by 29 descriptive variables [415] Table 6.2 which are standardized. Principal component of these variables gives the 3 z-scale values as given in Table VI [415]. The first scale z_1, is influenced by hydrophilicity, z_2 is influenced by size, $^1H\ NMR$, and some hydrophobicity scales, while z_3 incorporates information from pK_a, pI and $^1H\ NMR$ variables.

These z-scales values are further expanded by Partial Least square method (PLS) with 26 new descriptor variables [490, 869]. PLS correlates a latent variable with Y based on X, where X matrix contains 26 descriptors variables and Y matrix

Table 6.2 29 descriptive variables to characterize amino acids to generate first three score vectors of PCA

Descriptive variable no.	Properties
1	Molecular weight
2	$pK_{COOH}(COOH\ on\ C_\alpha)$
3	$pK_{NH_2}(NH_2 on C_\alpha,)$
4	pI,pH at isoelectric point
5	0 substituent van der Waals volume
6	$^1H\ \ NMR\ for\ C_\alpha\text{-}H(cation)$
7	$^1H\ \ NMR\ for\ C_\alpha\text{-}H(dipolar)$
8	$^1H\ \ NMR\ for\ C_\alpha\text{-}H(anion)$
9	$^{13}C\ \ NMR\ for\ C{=}O$
10	$^{13}C\ \ NMR\ for\ C_\alpha\text{-}H$
11	$^{13}C\ \ NMR\ for\ C{=}O\ in\ tetrapeptide$
12	$^{13}C\ \ NMR\ for\ C_\alpha\text{-}H\ in\ tetrapeptide$
13	$R_f\ for 1 - N - (4 - nitrobenzofurazono)$ amino acids in ethyl acetate/pyridine/water
14	Slope of plot 1 Vs mol% $H_2O\ in\ paper\ chromatography$
15	dG of transfer of amino acids from organic solvent to water
16	Hydration potential or free energy of transfer from vapor phase to water
17	$R_f saltchromatography$
18	log P, partition coefficient for amino acids in octanol/water
19	log D, partition coefficient at pH 7.1 for acetylamide derivatives of amino acids in octanol water
20	dG = RT in f; f = fraction buried/accessible amino acids in 22 proteins
21–29	HPLC retention times for nine combinations of three different DH and three eluent mixtures

contains 3 z-values [490, 869]. The goal of PLS is to align latent variables with previously calculated z-scale values. Multiple Linear Regression (MLR) uses 26 physicochemical descriptor variables (related to z_1–z_3) as predictors (X) and each variable as dependent (y). It generates a residual vector E which is orthogonal to z_1–z_3 and a coefficient h as illustrated in Eq. 6.28.

$$E = y - (z_1 \times h_1) - (z_2 \times h_2) - (z_3 \times h_3) \tag{6.28}$$

The first and second principal components score vectors of PCA values generated by the Residual matrix M, formed by 26 residual vector E, gave z_4 and z_5. This residual matrix M is generated when the information related to z_1–z_3 has been removed from the original matrix X [869] (Table 6.3).

Table 6.3 Descriptor scales z_1, z_2, and z_3 for amino acids (the first three score vectors of a principal component analysis of the amino acid data)

Amino acid	z1	z2	z3
Ala (A)	0.07	−1.73	0.09
Val (V)	−2.69	−2.53	−1.29
Leu (L)	−4.19	−1.03	−0.98
Ile (I)	−4.44	−1.68	−1.03
Pro (P)	−1.22	0.88	2.23
Phe (F)	−4.92	1.30	0.45
Trp (W)	−4.75	3.65	0.85
Met (M)	−2.49	−0.27	−0.41
LYS (K)	2.84	1.41	−3.14
Arg (R)	2.88	2.52	−3.44
His (H)	2.41	1.74	1.11
Gb (GI)	2.23	−5.36	0.30
Ser (S)	1.96	−1.63	0.57
CYS (C)	0.71	−0.97	4.13
Asn (N)	3.22	1.45	0.84
ASP (D)	3.64	1.13	2.36
Thr (T)	0.92	−2.09	−1.40
Tyr (Y)	−1.39	2.32	0.01
Gln (Q)	2.18	0.53	−1.14
Glu (E)	3.08	0.39	−0.07

6.7 Pseudo K-tuple Reduced Amino Acids Composition (PseKRAAC) Group Types

This encoder has been utilized in different types of protein sequence analysis tasks such as anti-cancer peptide prediction [155] and proteins conserved regions identification [1255]. PseKRAAC transforms protein sequences into statistical vectors to reduce dimensionality by dividing amino acids into different groups. PseKRAAC encoder has 19 different types (type1, type2, type3A, type3B, type4, type5, type6A, type6B, type6C, type7, type8, type9, type10, type11, type12, type13, type14, type15, type16) where each type contains different number of reduced amino acid compositions (RAAC) as shown in Table 6.4. Furthermore, this number of RAAC represents the groups for amino acids distribution. For example, type 1 can have 19 different distribution groups of amino acids as 2, 3, 4, 5, 6, 7, 8, 9, 10, 11, 12, 13, 14, 15, 16, 17, 18, 19, 20.

Table 6.5 illustrates distribution of amino acids in different RAAC groups of type1. Working paradigm of PseKRAAC encoder comprises of 3 main steps. Firstly, PseKRAAC type is selected from 19 different types and secondly reduced amino acid composition group is selected. Thirdly, sequences kmers are generated using g-gap or λ correlation strategy by sliding k-size window over the sequence where k ranges between 1 to 3. Both k-mers generation strategies are described in following subsections. Suppose a protein sequence P with L amino acid residues as follows:

Table 6.4 Group types with reduced amino acids composition types of PseKRAAC encoder

Type	Reduced amino acids compositions groups
Type 1	[2,3,4,5,6,7,8,9,10,11,12,13,14,15,16,17,18,19,20]
Type 2	[2,3,4,5,6,8,15,20]
Type 3A	[2,3,4,5,6,7,8,9,10,11,12,13,14,15,16,17,18,19,20]
Type 3B	[2,3,4,5,6,7,8,9,10,11,12,13,14,15,16,17,18,19,20]
Type 4	[5,8,9,11,13,20]
Type 5	[3,4,8,10,15,20]
Type 6A	[4,5,20]
Type 6B	[5]
Type 6C	[5]
Type 7	[2,3,4,5,6,7,8,9,10,11,12,13,14,15,16,17,18,19,20]
Type 8	[2,3,4,5,6,7,8,9,10,11,12,13,14,15,16,17,18,19,20]
Type 9	[2,3,4,5,6,7,8,9,10,11,12,13,14,15,16,17,18,19,20]
Type 10	[2,3,4,5,6,7,8,9,10,11,12,13,14,15,16,17,18,19,20]
Type 11	[2,3,4,5,6,7,8,9,10,11,12,13,14,15,16,17,18,19,20]
Type 12	[2,3,4,5,6,7,8,9,10,11,12,13,14,15,16,17,18,20]
Type 13	[4,12,17,20]
Type 14	[2,3,4,5,6,7,8,9,10,11,12,13,14,15,16,17,18,19,20]
Type 15	[2,3,4,5,6,7,8,9,10,11,12,13,14,15,16,20]
Type 16	[2,3,4,5,6,7,8,9,10,11,12,13,14,15,16,20]

$$P = R_1, R_2, R_3, R_4, R_5, ..., R_{L-3}, R_{L-2}, R_{L-1}, R_L \tag{6.29}$$

where R_1 represents amino acid residue at the sequence position 1, R_2 represents the amino acid residue at position 2 and so on. For each K-tuple of reduced amino acid cluster (RAAC), the feature vector of protein sequence has $(numberofgroups)^{kmer}$ dimensions.

6.7.1 g-Gap PseKRAAC

The g-gap PseKRAAC is used to represent a protein sequence as a feature vector by generating g gap-based kmers. Here, g represents gap between each K-tuple peptide and kmers are generated by using g gap value. Let's ACCDEFGKFLV is the given sequence, where g and k are set to 3 and 2, respectively. This encoder will first select type as type 1 and RAAC group is set to 3. Amino acids are divided into three groups as group 1 contains $\{C, M, F, I, L, V, W, Y\}$, group 2 has $\{A, G, T, S, P\}$ and group 3 contains $\{N, Q, D, E, H, R, K\}$. Each value in sequence that belongs to group 1 is encoded as 0 and group 2 is replaced with 1 and group 3 with 2. Hence, the sequence can be formulated as 1,0,0,2,2,0,1,2,0,0,0, and with k = 2, it generates bimers as $\{10, 20, 00\}$. In the next step it generates all possible pairs of group and bimer[0] represents the first group while bimer[1] denotes second group. Then, it calculates count of each amino acid and generates a $(numberofgroups)^{kmer}$ dimensional statistical vector(3^2 dimension in this example).

Table 6.5 RAAC groups of Type 1 PseKRAAC encoder

Group type 1	Distribution of amino acids
2	{'CMFILVWY' , 'AGTSNQDEHRKP'}
3	{'CMFILVWY' , 'AGTSP', 'NQDEHRK,}
4	{'CMFWY', 'ILV', 'AGTS', 'NQDEHRKP'}
5	{'WFYH'. 'MILV', 'CATSP', 'G', 'NQDERK'}
6	{'WFHY', 'MILV', 'CATS', 'P', 'G' , 'NQDERK'}
7	{'WFHY', 'MILV', 'CATS', 'P', 'G', 'NQDE', 'RK'}
8	{'WFHY', 'MILV', 'CA', 'NTS', 'P', 'G', 'DE', 'QRK'}
9	{'WFHY', 'MI', 'LV', 'CA', 'NTS', 'P', 'G', 'DE', 'QRK'}
10	{'WFY', 'ML', 'IV', 'CA', 'TS', 'NH', 'P', 'G', 'DE', 'QRK'}
11	{'WFY', 'ML', 'IV', 'CA', 'NH', 'TS', 'P', 'G', 'D', 'QE', 'RK'}
12	{'WFY', 'ML', 'IV', 'C', 'A', 'NH', 'TS', 'P', 'G', 'D', 'QE', 'RK'}
13	{'WFY', 'ML', 'IV', 'C', 'A', 'NH', 'T', 'S', 'P', 'G', 'D', 'QE', 'RK'}
14	{'WFY', 'ML', 'IV', 'C', 'A', 'NH', 'T', 'S', 'P', 'G', 'D', 'QE', 'R', 'K'}
15	{'WFY', 'ML', 'IV', 'C', 'A', 'N', 'H', 'T', 'S', 'P', 'G', 'DE', 'Q', 'R', 'K'}
16	{'W', 'FY', 'ML', 'IV', 'C', 'A', 'N', 'H', 'T', 'S', 'P', 'G', 'DE', 'Q', 'R', 'K'}
17	{'W', 'FY', 'H', 'ML','IV', 'C', 'A', 'N', 'T', 'S', 'P', 'G', 'D', 'E', 'Q', 'R', 'K'}
18	{'W', 'FY', 'H', 'M', 'L', 'IV', 'C', 'A', 'N', 'T', 'S', 'P', 'G', 'D', 'E', 'Q', 'R', 'K'}
19	'W', 'F', 'H', 'Y', 'M', 'L', 'IV', 'C', 'A', 'N', 'T', 'S', 'P', 'G', 'D', 'E', 'Q', 'R', 'K'
20	{'W', 'F', 'H', 'Y', 'M', 'I', 'L', V', 'C', 'A', 'N', 'T', 'S', 'P', 'G', 'D', 'E', 'Q', 'R', 'K'}

6.7.2 λ-Correlation PseKRAAC

λ-Correlation PseKRAAC generates kmer without any gap value and follows the working flow of generating feature vector. The λ-correlation PseKRAAC, also called parallel correlation PseKRAAC, is used to represent a protein sequence with a vector containing $(numberofgroups)^{kmer}$ components, where λ is an integer that represents the correlation between amino acids.

6.8 Local-Global Context Aware Encoding Methods

This encoder has been utilized in the development of computational predictors for two different tasks namely protein protein interaction prediction [1024] and viral host protein protein interaction prediction [56]. Local-Global Residue Context Aware Encoding (LGCAE) scheme is the fusion of 2 different modules (1) Local Residue Context Aware Sequence Encoding (LCAE) and (2) Global Residue

Context Aware Sequence Encoding (GCAE), paradigms of which are explained in following sub-sections.

6.8.1 Local Amino Acids Context Aware Encoding Method

Using the collection of protein sequences, local residue context aware encoder "LCAE" computes a dictionary D = {A,R,N,D,C,E,Q,G,H,I,L,K,M,F,P,S,T,W,Y,V} containing 20 unique amino acids. The distribution of 20 unique amino acids in protein sequences is reduced to 6 unique classes of amino acids using physicochemical properties of amino acids such as polarity and hydrophobicity. Instead of treating each amino acid as a distinct symbol in entropy calculation, the Shannon entropy of amino acid properties [489] categorizes them into 6 different classes: aliphatic (AVLIMC), polar (STNQ), positive (KR), negative (DE), aromatic (FWYH), and special (G,P). These classifications are based on specific conformational properties, as outlined in Table 6.6. With an aim to better compute the distribution of 6 different classes based on unique amino acids, iteratively, from all 6 classes, 3 amino acids classes are treated as one group whereas other classes are treated as individual groups. In this way, from six unique amino acid classes, 20 different patterns are acquired as shown in Table 6.7 where each pattern is comprised of four unique letters B, J, O, and U.

Furthermore, to better describe the process to generate statistical representation of protein sequences using 20 different patterns, lets consider a protein sequence $S = S_1.....S_n$ of length n where S_q represents a particular residue, initial level numerical representation is generated by iteratively mapping the sequence S to every group v_i expressed in terms of 4 unique letters (B, J, O, U) and replacing 4 unique letters with corresponding ASCII values. To optimize initial level numerical representation of protein sequences generated for each group v_i, it captures positional information of 4 unique letters to better track short range dependencies of residues. For all 4 letters, it computes their total occurrences in given protein sequence S represented as B_n, O_n,J_n, U_n as well their count in initial j entries (B_j, O_j,J_j, U_j) of sequence where j is iteratively updated until it matches the length n of sequence. Using total occurrences of 4 unique letters (B_n, O_n,J_n, U_n) and their iteratively increasing counts ((B_j, O_j,J_j, U_j)), it captures alteration in positional bits of all 4

Table 6.6 Amino acid (residue) categorization into 6 different classes

Descriptor	Property	Categorization
A1	Aliphatic amino acid	A,V,L,I,M,C
A2	Aromatic amino acid	F,W,Y,H
A3	Polar amino acid	S,T,N,Q
A4	Positive amino acid	K,R
A5	Negative amino acid	D,E
A6	Special conformations	G,P

Table 6.7 Twenty possible groups representing different combined patterns of residues which are described using 4 letters B, J, O, and U

V	B	J	O	U
v1	{A1, A2, A3}	A4	A5	A6
v2	{A1, A2, A4}	A3	A5	A6
v3	{A1, A2, A5}	A3	A4	A6
v4	{A1, A2, A6}	A3	A4	A5
v5	{A1, A3, A4}	A2	A5	A6
v6	{A1, A3, A5}	A2	A4	A6
v7	{A1, A3, A6}	A2	A4	A5
v8	{A1, A4, A5}	A2	A3	A6
v9	{A1, A4, A6}	A2	A3	A5
v10	{A1, A5, A6}	A2	A3	A4
v11	{A2, A3, A4}	A1	A5	A6
v12	{A2, A3, A5}	A1	A4	A6
v13	{A2, A3, A6}	A1	A4	A5
v14	{A2, A4, A5}	A1	A3	A6
v15	{A2, A4, A6}	A1	A3	A5
v16	{A2, A5, A6}	A1	A3	A4
v17	{A3, A4, A5}	A1	A2	A6
v18	{A3, A4, A6}	A1	A2	A5
v19	{A3, A5, A6}	A1	A2	A4
v20	{A4, A5, A6}	A1	A2	A3

unique letters in such a manner that odd and even position values fall in range of cosine and sine functions. Mathematical expressions of which are given in Eqs. 6.30 and 6.31 respectively.

$$XS_q(v_i) = \begin{cases} \cos\frac{\pi}{2} + \frac{\pi}{2}\frac{B_j}{B_n+1} & if\, S_q = B \\ \cos\frac{\pi}{2} + \frac{\pi}{2}\frac{J_j}{J_n+1} & if\, S_q = O \\ \cos\pi + \frac{\pi}{2}\frac{O_j}{O_n+1}, & if\, S_q = J \\ \cos\frac{3\pi}{2} + \frac{3\pi}{2}\frac{U_j}{U_n+1} & if\, S_q = U \end{cases} \tag{6.30}$$

$$YS_q(v_i) = \begin{cases} \sin\frac{\pi}{2} + \frac{\pi}{2}\frac{B_j}{B_n+1} & if \quad S_q = B \\ \sin\frac{\pi}{2} + \frac{\pi}{2}\frac{J_j}{J_n+1} & if \quad S_q = O \\ \sin\pi + \frac{\pi}{2}\frac{O_j}{O_n+1} & if \quad S_q = J \\ \sin\frac{\pi}{2} + \frac{\pi}{2}\frac{U_j}{U_n+1} & if \quad S_q = U \end{cases} \tag{6.31}$$

Using $XS_q(v_i)$ and $YS_q(v_i)$ obtained from Eqs. 6.30 and 6.31, 4 different normalized vectors can be computed for each group using mathematical expression given in Eqs. 6.32 and 6.33 respectively.

$$\begin{bmatrix} XN(v_i) = \frac{1}{len(seq)\times\sum XS_q(v_i)} \\ YN(v_i) = \frac{1}{len(seq)\times\sum YS_q(v_i)} \end{bmatrix} \tag{6.32}$$

In Eqs. 6.32 and 6.33, different mathematical formulas compute floating point values where in each mathematical formula, len(seq) denotes the number of residues present in the sequence, and $\sum XS_q(v_i)$, $\sum YS_q(v_i)$, $\sum A(v_i)$, and $\sum B(v_i)$ denote the sum of floating point values present in respective collections ($XS_q(v_i)$, $YS_q(v_i)$, $A(v_i)$, and $B(v_i)$).

$$\begin{bmatrix} A(v_i) = (XS_q(v_i) - XN(v_i))^2 \\ B(v_i) = (YS_q(v_i) - YN(v_i))^2 \\ AXN(v_i) = \frac{1}{len(seq)-1\times\sum A(v_i)} \\ BXN(v_i) = \frac{1}{len(seq)-1\times\sum B(v_i)} \end{bmatrix} \tag{6.33}$$

Then, LCAE vector with respect to each group can be computed by concatenating 4 normalized vectors (Eq. 6.34).

$$LCAE_{G1} = XN(v_i) \oplus YN(v_i) \oplus AXN(v_i) \oplus BXN(v_i) \tag{6.34}$$

$$\begin{bmatrix} LCAE = LCAE_{G1} \oplus LCAE_{G2} \oplus, ...LCAE_{G10} \\ LCAE\ Vector = Patterns \times 4 \end{bmatrix} \tag{6.35}$$

As we have 10 different groups, so comprehensive semantic relatedness and short range residue dependencies aware (LCAE) protein sequences representations can be generated by combining the representation of 10 different groups (Eq. 6.35). For each protein sequence, considering 4 normalization factors for every one out of 10 groups, LCAE generates a 40-dimensional vector (10×4) (Eq. 6.35).

6.8.2 *Global Residue Context Aware Encoding Generation Method*

Global residue context aware encoding (GCAE) handles positional invariance of residues by capturing their context in a more broader scope. First, using the collection of protein sequences, a dictionary D = {A,R,N,D,C,E,Q,G, H,I,L,K,M,F,P,S,T,W,Y,V } containing 20 unique amino acids is computed. Then for a given protein sequence S = S1,....,Sn; of length n where Si represents a particular amino acid present in dictionary D, GCAE generates a sparse 20 × n matrix A, where 20 unique amino acids act as row indices and sequence residues Sn act as column indices. Sparse matrix A looks like:

$$A = \begin{bmatrix} & S_1 & S_2 & \dots & S_n \\ A & a_{11} & a_{12} & \dots & a_{1n} \\ R & a_{21} & a_{22} & \dots & a_{2n} \\ \vdots & \vdots & \vdots & \vdots & \vdots \\ V & a_{20,1} & a_{20,2} & \dots & a_{20,n} \end{bmatrix} = a_{(i,j)} = \begin{Bmatrix} 1, if\, D(i) = S(j) \\ 0, others \end{Bmatrix} \tag{6.36}$$

Sparse matrix A is distributed with the values of 0 and 1 as every cell represented as $a_{(i,j)}$ gets the value of 1 if amino acid present in row index D(i) matches with amino acid present in column index S(j) and 0 otherwise. As basic working principal of global context aware encoder is to capture composition and transition of amino acids in protein sequences. Here in the matrix, length of each amino acid is equal to the length of particular protein sequence. Hence, instead of finding the composition and transition of amino acid by taking the full rows of the matrix, authors proposed to divide the rows of the matrix into L sub-rows/sub-regions and compute composition and transition with respect to each sub-region. To find appropriate number of sub-regions (L), a comprehensive experimentation is required.

The composition computes the frequency of '1", '11", and '111" in each row of sparse matrix to capture the proportion of various residues within sequence and their most dominant contexts.

$$Composition = Frequency\ of\ 1 \oplus Frequency\ of\ 11 \oplus Frequency\ of\ 111 \tag{6.37}$$

As composition computes frequencies of three different trends for each sub-vector ∈ L, therefore dimensions of composition vector can be computed as:

$$Composition\ Vector = 20 \times L \times 3 \tag{6.38}$$

Whereas, transition computes the frequency of '1" followed by '0" and '0" followed by '1" in each row of the sparse matrix to capture the dominant change in the context of the residue.

$$Transition = Frequency\ of\ 1 - 0 \oplus Frequency\ of\ 0 - 1 \tag{6.39}$$

Because, transition computes two different trends for each sub-vector ∈ L, hence dimensions of transition vector can be computed as:

$$TransitionVector = 20 \times L \times 2 \tag{6.40}$$

$$GCAE\ Vector = Composition\ Vector \oplus Transition\ Vector \tag{6.41}$$

By concatenating composition and transition vectors, GCAE protein sequence vector is generated. Considering L = 4, GCAE generates a 400-dimensional feature vector (composition vector = 240, transition vector = 160) for each protein sequence.

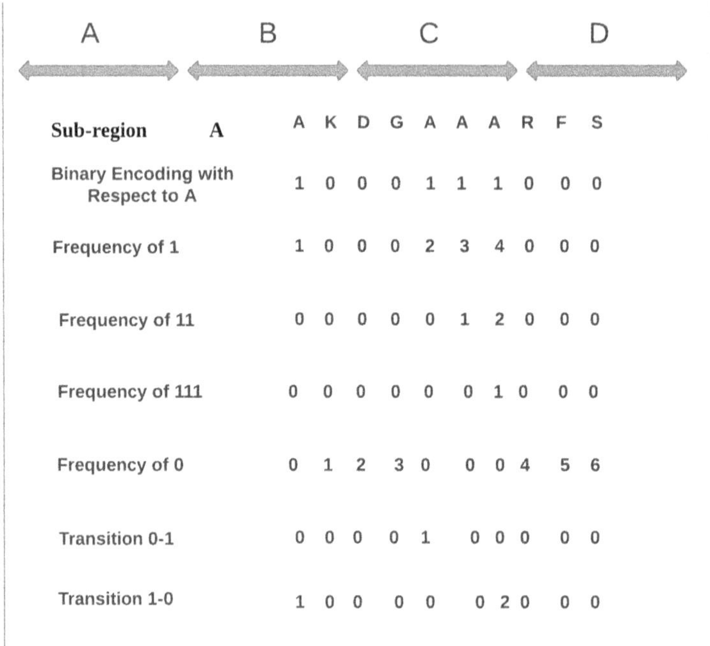

Fig. 6.7 Computing global residue context aware encoding for one amino acid using hypothetical sub-regions

To illustrate the working paradigm of global residue context aware encoding, a hypothetical example is demonstrated in Fig. 6.7.

Consider a hypothetical protein sequence of length 100 generated by combining 50 residues based viral protein sequence with 50 residues based host protein sequence. GCAE first generates a 2-dimensional matrix of size amino acid : 20 $\times$ size of sequence : 100. In this matrix, 20 amino acids act as rows indices and 100 amino acids present in sequence as column indices and each cell of every row gets the value of 1 or 0 depending on the match of row index with column index. As explained earlier, input sequence is divided into L:4 sub-regions where GCAE computes composition and transition of residues with respect to every sub-regions across entire rows of sparse matrix. To better illustrate the working paradigm of GCAE, Fig. 6.7 describes the process of learning representation for amino acid "A" using one of the 4 sub-regions. As shown by the Fig. 6.7, from 2-dimensional sequence matrix, for amino acid A, binary encoding of sub-region "AKDGAAARFS" looks like '1000111000". Using binary representation, GCAE computes composition by counting the frequency of '1", '11", and '111" whereas transition computes the frequency of 1 followed by 0 and 0 followed by 1 patterns. Statistical analysis indicates that this sub-region has four '1", six '0", two "11", one "111", one 0-1 transition, and two 1-0 transitions. As there are four sub-regions

(L = 4) so this sub-region will get a 400 dimensional feature vector (Amino Acid: 20 × L:4 × Composition:3 + Amino Acid: 20 × L:4 × Transition:2 = 400 dimensional vector).

6.8.3 *Fusion of Local and Global Amino Acids Context Aware Sequence Encoders*

Local-Global context aware sequence encoding is generated by fusing 40-dimensional local sequence vector with 400 dimensional global sequence vector. The concatenated feature space contains comprehensive short and long range positional as well as semantic information of residues important to discriminate interactive protein sequences into predefined classes.

6.9 Gap Based Amino Acids Distribution Encoding Methods

Simple distribution based encoding methods capture occurrence frequencies of 20 unique amino acids or different combinations of these amino acids such as bimer, trimer and higher order kmers. Researchers have extended simple distribution based encoding methods to capture long range dependencies of amino acids pairs. For example, bimers are utilized to capture information related to distribution of adjacent amino acids, and gap-based bimers can be adapted to capture long-range relationships between amino acid pairs in the sequence. The following sub-sections delve into different composition-based sequence encoding methods.

6.9.1 *K-spaced Composition Frequency (Kgap)*

Ghandi et al. [340] proposed Kgap sequence encoder that captures comprehensive distributional information of amino acids by considering different gaps among bimers in the protein sequences. This encoder has been widely used in diverse types of protein sequence analysis tasks such as identification of tyrosine nitration sites [234], phosphorylation sites prediction [1223], lysine formylation sites prediction [492], antifreeze proteins prediction [1010] and extracellular matrix proteins [30].

The encoding process of Kgap encoder can be categorized into two distinct phases. First, phase generates 400 bimers by utilizing all possible combinations of 20 unique amino acids ($20^2 = 400$). Second, phase computes distributional information of the generated bimers. Suppose a random bimer containing X and Y amino acids appears with different gaps in the sequence as XY, X*Y, X**Y, X***Y, and so on. In the sequence, * denotes a gap containing any random amino acid. The

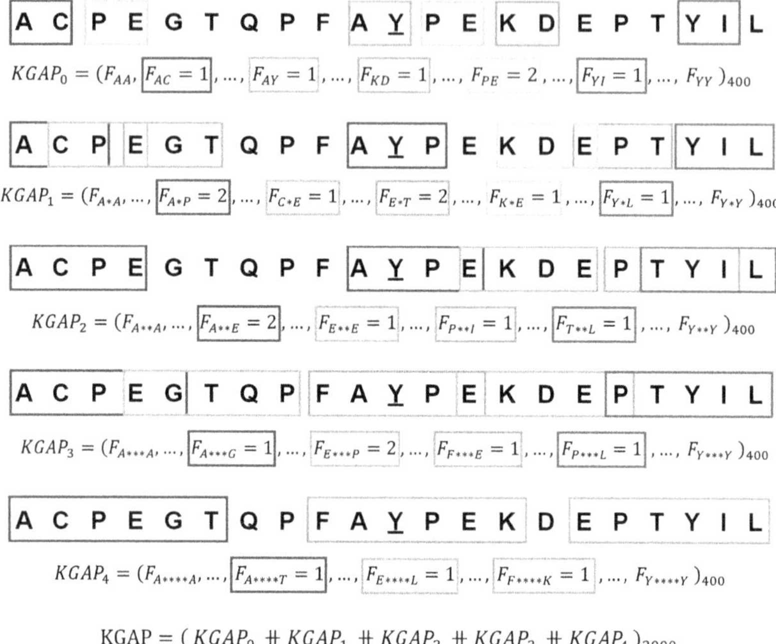

Fig. 6.8 Illustration of Kgap encoding

frequency of a bimer at various gap levels in a protein sequence is calculated by using gaps of differing lengths.

$$\text{bimers} \in \{AA, AC, AD, AE, ...YV, YW, YY\} \tag{6.42}$$

Equation 6.42 represents the generated set of all 400 possible bimers and Eq. 6.43 illustrates mathematical expression to compute gap based occurrence frequencies of bimers as:

$$KGAP_K = (F_{A(*^K)A}, F_{A(*^K)C}, F_{A(*^K)D}, ..., F_{Y(*^K)W}, F_{Y(*^K)Y})_{400} \tag{6.43}$$

In Eq. 6.43, F represents the occurrence frequency of a bimer, and $(*^K)$ represents the gap of length K in between the two amino acids in the bimer.

To clearly demonstrate Kgap sequence encoder process, bimers generated at first stage are provided in Eq. 6.42, and the rows of Fig. 6.8 describe a hypothetical protein sequence and bimers distribution at different gap values. In first row, the parameter K = 0 illustrates the occurrence frequencies of bimers without any gaps. For example, in the protein sequence, the bimer KD occurs once, hence $F_{KD} = 1$, PE occurs twice, hence $F_{PE} = 2$, bimer AA never occurs, hence $F_{AA} = 0$, and so on for all the 400 bimers. When parameter $K > 0$ is set, gaps are introduced in the bimers and value of K represents length of the gap. In second row, the parameter K = 1 illustrates occurrence frequencies of bimers with a gap of 1 between the two amino acids in each bimer. In the protein sequence, 1-gap-based bimer A*P occurs twice as ACP and AYP, hence $F_{A*P} = 2$. Similarly, E*T occurs twice as EGT and EPT, hence $F_{E*T} = 2$, and so on for all 400 1-gap-based bimers. In third row, the parameter K = 2 illustrates occurrence frequencies of bimers with a gap of 2 in-between the bimers. In the protein sequence, 2-gap-based bimer A**E occurs twice as ACPE and AYPE, hence $F_{A**P} = 2$, T**L occurs once as TYIL, hence $F_{T**L} = 1$, and so on for all 400 2-gap-based bimers. A similar process is used to compute the occurrence frequencies of bimers for higher values of K. Here, we utilize K in range 0 to 4 to generate the complete Kgap feature vector for each sequence. Since each value of K generates a 400 dimensional statistical vector, concatenating vectors from all 5 values of K generates a 2000 dimensional statistical vector.

6.9.2 *Composition of K-spaced Amino Acid Pairs (CKSAAP)*

The CKSAAP sequence encoder, developed by Chen et al. [163], is an extended version of the Kgap encoder discussed in Sect. 6.9.1. The Kgap encoder computes absolute occurrence frequencies of bimers. However, the length of protein sequences can vary based on the location of the extracted potential interaction site in the long protein chain. In Kgap encoder, the high variability of sequence length creates bias in the absolute occurrence frequencies of the bimers in the two classes. To mitigate this inherent bias, Chen et al. [163] introduced a sequence length based normalization of the feature vector in the CKSAAP encoder. This encoder has been successfully employed in a variety of protein sequence analysis studies such as, analysis of protein rigidity and flexibility [163], identification of protein phosphorylation and nitration sites [758, 1134, 1223], and detection of protein pupylation and succinylation [404, 405].

The working paradigm of this encoder is similar to the two-phase process of Kgap encoder. In addition, it normalizes the statistical vectors using sequence length at each gap-length. Adapting the Kgap vector defined in Eq. 6.43, the CKSAAP encoded vector can be mathematically formulated as:

$$CKSAAP_K = \left(\frac{F_{A(*^K)A}}{F_{total_K}}, \frac{F_{A(*^K)C}}{F_{total_K}}, \frac{F_{A(*^K)D}}{F_{total_K}}, \ldots, \frac{F_{Y(*^K)W}}{F_{total_K}}, \frac{F_{Y(*^K)Y}}{F_{total_K}}\right)_{400} \tag{6.44}$$

where,

$$F_{total_K} = \text{sequence length} - (K + 1), \tag{6.45}$$

Each row of Fig. 6.9 graphically illustrates the encoding of a hypothetical protein sequence of length 21 into the CKSAAP feature vector. The occurrence frequencies

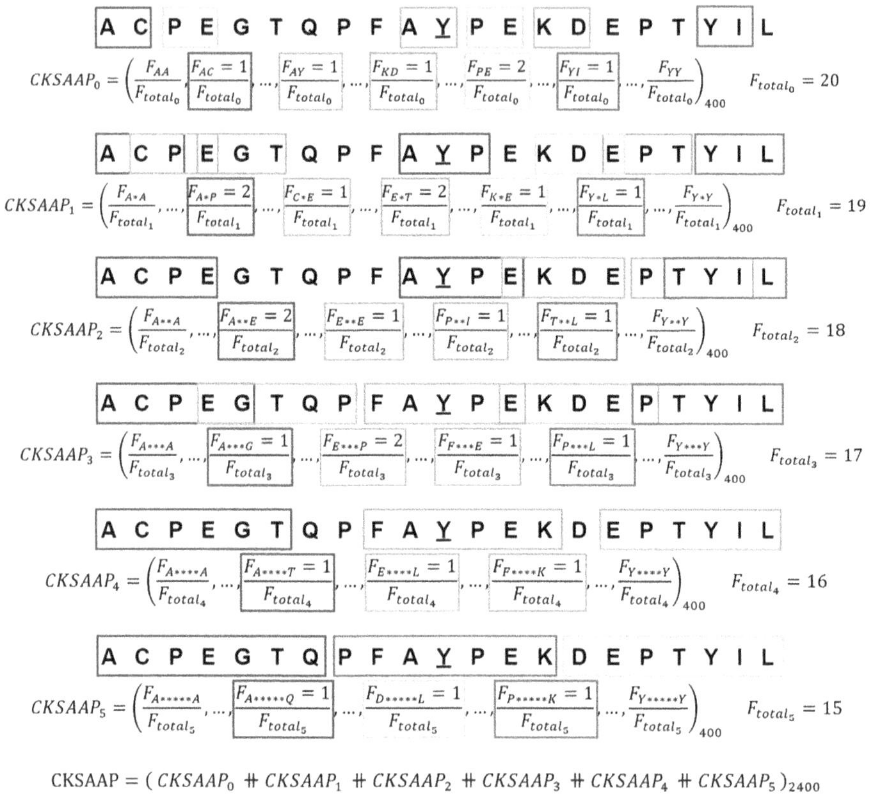

A C P E G T Q P F A Y P E K D E P T Y I L

$$CKSAAP_0 = \left(\frac{F_{AA}}{F_{total_0}}, \frac{F_{AC}=1}{F_{total_0}}, \ldots, \frac{F_{AY}=1}{F_{total_0}}, \ldots, \frac{F_{KD}=1}{F_{total_0}}, \ldots, \frac{F_{PE}=2}{F_{total_0}}, \ldots, \frac{F_{YI}=1}{F_{total_0}}, \ldots, \frac{F_{YY}}{F_{total_0}}\right)_{400} \quad F_{total_0} = 20$$

A C P E G T Q P F A Y P E K D E P T Y I L

$$CKSAAP_1 = \left(\frac{F_{A*A}}{F_{total_1}}, \ldots, \frac{F_{A*P}=2}{F_{total_1}}, \ldots, \frac{F_{C*E}=1}{F_{total_1}}, \ldots, \frac{F_{E*T}=2}{F_{total_1}}, \ldots, \frac{F_{K*E}=1}{F_{total_1}}, \ldots, \frac{F_{Y*L}=1}{F_{total_1}}, \ldots, \frac{F_{Y*Y}}{F_{total_1}}\right)_{400} \quad F_{total_1} = 19$$

A C P E G T Q P F A Y P E K D E P T Y I L

$$CKSAAP_2 = \left(\frac{F_{A**A}}{F_{total_2}}, \ldots, \frac{F_{A**E}=2}{F_{total_2}}, \ldots, \frac{F_{E**E}=1}{F_{total_2}}, \ldots, \frac{F_{P**I}=1}{F_{total_2}}, \ldots, \frac{F_{T**L}=1}{F_{total_2}}, \ldots, \frac{F_{Y**Y}}{F_{total_2}}\right)_{400} \quad F_{total_2} = 18$$

A C P E G T Q P F A Y P E K D E P T Y I L

$$CKSAAP_3 = \left(\frac{F_{A***A}}{F_{total_3}}, \ldots, \frac{F_{A***G}=1}{F_{total_3}}, \ldots, \frac{F_{E***P}=2}{F_{total_3}}, \ldots, \frac{F_{F***E}=1}{F_{total_3}}, \ldots, \frac{F_{P***L}=1}{F_{total_3}}, \ldots, \frac{F_{Y***Y}}{F_{total_3}}\right)_{400} \quad F_{total_3} = 17$$

A C P E G T Q P F A Y P E K D E P T Y I L

$$CKSAAP_4 = \left(\frac{F_{A****A}}{F_{total_4}}, \ldots, \frac{F_{A****T}=1}{F_{total_4}}, \ldots, \frac{F_{E****L}=1}{F_{total_4}}, \ldots, \frac{F_{P****K}=1}{F_{total_4}}, \ldots, \frac{F_{Y****Y}}{F_{total_4}}\right)_{400} \quad F_{total_4} = 16$$

A C P E G T Q P F A Y P E K D E P T Y I L

$$CKSAAP_5 = \left(\frac{F_{A*****A}}{F_{total_5}}, \ldots, \frac{F_{A*****Q}=1}{F_{total_5}}, \ldots, \frac{F_{D*****L}=1}{F_{total_5}}, \ldots, \frac{F_{P*****K}=1}{F_{total_5}}, \ldots, \frac{F_{Y*****Y}}{F_{total_5}}\right)_{400} \quad F_{total_5} = 15$$

$$\text{CKSAAP} = (CKSAAP_0 \mathbin{+\!\!+} CKSAAP_1 \mathbin{+\!\!+} CKSAAP_2 \mathbin{+\!\!+} CKSAAP_3 \mathbin{+\!\!+} CKSAAP_4 \mathbin{+\!\!+} CKSAAP_5)_{2400}$$

Fig. 6.9 Illustration of Composition of K-spaced amino acid pairs (CKSAAP) encoding

($F_{i(*^K)j}$) in the numerator are computed exactly like the Kgap encoder and normalization factor F_{total_K} varies with K. For K = 0, $F_{total_0} = 21 - (0 + 1) = 20$. Similarly, for K = 1, $F_{total_1} = 21 - (1 + 1) = 19$, and so on for any value of K. Parameter K in the range 0 to 5 is used to generate the full CKSAAP feature vector. Because each value of K produces a 400-dimensional statistical vector, concatenating vectors from all 6 values of K produces a 2400-dimensional statistical vector.

Although normalization potentially removes biases and variations introduced by the genetic sequencing processes and factors such as the length of the extracted sequences, it can adversely affect the biological variability of sequences [5, 1088]. Furthermore, Lovell et al. [665] showed that normalization into the simplex space (where, the components of the vector are only positive and sum to 1) can introduce spurious correlation between the relative compositions even though the absolute compositions are completely uncorrelated. Lovell et al. [665] implied that, wherever possible, both absolute abundance of the components and their relative abundance information should be analyzed using appropriate compositional features. Hence, in this study, both absolute and normalized composition encoders are utilized.

6.9.3 Adaptive Skip Dipeptide Composition (ASDC)

With an aim to control the dimensionality of generated statistical vectors, Wei et al. [1085] proposed the ASDC sequence encoder that captures compositional information from all possible adjacent and gap-based bimers in a sequence. In both Kgap and CKSAAP encoders, although long range dependencies of amino acids can be captured at large value of K, however it creates sparse high-dimensional feature spaces that adversely affect the performance of machine learning approaches for classification [239]. To mitigate this issue, Wei et al. [1085] proposed to compute the sum of occurrence frequencies of bimers at all possible values of K in a sequence. In this way rather than generating $400 \times (K_{max} + 1)$ dimensional vector, it generates a fixed 400-dimensional statistical vector. Furthermore, to address the bias in absolute occurrence frequencies of the bimers based on the length of the sequences, as described in Sect. 6.9.2, normalization is performed. This encoder has been utilized in a variety of protein sequence analysis tasks, such as, to identify cell-penetrating and anti-cancer peptides in proteins [141, 1085, 1086], and protein methylation sites [1084].

The encoding process of ASDC encoder can be categorized into three distinct phases. In first phase, similar to kgap and CKSAAP encoders, it generates the set of 400 bimers. In second phase, it computes the total occurrence frequency of each bimer considering all possible gap-values (K) that lie in range $0 \leq K \leq$ sequence_length-2. In the third phase, the total occurrence frequencies of the bimers are normalized to generate percentage composition. The ASDC feature vector can be mathematically defined as:

$$ASDC = (\frac{fv_{AA}}{F_{total}}, \frac{fv_{AC}}{F_{total}}, \frac{fv_{AD}}{F_{total}}, ..., \frac{fv_{YW}}{F_{total}}, \frac{fv_{YY}}{F_{total}})_{400} \tag{6.46}$$

In Eq. 6.46, combined occurrence frequency of each bimer at all possible K values can be computed as:

$$fv_{ij} = \sum_{K=0}^{L-2} F_{i(*^K)j} \tag{6.47}$$

where, $F_{i(*^K)j}$ represents the occurrence frequency of bimer with length of gap K between ith and jth amino acids. The normalization factor in the ASDC feature vector is formulated as:

A C P E G T Q P F A Y P E K D E P T Y I L

$$\text{ASDC} = (fv_{AA}, fv_{AC}, fv_{AD}, fv_{AE}, \ldots, fv_{YW}, fv_{YY})_{400} \quad \textit{where, } fv_{ij} = \frac{N_{ij}}{N}$$

$$N = \sum_{i,j\,\epsilon\,\{Amino\ acids\}} N_{ij} = N_{AA} + N_{AC} + N_{AD} + N_{AE} + \cdots + N_{YW} + N_{YY}$$

$$N_{AE} = (F_{AE} = 0) + \cdots + (F_{A**E} = 2) + \cdots + (F_{A*****E} = 1) + \cdots + ..$$

$$(F_{A(*)^{11}E} = 1) + \cdots + (F_{A(*)^{14}E} = 1) + \cdots + (F_{A(*)^{L-2}E} = 0)$$

Fig. 6.10 Illustration of Adaptive skip Dipeptide composition (ASDC) encoding

$$F_{total} = \sum_{i,j\, \in \text{ amino acids}} fv_{ij} \tag{6.48}$$

Figure 6.10 graphically describes the computation of total occurrence frequency of one bimer in the ASDC feature vector from a hypothetical protein sequence. The bimer AE appears in the sequence twice with gap K = 2, hence $F_{A**E} = 2$, once with gap K = 5, hence $F_{A*****E} = 1$, once with gap K = 11, hence $F_{A(*)^{11}E} = 1$, and once with gap K = 14, hence $F_{A(*)^{14}E} = 1$. For all other values of gap, it does not occur at all. Hence, in the ASDC encoded vector, the component fv_{AE} = 2+1+1+1 = 5. Similarly, the total frequencies are computed for all 400 bimers. The 400-dimensional vector is normalized by the sum of all fv_{ij} components in the ASDC feature vector.

6.9.4 PseAAC of Distance-Pair (DistancePair)

The Kgap, CKSAAP and ASDC encoders capture distribution information by considering bimers with different gap values. However, they do not capture composition of monomers. To address this shortcoming, Liu et al. [91] proposed DistancePair encoder that can capture comprehensive composition information of both monomers and gap-based bimers. This encoder is widely used in protein sequence analysis studies, such as identification of DNA-binding proteins [91, 657] and anti-cancer peptides [335].

The working paradigm of this encoder can be categorized into three distinct phases. In first phase, the set of monomers is selected from Table 6.8. The selection of monomers set reduce the unique vocabulary. It can be seen in Table 6.8, in case of 20 amino acids no reduction is applied while in other three sets monomers are reduced to 19, 14 and 13 amino acids.

Table 6.8 Physicochemical properties based amino acids grouping

Amino acids	A	C	D	E	F	G	H	I	K	L	M	N	P	Q	R	S	T	V	W	Y	X
20 amino acids	A	C	D	E	F	G	H	I	K	L	M	N	P	Q	R	S	T	V	W	Y	A
19 amino acids	A	C	D	E	F	G	H	I	K	L	M	N	P	Q	R	S	T	V	W	F	A
14 amino acids	A	C	D	E	F	G	H	I	H	L	I	N	P	H	H	S	T	I	W	W	A
13 amino acids	A	C	D	E	F	G	H	I	K	I	F	N	H	H	K	S	T	V	H	H	A

The working paradigm of PseAAC can be seen in three different phases. At the first phase, the set of bimers is generated based on selected set of monomers. In second phase, the occurrence frequencies of monomers and gap-based bimers at multiple gap-lengths in the sequence are computed. In third phase, the occurrence frequencies are normalized to encode the percentage composition. For any protein sequence, the DistancePair feature vector can be mathematically formulated as:

$$DP_d = \begin{cases} (\frac{F_A}{F_{total_d}}, \frac{F_C}{F_{total_d}}, \dots, \frac{F_Y}{F_{total_d}})_{20}, & d = 0 \\ \\ (\frac{F_{A(*^d)A}}{F_{total_d}}, \frac{F_{A(*^d)C}}{F_{total_d}}, \dots, \frac{F_{Y(*^d)Y}}{F_{total_d}})_{400}, & d > 0 \end{cases} \tag{6.49}$$

$$F_{total_d} = \begin{cases} \sum_{i \in Amino-acids} F_i, & d = 0, \\ \\ \sum_{i,j \in Amino-acids} F_{i(*^d)j}, & d > 0, \end{cases} \tag{6.50}$$

In Eqs. 6.49 and 6.50, the parameter d controls the level of encoding, F_i represents the occurrence frequencies of monomers, and $F_{i(*^d)j}$ represents the occurrence frequencies of bimers with (d-1) random intervening s in between the two amino acids in the bimer.

The rows of Fig. 6.11 graphically describes the encoding of a hypothetical protein sequence using DistancePair encoder by considering all 20 amino acids. In the first row, the parameter d = 0 is set that generates the DP_0 vector, which encodes the occurrence frequencies of 20 monomers. For example, amino acid A occurs twice, hence $F_A = 2$, amino acid P occurs four times, hence $F_P = 4$, and so on. For any value of $d > 0$, occurrence frequencies of bimers are computed exactly like the Kgap encoder, i.e., d = 1 computes occurrence frequencies of bimers without any gaps, d = 2 computes frequency for bimers with a gap of 1 between the two

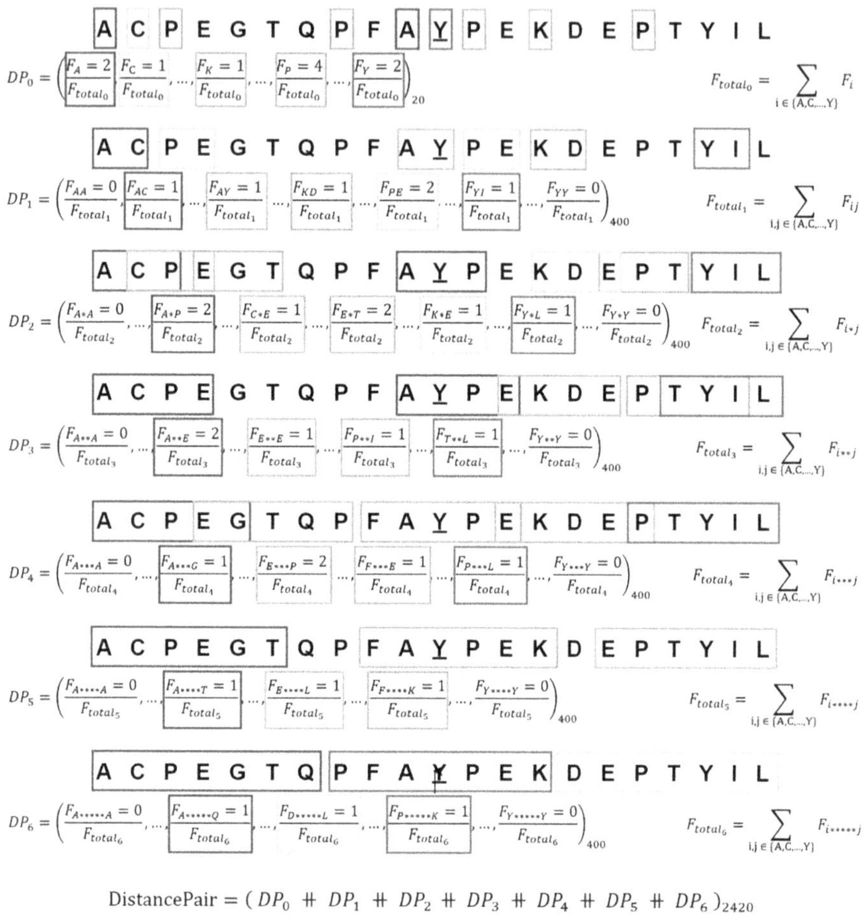

A C P E G T Q P F A Y P E K D E P T Y I L

$$DP_0 = \left(\frac{F_A = 2}{F_{total_0}}, \frac{F_C = 1}{F_{total_0}}, \ldots, \frac{F_K = 1}{F_{total_0}}, \ldots, \frac{F_P = 4}{F_{total_0}}, \ldots, \frac{F_Y = 2}{F_{total_0}}\right)_{20} \qquad F_{total_0} = \sum_{i \in \{A,C,\ldots,Y\}} F_i$$

A C P E G T Q P F A Y P E K D E P T Y I L

$$DP_1 = \left(\frac{F_{AA} = 0}{F_{total_1}}, \frac{F_{AC} = 1}{F_{total_1}}, \ldots, \frac{F_{AY} = 1}{F_{total_1}}, \ldots, \frac{F_{KD} = 1}{F_{total_1}}, \ldots, \frac{F_{PE} = 2}{F_{total_1}}, \ldots, \frac{F_{YI} = 1}{F_{total_1}}, \ldots, \frac{F_{YY} = 0}{F_{total_1}}\right)_{400} \qquad F_{total_1} = \sum_{i,j \in \{A,C,\ldots,Y\}} F_{ij}$$

A C P E G T Q P F A Y P E K D E P T Y I L

$$DP_2 = \left(\frac{F_{A*A} = 0}{F_{total_2}}, \ldots, \frac{F_{A*P} = 2}{F_{total_2}}, \ldots, \frac{F_{C*E} = 1}{F_{total_2}}, \ldots, \frac{F_{E*T} = 2}{F_{total_2}}, \ldots, \frac{F_{K*E} = 1}{F_{total_2}}, \ldots, \frac{F_{Y*L} = 1}{F_{total_2}}, \ldots, \frac{F_{Y*Y} = 0}{F_{total_2}}\right)_{400} \qquad F_{total_2} = \sum_{i,j \in \{A,C,\ldots,Y\}} F_{i*j}$$

A C P E G T Q P F A Y P E K D E P T Y I L

$$DP_3 = \left(\frac{F_{A**A} = 0}{F_{total_3}}, \ldots, \frac{F_{A**E} = 2}{F_{total_3}}, \ldots, \frac{F_{E**E} = 1}{F_{total_3}}, \ldots, \frac{F_{P**I} = 1}{F_{total_3}}, \ldots, \frac{F_{T**L} = 1}{F_{total_3}}, \ldots, \frac{F_{Y**Y} = 0}{F_{total_3}}\right)_{400} \qquad F_{total_3} = \sum_{i,j \in \{A,C,\ldots,Y\}} F_{i**j}$$

A C P E G T Q P F A Y P E K D E P T Y I L

$$DP_4 = \left(\frac{F_{A***A} = 0}{F_{total_4}}, \ldots, \frac{F_{A***G} = 1}{F_{total_4}}, \ldots, \frac{F_{E***P} = 2}{F_{total_4}}, \ldots, \frac{F_{F***E} = 1}{F_{total_4}}, \ldots, \frac{F_{P***L} = 1}{F_{total_4}}, \ldots, \frac{F_{Y***Y} = 0}{F_{total_4}}\right)_{400} \qquad F_{total_4} = \sum_{i,j \in \{A,C,\ldots,Y\}} F_{i***j}$$

A C P E G T Q P F A Y P E K D E P T Y I L

$$DP_5 = \left(\frac{F_{A****A} = 0}{F_{total_5}}, \ldots, \frac{F_{A****T} = 1}{F_{total_5}}, \ldots, \frac{F_{E****L} = 1}{F_{total_5}}, \ldots, \frac{F_{F****K} = 1}{F_{total_5}}, \ldots, \frac{F_{Y****Y} = 0}{F_{total_5}}\right)_{400} \qquad F_{total_5} = \sum_{i,j \in \{A,C,\ldots,Y\}} F_{i****j}$$

A C P E G T Q P F A Y P E K D E P T Y I L

$$DP_6 = \left(\frac{F_{A*****A} = 0}{F_{total_6}}, \ldots, \frac{F_{A*****Q} = 1}{F_{total_6}}, \ldots, \frac{F_{D*****L} = 1}{F_{total_6}}, \ldots, \frac{F_{P*****K} = 1}{F_{total_6}}, \ldots, \frac{F_{Y*****Y} = 0}{F_{total_6}}\right)_{400} \qquad F_{total_6} = \sum_{i,j \in \{A,C,\ldots,Y\}} F_{i*****j}$$

$$\text{DistancePair} = (DP_0 \mathbin{\#} DP_1 \mathbin{\#} DP_2 \mathbin{\#} DP_3 \mathbin{\#} DP_4 \mathbin{\#} DP_5 \mathbin{\#} DP_6)_{2420}$$

Fig. 6.11 Illustration of DistancePair encoding

amino acids, and so on. Each DP_d vector is normalized by sum of components of the respective vector, that converts the absolute occurrence frequencies into percentage composition. In this study, the parameter d in range 0 to 6 is utilized to generate DistancePair feature vector for each sequence. Since d = 0 generates a 20-dimensional statistical vector, and each value of $d > 0$ generates a 400 dimensional statistical vector, concatenating vectors from all 7 values of d generates a $20 + (400 \times 6) = 2420$ dimensional statistical vector.

6.10 Physicochemical Properties Based Amino Acids Grouping and Groups Distribution Based Encoding Methods

6.10.1 *Grouped Amino Acid Composition (GAAC)*

Grouped Amino Acid Composition encoding method has been widely used in multiple protein sequence analysis applications such as prediction of cell-penetrating Peptides [319], blood-brain barrier penetrating peptides [178], anticancer peptides [602], ion channel interacting peptides [1145]. This encoder categorizes 20 unique amino acids into 5 different groups namely: aromatic, aliphatic, negatively charged, positive charge and uncharged. Table 6.9 represents distribution of amino acids into their respective groups.

Furthermore, it makes use of five different groups $(G_1, G_2, G_3, G_4, G_5)$ for transforming protein sequence into another sequence in which amino acids are replaced with their respective groups. Groups based transformed sequence is segregated into unimers and distribution of 5 unique groups is computed. Equation 6.51 illustrates mathematical expression for computing final statistical vector. This encoder transforms protein sequences into 5 dimensional statistical vectors.

$$Encoding(Group_i) = \frac{Occurrence\ frequency\ of\ G_i}{Sequence\ Length}, \quad \text{where } G_i \in \{G_1, G_2, G_3, G_4, G_5\} \tag{6.51}$$

Figure 6.12 graphically represents working paradigm of GAAC encoder with a toy sequence.

6.10.2 *Grouped Di-Peptide Composition (GDPC)*

GDPC encoder is a variant of GAAC and has been utilized in different types of protein sequence analysis tasks such as identification of anticancer peptides [602], prediction of drugable [925] and cyclin [963] proteins. Similar to GAAC encoder, it also categorizes 20 unique amino acids into 5 different groups. Then it generates all

Table 6.9 Amino acids division for group-based encoders

Groups	Amino acids
Group-1: aliphatic	L, G, A, V, I, M
Group-2: aromatic	F, Y, W
Group-3: positive charge	R, K, H
Group-4: negative charged	D, E
Group-5: uncharged	S, T, C, P, Q, N

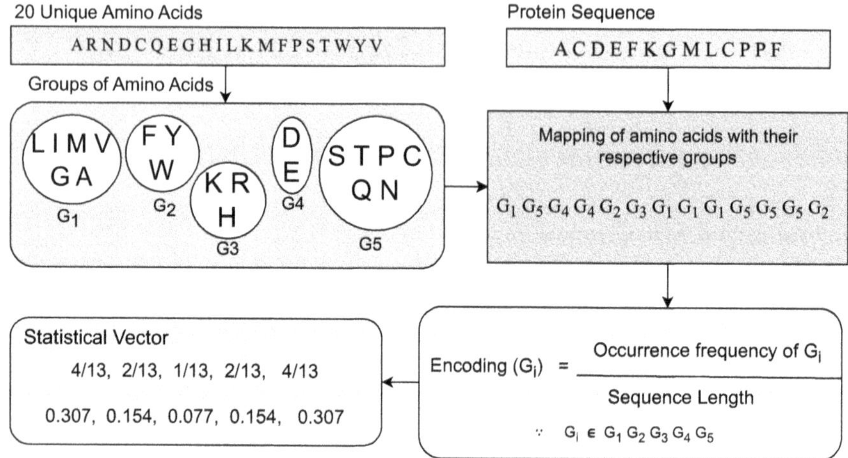

Fig. 6.12 Overall workflow of Grouped Amino Acid Composition (GAAC) encoder

possible pairs of 5 groups shown in Eq. 6.52. Furthermore, it segregates raw protein sequence into bimers by sliding a 2 size window with 1 stride size. In the next step, it computes occurrence frequencies of bimers with respect to group pairs.

$$\text{Groups Pairs} = \begin{cases} G_1G_1,\ G_1G_2,\ G_1G_3,\ G_1G_4,\ G_1G_5 \\ G_2G_1,\ G_2G_2,\ G_2G_3,\ G_2G_4,\ G_2G_5 \\ G_3G_1,\ G_3G_2,\ G_3G_3,\ G_3G_4,\ G_3G_5 \\ G_4G_1,\ G_4G_2,\ G_4G_3,\ G_4G_4,\ G_4G_5 \\ G_5G_1,\ G_5G_2,\ G_5G_3,\ G_5G_4,\ G_5G_5 \end{cases} \tag{6.52}$$

Equation 6.53 illustrates mathematical expression for computing group pairs occurrence frequencies.

$$Encoding(G_a.G_b) = \frac{\text{Occurrence frequency of } G_a.G_b}{\text{sequence length } - 2}, \quad \text{where a, b} \in \{1, 2, 3, 4, 5\} \tag{6.53}$$

This encoder transforms raw protein sequence into 25 dimensional statistical vector. Figure 6.13 graphically represents working paradigm of GDPC encoder with a hypothetical protein sequence 'ACDEFKGPFLVQNWMYI' with bimers as {AC, CD, DE, EF, FK, KG, GP, PF, FL, LV, VQ, QN, NW, WM, MY, YI}. In each bimer first amino acid or bimer[0] denotes to a group and second amino acid or bimer[1] also represents to a group. In other words, each bimers represents a group pair i.e.,

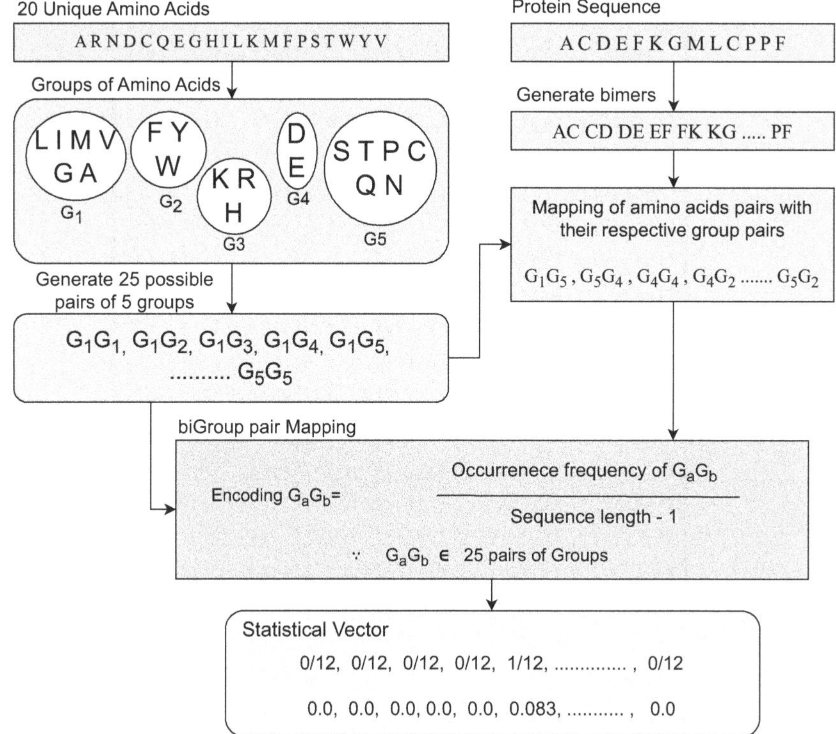

Fig. 6.13 Workflow of Grouped Di-Peptide Composition (GDPC) encoding method

if generated bimer is CD, C represents group 5 and D refers to group 4, then CD belongs to group G_5G_4.

6.10.3 *Grouped Tri-Peptide Composition (GTPC)*

GTPC encoder is a variant of GAAC and GDPC encoding methods. It has been utilized in different types of protein sequence analysis tasks including identification of ion channel interacting peptides [1145], anti-cancer peptide prediction [602], cyclin proteins prediction [963], drug target interaction prediction [518], RNA and protein bindings [641] and prediction of zyclin proteins [963].

Similar to GAAC and GDPC encoding methods, it categorizes 20 unique amino acids into 5 different groups. Then, it generates all possible tri-pairs of 5 groups as shown in Eq. 6.54.

$$
\text{Tri Groups} = \left\{ \begin{array}{l}
G_1G_1G_1,\ G_1G_1G_2,\ G_1G_1G_3,\ G_1G_1G_4,\ G_1G_1G_5 \\
G_1G_2G_1,\ G_1G_2G_2,\ G_1G_2G_3,\ G_1G_2G_4,\ G_1G_2G_5 \\
G_1G_3G_1,\ G_1G_3G_2,\ G_1G_3G_3,\ G_1G_3G_4,\ G_1G_3G_5 \\
G_1G_4G_1,\ G_1G_4G_2,\ G_1G_4G_3,\ G_1G_4G_4,\ G_1G_4G_5 \\
G_1G_5G_1,\ G_1G_5G_2,\ G_1G_5G_3,\ G_1G_5G_4,\ G_1G_5G_5 \\
G_2G_1G_1,\ G_2G_1G_2,\ G_2G_1G_3,\ G_2G_1G_4,\ G_2G_1G_5 \\
G_2G_2G_1,\ G_2G_2G_2,\ G_2G_2G_3,\ G_2G_2G_4,\ G_2G_2G_5 \\
G_2G_3G_1,\ G_2G_3G_2,\ G_2G_3G_3,\ G_2G_3G_4,\ G_2G_3G_5 \\
G_2G_4G_1,\ G_2G_4G_2,\ G_2G_4G_3,\ G_2G_4G_4,\ G_2G_4G_5 \\
G_2G_5G_1,\ G_2G_5G_2,\ G_2G_5G_3,\ G_2G_5G_4,\ G_2G_5G_5 \\
G_3G_1G_1,\ G_3G_1G_2,\ G_3G_1G_3,\ G_3G_1G_4,\ G_2G_1G_5 \\
G_3G_2G_1,\ G_3G_2G_2,\ G_3G_2G_3,\ G_3G_2G_4,\ G_3G_2G_5 \\
G_3G_3G_1,\ G_3G_3G_2,\ G_3G_3G_3,\ G_3G_3G_4,\ G_2G_3G_5 \\
G_3G_4G_1,\ G_3G_4G_2,\ G_3G_4G_3,\ G_3G_4G_4,\ G_3G_4G_5 \\
G_3G_5G_1,\ G_3G_5G_2,\ G_3G_5G_3,\ G_3G_5G_4,\ G_3G_5G_5 \\
G_4G_1G_1,\ G_4G_1G_2,\ G_4G_1G_3,\ G_4G_1G_4,\ G_4G_1G_5 \\
G_4G_2G_1,\ G_4G_2G_2,\ G_4G_2G_3,\ G_4G_2G_4,\ G_4G_2G_5 \\
G_4G_3G_1,\ G_4G_3G_2,\ G_4G_3G_3,\ G_4G_3G_4,\ G_4G_3G_5 \\
G_4G_4G_1,\ G_4G_4G_2,\ G_4G_4G_3,\ G_4G_4G_4,\ G_2G_4G_5 \\
G_4G_5G_1,\ G_4G_5G_2,\ G_4G_5G_3,\ G_4G_5G_4,\ G_4G_5G_5 \\
G_2G_1G_1,\ G_2G_1G_2,\ G_2G_1G_3,\ G_2G_1G_4,\ G_2G_1G_5 \\
G_2G_2G_1,\ G_2G_2G_2,\ G_2G_2G_3,\ G_2G_2G_4,\ G_2G_2G_5 \\
G_2G_3G_1,\ G_2G_3G_2,\ G_2G_3G_3,\ G_2G_3G_4,\ G_2G_3G_5 \\
G_2G_4G_1,\ G_2G_4G_2,\ G_2G_4G_3,\ G_2G_4G_4,\ G_2G_4G_5 \\
G_2G_5G_1,\ G_2G_5G_2,\ G_2G_5G_3,\ G_2G_5G_4,\ G_2G_5G_5
\end{array} \right\} \tag{6.54}
$$

In the next step, it segregates protein sequence into trimers and computes occurrence frequencies of trimers with respect to tri groups. Equation 6.55 illustrates mathematical expression to compute groups occurrence frequencies with respect to sequence trimers.

$$
Encoding(G_a.G_b.G_c) = \frac{\text{Occurrence frequency of } G_a.G_b.G_c}{\text{sequence length - 2}}, \quad \text{where a, b, c} \in \{1, 2, 3\} \tag{6.55}
$$

Let's consider a toy protein sequence 'ACDEFKGPFLVQNWMYI' with trimers as ACD, CDE, DEF, EFK, FKG, KGP, GPF, PFL, FLV, LVQ, VQN, QNW, NWM, WMY, MYI. In each trimer first amino acid or trimer[0] denotes a group and second amino acid or trimer[1] also represents a group and third amino acid or trimer[2] denotes a group. All amino acids of a trimer may belong to same group or two or three different groups. Trimer[0], trimer[1] and trimer[2] represent group tri-pairs i.e., if generated trimer is CDC, C belongs to group 5 and D denotes group 4, then

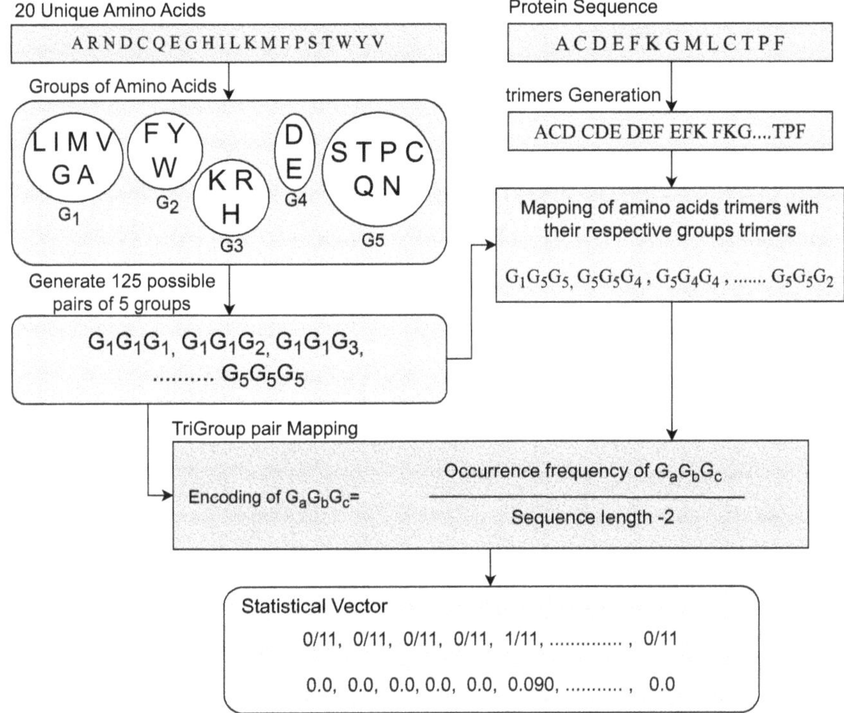

Fig. 6.14 Overall workflow of Grouped Tri-Peptide Composition (GTPC) encoder

it is included in group $G_5G_4G_5$. Similarly, it computes the count of other group pairs and normalizes it with sequence length − 2 to generate 5^3 dimensional final statistical vector. Figure 6.14 graphically represents encoding process employed by GTPC encoder using a toy sequence 'ACDEFKGMLCPPF'.

6.10.4 Conjoint Triad (CTriad)

Shen et al. [906] proposed CTriad encoder that has been widely used to develop computational predictors for nuclear receptor prediction [1045], RNA/Protein interaction prediction [1048], protein protein interaction prediction [251, 906, 1056], identification of amyloid proteins [148] and identification of Arginine Methylation Sites [1225]. CTriad is another variant of group based encoders that transforms protein sequences into statistical vectors by computing groups based distribution of amino acids. This encoder categorize 20 unique amino acids into seven different groups that are illustrated in Table 6.10 (Fig. 6.15).

Similar to GTPC encoder, it generates 7^3 triplets of groups as illustrated in Eq. 6.54. In the next step, it generates trimers of raw sequence and computes occur-

Table 6.10 Amino acid categorization into 7 different groups

Groups	Amino acids
Group-1	V, G, A
Group-2	F, L, I, P
Group-3	S, T, Y, M
Group-4	Q, N, W, H
Group-5	R,K
Group-6	D,E
Group-7	C

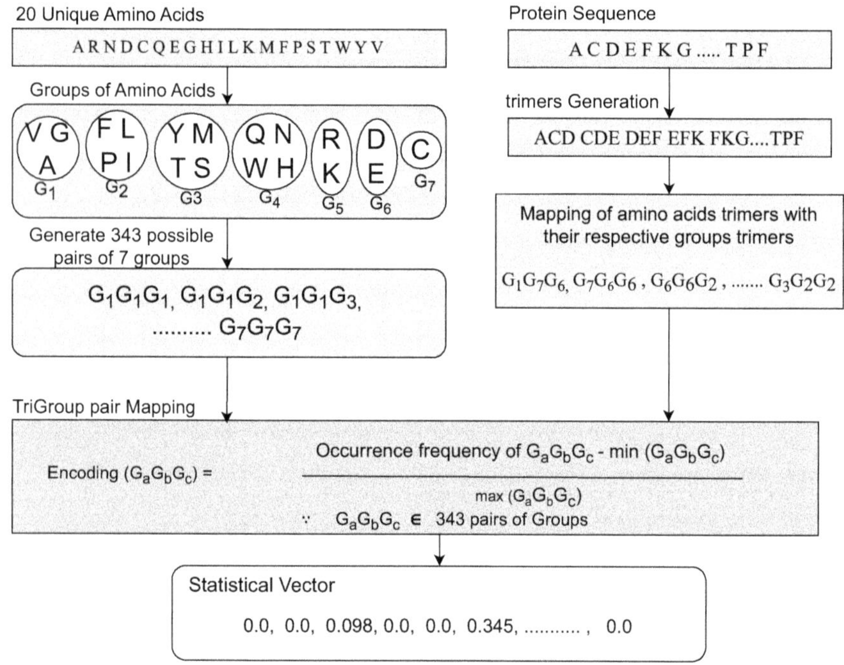

Fig. 6.15 Workflow of Conjoint Triad (CTriad) method

rence frequencies of trimers with respect to generated group pairs. Equation 6.56 illustrates mathematical expression to compute groups occurrence frequencies with respect to sequence trimers.

$$Encoding(G_a.G_b.G_c) = \frac{\text{Occurrence frequency of } G_a.G_b.G_c - min(G_a.G_b.G_c)}{max(G_a.G_b.G_c)},$$
$$\text{where a, b, c} \in \{1, 2, 3, 4, 5\} \tag{6.56}$$

Consider a toy protein sequence ACDEFKGPFLVQNWMYI and its trimers ACD, CDE, DEF, EFK, FKG, KGP, GPF, PFL, FLV, LVQ, VQN, QNW, NWM, WMY, MYI. In each trimer first amino acid or trimer[0] denotes a group and second amino acid or trimer[1] also represents a group and third amino acid or trimer[2] denotes a group. All amino acids of a trimer may belong to same group or two or three different groups.

Letters of triplets represent the triplet to which that particular trimer belongs. Let's if GPC is one of the generated trimer, G denotes group 1, P indicates group 2 and C belongs to group 7. In simple words, GPC shows $G_1G_2G_7$ and after generating trimers, it computes frequency occurrence of each triplet. A final 7^3 dimensional feature vector is generated by CTriad encoder.

6.10.5 k-Spaced Conjoint Triad (KSCTriad)

This encoder has been utilized in different types of protein sequence analysis tasks such as nuclear receptor prediction [1045], RNA/Protein interaction prediction [1048]. KSCTriad is a variant of CTriad encoder and has been utilized in different types of protein sequences analysis tasks such as peptide classification [943] and protein protein interaction prediction [9]. Similar to CTriad encoder it categorizes 20 unique amino acids into 7 different groups as illustrated in Table 6.10. It generates 7^3 triplets of groups that are briefly describe in Sect. 6.10.3 encoder. Further, it generates trimers with different gap values (say n) between amino acids. Primarily, motivation behind generation of gap based trimers is to estimate distribution of amino acid pairs with different gaps. For hypothetical sequence $A_0A_1A_2A_3A_4......A_l$ and gap value is set as n, the encoder generates n different types of samples having trimers that are briefly described in the following mathematical expression 6.57.

$$\begin{cases} A_0A_1A_2,\ A_1A_2A_3,\ A_2A_3A_4,\ ...,\ A_{l-2}A_{l-1}A_l \quad nogap \\ A_0A_2A_4,\ A_1A_3A_5,\ A_2A_4A_6,\ ...,\ A_{l-4}A_{l-2}A_l \quad gap = 1 \\ A_0A_3A_5,\ A_1A_4A_7,\ A_2A_5A_8,\ ...,\ A_{l-5}A_{l-3}A_l \quad gap = 2 \\ \qquad . \qquad\qquad\qquad . \\ \qquad . \qquad\qquad\qquad . \\ \qquad . \qquad\qquad\qquad . \\ A_0A_{n+1}A_{2n+1},\ ,\ ...,\ ,\ A_{l-(2n+1)}A_{l-(n+1)}A_l \quad gap = n \end{cases} \tag{6.57}$$

For each gap value based generated trimers are mapped to triplets of groups. Equation 6.58 illustrates mathematical expression for computing distribution of groups triplets.

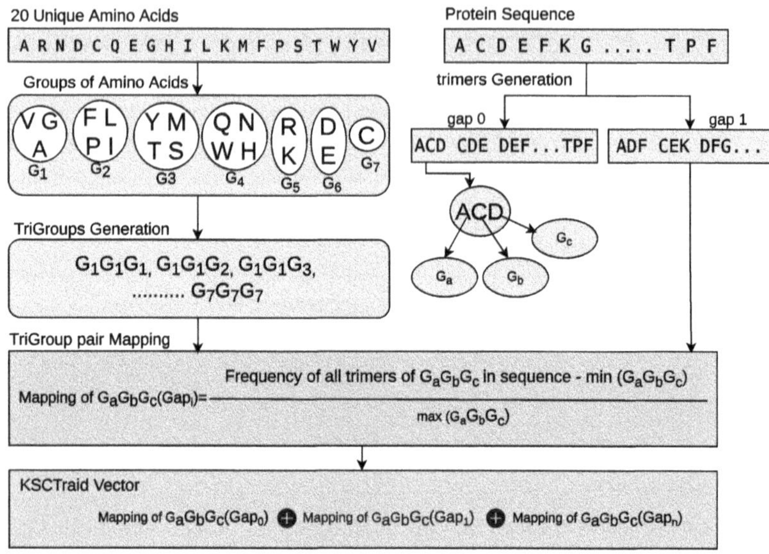

Fig. 6.16 Workflow of k-Spaced Conjoint Triad (KSCTriad) encoder

$$Triplet[i] = \frac{Count\ of\ Triplet[i] - min(Triplet[1], Triplet[2], ..., Triplet[n])}{max(Triplet[1], Triplet[2], ..., Triplet[n])} \tag{6.58}$$

After generating statistical vectors of each gap value, it concatenates vectors of all gap values and generate final statistical vector of dimensions $(n + 1) \times 7^3$, where n represents gap value. Figure 6.16 provides graphical illustration of overall working paradigm of KSCTraid encoder.

6.10.6 Composition of k-Spaced Amino Acid Group Pairs (CKSAAGP)

CKSAAGP is an advance version of CKSAAP encoder and has been utilized in diverse types of protein sequence analysis applications such as anti-cancer peptide prediction [155], PE_PGRS proteins prediction [611], identification of plant resistance proteins [1065] and major histocompatibility protein prediction [161]. Motivation behind development of this encoder was to map gap based bimers of protein sequences into different groups and compute distribution of groups.

Figure 6.17 graphically represents overall working paradigm of CKAAGP encoder. To compute distribution of amino acid groups, it generates bimers of

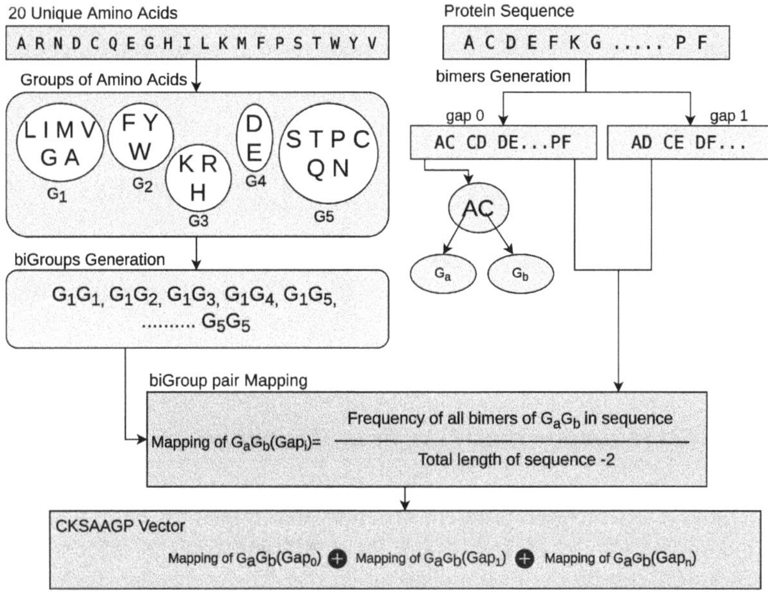

Fig. 6.17 Overall workflow of Composition of k-Spaced Amino Acid Group Pairs (CKSAAGP) encoder

protein sequence with different gap values. Consider a protein $A_0A_1A_2A_3A_4......A_l$ and gap value is set as n, CKSAAGP generates n different types of bimers samples that are briefly described in mathematical expression 6.59.

$$\begin{cases} A_0A_1,\ A_1A_2,\ A_2A_3,\ ...,\ A_{l-1}A_l & nogap \\ A_0A_2,\ A_1A_3,\ A_2A_4,\ ...,\ A_{l-2}A_l & gap = 1 \\ A_0A_3,\ A_1A_4,\ A_2A_5,\ ...,\ A_{l-3}A_l & gap = 2 \\ . & . \\ . & . \\ . & . \\ A_0A_{n+1},\ A_1A_{n+2},\ A_2A_{n+3},\ ,\ ...,\ A_{l-(n+1)}A_l & gap = n \end{cases} \tag{6.59}$$

It can be seen from Eq. 6.59, at gap value 0 bimers are generated with adjacent amino acids, at gap value 1 bimers are generated by skipping 1 amino acid between first letter of bimer and second letter of bimer and at gap n bimers are generated by skipping n amino acids between first letter of bimer and second letter of bimer.

As shown in Fig. 6.17, based on two different properties namely: volume and dipole moment 20 unique amino acids are categorized into 5 different groups. Further, it generates 25 pairs of groups by taking all possible combinations of 5 groups. In the next step, for each gap based bimers, it maps bimers to group pairs and transforms bimers samples into statistical feature space by computing occurrence frequency of each paired unit of group and normalizing it with the sequence length – kmer size – 1. Equation 6.60 illustrates mathematical expression for computing statistical value of each group pair.

$$Encoding(G_aG_b) = \frac{Occurrence\ frequency\ of\ G_aG_b}{Sequence\ length - kmer\ size\ - 1} \tag{6.60}$$

In the above equation, G_aG_b represents to a group pair. In this way, n+1 different statistical vectors are generated and $5^2 \times (n + 1)$ dimension feature space is generated.

6.10.7 CTD (Composition/Transition/Distribution)

CTD includes a set of three different encoders that transform protein sequences into numerical vectors by making use of three different types of information i.e., composition (C), transition (T) and distribution (D). These encoders include CTD-C , CTD-T, and CTD-D, and these encoders provide a quantitative representation of how amino acids with certain properties are distributed throughout the sequence. Researchers have utilized these encoding methods with 13 different types of physicochemical properties including solvent accessibility, normalized van der Waals Volume, charge, polarity, polarizability, secondary structures and hydrophobicity. Table 6.11 represents amino acids classification into three groups based on 13 distinctive physical and chemical properties.

CTD encoders have been applied in different types of protein sequence analysis applications such as identification of folding proteins [277, 278], enzyme family prediction [126], RNA protein interaction prediction [395], protein structure prediction [995] and identification of anti-cancer peptides [1086].

CTD features are often calculated using global statistics across the entire protein or peptide sequence. This approach may mask or dilute localized patterns or regions with specific amino acid distributions. It may be challenging to capture and differentiate specific patterns in regions with high sequence variability or when considering sequences with diverse lengths. CTD features are typically calculated based on specific property thresholds or predefined groups of amino acids. This can limit their generalizability to diverse protein families or sequences with unique properties not captured by the predefined categories. The effectiveness of CTD features may vary across different protein contexts, and their performance may be sensitive to the choice of property thresholds or groups. To mitigate these drawbacks, researchers often combine CTD features with other different types encoding methods. Following subsections briefly describe the working paradigm of CTD encoders.

Table 6.11 Amino acids classification with respect to 13 physicochemical properties

Physicochemical properties	Group 1	Group 2	Group 3
Secondary structure	Helix M,E,A,K,R,H,L,Q	Strand; C,F,I,T,V,W,Y	Coil; S,D,G,P,N
Hydrophobicity_ARGP820101	Polar; Q,S,T,N,G,D,E	Neutral; R,A,H,C,K,M,V	Hydrophobicity; L,Y,P,F,I,W
Hydrophobicity_ZIMJ680101	Polar; Q,N,G,S,W,T,D,E,R,A	Neutral; H,M,C,K,V	Hydrophobicity; L,P,F,Y,I
Hydrophobicity_PONP930101	Polar; K,P,D,E,S,N,Q,T	Neutral; G,R,H,A	Hydrophobicity; Y,M,F,W,L,C,V,I
Hydrophobicity_CASG920101	Polar; K,D,E,Q,P,S,R,N,T,G	Neutral; A,H,Y,M,L,V	Hydrophobicity; F,I,W,C
Hydrophobicity_ENGD860101	Polar; R,D,K,E,N,Q,H,Y,P	Neutral; S,G,T,A,W	Hydrophobicity; C,V,L,I,M,F
Hydrophobicity_FASG890101	Polar; K,E,R,S,Q,D	Neutral; N,T,P,G	Hydrophobicity; A,Y,H,W,V,M,F,L,I,C
Hydrophobicity_PRAM900101	Polar; N,Q,D,E,K,R	Neutral; Y,P,H,S,T,A,G	Hydrophobicity; M,F,I,L,C,W,V
Normalized van der Waals volume	0–2.78; T,S,P,A,G,D	2.95–94.0; Q,L,V,N,E,I	4.03–8.08; M,H,K,F,R,Y,W
Solvent accessibility	Buried; W,V,I,C,G,F,A,L	Exposed; Q,E,D,N,K,P	Intermediate; H,Y,M,S,P,T
Polarizability	0–1.08; G,A,S,D,T	0.128–120.186 G,P,N,V,E,Q,I,L	0.219–0.409 K,M,G,F,R,Y,W
Charge	Positive; K,R	Neutral; Q,G,H,I,A,N,C,L,M,F,P,S,T,W,Y,V	Negative; E,D
Polarity	4.9–6.2; L,I,F,W,C,M,V,Y	8.0–9.2; P,A,T,G,S	10.4–13.0; H,Q,R,K,N,E,D

6.10.7.1 CTDC

Dubchak et al. [277] proposed CTDC encoder and it has been extensively utilized in diverse types of protein sequence analysis applications such as prediction of lactylation site [485], identification of umami peptides [147], prediction of small secreted peptides [584], identification of protein coupled receptors [370], and therapeutic peptides prediction [61].

This encoder takes raw protein sequence and transforms it into another sequence in which each amino acid is represented by a group to which it belongs i.e.,1, G1, G2 or G3 based on the specific physicochemical property. Consider a toy protein sequence "RKEDQNGASTPHYCLVIMFW". Utilizing the charge property, the amino acids in a protein sequence are categorized into three distinct groups. The first group encompasses amino acids with a positive charge, the second group comprises those with a negative charge, and the third group includes neutral amino acids. After representing amino acids with their respective groups toy protein sequence is transformed to "$G_1G_1G_3G_3G_2G_2G_2G_2G_2G_2G_2G_2G_2G_2G_2G_2G_2G_2G_2G_2$". Similarly, for each physicochemical property it transforms raw protein sequence into new sequence. It is important to note, in each physicochemical property categorization of amino acids into predefined groups varies.

On the basis of transformed sequence, it computes normalized occurrence frequency of each group. Equation 6.61 illustrates mathematical expression to compute normalized occurrence frequency of each group.

$$Encoding(G_i) = \frac{\text{Occurrence frequency of } G_i}{\text{sequence length}}, \text{ where i } \epsilon \; \{1, 2, 3\} \tag{6.61}$$

In above expression G_i represents to 3 different Groups (G_1, G_2, G_3) that denotes 20 unique amino acids in the raw protein sequence. The whole process is repeated for all physicochemical properties. For each property encoder generates a 3 dimensional vector. Dimension of final statistical vector can be computed as number of physicochemical properties $\times$ 3. Figure 6.18 graphically depicts the working paradigm of CTDC encoder.

6.10.7.2 CTDT

CTDT encoder computes amino acids transition information in protein sequence and has been widely utilized in diverse types of protein sequence analysis tasks including identification of non classical secreted proteins [1203], SNARE proteins [593], anticancer peptides prediction [176], proteins sub-cellular location prediction [594].

Similar to CTDC encoder, CTDT also makes use of 13 different types of physicochemical properties and categorizes amino acids into 3 different groups G_1, G_2 and G_3. Furthermore, it generates three pairs of groups ($G_1.G_2$ or $G_2.G_1$), ($G_1.G_3$ or $G_3.G_1$), ($G_2.G_3$ or $G_3.G_2$). Here, $G_1.G_2$ and $G_2.G_1$ are considered

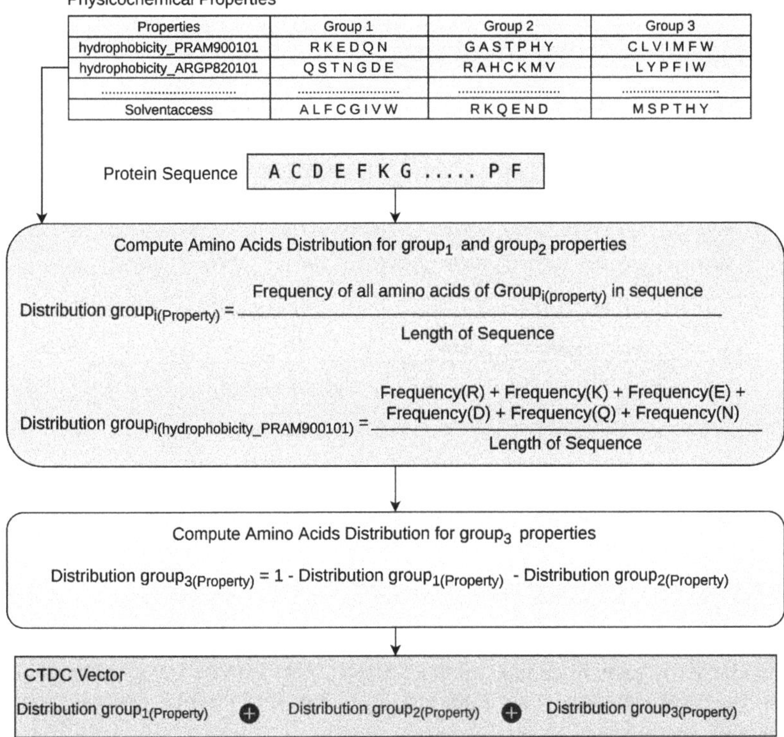

Fig. 6.18 Graphical illustration of composition based CTD encoding method

similar as the prime objective of this CTDT is to count transitions of amino acids from one group to the other group.

In next step, to compute transition information of amino acids, it generates bimers of protein sequence and maps each bimer to pairs of groups. Finally, it computes occurrence frequencies of pairs of groups and normalize with sequence length -1. Equation 6.62 represents mathematical expression for computing occurrence frequencies of group pairs.

$$Encoding(G_a.G_b) = \frac{\text{Occurrence frequency of } G_a.G_b}{\text{sequence length -1}}, \text{ where a,b } \epsilon \{1, 2, 3\} \tag{6.62}$$

In above expression $(G_a.G_b)$ represents to 3 different Groups ($G_1.G_2$ or $G_2.G_1$), ($G_1.G_3$ or $G_3.G_1$), ($G_2.G_3$ or $G_3.G_2$.) that represent 20 different amino acids. The whole process is repeated for all physicochemical properties. For each property this encoder generates a 3 dimensional vector. Dimension of final statistical vector can be computed as number of physicochemical properties $\times$ 3. Figure 6.19 graphically depicts the working paradigm of CTDT encoder.

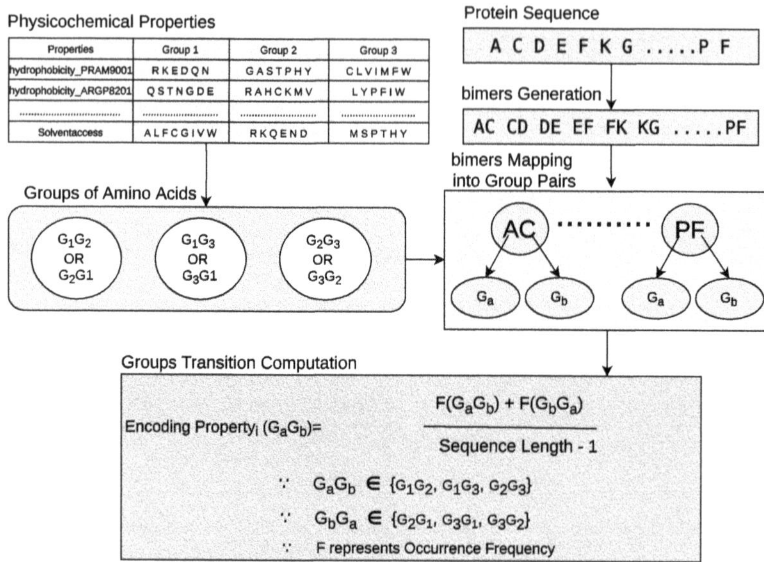

Fig. 6.19 Graphical illustration of transition based CTD encoding method

Consider a toy protein sequence "RKEDQNGASTPHYCLVIMFW" and generated bimers "RK, KE, ED, DQ, QN, NG, GA, AS, ST, TP, PH, HY, YC, CL, LV, VI, IM, MF, FW". Using charge property, bimers of protein sequence are mapped into group pairs. "G_1G_1, G_1G_3, G_3G_3, G_3G_2, G_2G_2, G_2G_2, G_2G_2, G_2G_2, G_2G_2, G_2G_2, G_2G_2, G_2G_2, G_2G_2, G_2G_2, G_2G_2, G_2G_2, G_2G_2, G_2G_2, G_2G_2". As this encoder computes transition of amino acids into different group pairs so pairs in which both letters belongs to similar group discarded and remaining group pairs are "G_1G_3, G_3G_2,". Next step is to count number of group pairs in each category of group.

$(G_1.G_2$ or $G_2.G_1)$: 2

$(G_1.G_3$ or $G_3.G_1)$: 3

$(G_2.G_3$ or $G_3.G_2.) : 4$

Further, it normalizes number of counts with sequence length that is 20.

Final statistical vector is 0.105, 0.158, 0.210.

6.10.7.3 CTDD

CTDD (Distribution) encodes distribution of each group of the amino acid for a considered property by the five chain lengths within which the first, 25%, 50%, 75%, and 100% of the amino acids with a certain group are contained for a considered property.The computation is performed for every physicochemical properties in order to generate the feature vector from CTDD feature descriptor [360]. It can be calculated as,

$$CTDT(G_i[P_i]) = \frac{N_{G_i}[P_i] \times (1, 25, 50, 75, 100)}{N} \tag{6.63}$$

Where N = total length of protein sequence, $N_{G_i}[P_i]$ represents number of amino acids of a group, and P_i is the physicochemical property. Table 6.11 represents the division of 20 amino acids into 3 groups based on the 13 physical and chemical properties.

Consider a protein sequence, 'RKEDQNGASTPHYCLVIMFW', based on the physicochemical property hydrophobicity it can be encoded as '11111122222223333333'. The encoded model has 6 amino acids of group 1, and 7 amino acids of group 2 and group 3 respectively. For group 1, the first residue coincides with the beginning of the chain, hence 0. 25% residue of the group 1 (rounded to 2) are present until second position of the protein sequence , 50% residue of the group 1 are present until third position of the protein sequence, 75% residue of the group 1 are present until fourth position of the protein sequence, and 100% residue of group 1 is present until sixth position. Hence,

$$CTDD_{25}(G_1[P_{hydrophobicity}]) = \frac{2}{20} \times 100 = 10\%$$

$$CTDD_{50}(G_1[P_{hydrophobicity}]) = \frac{3}{20} \times 100 = 15\%$$

$$CTDD_{75}(G_1[P_{hydrophobicity}]) = \frac{4}{20} \times 100 = 20\%$$

$$CTDD_{100}(G_1[P_{hydrophobicity}]) = \frac{6}{20} \times 100 = 30\%$$

Similar computation will be carried out for group 2 and group 3 respectively, and can be calculated as,

$$CTDD_{1}(G_2[P_{hydrophobicity}]) = \frac{7}{20} \times 100 = 35\%$$

$$CTDD_{25}(G_2[P_{hydrophobicity}]) = \frac{8}{20} \times 100 = 40\%$$

$$CTDD_{50}(G_2[P_{hydrophobicity}]) = \frac{10}{20} \times 100 = 50\%$$

$$CTDD_{75}(G_1[P_{hydrophobicity}]) = \frac{11}{20} \times 100 = 55\%$$

$$CTDD_{100}(G_1[P_{hydrophobicity}]) = \frac{13}{20} \times 100 = 65\%$$

The five chain length value for group 3 are 70%, 75%, 80%, 90%, 100% respectively [360, 780]. Hence, using 13 physicochemical properties, CTDD will 7×(3×5) = 105

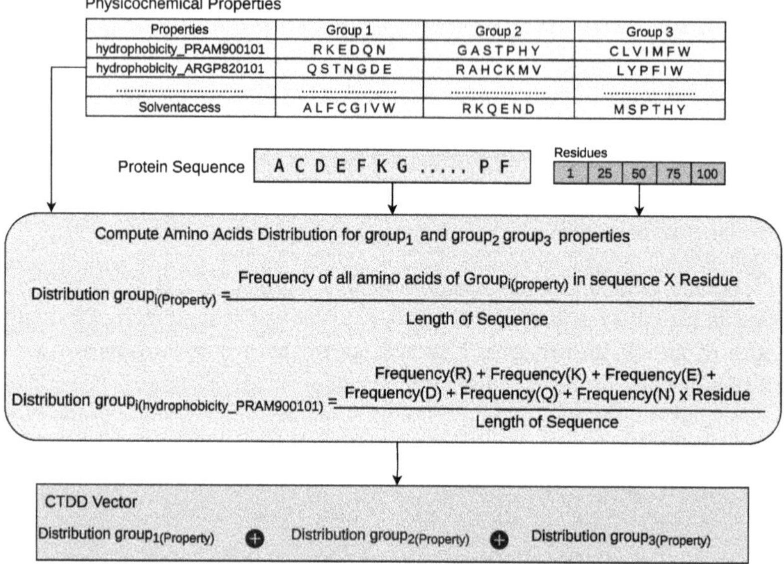

Fig. 6.20 Graphical illustration of distribution based CTD encoding method

dimensional sequence vector [277, 360]. Figure 6.20 graphically depicts the working paradigm of CTDD encoder.

6.11 Physicochemical Properties and Sequence Order Based Encoding Methods

6.11.1 Sequence Order Coupling Number (SocNumber)

Chou et al. [187] proposed sequence order coupling number encoder for transforming raw protein sequences into statistical vectors by extracting amino acids sequence order information at different levels. This encoder has been widely utilized in different types of protein sequence analysis tasks such as protein sub-cellular location prediction [187] and protein protein interaction prediction [1024].

As shown in Fig. 6.21, this encoder makes use of lag based bimers of amino acids and their pre-computed physicochemical distances (Schneider-Wrede [888] and Grantham [362]) for capturing amino acids sequence order patterns. Tables 6.12 and 6.13 illustrates physicochemical distances between pairs of 20 amino acids computed by Schneider-Wrede [888] and Grantham [362]. Primarily, Schneider-Wrede [888] and Grantham [362] computed Manhattan distance between amino acids by utilizing values of 4 different physicochemical properties namely, hydrophobicity, hydrophilicity, polarity, and side chain volume.

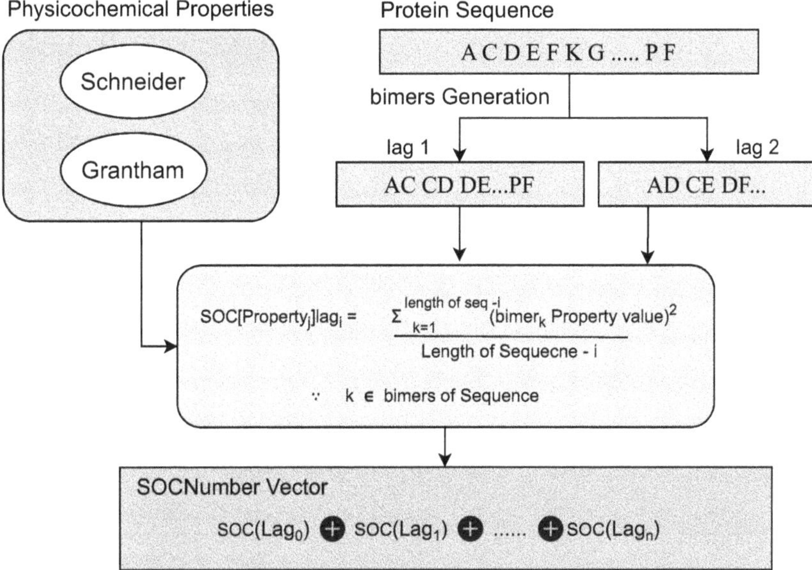

Fig. 6.21 Overall workflow of Sequence Order Coupling Number(SocNumber) encoder

To elaborate working paradigm of SOCNumber encoder, consider a sequence of protein comprises of L amino acids, $R_1 R_2 R_3 R_4....R_L$, where R_1 is amino acid at sequence location 1, R_2 is amino acid at sequence location 2, R_3 is amino acid at sequence location 3 and so on. Following mathematical expression illustrates process of computing lag based sequence order coupling number of amino acids.

$$\left.\begin{aligned} \tau_1 &= \frac{1}{L-1}\sum_{i=1}^{L-1}(J_{i,i+1}) \\ \tau_2 &= \frac{1}{L-2}\sum_{i=1}^{L-2}(J_{i,i+2}) \\ &\vdots \\ \tau_n &= \frac{1}{L-n}\sum_{i=1}^{L-n}(J_{i,i+n}),\ n < L \end{aligned}\right\} \text{where i} \in \{R_1, R_2, ...R_L\}$$

In above expression L denotes total number of amino acids in the sequence and i represents amino acids of the sequence. $J_{i,i+1}$, $J_{i,i+2}$ and $J_{i,i+n}$ refer to physicochemical distance between amino acids pairs $(R_i, R_i + 1)$, $(R_i, R_i + 2)$ and $(R_i, R_i + n)$ respectively. τ_1 computes first rank sequence order coupling number between adjacent amino acids at lag value 1. τ_2 computes second rank coupling number between amino acid pairs at lag value 2. Similarly, τ_n computes nth rank coupling number between amino acid pairs at lag value n. Here lag value defines

Table 6.12 Distance or content matrix computed by Schneider and Wrebe on the basis of 4 different physicochemical properties i.e., hydrophobicity, hydrophilicity, polarity, chain mass

Name	A	C	D	E	F	G	H	I	K	L	M	N	P	Q	R	S	T	V	W	Y
A	0	0.11	0.81	0.82	0.54	0.20	0.69	0.40	0.89	0.40	0.37	0.318	0.19	0.37	1	0.09	0.22	0.27	0.73	0.55
C	0.11	0	0.84	0.83	0.43	0.32	0.66	0.30	0.88	0.30	0.27	0.324	0.15	0.34	1	0.17	0.23	0.16	0.63	0.45
D	0.72	0.74	0	0.12	0.92	0.69	0.43	0.84	0.24	0.84	0.81	0.56	0.65	0.58	0.29	0.66	0.64	0.79	1	0.83
E	0.79	0.78	0.13	0	0.93	0.77	0.40	0.86	0.14	0.85	0.83	0.599	0.68	0.59	0.23	0.72	0.68	0.82	1	0.83
F	0.50	0.40	0.97	0.91	0	0.69	0.66	0.12	0.90	0.13	0.16	0.541	0.42	0.45	1	0.54	0.49	0.25	0.20	0.17
G	0.20	0.31	0.77	0.80	0.72	0	0.76	0.59	0.89	0.59	0.55	0.381	0.32	0.46	1	0.15	0.27	0.46	0.92	0.72
H	0.89	0.83	0.62	0.54	0.90	1	0	0.84	0.56	0.84	0.82	0.754	0.77	0.71	0.69	0.86	0.83	0.83	0.98	0.82
I	0.40	0.29	0.94	0.89	0.13	0.59	0.65	0	0.89	0.01	0.05	0.457	0.31	0.38	1	0.44	0.39	0.13	0.33	0.21
K	0.88	0.87	0.27	0.14	0.95	0.9	0.43	0.89	0	0.89	0.87	0.667	0.75	0.63	0.15	0.82	0.75	0.88	1	0.84
L	0.40	0.29	0.94	0.89	0.13	0.59	0.65	0.01	0.89	0	0.06	0.452	0.30	0.37	1	0.44	0.39	0.13	0.34	0.20
M	0.38	0.27	0.93	0.87	0.18	0.56	0.64	0.05	0.88	0.06	0	0.447	0.28	0.37	1	0.41	0.35	0.12	0.39	0.25
N	0.42	0.42	0.83	0.83	0.76	0.51	0.78	0.61	0.89	0.60	0.58	0	0.26	0.17	1	0.36	0.36	0.50	0.94	0.64
P	0.22	0.17	0.85	0.83	0.51	0.37	0.69	0.36	0.87	0.35	0.32	0.231	0	0.22	1	0.19	0.16	0.24	0.72	0.48
Q	0.51	0.46	0.90	0.86	0.67	0.64	0.76	0.53	0.88	0.51	0.50	0.181	0.27	0	1	0.46	0.38	0.46	0.83	0.52
R	0.91	0.90	0.30	0.22	0.97	0.92	0.49	0.92	0.14	0.92	0.90	0.69	0.79	0.66	0	0.86	0.80	0.91	1	0.85
S	0.1	0.18	0.80	0.81	0.62	0.17	0.71	0.47	0.88	0.47	0.44	0.289	0.18	0.35	1	0	0.17	0.34	0.82	0.61
T	0.25	0.26	0.83	0.81	0.60	0.31	0.73	0.45	0.86	0.45	0.40	0.315	0.15	0.32	1	0.18	0	0.34	0.81	0.59
V	0.27	0.16	0.9	0.86	0.26	0.47	0.64	0.13	0.88	0.13	0.12	0.38	0.21	0.33	1	0.32	0.30	0	0.47	0.31
W	0.65	0.56	1	0.93	0.19	0.82	0.67	0.30	0.89	0.30	0.34	0.631	0.55	0.53	0.96	0.68	0.63	0.41	0	0.20
Y	0.58	0.47	1	0.93	0.20	0.78	0.67	0.23	0.90	0.21	0.26	0.512	0.44	0.40	0.99	0.61	0.55	0.32	0.24	0

Table 6.13 Distance or content matrix computed by Grantham on the basis of 4 different physicochemical properties i.e., hydrophobicity, hydrophilicity, polarity, chain mass

Name	A	R	N	D	C	Q	E	G	H	I	L	K	M	F	P	S	T	W	Y	V
A	0	112	111	126	195	91	107	60	86	94	96	106	84	113	27	99	58	148	112	64
R	112	0	86	96	180	43	54	125	29	97	102	26	91	97	103	110	71	101	77	96
N	111	86	0	23	139	46	42	80	68	149	153	94	142	158	91	46	65	174	143	133
D	126	96	23	0	154	61	45	94	81	168	172	101	160	177	108	65	85	181	160	152
C	195	180	139	154	0	154	170	159	174	198	198	202	196	205	169	112	149	215	194	192
Q	91	43	46	61	154	0	29	87	24	109	113	53	101	116	76	68	42	130	99	96
E	107	54	42	45	170	29	0	98	40	134	138	56	126	140	93	80	65	152	122	121
G	60	125	80	94	159	87	98	0	98	135	138	127	127	153	42	56	59	184	147	109
H	86	29	68	81	174	24	40	98	0	94	99	32	87	100	77	89	47	115	83	84
I	94	97	149	168	198	109	134	135	94	0	5	102	10	21	95	142	89	61	33	29
L	96	102	153	172	198	113	138	138	99	5	0	107	15	22	98	145	92	61	36	32
K	106	26	94	101	202	53	56	127	32	102	107	0	95	102	103	121	78	110	85	97
M	84	91	142	160	196	101	126	127	87	10	15	95	0	28	87	135	81	67	36	21
F	113	97	158	177	205	116	140	153	100	21	22	102	28	0	114	155	103	40	22	50
P	27	103	91	108	169	76	93	42	77	95	98	103	87	114	0	74	38	147	110	68
S	99	110	46	65	112	68	80	56	89	142	145	121	135	155	74	0	58	177	144	124
T	58	71	65	85	149	42	65	59	47	89	92	78	81	103	38	58	0	128	92	69
W	148	101	174	181	215	130	152	184	115	61	61	110	67	40	147	177	128	0	37	88
Y	112	77	143	160	194	99	122	147	83	33	36	85	36	22	110	144	92	37	0	55
V	64	96	133	152	192	96	121	109	84	29	32	97	21	50	68	124	69	88	55	0

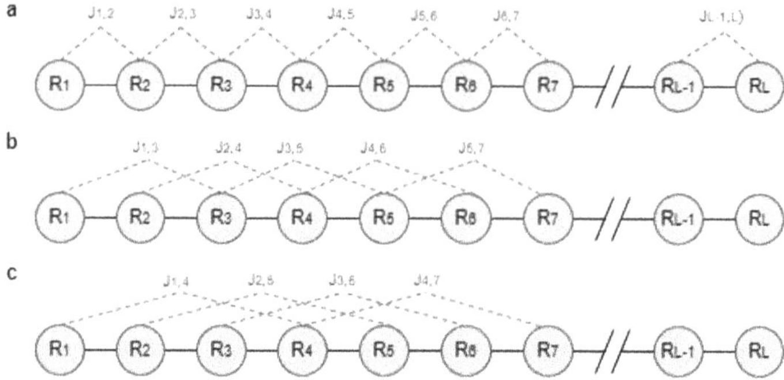

Fig. 6.22 (**a**) The first rank reflects the correlation between most adjacent residues. (**b**) The second rank, reflects the correlation between most adjacent plus one. (**c**) The third rank, reflects the correlation between most adjacent plus two. This figure is adapted from Ref. [187]

the rank of coupling number between amino acid pairs. Primarily, sequence order coupling number is physicochemical distance between sequence amino acids pairs. Given mathematical expression and Fig. 6.22, it is evident that τ_1 calculates the distance at lag 1 between two adjacent amino acids. Likewise, τ_2 computes the distance at lag 2 between pairs of amino acids by skipping one amino acid. For τ_n, the calculation involves the distance at lag n between pairs of amino acids, with n amino acids skipped between them. Furthermore, lag is a hyper-parameter that needs to be set according to nature and complexity of the task. To generate final statistical vector SOCNumber encoder concatenates all values of computed τ for both distance matrices provided by Schneider-Wrede [888] and Grantham [362].

6.11.2 Quasi Sequence (QS) Order

Quasi Sequence (QS) Order encoder is advance version of SocNumber, which computes both occurrence distribution of amino acids and their physicochemical properties distance based on distance values of amino acid pairs provided by Schneider-Wrede [888] and Grantham [362]. This encoder has been utilized in different types of protein sequence analysis tasks such as nuclear proteins sub-nuclear location prediction [1143], plant proteins sub-cellular locations prediction [863], membrane protein types prediction [124] and proteins classification into their functional families [780].

Figure 6.23 graphically represents working paradigm of QS Order encoder. To understand working of QS order encoder, consider a protein sequence $R_1R_2R_3R_4....R_L$, where R_i represents 20 unique amino acids. Similar to SocNumber encoder, QS Order encoder uses following mathematical expression to compute lag based sequence order coupling number of amino acids pairs.

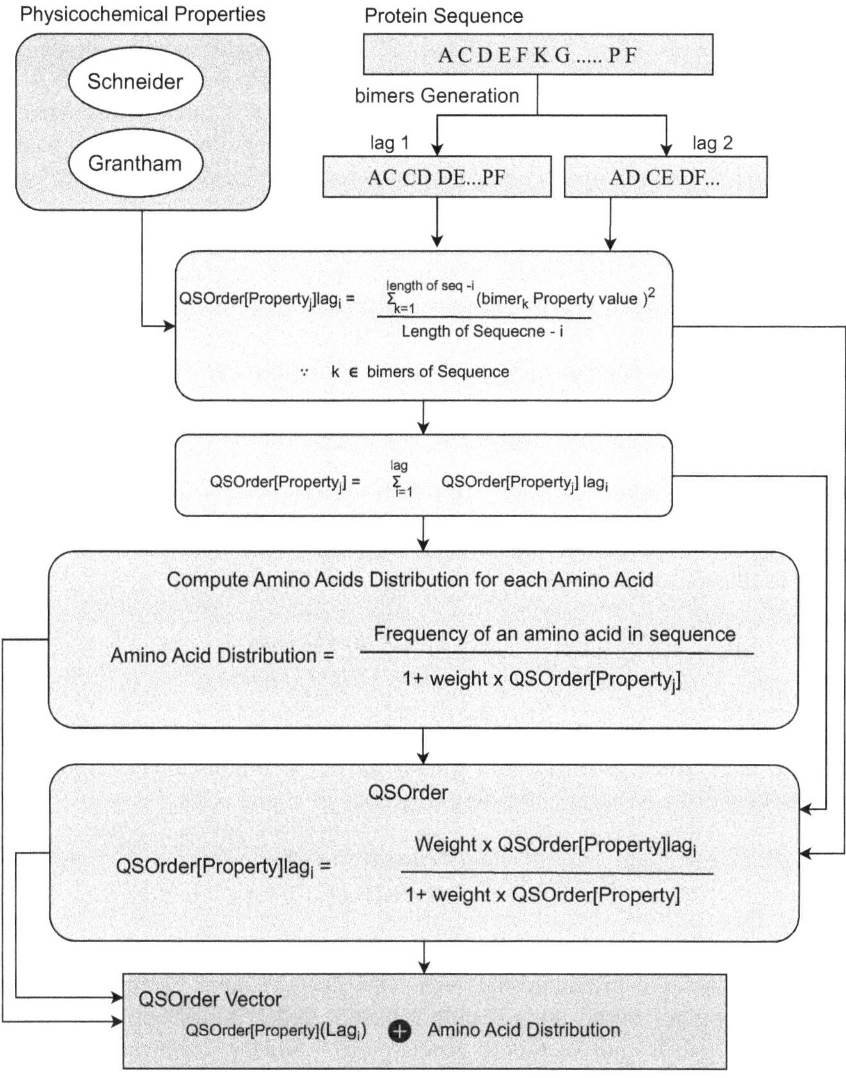

Fig. 6.23 Overall workflow of Quasi Sequence(QS) Order encoding method

$$\left.\begin{aligned} \tau_1 &= \tfrac{1}{L-1} \sum_{i=1}^{L-1} (J_{i,i+1}) \\ \tau_2 &= \tfrac{1}{L-2} \sum_{i=1}^{L-2} (J_{i,i+2}) \\ &\vdots \\ \tau_n &= \tfrac{1}{L-n} \sum_{i=1}^{L-n} (J_{i,i+n}), \ n < L \end{aligned}\right\} \text{where } J_i \in \{R_1, R_2, \ldots R_L\}$$

In above expression L denotes length of protein sequence and J_i represents to 20 unique amino acids. $(J_{i,i+1})$, $(J_{i,i+2})$ and $(J_{i,i+n})$ are physicochemical properties distances between amino acids pairs $(R_i, R_i + 1)$, $(R_i, R_i + 2)$ and $(R_i, R_i + n)$ respectively. More detailed information about different τ is already discussed in Sect. 6.11.1 to generate final statistical vector this encoder concatenates all values of computed τs for both distance matrices provided by Schneider-Wrede [888] and Grantham [362].

After computing amino acids sequence order information at different lag values, it computes overall sequence order information by calculating sum of amino acids sequence order information at all lag values. Equation 6.64 illustrates mathematical formulation to compute overall sequence order information with lag value n.

$$OSOIproperty_i = \sum_{i=1}^{lag_n} \tau_i \tag{6.64}$$

After computing overall sequence order information, it normalizes τ_i with a weight factor and combines sum of amino acids sequence order information at all lag values, as illustrated in Eq. 6.65.

$$NSOI[property_i]_{lagi} = \frac{w \times En[D_i]_{lagi}}{1 + w \times OSOIproperty_i} \tag{6.65}$$

Furthermore, it computes occurrence frequency of each amino acid in given protein sequence and normalizes it with $w \times Encoding[D_i] + 1$. Equation 6.66 illustrates mathematical expression for computing frequency of amino acids.

$$En[A_i] = \frac{Frequency\, of\ A_i}{w \times Encoding[D_i] + 1} \tag{6.66}$$

Finally, it generates two different vector spaces; one ($En[A_i]$) contains information about amino acids distribution and other ($En[D_i]_{lagi}$) other captures physicochemical properties based amino acids sequence order information. Finally, it concatenates both feature vectors to generate $(20 + lag) \times 2$ dimensional statistical vector. Equation 6.67 provides mathematical expression of final amino acid encoding generated by QS order encoder.

$$Sequence\ Encoding = En[A_i]||En[D_i]_{lagi} \tag{6.67}$$

6.11.3 Pseudo-Amino Acid Composition (PAAC)

Kuo- Chen Chou [188] proposed PAAC encoder that has been extensively utilized in diverse types of protein sequence analysis tasks such as identification of

Table 6.14 Physicochemical values used in the APAAC encoder for hydrophobicity, hydrophilicity and side chain mass

AccNo	Hydrophobicity	Hydrophilicity	SideChainMass
A	0.62	−0.5	15
R	−2.53	3	101
N	−0.78	0.2	58
D	−0.9	3	59
C	0.29	−1	47
Q	−0.85	0.2	72
E	−0.74	3	73
G	0.48	0	1
H	−0.4	−0.5	82
I	1.38	−1.8	57
L	1.06	−1.8	57
K	−1.5	3	73
M	0.64	−1.3	75
F	1.19	−2.5	91
P	0.12	0	42
S	−0.18	0.3	31
T	−0.05	−0.4	45
W	0.81	−3.4	130
Y	0.26	−2.3	107
V	1.08	−1.5	43

lysine malonylation sites [193], non-classically secreted proteins prediction [1038], proteins structural classes prediction [1116], infectious diseases associated host genes prediction [74], identification of phosphorylation site [669], bitter peptide prediction [146] and classification of human coding variants [624].

PAAC encoder transforms variable or fixed-length protein sequences into fixed length statistical vectors by computing physicochemical properties and amino acids distribution based features. Researchers have utilized this encoder with three different types of physicochemical properties namely; hydrophobicity, hydrophilicity and side chain mass. Table 6.14 illustrates 20 unique amino acids physicochemical values for all 3 properties.

As PAAC encoder makes use of 3 different types of properties so it normalizes physicochemical values of amino acids. Standardizing the physicochemical values of amino acids ensures that the resulting feature vectors are consistent and comparable across different protein sequences. In the absence of normalization, the absolute values of physicochemical properties can exhibit substantial variations across different amino acids which poses challenges for direct comparisons. By normalizing the values, the relative differences between amino acids are preserved to allow more meaningful comparisons.

Equation 6.68 illustrates mathematical expression for normalizing amino acids physicochemical values.

$$f(x) = \begin{cases} Mean[p_i] = \frac{\sum_{k=1}^{20} p_i[AA_k]}{20} \\ S[p_i] = \sqrt{\frac{(\sum_{k=1}^{20} (p_i[AA_k] - Mean[p_i])^2)}{20}} \\ p_i[AA_k] = \frac{p_i[AA_k] - Mean[p_i]}{S[p_i]}, \quad AA_k] \in \{1., 2, 3, \cdots, 20\} \\ p_i \in \{\text{hydrophobicity, hydrophilicity, side chain mass}\} \end{cases} \tag{6.68}$$

Here, p_i represents the physicochemical property based value of a amino acid (AA_k) which is either hydrophobicity, hydrophilicity or side chain mass. In Eq. 6.68, Mean[p_i] is the mean of 20 amino acids in each property and S[p_i] is the standard deviation, where both can be computed using Eq. 6.68.

Figure 6.24 elucidates the overall encoding paradigm utilized by the PAAC encoder. Specifically, it produces lag-based bimers to calculate the physicochemical values of amino acids. Particularly, lag values entail the skipping of multiple amino acids between consecutive bimers of amino acids.

Consider a protein sequence $R_1 R_2 R_3 R_4 \ldots R_L$. The sequence order effect reflected by the protein sequence can be represented with a set of sequence order correlation factors, which can be formulated as,

$$\begin{aligned} \theta_1 &= \frac{1}{L-1} \sum_{i=1}^{L-1} \Theta(R_i, R_{i+1}) \\ \theta_2 &= \frac{1}{L-2} \sum_{i=1}^{L-2} \Theta(R_i, R_{i+2}) \\ \theta_3 &= \frac{1}{L-3} \sum_{i=1}^{L-3} \Theta(R_i, R_{i+3}) \\ &\vdots \\ \theta_\lambda &= \frac{1}{L-\lambda} \sum_{i=1}^{L-\lambda} \Theta(R_i, R_{i+\lambda}) \end{aligned} \tag{6.69}$$

Here, λ $(\lambda < L)$,θ_1 is the first tier called the first-tier correlation factor that reflects sequence order correspondence among most adjacent residues along a protein chain, θ_2 is second-tier correlation factor that reflects the sequence order correspondence between all the second most contiguous residues along a protein chain. θ_3 is the third tier called the third-tier correlation factor that reflects the sequence order correspondence between all the third most contiguous residues along a protein chain [188].

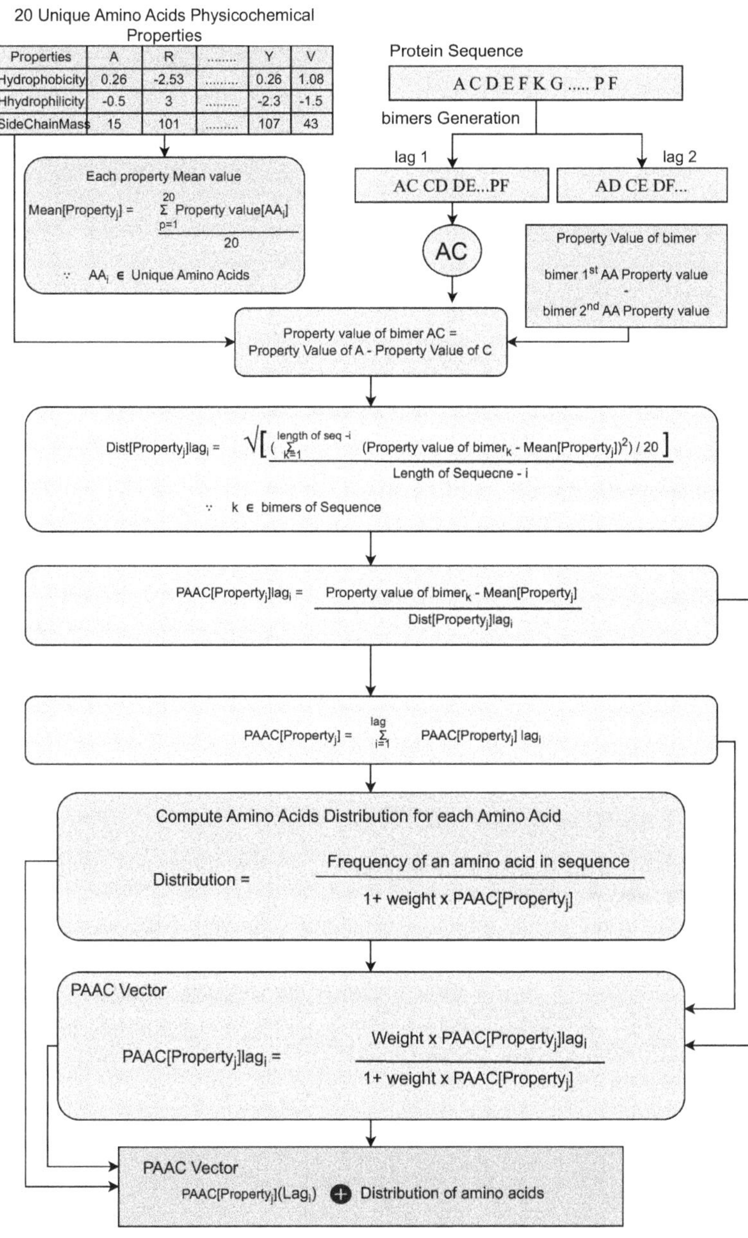

Fig. 6.24 Overall workflow of Pseudo-Amino Acid Composition (PAAC) encoding method

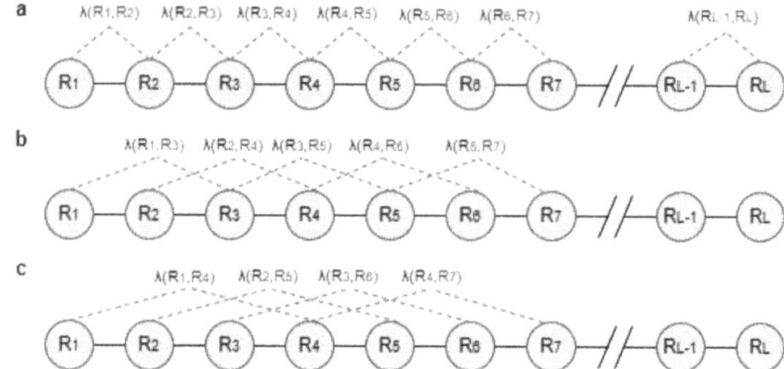

Fig. 6.25 (**a**) The first tier reflects the coupling mode between all the most adjacent residues. (**b**) The second tier, reflects the coupling mode between adjacent plus one. (**c**) The third tier,reflects the coupling between most adjacent plus two. This figure is adapted from [188]

Hence, correlation function for each physicochemical property can be defined as,

$$\Theta(R_i, R_j) = [P(R_i) - P(R_j)]^2, \tag{6.70}$$

Were $P(R_i)$,and $P(R_j)$ are properties of Amino acid R_i, and R_j after normalization. The correlation function provides averaged value for amino acids across 3 different physicochemical properties.

$$\Theta(R_i, R_j) = \frac{1}{n}\sum_{n=1}^{n}[P_k(R_j) - P_k(R_i)]^2 \tag{6.71}$$

where $P_k(R_i)$,and $P(R_j)$ are the Kth property in the amino acid property set of amino acid R_i, and R_j respectively.

Figure 6.25a–c represents effect of sequence order for different value of λ of protein can be reflected by the sequence correlation factors θ_1,θ_2,$\theta_3, \ldots, \theta_\lambda$.The amino acid composition formulation includes the discrete numbers provided by the sequence order correlation factors.Hence, rather than using the vector of 20-D defined by 20 components, vector of (20+λ)-D is used to represent a protein **X** as,

$$\mathbf{X} = \begin{bmatrix} x_1 \\ x_2 \\ \vdots \\ x_{20} \\ x_{20+1} \\ \vdots \\ x_{20+\lambda} \end{bmatrix} \tag{6.72}$$

where,

$$x_n = \frac{f_n}{\sum_{i=1}^{20} f_i + w \times \sum_{j=1}^{\lambda} \theta_j}, (1 < n < 20) \quad (6.73)$$

$$= \frac{w \times \theta_{n-20}}{\sum_{i=1}^{20} f_i + w \times \sum_{j=1}^{\lambda} \theta_j}, (20 + 1 < n < 20 + \lambda) \quad (6.74)$$

Where f_i is the normalized occurrence frequency of amino acid i in the protein and w shows the weighting factor for the sequence-order effect (w = 0.5 or 0.1). θ_j is the j-tier correlation factor as defined in Eq. 6.70 for protein **X**. Equation 6.73 represent 20 components represent the effect of amino acid composition, and Eq. 6.74 represent the (20+λ) component, represent the effect of sequence order.This set of 20+λ components is called pseudo-amino acid composition for protein **X** [188].

6.11.4 Amphiphilic Pseudo-Amino Acid Composition (APAAC)

Chou et al. [188, 189] proposed APAAC encoder that makes use of pre-computed physicochemical values of 3 different properties namely hydrophobicity, hydrophilicity and side chain mass [188, 189]. This encoder has been widely used in different types of protein sequence analysis tasks including protein solubility prediction [706, 1073], identification of major histocompatibility complex [161], protein-protein interactions prediction [1056], viral host protein-protein interactions prediction [56]. Table 6.14 illustrates each physicochemical property related 20 float values associated with 20 unique amino acids. The physicochemical values are computed based on diverse types of information related to protein folding, and protein's interactions with the environment and other molecules. For each of the three quantitative properties, the values of its corresponding amino acids are normalized to zero mean and unit standard deviation through Eq. 6.75. Mean normalization and standardization can improve the performance of machine learning algorithms by ensuring that the feature values have similar ranges and variances. Algorithms that rely on distance metrics or gradient-based optimization, such as support vector machines or neural networks, benefit from having features with similar scales. Normalizing the values helps prevent certain features from dominating the learning process or biasing the model's decisions.

$$f(x) = \begin{cases} Mean[p_i] = \dfrac{\sum_{k=1}^{20} p_i[AA_k]}{20} \\ S[p_i] = \sqrt{\dfrac{(\sum_{k=1}^{20} (p_i[AA_k] - Mean[p_i])^2)}{20}} \\ p_i[AA_k] = \dfrac{p_i[AA_k] - Mean[p_i]}{S[p_i]}, \quad AA_k] \in \{1., 2, 3, \cdots, 20\} \\ p_i \in \{\text{hydrophobicity, hydrophilicity, side chain mass}\} \end{cases} \tag{6.75}$$

In Eq. 6.75 p_i represents the physicochemical property based value of a amino acid (AA_k) which is either hydrophobicity, hydrophilicity or side chain mass. Mean[p_i] is the mean of 20 amino acids in each property and S[p_i] is the standard deviation.

In each physicochemical property, using normalized values of all 20 amino acids, order of amino acids within protein sequences is captured using lag-based phenomenon.

Figure 6.26 illustrates overall workflow of Amphiphilic Pseudo-Amino Acid Composition (APAAC) encoder. For instance, consider a raw sequence S = $R_1, R_2, R_3, R_4, \cdots, R_L$, where $R_{1,\cdots,L}$ denotes repetitive patterns of 20 unique amino acids. If lag = 1, then two most contiguous amino acids i.e., $S_{lag1} = R_1R_2, R_2R_3, R_3R_4, R_4R_5$, are taken, for lag = 2, second-most contiguous amino acids, i.e., $S_{lag2} = R_1R_3, R_2R_4, R_3R_5$ are taken by skipping 1 amino acid, and for lag = 3, third-most contiguous amino acids are taken by skipping 2 amino acids i.e., $S_{lag3} = R_1R_4, R_2R_5$, and so on. After generating bigrams at different lag values i.e., $S_{lag1}, S_{lag2}, S_{lag3}$. Iteratively for each lag value, bigrams are taken and in each bigram, physicochemical values of both amino acids are multiplied using a correlation function shown in Eq. 6.76.

$$\begin{gathered} P_i[B] = p_i(AA_j).p_i(AA_k), \\ p_i \in \{\text{hydrophobicity, hydrophilicity, side chain mass}\} \end{gathered} \tag{6.76}$$

After computing correlation, for each property, across all lags, a single float is computed by averaging property values across all the lag-based amino acid bigrams.

$$Enc[p_i] = \sum_{l=1}^{lag} \frac{P[B]}{seq\ len - lag_l}. \tag{6.77}$$

Furthermore, both types of sequence order and amino acid distributional information can be captured using Eq. 6.78.

$$Enc\ [AA] = \frac{frequency\ of\ AA\ in\ protein\ sequence}{1 + w \times Enc[p_i]}, \tag{6.78}$$

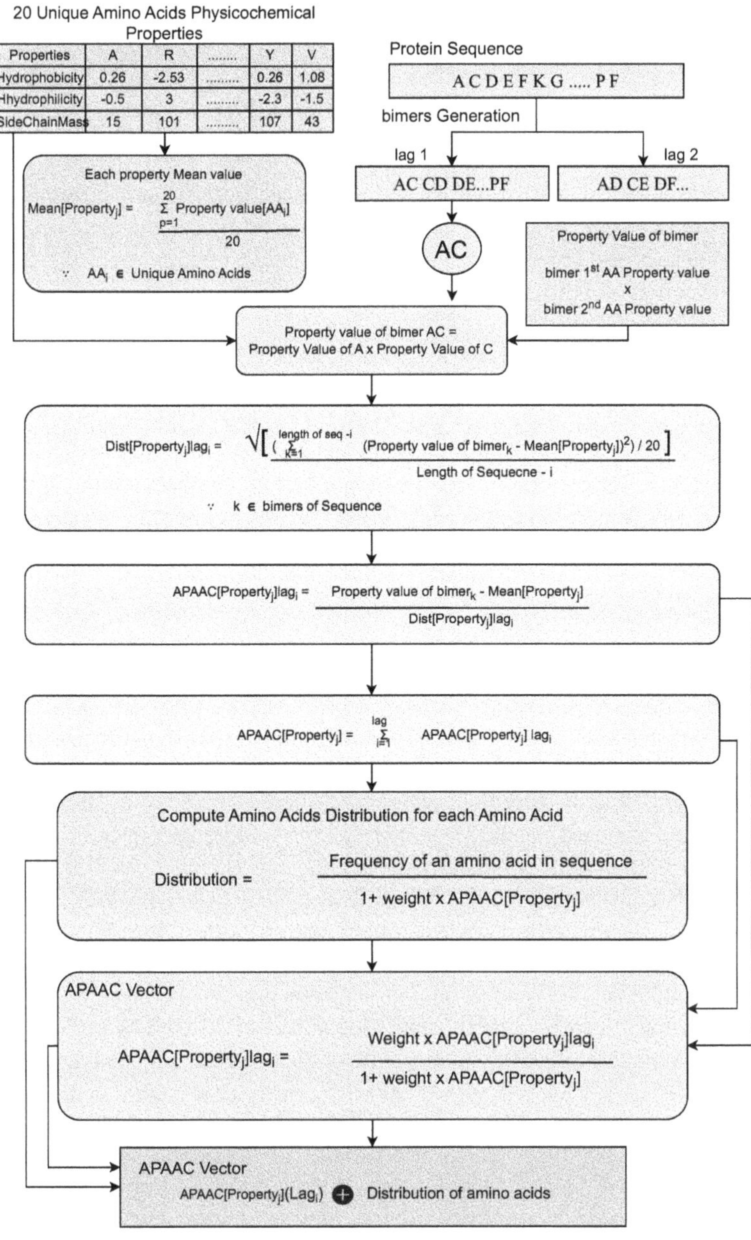

Fig. 6.26 Overall workflow of Amphiphilic Pseudo-Amino Acid Composition (APAAC) encoder

here, w is a weight parameter that varies from 0.1 to 1. Similarly, normalization is applied on the original sequence order information by using Eq. 6.79,

$$Enc[p_i]\, lag_i = \frac{w \times Enc[p_i]lag_i}{1 + w \times Enc[p_i]}. \tag{6.79}$$

Once the amino acid distribution and sequence order related information are encoded, the final statistical representation is obtained by concatenating the amino acid distributions and correlations among amino acids, that represent the sequence order information of a protein sequence.

$$Encoding\ [seq]P_i = Enc[AA] \parallel Enc\ [p_i]_{lag_i} \tag{6.80}$$

The dimension of the final statistical vector for a single physicochemical property is 20 + lag-D vector and for 3 physicochemical properties, the final statistical vector is (20 + lag) × 3 dimensional vector. In which, first 20 numbers are the normalized amino acid frequencies and the next following discrete numbers reminisce the amphiphilic amino acid correlations along a protein chain.

6.12 Binary Descriptor

Binary descriptor also known as one-hot encoding method generates statistical representation of a protein sequence by substituting amino acids with 0s and 1s. Each amino acid in a protein sequence is represented by a 20-dimensional binary vector composed solely of 0_s and 1_s [648]. More precisely, in this representation ith amino acid is indicated by setting the ith bit to "1" and all other bits to "0". For instance [A,C,D,E,F,G,H,I,K,L,M,N,P,Q,R,S,T,V,W,Y] represents sequence of 20 unique amino acids, then one hot encoding for 'A' will be '10000000000000000000', for 'C' it will be '01000000000000000000', for 'D' '00100000000000000000' and so on. Furthermore, physicochemical property value of each amino acid can be directly appended at end of 20-bit string. For instance, the 'hydrophobicity" value of amino acid of type 'A' is 1.8, and hence it can be encoded as, '100000000000000000001.8', which facilitates to discriminate between different amino acids. Contrarily, some protein sequences contain an unknown amino acid type which necessitates an additional bit and resulting in a 21-dimensional binary encoding generates high-dimensional feature vector which increases model complexity. Researchers have proposed diverse types of strategies to handle high dimension of vector. Primarily these strategies make use of physiochemical properties to form groups of amino acids to generate one hot vector encoding of groups as described in following subsection. Binary encoders have been used for proteomic sequence analysis tasks such as protein secondary structure prediction [1168], protein sequence classification [1049] and protein sequence back-translation [1093].

6.12.1 Binary (3-bit)

This encoder is capable of generating 7 distinct types statistical representations of protein sequences. Primarily, generated statistical representations differ based on disparate groupings of amino acids. This encoder makes use of 7 different physicochemical properties to categorise 20 unique amino acids into 3 different groups. Table 6.15 illustrates specific amino acid distribution into 3 different groups corresponding to 7 types. After grouping amino acids, it generates 3-bit one hot-encoding of each amino acid depending on its presence in particular group. Amino acids belonging to group 1 are represented by $\{1, 0, 0\}$ 3-bit encoding, while group 2 and 3 amino acids are denoted by $\{0, 1, 0\}$ and $\{0, 0, 1\}$, respectively. Finally, amino acids in sequences are replaced with their binary encoding to generate (sequence length $\times$ 3) dimensional statistical vector. Figure 6.27 graphically illustrates complete working paradigm of transforming toy sequence $A, C, D, E, F, K, G, ..., P, F$ to binary 3-bit (Type 1) based statistical representation.

Table 6.15 Amino acids division for 3-bit binary encoder

Binary 3-bit	Group 1	Group 2	Group 3
Type 1	Q,R,K,N,D,E	G,A,H,Y,P,T,S	L,M,C,V,F,W,I
Type 2	G,D,A,P,T,S	N,Q,V,I,L,E	R,M,C,F,W,K,H,Y
Type 3	Q,R,K,N,D,E	Y,G,A,P,H,T,S	L,M,C,V,F,W,I
Type 4	G,D,A,T,S	N,Q,I,C,V,L,P,E	R,M,K,F,H,Y,W
Type 5	R,K	G,I,NA,T,Q,S,L,C,M,P,Y,H,F,V,W	E,D
Type 6	E,Q,A,H,K,L,R,M	I,F,C,W,Y,V,T	D,N,P,S,G
Type 7	F,A,L,W,C,V,I,G	P,Q,E,K,N,D	H,T,M,S,R,Y

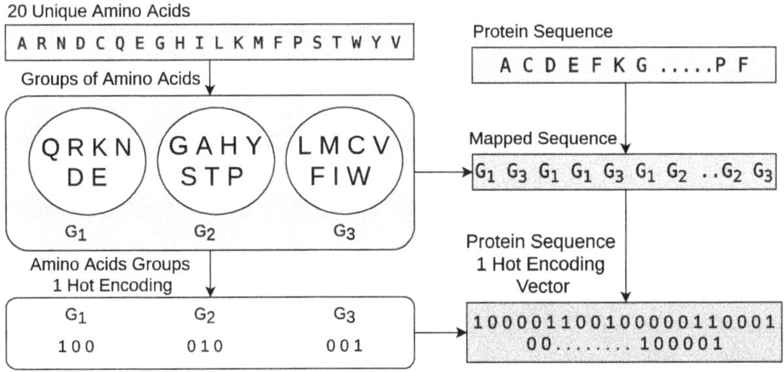

Fig. 6.27 Overall workflow of Binary (3bit-type1) encoder

Table 6.16 Amino acids division for 6-bit binary encoder

Binary 6-bit	Group 1	Group 2	Group 3	Group 4	Group 5	Group 6
Type 1	K,H,R,	D,E,N,Q	C	S,G,T,P,A	V,M,I,L	Y,F,W

6.12.2 Binary (6-bit)

The overall working paradigm of 6-bit binary encoder is similar to 3-bit binary encoder, with key difference being segregation of amino acids into 6 distinct groups instead of 3 as shown in Table 6.16. Among these groups, group 1 contains 3 amino acids K, R, H while group 2 comprises amino acids D, E, N and Q. Group 3 contains only one amino acid i.e. C and group 4 4 amino acids S, G, P, T, A. Group 5 consists of 3 amino acids V, I, M, and L, and Group 6 includes 3 amino acids Y, F, and W. Based on amino acid distribution in each group, encoder generates 6-bit binary encoding depending on presence of amino acids in a particular group. It replaces each amino acid of raw sequence with a corresponding 6-bit binary representation and generates (sequence length $\times$ 6) dimensional feature vector. Let's assume toy sequence AGTSF, binary (6-bit) encoder generates following statistical vector: $\{0, 0, 0, 1, 0, 0, 0, 0, 0, 1, 0, 0, 0, 0, 0, 1, 0, 0, 0, 0, 0, 1, 0, 0, 0, 0, 0, 0, 0, 1\}$.

6.12.3 Binary (5-Bit Type1)

The working paradigm of binary (5-bit Type-1) encoder is comparable to binary (6-bit) except it categorizes amino acids into 5 groups such as G1 {I,V,M,A,G,L}, G2 {Y,F,W}, G3 {R,H,K}, G4 {D,E} and G5 {T,S,C,P,N,Q}. Then, it computes binary vector for each amino acid based on the presence of amino acid in particular group. Next, it replaces amino acids with their respective binary vector to generate (sequence length $\times$ 5) dimensional feature vector.

6.12.4 Binary (5-Bit Type2)

Binary 5-bit type 2 encoder generates 5 groups with unique amino acids distribution as shown in Table 6.17. The working paradigm of this encoder comprises 3 steps. First, encoder generates 5-bit binary encodings for each amino acid. Each bit in the encoding corresponds to a specific group and presence of the amino acid in that particular group is denoted by value 1, while the absence is represented by 0. For instance, 5 bit type 2 binary encoding of A can be formulated as 00011 which denotes that A belongs to groups 4 and 5. Lastly, encoder generates (sequence length $\times$ $\times$5) dimensional feature vector of raw sequence using feature mapping.

Table 6.17 Amino acids division for 7-bit OPF encoder

OPF 7-bit	Group 1	Group 2	Group 3	Group 4	Group 5	Group 6	Group 7
Type 1	A,C,F,G,H, I,L,M,N,P, Q,S,T,V,W,Y	C,F,I,L,M, V,W	A,C,D,G,P, S,T	C,F,I,L,M, V,W,Y	A,D,G,S,T	D,G,N,P,S	A,C,F,G,I, L,V,W
Type 2	D,E	A,G,H,P,S, T,Y	E,I,L,N,Q,V	A,G,P,S,T	C,E,I,L,N, P,Q,V	A,E,H,K,L, M,Q,R	H,M,P,S,T, Y
Type 3	K,R	D,E,K,N,Q, R	F,H,K,M,R, W,Y	D,E,H,K,N, Q,R	F,H,K,M,R, W,Y	C,F,I,T,V, W,Y	D,E,K,N,R, Q

6.12.5 Overlapping Property Features (OPF) (7-Bit)

Overall working paradigm of OPF 7bit encoder is similar to binary 5-bit type 2 encoder but OPF 7-bit encoder can generate 7 groups with 3 different types of amino acids distribution as shown in Table 6.17. Table 6.17 illustrates that OPF (7-bit) encoder can generate 3 different types of feature vectors against raw sequence using 3 types of amino acids distribution. First, it generates 7-bit binary encodings of each amino acid. Each bit represents a group and presence of amino acid in that particular group can be represented as 1, otherwise 0. In the last step, it generates (sequence length $\times$ 7) dimensional feature vector.

6.12.6 OPF 10-Bit

The working paradigm of OPF 10-bit encoder is similar to ODF 7-bit encoder and can be summarized in 3 different steps. First, it generates 10 groups that contain amino acids as illustrated in Table 6.18. Next, it generates 10-bit binary encoding of each amino acid based on amino acid presence in each group. Lastly, it generates (sequence length $\times$ 10) dimensional statistical vector by replacing each amino acid with their respective binary encodings.

6.13 Physicochemical Properties Mapping Based Encoding Methods

6.13.1 Dipole Moment and Alpha Mapping

Dipole moment refers to Vanderwaal forces between the non-polar molecules connected with polar molecules of permanent dipoles that sum all the partial atomic charges. Experimentally calculated dipole moment and alpha mapping values associated with amino acids are shown in Table 6.20. In protein sequence amino acids are replaced with a unique number of alpha and dipole moment values of amino acids for numerical representation of the sequence. This mapping method has been utilized in different types of protein sequence analysis tasks such as exon prediction [231], protein-coding region identification [225] and protein backbone geometry calculation [663].

Table 6.18 Amino acids division for 10-bit OPF encoder

OPF 10-bit	Group 1	Group 2	Group 3	Group 4	Group 5	Group 6	Group 7	Group 8	Group 9	Group 10
Amino acids	F,Y,H,W	D,E	K,R,H	S,D,E,T,Q, C,H,N,R,K, Y,W	G,A,C,T,I, V,L,K,H,Y, M,F,W	I,V,L	A,G,S,C	K,H,R,D	P,N,D,T,Q, C,H,A,S,V, G	P

6.13.2 *Electron Ion Interaction Potential (EIIP)*

EIIP represents energy of valence electrons of amino acids. Oscillations between bio-molecules produce electromagnetic affinity which is known as EIIP. Statistical representation of a given protein sequence is generated by placing experimentally calculated EIIP values (Table 6.20) of amino acids in the corresponding protein sequence. This mapping method has been utilized in different types of protein sequence analysis tasks such as exon prediction [231], and hotspot location identification [831, 832, 864].

6.13.3 *AAindex (AAindex)*

This encoder has been utilized in different types of protein sequence analysis tasks such as outer membrane prediction [1252], and tumor T-cell antigens prediction [419]. AAindex make use of AAindex database that facilitates 531 different types of physical and biochemical values of amino acids at single platform [308]. It transforms raw protein sequences into statistical vectors by substituting sequence amino acids with their physical and biochemical properties values. However, type and number of properties are selected according to nature and complexity of task. Furthermore, dimension of generated statistical vector is equal to original length of protein sequence × number of selected properties. Figure 6.28 graphically represents working paradigm of AAindex encoding method where a toy sequence is transformed into statistical vector using two different physicochemical properties namely ANDN920101 and FASG890101.

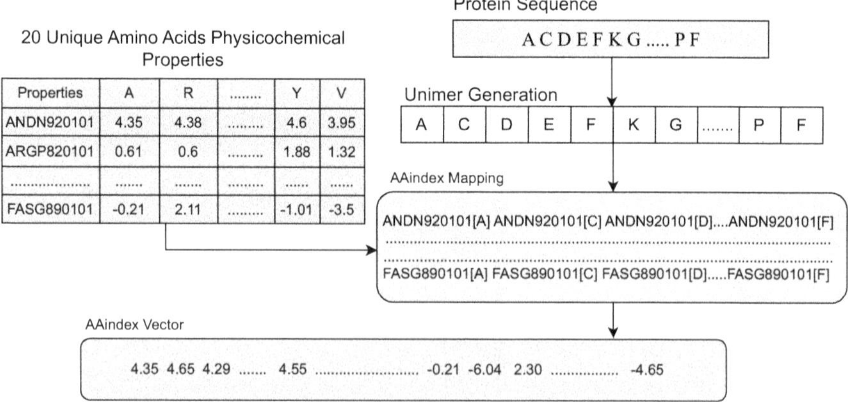

Fig. 6.28 Overall workflow of AAindex encoder

6.13.4 AESNN3

AESNN3 encoder was proposed by Lin et al. [629] and has been widely utilized in diverse types of protein sequence analysis applications such as phosphorylation site detection [660], protein structure alignments [629] and prediction of kinase-specific phosphorylation sites in proteins [154]. Lin et al. [629] trained a neural network model on extensive data related to protein structure alignment and extracted the last layer's three neuron values for each amino acid. The authors claimed that these neuron values effectively preserve diverse types of properties associated with amino acids. Further, they linearly transformed amino acids 3-dimensional values into range of $\{-1, 1\}$. All unique amino acids transformed values are shown in Table 6.19. To transform raw protein sequences into statistical vectors, sequence amino acids are replaced with precomputed values. Given a toy protein sequence "RCDNA," the statistical vector resulting from the replacement of precomputed values for each amino acid in the sequence is represented as follows: 0.28, −0.99, −0.22, 0.34, 0.88, 0.35, 0.74, −0.72, −0.35, 0.77, −0.24, 0.59, −0.99, −0.61, 0.00.

6.13.5 Hydropathy Index

Hydropathy index (HI) quantifies energy required by amino acid to convert cyclohexane to water. A positive HI index represents hydrophobic amino acids while negative HI index refers to hydrophilic amino acids. Researchers have conducted experiments to empirically calculate amino acids hydropathy index values shown in Table 6.20. This mapping method has been utilized in different types of protein sequence analysis tasks such as exon prediction [231], and homosepian cancerous gene identification [227, 229].

6.13.6 P-adic Mapping

To develop computational predictor for hotspot [1141] and exon prediction [231], Dragovich et al. [270] proposed p-adic encoding method. Motivation behind development of this encoding method was to transform raw protein sequences into statistical vectors by acquiring amino acids codons distribution. In DNA sequence three different nucleotides combinations represent multiple codons and these codons refers to amino acids that form protein sequence. Considering protein sequence relation with DNA sequence, authors proposed a unique strategy that makes use of nucleotides and codons to represent each amino acid with an integer value. Specifically to compute amino acids integer values, authors assigned a single unique integer value to each nucleotide as C = 1, A = 2, T = U = 3, and G = 4. Furthermore, these nucleotides values along with codons are used to derived amino acids unique

Table 6.19 AESNN3 encoding of amino acids

Amino acids	AESNN3 encoding	Amino acids	AESNN3 encoding	Amino acids	AESNN3 encoding	Amino acids	AESNN3 encoding
A	0.28, −0.99, −0.22	Q	0.12, −0.99, −0.99	L	−0.92, 0.31, −0.99	S	0.99, 0.40, 0.37
R	0.28, −0.99, −0.22	E	0.59, −0.55, −0.99	K	−0.63, 0.25, 0.50	T	0.42, 0.21, 0.97
N	0.77, −0.24, 0.59	G	−0.79, −0.99, 0.10	M	−0.80, 0.44, −0.71	W	−0.13, 0.77, −0.90
D	0.74, −0.72, −0.35	H	0.08, −0.71, 0.68	F	0.87, 0.65, −0.53	Y	0.59, 0.33, −0.99
C	0.34, 0.88, 0.35	I	−0.77, 0.67, −0.37	P	−0.99, −0.99, −0.99	V	−0.99, 0.27, −0.52

Table 6.20 Amino acids mapping

Amino acid	Symbol	Codons	Genetic Code Context (GCC)	Alpha	EIIP	p-Adic	Dipole moment	HI	CPNR	10-Bit binary encoding
Alanine	A	GCA,GCC,GCG,GCT	0.61 + 88.3i	1.409	0.0373	41	5.937	1.8	37	1000011000
Cysteine	C	TGC, TGT	1.07+ 112.4i	1.069	0.0829	34	10.74	2.5	11	1000010000
Aspartic	D	GAG, GAT	0.46 + 110.8i	0.192	0.1263	42	29.49	−3.5	59	1011110000
Glutamic	E	GAA, GAG	0.47+ 140.5i	0.175	0.0058	42	42.52	−3.5	61	1011100000
Phenylalanine	F	TTC,TTT	2.02 + 189i	1.966	0.0946	33	5.98	2.8	3	1000000010
Glycine	G	GGA, GGC, GGT, GGG	0.07 + 60i	1.058	0.0050	44	0.0	−0.4	43	1000011000
Histidine	H	CAC, CAT	0.61 + 152.6i	0.558	0.0242	12	20.44	−3.2	17	1101100010
Isoleucine	I	ATA,ATC,ATT	2.22 + 168.5i	1.990	0.0000	23	3.371	4.5	53	1000000100
Lysine	K	AAA, AAG, TTA, TTG, CTA, CTC	1.15 + 175.6i	0.181	0.0371	22	50.02	−3.9	67	1101100000
Leucine	L	CTG,CTT	1.53 + 168.5i	1.702	0.0000	13	3.782	3.8	23	1000000100
Methionine	M	ATG	1.18 + 162.2i	1.501	0.0823	23	8.589	1.9	1	1000000000
Asparagine	N	AAC, AAT	0.06 + 125.1i	0.434	0.0036	22	18.89	−3.5	41	1001010000
Proline	P	CCA, CCC, CCG, CCT	1.95+ 122.2i	0.519	0.0198	11	7.916	−1.6	7	1000010001
Glutamine	Q	CAA, CAG, AGA, AGG, CGA, CGC, CGG	148.7i	1.058	0.0761	12	0.0	−3.5	29	1011000000
Arginine	R	CGT, AGC, AGT, TCA, TCC	0.60 + 181.2i	0.240	0.0959	14	37.5	−3.8	47	1101100000
Serine	S	TCG, TCT	0.05+ 88.7i	0.774	0.0829	31	9.836	−0.8	31	1001011000
Threonine	T	ACA, ACC, ACG, ACT	0.05+ 118.2i	0.828	0.0941	21	9.304	−0.7	13	1001010000
Valine	V	GTA, GTC, GTG, GTT	1.32+ 141.4i	1.694	0.0057	43	2.692	4.2	19	1000010100
Tryptophan	W	TGG	2.65 + 227i	1.314	0.0548	34	10.73	−0.9	2	1001000010
Tyrosine	Y	TAG, TAT	1.88+ 193i	0.979	0.0516	32	10.41	−1.3	5	1001000010

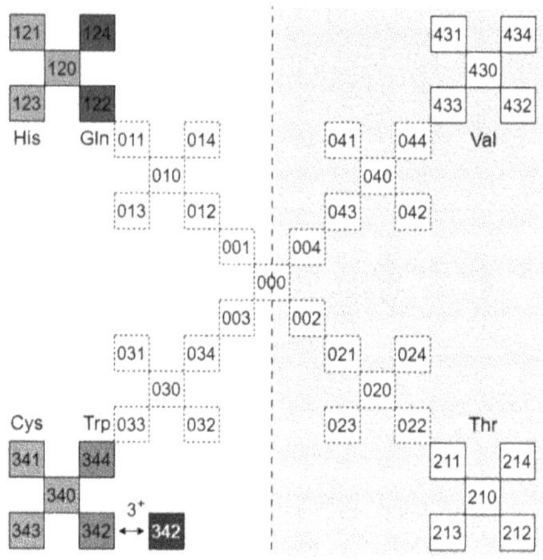

Fig. 6.29 Graphical architecture of p-adic Mapping encoder

values. Table 6.20 illustrates that several codons correspond to the same amino acid, with the notable characteristic that the first and second nucleotides in each set of codons associated with a particular amino acid are identical. For example 4 different codons (GCA, GCC, GCT, GCG) represent alanine (A) amino acid and in all these codons first nucleotide G and second nucleotide C are same. To derive amino acids integer values, authors proposed to concatenate codon first and second nucleotides integer values. Such as alanine (A) amino acid, p-adic value is 41. In this integer value, 4 refers to codon first nucleotide G and 1 represents C second nucleotide of codon. Finally, it make use of amino acids derived values to transform protein sequence into statistical vector. A toy sequence "APMSTVEF" can be transformed into statistical vector $\{41, 11, 23, 31, 21, 43, 42, 33\}$ by replacing amino acids values with unique integers of amino acids shown in Table 6.20. Amino acids are functionally linked to each other in p-adic space that is constructed using p-adic value of amino acids as shown in Fig. 6.29.

6.14 Correlation Based Encoding Methods

Correlation quantifies how two variables vary together or change in relation to each other. On the other hand, auto-correlation or serial correlation, measures the relationship between variables separated with certain distance. Auto-correlation descriptors such as Moran, Geary, and Moreau-Broto are statistical representations of different correlation patterns within a protein sequence estimated based on amino acids physicochemical properties. These descriptors provide information about the arrangement as well as distribution of unique amino acid properties along the sequence. Auto-correlation descriptors calculate the correlation among the values of a given physicochemical property at various places within the sequence. These descriptors capture crucial structural and functional aspects of proteins by evaluating how physicochemical characteristics fluctuate and interact along the sequence. They can be used in proteomics sequence analysis to compare protein sequences [375], anti-cancer peptides identification [1086], forecast protein structures [995] and explore protein-protein interactions [1056].

Initially, eight different amino acid properties have been used by auto-correlation descriptors including hydrophobicity, polarizability, average flexibility, free energy of amino acid solution in water, amino acid residue volumes, residue accessible surface areas, relative mutability, and steric parameters [512]. However, Guo et al. [375] found that there exist four kinds of interactions between two proteins namely; electrostatic interactions, hydrophobic interactions, steric interactions, and hydrogen bonds interactions. Therefore researchers have started using six physicochemical properties in auto-correlation based sequence descriptors including; hydrophobicity, volume of side chains, polarity, polarizability, solvent accessible surface area, and net charge index of side chains to effectively characterize different interactions.

Auto-correlation based sequence descriptors utilize amino acid indices (AAidx) matrix, which describes 544 physicochemical properties of 20 different amino acids [512, 965]. Primarily, Amino acid indices (AAidx) are numerical numbers that represent amino acid physicochemical characteristics. They offer a quantitative assessment of unique characteristics including hydrophobicity, charge, polarity, etc. Every amino acid has an AAidx value for each of its properties. Typically, these values are obtained from theoretical or experimental data and are preserved in databases. In proteomics sequence analysis, researchers can use AAidx values to quantify and compare different physicochemical properties of amino acids, allowing them to calculate correlation based descriptors to analyze protein sequences.

Physicochemical properties of amino acids can have inherent biases due to differences in measurement techniques or scales used by different researchers. Mean normalization and standardization help to remove such biases and normalize the values based on a common reference point. This allows for a more objective representation of the physicochemical properties, reducing the influence of any particular experimental or methodological bias. Before performing calculations, all three auto-correlation based descriptors standardize the value of each physicochemical property using Eq. 6.81.

$$P_{i,j} = (P_{i,j} - \bar{P_j})/\sigma_j \tag{6.81}$$

Here, $P_{i,j}$ is the jth physicochemical property of ith amino acid, $\bar{P_j}$ is the average value of jth physicochemical property of the amino acid from position 1 to N in the protein sequence of length N and σ_j is the standard deviation of the value the jth physicochemical property of the 20 amino acids. Equations 6.82 and 6.83 illustrate mathematical formulation to calculate P_j and σ_j, respectively, where σ_j is calculated using Eq. 6.82.

$$\bar{P}_j = \frac{\sum_{i=1}^{N} P_{i,j}}{N} \tag{6.82}$$

$$\sigma_j = \sqrt{\sum_{i=1}^{20} \frac{(P_{i,j} - \bar{P}_j)^2}{20}} \tag{6.83}$$

6.14.1 Moran Auto-correlation

Moran Auto-correlation was proposed by Patrick Alfred Pierce Moran to measure spatial auto-correlation, which is the degree to which data values in spatially adjacent locations are similar to each other. Moran auto-correlation calculates a correlation coefficient that ranges from -1 to 1. A positive value indicates that similar data values are present at spatially adjacent locations and a negative value indicates that dissimilar values are present at spatially adjacent locations. A value close to 0 suggests no significant auto-correlation, indicating randomness or independence. It has been used for different applications including protein-protein interaction prediction [4, 1109], drug-target interactions prediction [1066], prediction of membrane protein type [305], secondary structural content of proteins [627], identification of Antioxidant proteins [425], prediction of cyclin protein [963] and in the spatio-temporal analysis of virus outbreaks [690]. Figure 6.30 graphically illustrates complete working paradigm of Moran auto-correlation encoding method.

Moran auto-correlation uses the covariance of the property values of the amino acids with their lags in order to compute the statistical vector of protein sequence of size N using Eq. 6.84:

$$M(j, lag) = \frac{(1/N - lag) \times \sum_{i=1}^{N-lag} (P_{i,j} - \bar{P}_j) \times (P_{i+lag,j} - \bar{P}_j)}{1/N \times \sum_{i=1}^{N} (P_{i,j} - \bar{P}_j)} \tag{6.84}$$

Here, j = 1,2,...n, lag = 1,2,...m, $P_{i,j}$ and $P_{i+lag,j}$ represent the value of jth physicochemical property of amino acids at the i-position and i+lag position respectively within a protein sequence. Furthermore, $\bar{P}_j$ is the average value of jth physicochemical property of the amino acid from position 1 to N in the protein sequence, while lag denotes distance between two amino acids. Most researchers either use shortest protein sequence length or 30 as a lag value.

An intuitive analysis of nominator indicates that Moran auto-correlation computes the covariance based on deviations of amino acids property values with respect to their property means. covariance is measured as the average of products of deviations of amino acids property values. Furthermore, products of deviations takes directionality of deviations into account, like when both property values of amino acids separated at lagged distance have positive or negative deviations, their product

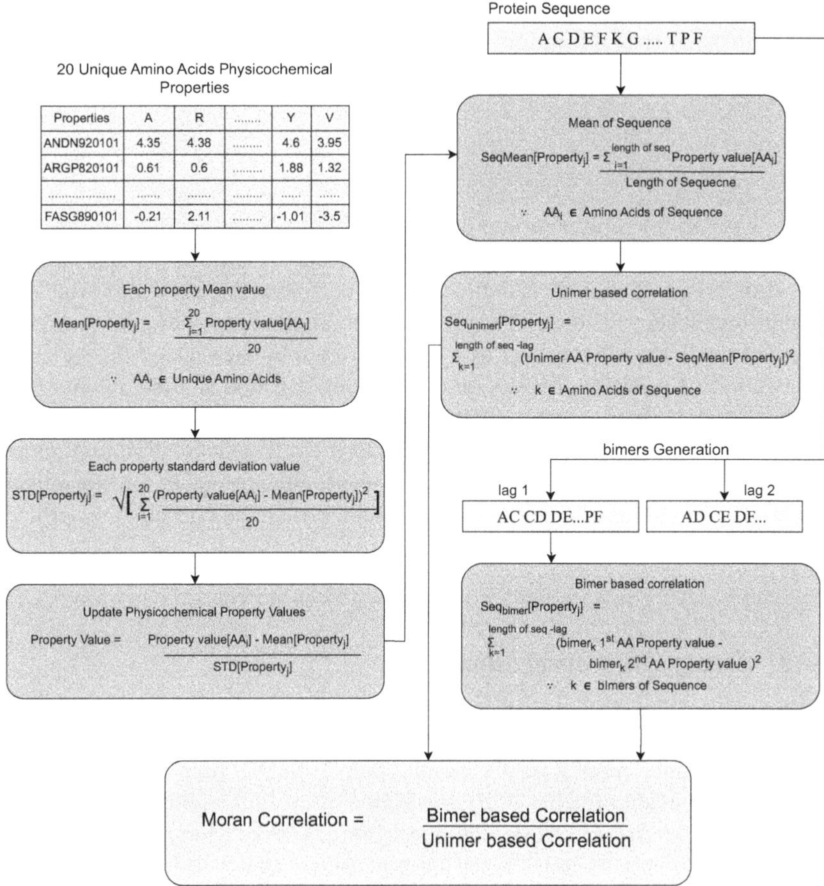

Fig. 6.30 Overall workflow of Moran Correlation encoding method

is positive, otherwise negative. The denominator helps to compute the average joint variability among the property values of amino acids separated with a lag distance.

The dimension of statistical vector produced by Moran auto-correlation for protein sequence of size N, using X number of properties can be calculated using Eq. 6.85.

$$MoranDim.vector = N \times X \tag{6.85}$$

One of the possible alternative to Moran auto-correlation is correlation coefficient. Moran's auto-correlation coefficient is used to detect spatial auto-correlation, which is the degree to which amino acids physicochemical properties are similar to neighbouring amino acids physicochemical properties within a sequence. Furthermore, correlation coefficient measures linear relationship between amino acids, regardless of their spatial location.

Mathematically, the correlation coefficient (r) can be computed using Eq. 6.86.

$$r = \frac{\text{Cov}(X, Y)}{\sigma_X \cdot \sigma_Y} \tag{6.86}$$

Here, Cov(X, Y) represents the covariance between variables X and Y while σX and σY represent the standard deviations of X and Y, respectively. This division allows comparison of strength and direction of the relationship between variables on a standardized scale (-1 to 1). The correlation coefficient provides a more effective measure of the linear relationship, as it is not affected by the scale of the variables.

The choice of measure depends on the research question and the nature of data being analyzed. If the research question involves understanding the spatial patterns of a phenomenon, then Moran's auto-correlation coefficient is more appropriate. On the other hand, if the research question involves understanding the strength and direction of the relationship between two variables, then the correlation coefficient is more appropriate.

6.14.2 *Geary Auto-correlation*

Geary auto-correlation was proposed by Geary in 1951 [336, 1144] in order to compute spatial auto-correlation. Geary auto-correlation has been utilized for identification of antioxidant proteins [425], prediction of cyclin proteins [963, 1253] and spatio-temporal analysis of virus outbreaks [690]. Unlike Moran auto-correlation that utilizes covariance of amino acids attribute values with their lags, Geary's auto-correlation calculates square of difference between properties of an amino acid and its lag in a sequence.

$$G(j, lag) = \frac{\frac{1}{2\times(N-lag)} \times \sum_{i=1}^{N-lag}((P_{i,j}) - (P_{i+lag}, j))^2}{\frac{1}{N-1} \times \sum_{i=1}^{N}((P_{i,j}) - (\bar{P_j}))^2} \tag{6.87}$$

Parameters used in Eq. 6.87 has been already explained in Moran auto-correlation. Figure 6.31 graphically illustrates complete working paradigm of Geary auto-correlation method. The paradigm of squaring the difference in Geary auto-correlation provides multiple benefits. Geary auto-correlation ensures that results are always positive and in the auto-correlation analysis, positive value reflects spatial dependency. This helps to identify protein sequence regions where amino acids reveal consistent dependencies or patterns related to specific physicochemical property. Furthermore, the squaring paradigm puts more weights to the larger differences among the property values of amino acid and its lagged version. This allows identification of discriminative patterns in protein sequence that might be

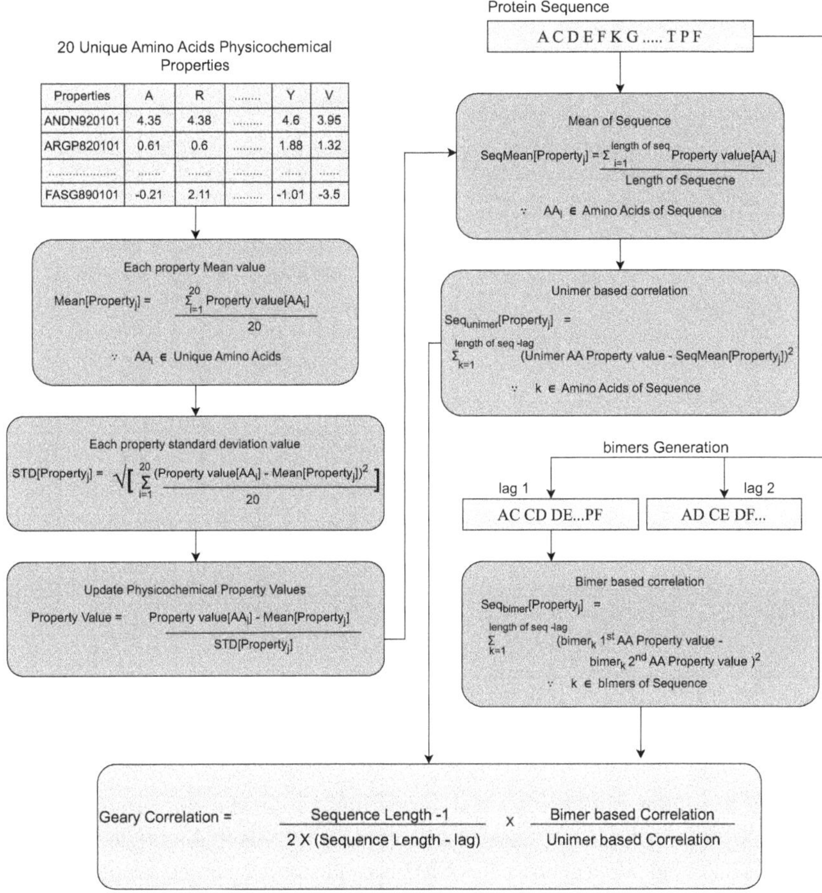

Fig. 6.31 Overall workflow of Geary Correlation encoding method

associated to particular functional or structural attributes of protein. This paradigm mitigates the possibility of cancellation that may arise due to the summation of positive and negative differences. The process of squaring the differences results in contribution from all values towards the measure. This prevents any cancellations and guarantees that the auto-correlation accurately reflects the spatial pattern that underlies it.

6.14.3 Normalized Moreau-Broto Auto-correlation (NMBroto)

The NMBroto descriptor also called auto-correlation of a Topological Structure (ATS) was proposed by Moreau and Broto [425]. It captures the structure of

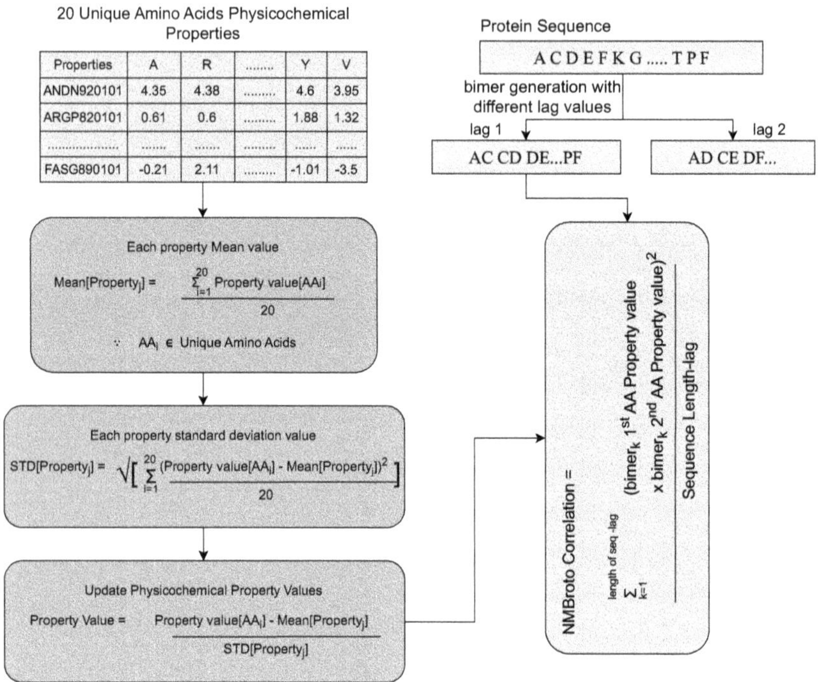

Fig. 6.32 Overall workflow of NMBroto Correlation encoding method

molecules based on physicochemical properties of constituent atoms and their neighbours. It describes how a specific property of atoms is distributed along the molecule's topological structure. The NMBroto descriptor can be adapted for protein sequences based on the constituent amino acid residues and their standardized AAindex properties. With adoption of lag that represents amino acid residues at a distance from each other, it can also capture long range topological structure of sequences. The NMBroto descriptor has been utilized in prediction of protein helix content [437], identification of antioxidant proteins [425], prediction of cyclin proteins [963] and lipocalin proteins [1253] and identification of acetylation sites in proteins [1172]. Figure 6.32 graphically illustrates complete working paradigm of NMBroto encoder.

Unlike Moran and Geary, Moreau-Broto auto-correlation is computed based on amino acids properties values and their lags in a protein sequence as shown in Eq. 6.88.

$$MBj, lag = \sum_{i=1}^{N-lag} P_{i,j} \times P_{i+lag,j} \tag{6.88}$$

The Normalized Moreau-Broto auto-correlation coefficient can be mathematically defined as illustrated in Eq. 6.89.

$$NMB_{j,lag} = \frac{MB_{j,lag}}{N - lag} \tag{6.89}$$

The parameters used in Eqs. 6.88 and 6.89 are already explained in previous subsections. It is important to mention that Moreau-Broto auto-correlation investigates a different aspect of auto-correlation. Considering, covariances of amino acid property values or taking their squared differences lack to capture non-linear relations among protein sequences, Moreau-Broto utilizes property values of amino acid and its lag version tries to capture this information.

6.15 Covariance Based Encoding Methods

6.15.1 Auto-Covariance (AC)

Auto-Covariance (AC) is a statistical measure given by Wold et al. [1098] in 1993 with an aim to transform protein sequences with variable lengths and compositions into the standardized and uniform matrices. These matrices have consistent dimensions and properties across different protein sequences. AC takes average interaction of amino acids which are separated with a lag distance throughout the complete sequence. AC considers seven different physicochemical properties of amino acids including; hydrophilicity, hydrophobicity, net charge index (NCI) of side chains of amino acids, polarizability, polarity, volumes of side chains of amino acids and solvent-accessible surface area (SASA) to effectively characterize their different interactions. AC descriptor has been used in different genomics and proteomics sequence analysis applications including protein-protein interaction prediction [373, 375, 1208], cancer treatment [788] and N7-methylguanosine sites prediction [1251].

Statistical representation of a protein sequence of length N using AC is computed using Eq. 6.90.

$$\begin{aligned} AC_{j,lag} = \frac{1}{N - lag} \sum_{i=1}^{N-lag} (P_{i,j} - \frac{1}{N} \times \sum_{i=1}^{N} P_{i,j}) \times \\ (P_{i+lag,j} - \frac{1}{N} \times \sum_{i=1}^{N} P_{i,j}) \end{aligned} \tag{6.90}$$

Here, lag (1,2,...m) is the distance among amino acids, $P_{i,j}$ represents jth physicochemical property value of ith amino acid and $P_{i+lag,j}$ denotes jth physicochemical property value of amino acid present at i+lag position. The dimensions of AC descriptors are equal to the value obtained by multiplying the maximum possible lag value with the number of properties accounted ($maximum\_lag \times numberofproperties$).

6.15.2 Cross-Covariance (CC)

Cross covariance was initially proposed by Biagio et al. [868] to effectively analyze different segments of images. Later on researchers adapted it to characterize protein sequences for various genomics and proteomics sequence analysis tasks including; identification of DNA-Binding proteins [654], protein-protein interaction [375] and response prediction to antiviral therapy in humans [63]. Cross covariance generates statistical representation of a protein sequence by measuring the correlation of different properties among amino acids separated with lag distance along the sequence. For a given protein sequence of length N, fixed length statistical vector $\mathbf{P}'$ can be formulated using Eq. 6.91.

$$\mathbf{P}' = \left[\psi_1 \ldots \psi_{X*(X-1)*lag}\right] \tag{6.91}$$

In Eq. 6.91, X represents number of properties of amino acids and ψ_u can be computed using Eq. 6.92.

$$\begin{aligned} \psi_u &= CC\,(\mathrm{j_m}, \mathrm{j_n}, lag) \\ &= \sum_{i=1}^{N-lag} \frac{\left(P_{i,j_m} - \overline{P_{i_m}}\right)\left(P_{i+lag,j_n} - \overline{P_{i_n}}\right)}{(N-lag)} \end{aligned} \tag{6.92}$$

The parameters of Eq. 6.92 are quite similar to auto-correlation and auto-covariance based methods and have already been explained in previous subsections. Few different parameters such as j_m, j_n represent two different physicochemical properties. Moreover, P_{i_m} and P_{i_n} are mean value of amino acids for two different physicochemical properties m and n, respectively.

6.15.3 Auto-Cross-Covariance (ACC)

Unlike cross covariance, ACC transforms protein sequences of different lengths into fixed-length vectors by computing correlation of same property among amino acids separated with a lag distance. ACC has been used to predict DNA-Binding proteins [654], protein-protein interaction [375] chromatography retention behavior of oligonucleotides [1222], antigenic drift in H3 influenza A viruses in Swine [1185] and complex traits in maize [65]. For a given protein sequence of length N, fixed length statistical vector $\mathbf{P}'$ can be formulated using Eq. 6.93.

$$\mathbf{P}' = \left[\psi_1 \ldots \psi_{X \times lag}\right] \tag{6.93}$$

Here, X represents number of properties of amino acids and ψ_u can be computed using Eq. 6.94.

$$\begin{aligned} \psi_u &= ACC\,(j, lag) \\ &= \sum_{i=1}^{N-lag} \frac{\left(P_{i,j} - \overline{P_i}\right)\left(P_{i+lag,j} - \overline{P_i}\right)}{(N - lag)} \end{aligned} \tag{6.94}$$

6.16 Special Gap Based Encoding Methods

As described in previous sections, diverse types of gap based encoding methods have been proposed. Primarily, main motivation behind these methods is to generate gap based bimers and compute their occurrence frequencies. Following the success of gap based encoding methods, researchers have proposed special type of gap based encoding methods that generate k-mers with diverse types of gaps. Furthermore, these methods can be utilized for DNA, RNA or protein sequence analysis tasks. Following subsections briefly describe working paradigm of these encoding methods.

6.16.1 MonoMonoKGap

This encoder has been utilized in different types of protein and DNA/RNA sequence analysis tasks such as presynaptic and postsynaptic neurotoxins identification [1241], lncRNA-protein interaction identification [1235], Parkinson's disease identification [416] and anticancer peptides prediction [1230]. The MonoMonoKGap is a variant of CKSNAP encoder that generates statistical representation of raw sequences by capturing gap based distribution of nucleotides or amino acids. Working paradigm of this encoder can be summarized in 4 different steps. First, this encoder generates m^2 bimers by taking all possible pairs of basic nucleotides or amino acids where m denotes basic nucleotides (4) or amino acids (20). Next, it segregates raw sequences into kmers by iteratively sliding (n+2) size window over the whole sequence. Here, n denotes gap value, if the gap value is 3 then sequences are segregated into 3 different sizes k-mers. Specifically, first it segregate sequences into 3 size kmers by taking n = 1, then it generates 4 size kmers by taking n = 2, and lastly it generates 5 size kmers by taking n = 3. Mainly, this encoder generates different size kmers from n = 1 up to given value of gap with an increment of 1. To briefly understand k-mer generation process, consider a generic sequence $A_0A_1A_2A_3A_4.........A_l$, where A_l can be any of nucleotides or amino acids, different types of kmers for gap value n are shown in Eq. 6.95.

$$
\begin{cases}
A_0A_1A_2,\ A_1A_2A_3,\ A_2A_3A_4,\ ...,\ A_{l-2}A_{l-1}A_l & gap=1 \\
A_0A_1A_2A_3,\ A_1A_2A_3A_4,\ ...,\ A_{l-3}A_{l-2}A_{l-1}A_l & gap=2 \\
A_0A_1A_2A_3A_4,\ A_1A_2A_3A_4A_5,\ ...,\ A_{l-4}A_{l-3}A_{l-2}A_{l-1}A_l & gap=3 \\
\vdots & \vdots \\
A_0A_1A_2..A_{n+1},\ A_1A_2..A_{n+2},\ A_2A_3..A_{n+3},\ ....,\ A_{l-(n+1)}..A_{l-2}A_{l-1}A_l & gap=n
\end{cases}
\tag{6.95}
$$

After generating kmers, next it computes occurrence frequencies of unique bimers. To accomplish this, encoder first maps generated kmers into bimers by taking their first and last letters. Figure 6.33 illustrates working paradigm of MonoMonoKGap encoder with a toy sequence AGTATC and gap value 2. It can be seen in Fig. 6.33, encoder generates both 3 size and 4 size kmers. Then it maps generated kmers on bimers and computes bimers occurrence frequencies. Let's for kmer AGT, it counts all 3 size kmers that have A as first and T as last letter with gap value 1 between them (A_T) and maps the count on feature space of bimers. Similarly, if kmer is AGTA, this encoder computes occurrence frequency of 4-size kmers that begin and

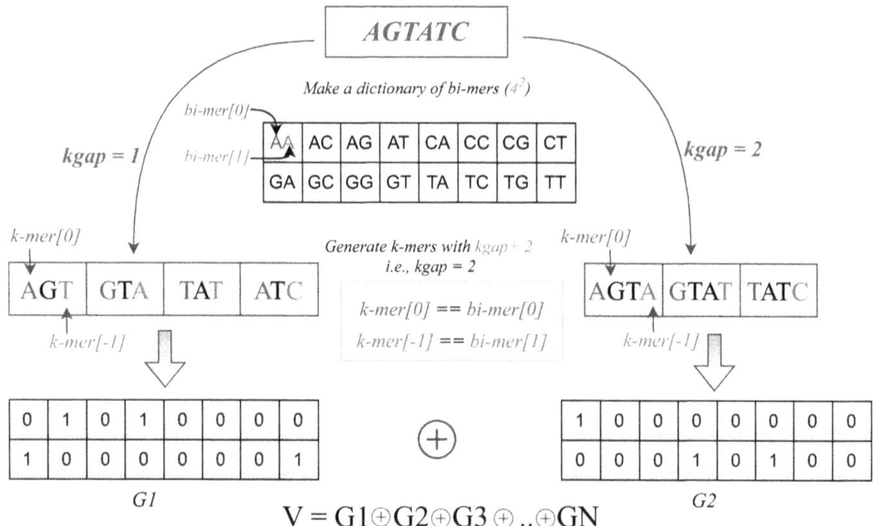

Fig. 6.33 Working of MonoMonoKGap encoder

end with A having gap value 2 to map on bimers(A__A). In this way, two vectors are generated and concatenated to create a $m^2 \times m$ dimensional statistical vector.

To more briefly understand the working paradigm of this encoder for DNA sequences, consider a toy sequence "AGTATC" as illustrated in Fig. 6.33. First of all, it generates a dictionary of all possible pairs of four nucleotides (A, C, G, T) that can be formulated as [AA, AC, AG, AT, CA, CC, CG, CT, GA, GC, GG, GT, TA, TC, TG, TT]. In the next step, it selects gap value n = 2 and segregates raw sequence into k-mers by iteratively sliding a n+2 window. Particularly, it generates two sets of k-mers, each with a different size, by considering gap values from 1 to 2. Specifically, for n = 1, it generates 3-mers by taking n value 1 and sliding (1+2) 3 size window which are AGT, GTA, TAT, and ATC. Then, it generates 4-mers as AGTA, GTAT, TATC by sliding 4 size window (n = 2) over the raw sequence. In the next step, it maps these k-mers on bi-mers and generates statistical vectors for all three size k-mers. For this, it takes the initial and last letter of k-mers for each size iteratively and maps their count on bi-mers dictionary. It first takes generated 3-mers (AGT, GTA, TAT, ATC) and maps them on bimers by taking initial and last letter of 3-mers. For first 3-mer "AGT", it counts all 3-mers that have A as first and C as last letter with gap value 1 between them (A_T) and map count 1 at AT in feature space of bi-mers. Similarly, for GTA it computes the occurrence frequency of 3-mers that have G as first letter and A as last letter (G_A). It means that it considers AGT as AT and GTA as GA to map on feature space of bi-mers. Similarly, for TAT and ATC, it maps 1 at TT and AC in the feature space because their occurrence frequency is 1. The statistical vector for 3-mers can be formulated as $\{0, 1, 0, 1, 0, 0, 0, 0, 1, 0, 0, 0, 0, 0, 0, 1\}$. After generating statistical vectors for 3-mers, it moves to 4-mer and for first 4-mer "AGTA", this encoder computes occurrence frequency of 4-mers that have A at the beginning and A at the end with gap value 2 to map on bimers(A__A) in feature space. It repeats the process for other 4-mers and statistical vector for 4-mers can be derived as $\{1, 0, 0, 0, 0, 0, 0, 0, 0, 0, 0, 1, 0, 1, 0, 0\}$. The last step concatenates all three vectors and generates a 3×4^2 vector representation of given toy sequence as $\{0, 1, 0, 1, 0, 0, 0, 0, 1, 0, 0, 0, 0, 0, 0, 1\} \oplus \{1, 0, 0, 0, 0, 0, 0, 0, 0, 0, 0, 1, 0, 1, 0, 0\}$. For protein sequences, the only variation lies in the length of generated dictionary. With 20 unique amino acids (A, C, D, E, F, G, H, I, K, L, M, N, P, Q, R, S, T, V, W, Y), a dictionary of 20^2 dimensions is generated, such as $\{AA, AC, AD, AE, ...YT, YV, YW, YY\}$. Similarly, for a fixed gap value, it generates k-mers, computes their occurrence frequencies, and maps them on the feature space of bi-mers. In this way, a $n \times 20^2$ dimensional statistical vector is generated for protein sequence, where n refers to gap value.

6.16.2 MonoDiKGap

This encoder has been utilized in different types of protein and DNA/RNA sequence analysis tasks such as lncRNA-protein interaction prediction [1235], DNA modification sites identification [1176], Parkinson's disease identification [416]. Working Criterion of MonoDiKGap encoder can be summarized in 4 different steps. First, this encoder generates m^3 trimers by taking all possible combinations of basic nucleotides or amino acids where m represents basic nucleotides and amino acids. Secondly, it generates kmers by sliding (n+3) size window iteratively over raw sequence, where n is defined as gap value. This encoder starts working by taking n equal to 1 as and generates 4 size kmers in first iteration. Then, with increment 1 in n, it generates 5 size kmers for second iteration. In this way, it generates different samples of n different size of kmers and each size kmer represents a particular gap value. Equation 6.96 illustrates a generic sequence $A_0A_1A_2A_3A_4......A_l$, where A_l can be any nucleotide or amino acid. It can be seen in Eq. 6.96, for n gap value, n different types of kmer sizes samples are generated.

$$\begin{cases} A_0A_1A_2A_3,\ A_1A_2A_3A_4,\ A_2A_3A_4A_5,\ ...,\ A_{l-3}A_{l-2}A_{l-1}A_l & gap{=}1 \\ A_0A_1A_2A_3A_4,\ A_1A_2A_3A_4A_5,\ ...,\ A_{l-4}A_{l-3}A_{l-2}A_{l-1}A_l & gap{=}2 \\ A_0A_1A_2A_3A_4A_5,\ A_1A_2A_3A_4A_5A_6,\ ...,\ A_{l-5}A_{l-4}A_{l-3}A_{l-2}A_{l-1}A_l & gap{=}3 \\ \vdots & \vdots \\ A_0A_1A_2..A_{n+2},\ A_1A_2..A_{n+3},\ A_2A_3..A_{n+4},\ ....,\ A_{l-(n+2..)}..A_{l-2}A_{l-1}A_l & gap{=}n \end{cases} \tag{6.96}$$

After generating kmers, next it computes occurrence frequencies of m^3 trimers. In the next step, all generated kmers are mapped to trimers by taking only one initial and last two letters of generated kmers. Let's if gap value is 2, two different size kmer samples are generated, one is 4-size kmers and the other is 5-size kmer. First, generate a feature vector of 4-size kmers by mapping on trimers and if generated 4-size kmer is ATCG, this encoder computes occurrence frequency of 4-size kmers that have A as initial letter and CG at the end with one gap value between them. After 4-mers, it counts occurrence frequency of all 5 size kmers and maps it on dictionary of trimers. In the last step, it concatenates both feature vectors and generates a final statistical vector of $m^3 \times n$ dimensions.

6.16.3 MonoTriKGap

This encoder has been utilized in different types of protein and DNA/RNA sequence analysis tasks such as Pre-miRNA identification [1173], Parkinson's disease identification [416], lncRNA-protein interaction identification [1235] and Immunoglobulin prediction [353]. Working paradigm of MonoTriKGap encoder consists of 4 different steps. First it generates m^4 4mers by taking all possible pairs of basic nucleotides or amino acids (m indicates basic nucleotides or amino acids). Next, it segregates raw sequence into kmers by sliding (n+4) size window for n times. Here, n is the gap value that ranges between 1 to n, which changes with an increment of 1 for each iteration. Then, it generates 5 size kmers with initial gap value of 1 and iteratively generates samples with different size kmers for up to n gap values. Let's if gap value is set as n for raw sequence $A_0A_1A_2A_3A_4......A_l$, the encoder generates different samples with n kmers types as shown in Eq. 6.97.

$$\begin{cases} A_0A_1A_2A_3A_4,\ A_1A_2A_3A_4A_5,\ A_2A_3A_4A_5A_6,\ ...,\ A_{l-4}A_{l-3}A_{l-2}A_{l-1}A_l & gap = 1 \\ A_0A_1A_2A_3A_4A_5,\ A_1A_2A_3A_4A_5A_6,\ ...,\ A_{l-5}A_{l-4}A_{l-3}A_{l-2}A_{l-1}A_l & gap = 2 \\ A_0A_1A_2A_3A_4A_5A_6,\ A_1A_2A_3A_4A_5A_6A_7,\ ...,\ A_{l-6},\ A_{l-5}A_{l-4}A_{l-3}A_{l-2}A_{l-1}A_l & gap = 3 \\ . & . \\ . & . \\ . & . \\ A_0A_1A_2..A_{n+3},\ A_1A_2...A_{n+4},\ A_2A_3...A_{n+5},\ ...,\ A_{l-(n+3)}..A_{l-2}A_{l-1}A_l & gap = n \end{cases} \tag{6.97}$$

Next, it counts occurrence frequencies of m^4 4mers for n types of kmers. To achieve this, it maps generated kmers on 4mers by considering one letter from beginning and three letters from end with n gaps between them. Let's if gap value is 2 as illustrated in Fig. 6.34, two types of kmer samples are created namely: 5mers and 6mers. The encoder maps all unique samples of generated 5mers on 4mers by using X_XXX structure and generates a feature vector. Similarly, this encoder generates a feature space of 6mers by mapping them on 4mers. Both feature vectors are concatenated to generate a $m^4 \times n$ dimensional statistical representation.

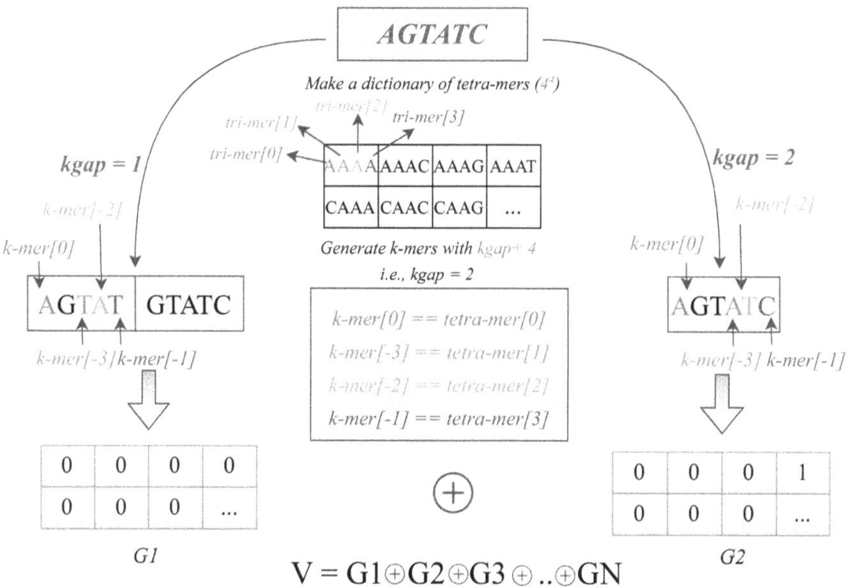

Fig. 6.34 Working of MonoTriKGap encoder

6.16.4 DiMonoKGap

This encoder has been utilized in different types of protein and DNA/RNA sequence analysis tasks such as lncRNA-protein interaction identification [1235] and Parkinson's disease identification [416]. DiMonoKGap encoder, first, generates a unique vocabulary of m^3 trimers by taking basic nucleotides or amino acids, where m is the number of basic amino acids or nucleotides. Then, it manually assigns a gap value to generate different samples with n kmer types by iteratively sliding n+3 size window over raw sequence. Here, n denotes gap value and lets if n is set as 3, encoder generates samples of 3 different kmers types namely: 4mers, 5mers and 6mers corresponding to n vlaue 1, 2 and 3, respectively. For a generic DNA/RNA or amino acid sequence, $A_0A_1A_2A_3A_4......A_l$ and gap is set as n, following mathematical expression 6.98 illustrates different types of kmer samples for n gap values.

$$\begin{cases} A_0A_1A_2A_3,\ A_1A_2A_3A_4,\ A_2A_3A_4A_5,\ ...,\ A_{l-3}A_{l-2}A_{l-1}A_l & gap=1 \\ A_0A_1A_2A_3A_4,\ A_1A_2A_3A_4A_5,\ ...,\ A_{l-4}A_{l-3}A_{l-2}A_{l-1}A_l & gap=2 \\ A_0A_1A_2A_3A_4A_5,\ A_1A_2A_3A_4A_5A_6,\ ...,\ A_{l-5}A_{l-4}A_{l-3}A_{l-2}A_{l-1}A_l & gap=3 \\ . & . \\ . & . \\ . & . \\ A_0A_1A_2..A_{n+2},\ A_1A_2...A_{n+3},\ A_2A_3...A_{n+4},\ ...,\ A_{l-(n+2)}..A_{l-2}A_{l-1}A_l & gap=n \end{cases} \tag{6.98}$$

It can be seen in Fig. 6.35 after generating different size kmers, it maps each size kmers on unique vocabulary of trimers, one-by-one. To accomplish this, it computes frequencies of kmers by taking two initial letters and one ending letter from kmers. It computes frequencies of all n types kmer samples and maps these values on unique vocabulary of trimers. In this way, n feature vectors are generated and this encoder concatenates both feature vectors as $G_1 \oplus G_2$. $m^3 \times n$ dimension vector is generated by DiMonoKGap encoder.

$$num(DiMono) = (4) \times (4 \times 4) \times n = 4^3 \times n \tag{6.99}$$

6.16.5 DiDiKGap

This encoder has been utilized in different types of protein and DNA/RNA sequence analysis tasks such as lncRNA-protein interaction identification [1235] and Parkinson's disease identification [416]. Figure 6.36 graphically illustrates the working paradigm of DiDIKGap encoder. DiDiKGap encoder works similar to TriMonoKGap encoder by generating a unique vocabulary of m^4 4mers and sliding (n+3) size window over raw sequence, iteratively. The sequence is segregated into n

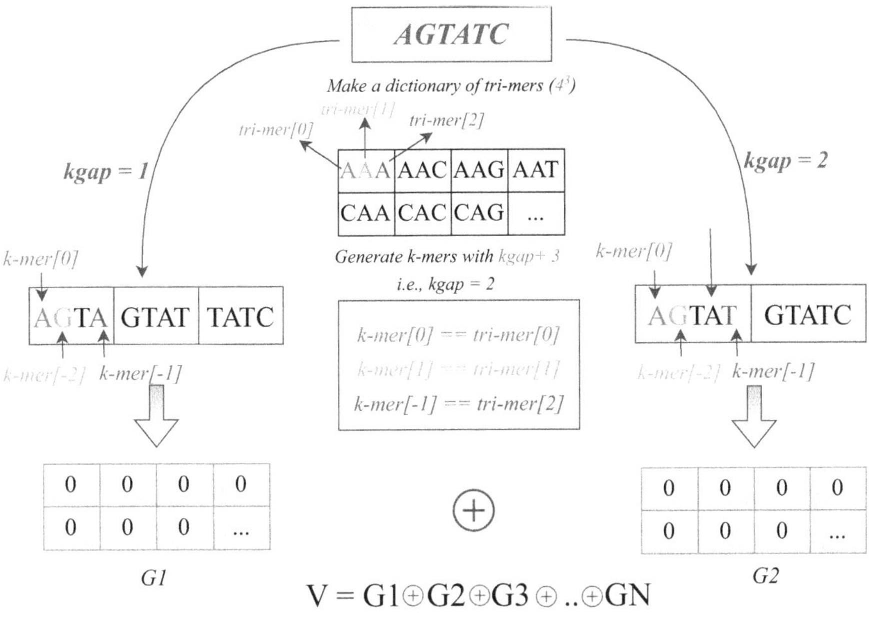

Fig. 6.35 Working of DiMonoKGap encoder

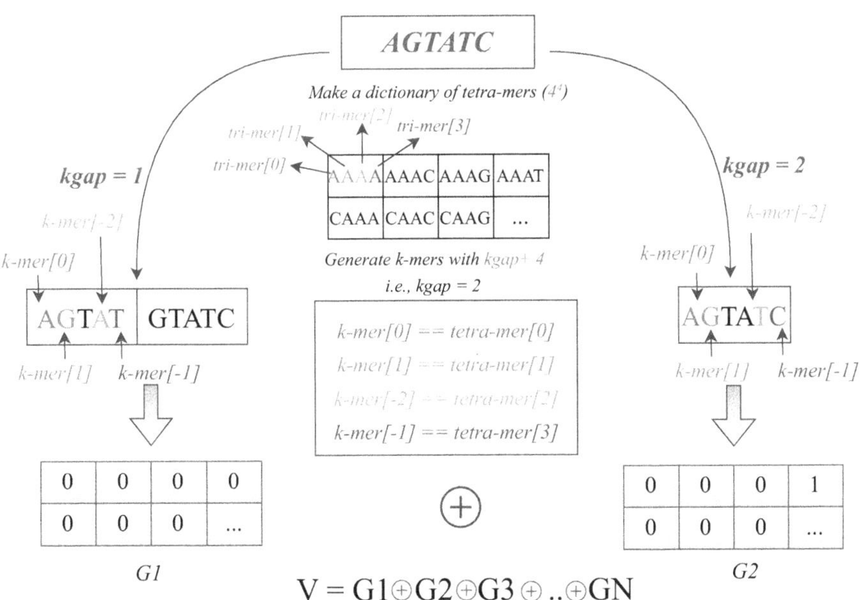

Fig. 6.36 Working of DiDiKGap encoder

different size kmer samples. Let's assume a sequence $A_0A_1A_2A_3A_4......A_l$, where A_l can be any nucleotide or amino acid. Then, it generates n different type of kmers samples as shown in following mathematical expression 6.100.

$$\begin{cases} A_0A_1A_2A_3A_4,\ A_1A_2A_3A_4A_5,\ A_2A_3A_4A_5A_6,\ ...,\ A_{l-4}A_{l-3}A_{l-2}A_{l-1}A_l & gap = 1 \\ A_0A_1A_2A_3A_4A_5,\ A_1A_2A_3A_4A_5A_6,\ ...,\ A_{l-5}A_{l-4}A_{l-3}A_{l-2}A_{l-1}A_l & gap = 2 \\ A_0A_1A_2A_3A_4A_5A_6,\ A_1A_2A_3A_4A_5A_6A_7,\ ...,\ A_{l-6},\ A_{l-5}A_{l-4}A_{l-3}A_{l-2}A_{l-1}A_l & gap = 3 \\ . & . \\ . & . \\ . & . \\ A_0A_1A_2..A_{n+3},\ A_1...A_{n+4},\ A_2...A_{n+5},\ ...,\ A_{l-(n+3)}..A_{l-2}A_{l-1}A_l & gap = n \end{cases} \tag{6.100}$$

After generating kmers, this encoder computes occurrence frequencies of different types of kmers to map it on feature space of 4mers. It computes frequencies by taking two letters from beginning and two from the end. It generates n feature vectors and concatenates for a final statistical representation of $m^4 \times n$ dimensions.

6.16.6 DiTriKGap

This encoder has been utilized in different types of protein and DNA/RNA sequence analysis tasks such as lncRNA-protein interaction identification [1235], Parkinson's disease identification [416]. DiTriKGap encoder generates a unique vocabulary of m^5 5mers by selecting four unique nucleotides and amino acids. Raw sequence is segregated into kmer by sliding window of size n+5 for n times, Here n denotes gap value and for a range of 1 to n, n different size kmer samples are generated. Equation 6.101 describes the generation of different types of kmers for sequence $A_0A_1A_2A_3.....A_l$, where A_l can be any nucleotide or amino acid.

$$\begin{cases} A_0A_1A_2A_3A_4A_5,\ A_1A_2A_3A_4A_5A_6,\ A_2A_3A_4A_5A_6A_7,\ ...,\ A_{l-5}A_{l-4}A_{l-3}A_{l-2}A_{l-1}A_l & gap = 1 \\ A_0A_1A_2A_3A_4A_5A_6,\ A_1A_2A_3A_4A_5A_6A_7,\ ...,\ A_{l-7}A_{l-6},\ A_{l-5}A_{l-4}A_{l-3}A_{l-2}A_{l-1}A_l & gap = 2 \\ A_0A_1A_2A_3A_4A_5A_6A_7,\ A_1A_2A_3A_4A_5A_6A_7A_8,\ ...,\ A_{l-7}A_{l-6}A_{l-5}A_{l-4}A_{l-3}A_{l-2}A_{l-1}A_l & gap = 3 \\ . & . \\ . & . \\ . & . \\ A_0A_1A_2..A_{n+4},\ A_1...A_{n+5},\ A_2...A_{n+6},\ ...,\ A_{l-(n+4)}..A_{l-2}A_{l-1}A_l & gap = n \end{cases} \tag{6.101}$$

Figure 6.37 illustrates an example for generating dictionary of m^5 unique 5mers and kmers with different gap values. To generate a statistical vector, it maps different size kmers on 5mers by taking two initial and three letters from the end of sub-sequences. Let's if gap is set as 2, 6mers and 7mers are two different types of generated kmers. It computes the frequencies of 6mers to generate a statistical vector. If the generated 6mer is ATCGTA, the encoder computes occurrence frequencies

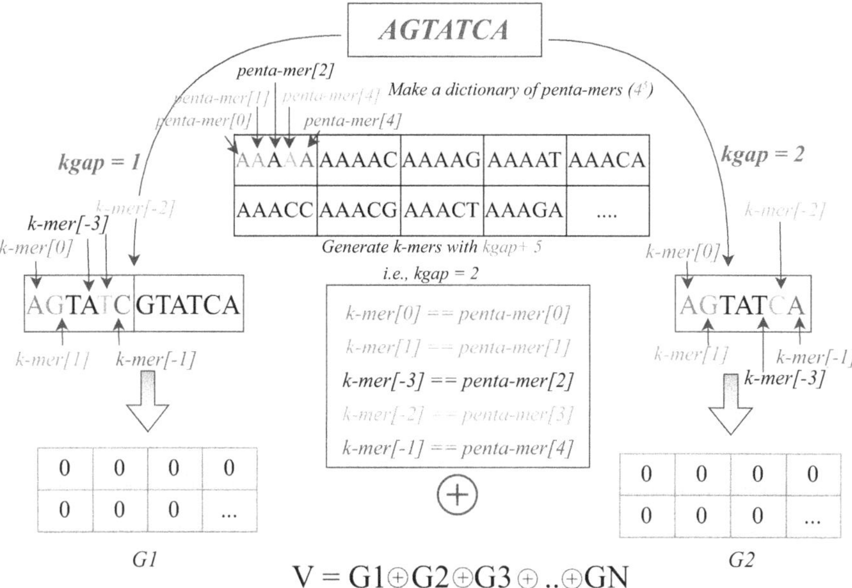

Fig. 6.37 Working of DiTriKGap encoder

of 6mers with AT_GTA structure and updates ATGTA value in feature space. It generates statistical vector for 7mers in similar manner. Finally, it concatenates feature spaces of 6mers and 7mers, to generate $m^5 \times n$ dimensional feature vector.

6.16.7 *TriMonoKGap*

This encoder has been utilized in different types of protein and DNA/RNA sequence analysis tasks such as lncRNA-protein interaction identification [1235] and Parkinson's disease identification [416]. TriMonoKGap follows footsteps of MonoTriKGap and DiDiKGap encoders. Equation 6.102 represents all possible size kmers that can be generated by sliding n+4 size window over raw sequence $A_0A_1A_2A_3....A_l$, A_l for n different gap values.

$$\begin{cases} A_0A_1A_2A_3A_4,\ A_1A_2A_3A_4A_5,\ A_2A_3A_4A_5A_6,\ ...,\ A_{l-4}A_{l-3}A_{l-2}A_{l-1}A_l & gap = 1 \\ A_0A_1A_2A_3A_4A_5,\ A_1A_2A_3A_4A_5A_6,\ ...,\ A_{l-5}A_{l-4}A_{l-3}A_{l-2}A_{l-1}A_l & gap = 2 \\ A_0A_1A_2A_3A_4A_5A_6,\ A_1A_2A_3A_4A_5A_6A_7,\ ...,\ A_{l-6},\ A_{l-5}A_{l-4}A_{l-3}A_{l-2}A_{l-1}A_l & gap = 3 \\ . & . \\ . & . \\ . & . \\ A_0A_1A_2..A_{n+3},\ A_1A_2...A_{n+4},\ A_2A_3...A_{n+5},\ ...,\ A_{l-(n+3)}..A_{l-2}A_{l-1}A_l & gap = n \end{cases} \tag{6.102}$$

The only difference in MonoTri, DiDi, and TriMono Kgap is the characteristic structure of kmers that are mapped on vocabulary of m^4 unique 4mers. TriMonoKGap encoder computes the frequencies of 4mers by mapping different size kmers. For each size kmer, it selects three initial and one last letter with different gap values. With n = 1, characteristic structure looks like XXX_X, XXX__X with n = 2, and so on. After computing frequencies and generating statistical representation with distinct kmer size samples, all feature vectors are concatenated to generate a $m^4 \times n$ dimension feature vector.

6.16.8 TriDiKGap

This encoder has been utilized in different types of protein and DNA/RNA sequence analysis tasks such as lncRNA-protein interaction identification [1235] and Parkinson's disease identification [416]. Working criterion of TriDiKGap consists of 4 different steps. First, it generates a unique dictionary of m^5 5mers by taking all basic amino acids and nucleotides. In the next step, it segregates raw sequence into different size kmer samples by sliding n+5 size window. Equation 6.103 illustrates different n size kmer samples that can be generated by using n gap value for the given sequence in $A_0A_1A_2A_3A_4.....A_l$, where A_l represents any of basic nucleotides and amino acids.

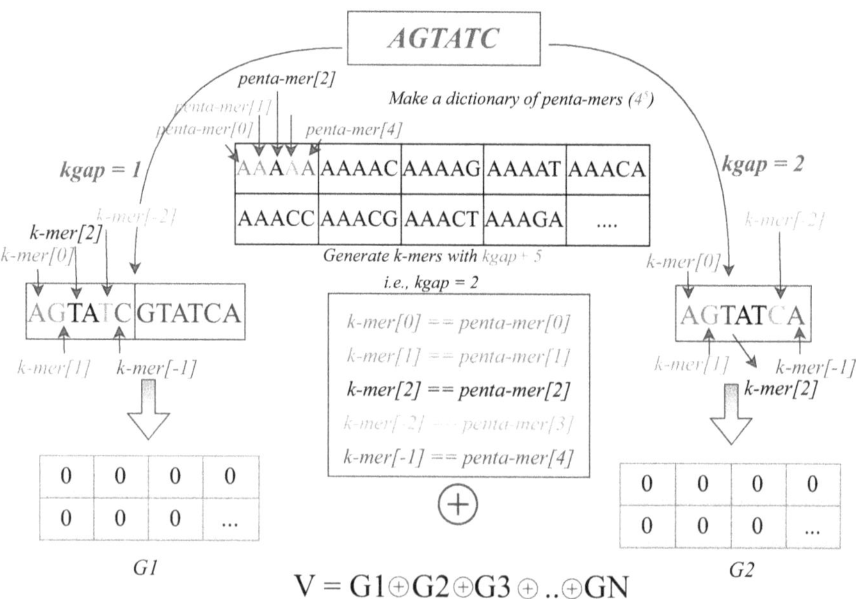

Fig. 6.38 Working of TriDiKGap encoder

$$\begin{cases} A_0A_1A_2A_3A_4A_5,\ A_1A_2A_3A_4A_5A_6,\ A_2A_3A_4A_5A_6A_7,\ ....,\ A_{l-5}A_{l-4}A_{l-3}A_{l-2}A_{l-1}A_l & gap = 1 \\ A_0A_1A_2A_3A_4A_5A_6,\ A_1A_2A_3A_4A_5A_6A_7,\ ....,\ A_{l-7}A_{l-6},\ A_{l-5}A_{l-4}A_{l-3}A_{l-2}A_{l-1}A_l & gap = 2 \\ A_0A_1A_2A_3A_4A_5A_6A_7,\ A_1A_2A_3A_4A_5A_6A_7A_8,\ ....,\ A_{l-7}A_{l-6}A_{l-5}A_{l-4}A_{l-3}A_{l-2}A_{l-1}A_l & gap = 3 \\ \vdots & \vdots \\ A_0A_1A_2..A_{n+4},\ A_1...A_{n+5},\ A_2...A_{n+6},\ ....,\ A_{l-(n+4)}..A_{l-2}A_{l-1}A_l & gap = n \end{cases} \tag{6.103}$$

This encoder computes the frequencies of m^5 unique 5mers to generate a statistical representation of the sequence. It calculates the count of kmers by taking three initial and last two letters of the kmer with n gaps between them, for all n types of kmer based samples. Figure 6.38 illustrates a toy example of a small sequence with gap value 2, this encoder generates two different types of kmers namely; 6mers and 7mers. Next it generates feature vector of each kmers types by computing frequencies of 6mers and 7mers with characteristic structures XXX_XX and XXX__XX, respectively. These two feature spaces are concatenated to generate final statistical representation of $m^5 \times 2$ dimension for gap equal to 2. In this way, a $m^5 \times n$ dimension feature vector is generated by TriDiKGap encoder.

Chapter 7
CRISPR System and AI Applications

7.1 Introduction

According to the World Health Organization (WHO), more than 10,000 diseases have emerged with unique characteristics, causes, and symptoms [207]. These diseases can be placed into four categories namely, infectious, non-communicable, genetic, and others [114, 430, 527]. Infectious diseases are caused by microorganisms such as bacteria, viruses, or parasites and can spread directly or indirectly from one person to another [114]. Genetic diseases arise from mutations or alterations in an individual's DNA. These mutations can be inherited from parents or occur spontaneously [527]. Non-communicable diseases arise from genetic, physiological, environmental, and behavioral factors [430]. These chronic conditions develop gradually and are not typically passed from person to person [430]. Additionally, other diseases include various conditions arising from different mechanisms, such as injuries caused by external factors and congenital anomalies resulting from developmental malformations.

Among the four major disease categories, infectious and non-communicable diseases have a wider range of treatment possibilities. These options encompass traditional medications and cutting-edge therapies, including Proteolysis Targeting Chimeras (PROTACs) and RNA-based approaches [1068]. However, these treatments are often ineffective for genetic diseases such as cancers, autoimmune disorders, hereditary conditions, and nervous system disorders [1018]. This is because these diseases involve complex genetic mutations and pathways which are not easily targeted by conventional or even some modern therapies. Additionally, the variability in genetic makeup among individuals can result in diverse responses to treatment, making it difficult to develop universally effective therapies.

Following the need of a more effective treatment for genetic diseases, Doudna et al. [75, 269] proposed a unique system for DNA sequence editing named Clustered Regularly Interspaced Short Palindromic Repeats (CRISPR). CRISPR offers multiple advantages including, speed, flexibility, cost-effectiveness, and the

M. Nabeel Asim et al., *Artificial Intelligence for Molecular Biology*,
https://doi.org/10.1007/978-3-031-90450-9_7

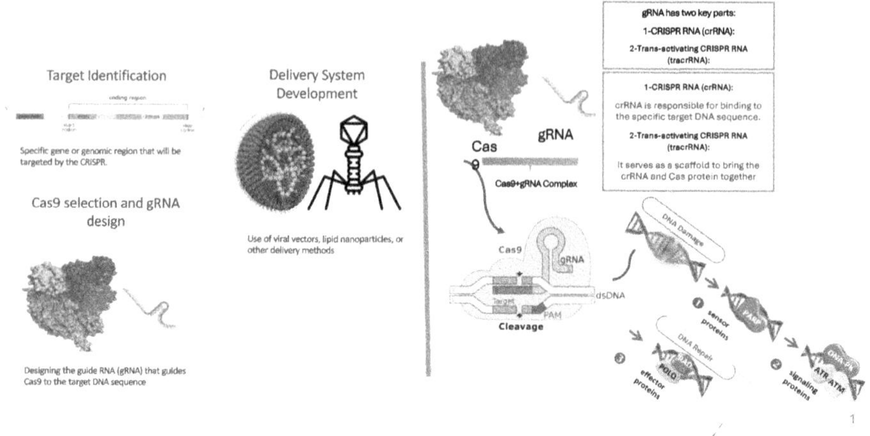

Fig. 7.1 Adaptation of CRISPR system for DNA editing. In the very first step, target regions are identified with the selection of an appropriate CRISPR system. Afterward, CRISPR, trans-activating, and gRNA are designed to help Cas proteins cleave at the desired site. The complete complex is then delivered inside the cell with the help of a vector (virus)

capacity to manipulate multiple genomic locations simultaneously [25, 460]. The effectiveness of CRISPR system has been proven by various clinical studies related to diseases such as inherited eye disease Leber congenital amaurosis (LCA), Duchenne muscular dystrophy (DMD), and genetic lung and liver diseases [912]. In addition, the CRISPR system has been approved as the first gene therapy for Hemoglobinopathies (sickle cell disease) [242].

Figure 7.1 graphically represents a generic muti-step process of CRISPR for the treatment of genetic diseases [1103]. The multi-step process initiates with the identification of problematic regions in the DNA such as disease-related genes identification [633]. Following characteristics of desired cleavage regions, the next step is to design a CRISPR system which includes deep analysis of available diverse types of Cas Proteins, and design of guide RNA. gRNA is itself made up of two different parts i.e., CRISPR RNA, guides the Cas protein to the specific location in the genome where the cut is to be made, and trans-activating RNA which is necessary for binding of crRNA and the Cas protein, helping to form the active CRISPR-Cas9 complex [725]. The active CRISPR complex is sent into living organisms through a delivery system such as viral mechanisms or lipid nanoparticles [633]. In the CRISPR complex, guide RNA contains instructions about cleavage regions and Cas proteins cleave those regions. Sometimes this complex does not make cleavage at desired locations due to multiple reasons such as the weak design of guide RNA, off-target effects, poor delivery efficiency, chromatin accessibility, and cellular repair mechanisms. To make sure whether the designed complex will make cleavage at desired locations or not, researchers perform a deep analysis of this complex with the possibility of making cleavage at desired locations or wrong locations. Furthermore, if the CRISPR complex initiates cleavage at incorrect positions, a specialized process can be employed to halt it by introducing anti-CRISPR (acr) proteins and anti-CRISPR-associated proteins [1170], which inhibit the CRISPR complex from making further cuts in the DNA. Finally, after DNA is cleaved, it has natural processes to repair the cleaved DNA with two different types of DNA repair mechanisms i.e., Non-Homologous End Joining (NHEJ), and Homology-Directed Repair (HDR) [744]. After, this rebuilding process there is a need to perform a genetic analysis of mutations. In a nutshell, this whole process contains 10 distinct types of tasks including CRISPR arrays [249], CRISPR loci [857], CRISPR-Cas systems [790], acr proteins and their activity [589], aca proteins [1158], CRISPR operons [1148, 1164], Cas protein [1194], Off-target activity [143], On-target activity [550], editable target regions [70], and gene editing outcomes [642]. A brief biological foundation of all 10 tasks is given in Sect. 7.5.

Notably, all these tasks are usually performed through wet-lab experiments that are expensive, time-consuming, and error-prone. Following the success of artificial intelligence (AI) in diverse fields and with an aim for transitioning from wet-lab to AI-driven applications for CRISPR-based therapies development, researchers are trying to develop AI-driven applications for all 10 tasks [143, 222, 550, 642, 790, 1164]. Although several AI-driven applications have been developed for CRISPR systems there is still a lot of room for the development of new applications. To accelerate and expedite the development of AI-driven applications for all 10 tasks, apart from the development of task-specific applications, in the last 3 years, 13 review articles have been published [94, 258, 407, 458, 519, 538, 569, 773, 826, 904, 910, 994, 1060]. The primary focus of these articles was to summarise existing AI predictors within the context of CRISPR. The focus of these review articles is often constrained to only a single task of CRISPR and fail to bridge the gap between the broader landscape of CRISPR and AI predictors trends.

With an aim to bridge the gap between both fields and provide a unique platform encompassing biological foundations and AI advancements related to all 10 tasks, the contributions of this manuscript are manifold. (1) First, it equips AI researchers with biological foundations of 10 distinct tasks of CRISPR. (2) It presents details of the existing 80 public datasets related to 10 distinct tasks and provides overview of 10 public CRISPR databases for the development of new datasets. (3) In the context of all 10 tasks, it provides an in-depth analysis of the representation learning methods and classifiers/regressors employed in the existing AI predictors. (4) It discusses experimental settings and evaluation strategies utilized to evaluate existing AI-driven applications across 10 distinct tasks. (5) Finally, it provides performance values of 50 predictive pipelines across 80 public benchmark datasets of 10 distinct tasks. AI researchers can utilize this information to find predictors' architectural details and current state-of-the-art performance values of predictors for each task. Besides this, partial content of this chapter is published in [3].

7.2 Examining CRISPR Tasks Through the Lens of AI Researcher

AI researchers often lack a deep understanding of the biological foundations of CRISPR and generally show little interest in the development of AI-driven CRISPR applications. In addition, the alignment of CRISPR tasks with AI paradigms may require extensive effort for understanding CRISPR tasks background knowledge. However, AI researchers can be facilitated by aligning CRISPR tasks with their familiar AI paradigms—such as binary classification, multi-class classification, or regression. In this context, we aligned 10 different CRISPR-related tasks from an AI perspective with their associated AI paradigms. Figure 7.2 visually illustrates 10 CRISPR tasks alignment with 4 AI paradigms namely binary classification, multi-class classification, regression, and reinforcement learning (RL) based optimization.

Figure 7.3 provides an overview of AI predictive pipeline for all 4 paradigms. This pipeline begins with the creation of novel benchmark datasets from publicly available sources. A high level analysis of Fig. 7.3 reveals that within aforementioned 4 AI-paradighms development of AI-driven CRISPR applications comprises 4 different components. These components include the development or utilization of existing benchmark datasets, transformations of raw sequences into numerical vectors, utilization of classifier or regressor, and evaluation measures. The datasets are usually developed by acquiring sequences and associated information from public databases. Detailed information about commonly used CRISPR databases and existing benchmark datasets can be found in Sects. 7.5 and 7.5.2.10. Sequence data cannot be directly used in ML and DL classifiers or regressors due to their

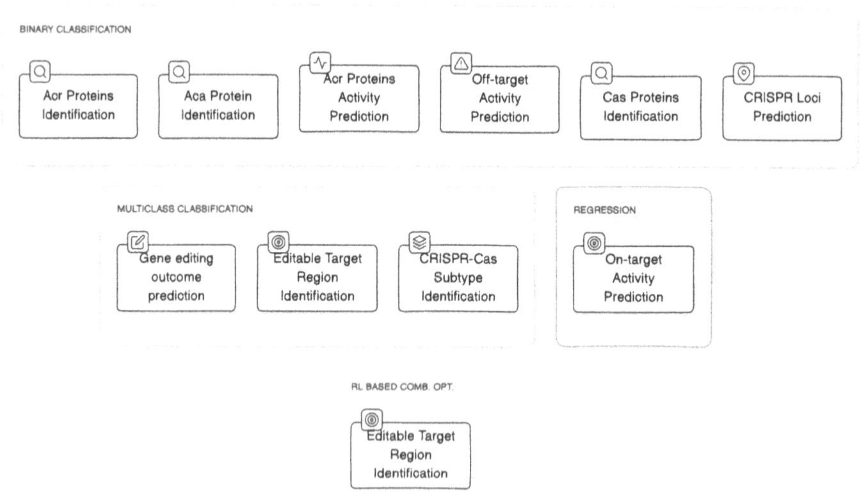

Fig. 7.2 Categorization of CRISPR related applications/tasks

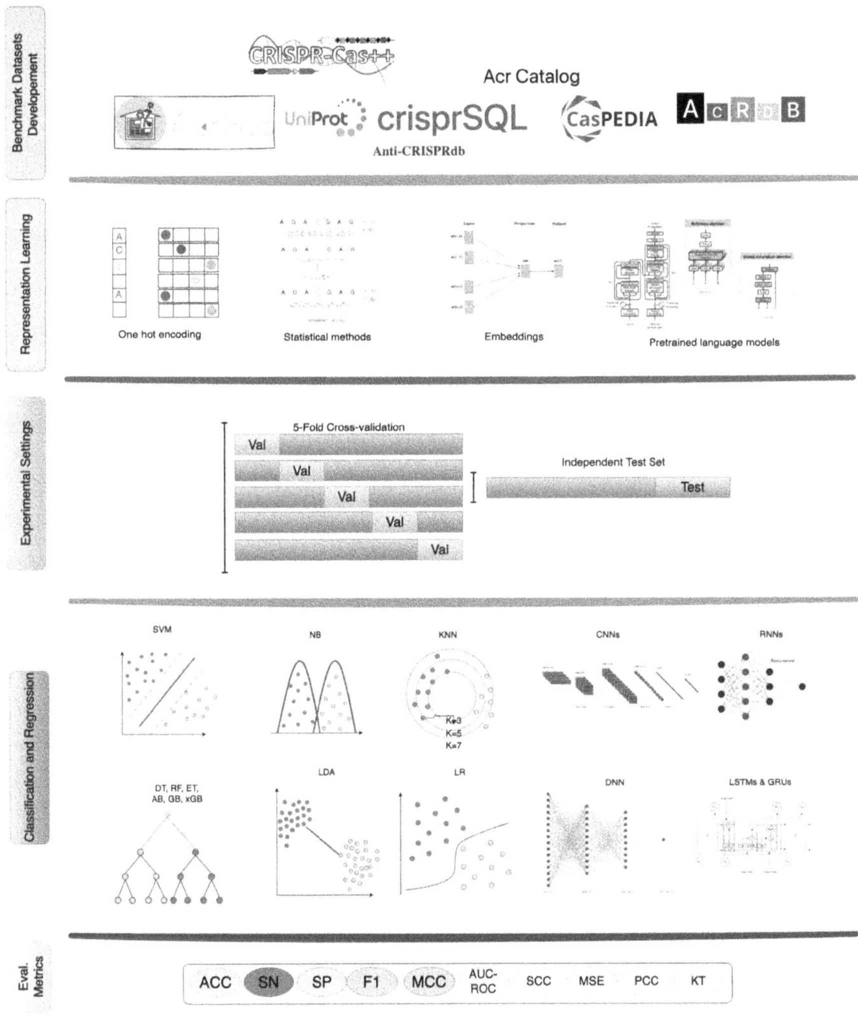

Fig. 7.3 Components of AI predictive pipelines for CRISPR tasks involve several key steps. Initially, datasets are generated from publicly available databases. These sequences are then transformed into numerical vectors using statistical methods, one-hot encoding (OHE), or embedding techniques. The resulting datasets are divided into training and testing sets through cross-validation or independent testing. Subsequently, AI predictors are trained and tested, with evaluation scores calculated to determine their predictive performance

dependence on numerical vectors, representation learning methods are employed to convert sequences into numerical vectors. Section 7.5.3 elaborates distinct types of sequence representation learning methods that are employed in existing AI-driven CRISPPR applications. ML and DL predictors that are utilized in existing AI-driven CRISPR applications are described in Sect. 7.5.4. Finally, in order to assess predictive pipelines, numerous evaluation measures are used which are comprehensively discussed in Sect. 7.5.5.

7.3 A Look into CRISPR and AI Focused Review Studies

To enhance the integration of AI approaches into CRISPR, 13 distinct review articles have been published in the last four years. The primary focus of these articles is to summarise insights of existing AI predictors that have been developed to empower CRISPR. Table 7.1 provides a comparative overview of these reviews, including the number of research articles they covered, their overall scope, and limitations.

A high-level analysis of the scope of 13 existing review articles in Table 7.1 reveals that these articles can be categorized into three distinct groups based on their focus on AI predictors for CRISPR. Two reviews delve exclusively into ML driven CRISPR methods [773, 826], while another two explore DL methods [538, 569]. Notably, the remaining nine reviews covered both ML and DL applications within CRISPR [94, 258, 407, 458, 519, 910, 994, 1060].

Leveraging insights from Fig. 7.5 and the CRISPR summarized in the introduction section, a comprehensive review article should focus on the analysis of developed AI-driven applications for 10 distinct CRISPR tasks. The focused CRISPR tasks are on/off-target activity prediction [143, 550, 1204], CRISPR array [249], loci, and system identification [790], acr and aca prediction [590], acr activity prediction [222], gene editing outcome prediction [152], and CRISPR operons identification [567]. Within this scope, an in-depth analysis of existing review articles reveals that 8 review papers emphasize on the design of CRISPR systems and the on/off-target activity prediction combined with additional topics such as cancer treatment or the usage of nanovectors [94, 407, 519, 538, 773, 910, 994, 1060]. Two review papers incorporate 3 different tasks including acr proteins, gene editing outcomes, Cas9 activity [258, 569]. One review focuses on drug discovery [519] and one combines CRISPR and the development of biosensors [458]. It is noteworthy to mention that, out of all review papers, only Sharma et al. [904] cover 6 distinct tasks related to CRISPR i.e., PAM prediction, gRNA designing, on/off-target activity prediction, and Prime editing and pegRNA designing.

The overall objective of existing review papers was to consolidate AI predictors related to all tasks of CRISPR into a single platform. Additionally, these review papers attempted to bridge the gap between AI researchers and the complex biology associated with the various facets of CRISPR. However, there are two significant problems with the existing review papers. First, none of the review papers provide a complete and comprehensive picture of all CRISPR-related tasks. Secondly, these reviews inadequately capture the current landscape of datasets, feature extraction methods, and ML and DL models across various CRISPR tasks. Consequently, the bridge between AI and CRISPR remains incomplete due to the limitations of the current literature. For example, Wang et al. [1060] offers a survey of ML and DL models used to predict CRISPR gRNA on and off-target activities. However, they

Table 7.1 CRISPR and machine learning/deep learning related reviews

Article	Year range	Papers	Scope	Shortcomings
[258]	2017–2022	57	This study focuses on ML techniques to predict CRISPR/Cas9 sgRNA activity (on/off-target cleavage), to assist sgRNA design and identify current research trends.	The study is limited to a systematic mapping, excluding comparisons of methods or results.
[569]	2019–2023	54	This review article focuses on the applications of DL in multiple aspects of CRISPR-Cas, the prime focus is on gRNA activity prediction, CRISPR-Cas editing outcomes, design of High-Activity gRNAs, Automated System Implementation, Nucleic Acid Detection, Anti-CRISPR Protein Identification, Cas9 Variant Activity Prediction, Transcription Factor Binding Prediction	Not all topics are equally focused on. ML models, feature representation methods, and publicly available CRISPR-Cas associated benchmark datasets are not discussed.
[458]	2016–2019	11	Future of CRISPR-based biosensors, genome engineering, discovery of CRISPR, conventional biosensors, IoT, Big Biomedical Data, Cloud Computing Systems, integration of AI in CRISPR-based biosensors	There is no discussion on the use of AI in CRISPR.
[94]	–	–	Applicability of CRISPR/Cas9 in cancer research, CRISPR/Cas9 in drug resistance, CRISPR clinical trials, on/off-target gRNA activity prediction	The focus in biological/biochemical aspects is much bigger than on AI
[826]	till 2022	–	ML models in cancer, limited CRISPR details, drug discovery through AI/ML, precision and genomic medicine, different ML Models	Deep Learning is not described in detail and there is only a small discussion of CRISPR
[994]	2014–2022	15	CRISPR for breast cancer treatment, AI/ML for therapy strategy, on/off-target effects of gRNA	Specific focus on Triple Negative Breast Cancer, no other fields than on/off target effects are dealt with
[519]	2017–2022	21	A perspective on AI in CRISPR/Cas9 modification, gRNA design, clinical trials. It explores how AI can enhance CRISPR's precision and effectiveness in treating genetic diseases, particularly cancer, while also examining the current limitations and future possibilities of this approach.	This perspective study does not discuss any details of benchmark datasets, feature engineering approaches, and ML or DL methods.
[773]	–	–	ML effects on CRISPR gene editing, data labeling pitfalls, data selection, feature engineering, gRNA design and effects prediction	Only on/off-target activity prediction is discussed
[910]	2014–2022	49	ML/DL models in CRISPR/Cas9, on/off target activity prediction, data preprocessing, gRNA encoding	Only on/off-target activity prediction is discussed

(continued)

Table 7.1 (continued)

Article	Year range	Papers	Scope	Shortcomings
[407]	2017–2021	9	AI in designing gene delivery vehicles, improving CRISPR/Cas, nanobots and mRNA vaccine carriers development	No other fields than on/off target effects are dealt with
[538]	2015–2021	20	On-target activity prediction, gRNA design, DL tools evaluation, comparison of learning based (DL) and hypothesis driven tools	No other fields than on target effects are dealt with
[1060]	till 2019	20	ML/DL algorithms for on/off target prediction, gRNA design, challenges in CRISPR activity and specificity prediction	No other fields than on target effects are dealt with
[904]	till 2023	–	ML and DL models in PAM prediction, gRNA designing, on/off-target activity prediction, and Prime editing and pegRNA designing	Important details related to datasets, representation learning, and ML and DL models are missing. In addition, only 5 different of AI in CRISPR are covered.

only examine 10 distinct models for this purpose and do not delve deeply into trends regarding datasets and feature representation methods. Furthermore, they neglect other important aspects of CRISPR, such as gene editing outcomes, aca and acr proteins prediction. Dimarou et al. [258], focus on creating a catalog of ML and DL applications in CRISPR/Cas9 gRNA design, without delving into the specific technical details of ML/DL methods. Similarly, Khoshandam et al. [519] offer a generic perspective on the applications of AI in CRISPR. Particularly, only Lee et al. [569] and Sharma et al. [904] delve into the diverse applications of DL across different aspects of CRISPR, including on/off-target activity, automated systems, Cas9 variants, PAM prediction, and peGRNA design. However, both of these articles also fail to capture the current trends in feature extraction methods and AI algorithms used in CRISPR research. Moreover, crucial topics like CRISPR arrays, operons, and aca proteins are conspicuously absent. Furthermore, details regarding the datasets used in the research articles are rarely discussed. This limited scope hinders a comprehensive understanding of the current landscape of AI predictors in CRISPR research. While existing review articles provide some overview of CRISPR-associated predictive pipelines, there is a pressing need to consolidate diverse information into a unified platform that offers comprehensive insights, patterns, and trends in CRISPR-associated predictive pipelines.

7.4 Methodology

This section explains different stages of preferred reporting items for systematic review and meta-analysis (PRISMA) strategy [900], which is used to gather relevant papers on the applications of AI in different CRISPR tasks. Figure 7.4 illustrates a visual representation of various stages from PRISMA that are summarized in the following subsections.

7.4.1 Search Strategy

Figure 7.4 illustrates the identification stage with different combinations of keywords that are utilized to search research articles. The keyword block has two key operators i.e. 'AND' and 'OR'. We leverage these operators to connect keywords and build search queries such as, *'CRISPR AND ARTIFICIAL INTELLIGENCE'*, *'CRISPR AND MACHINE LEARNING'*, *'DEEP OR MACHINE AND LEARNING*

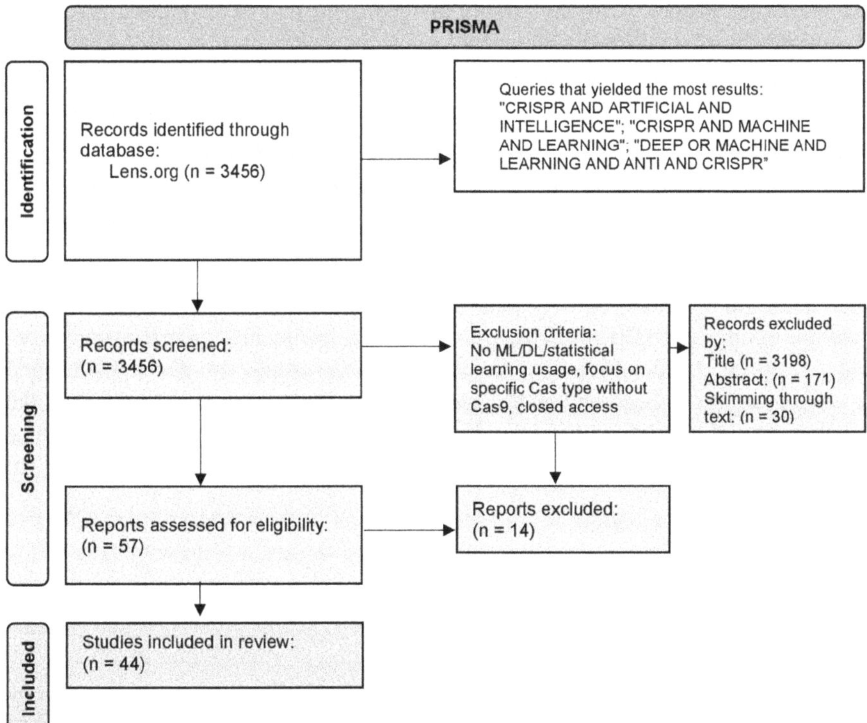

Fig. 7.4 This figure depicts the workflow of searching and screening articles, with 'n' representing the number of papers at each stage

AND EFFICACY AND CRISPR' and *'MACHINE OR DEEP LEARNING AND CRISPR'*. These queries are utilized in literature search engines like Lens (https://www.lens.org/) and Google Scholar (https://scholar.google.com/) for literature searches from Jan 2020 to Dec 2023. With the help of these queries, a substantial number of 3456 research articles are retrieved which are screened further.

7.4.2 *Screening Strategy*

With an aim to retain papers related to CRISPR-Cas, title, and abstract-based screenings are carried based on following criteria:

- Studies, that make use of ML or DL techniques.
- Studies exclusively focus on CRISPR problems.
- Studies with open access.

Following the preliminary title based screening, 3198 papers are sifted out. In a subsequent step, these 258 papers are screened by abstract, resulting in 87 papers for full-text screening. An additional 27 papers are discarded after scrutinizing the full text of the papers. After thoroughly reviewing the text of the papers, 50 papers are ultimately selected for the literature review.

7.5 Background of CRISPR Tasks and Benchmark Datasets for Development of AI Predictive Pipelines

This section provides a concise overview of 10 distinct CRISPR tasks. These tasks encompass on/off-target activity prediction, acr proteins prediction, gene editing outcome prediction, CRISPR arrays analysis, acr associated proteins prediction, and Cas proteins identification. Additionally, it presents sample statistics and details of various public benchmark datasets pertinent to each task, facilitating the development of innovative AI tools. This section also discusses the distribution of AI predictors for each task and the types of datasets utilized in their experimental setups.

7.5.1 *Basics of CRISPR*

CRISPR originates from bacteria and develops inside the bacterial genome as a defense mechanism through past encounters with the foreign genetic material of viruses or plasmids [438, 471]. The overall process of viral infection and bacterial response is shown in Fig. 7.5. Particularly, this bacterial defense mechanism against viral sequences is adopted in genetic engineering, where synthetic guide RNAs

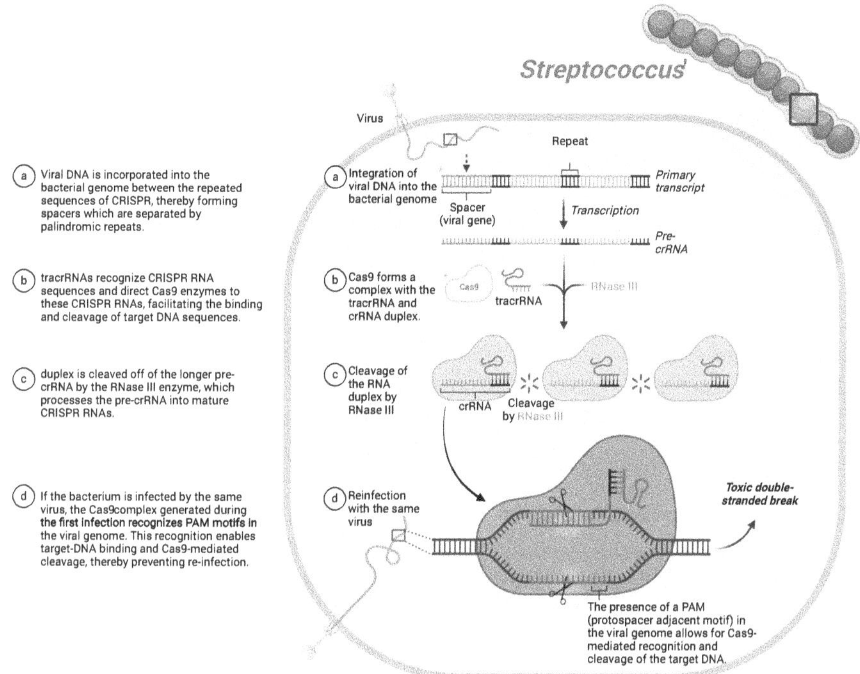

Fig. 7.5 Viral DNA is integrated into the bacterial genome, forming spacers. tracrRNAs guide Cas9 to process and mature crRNAs. Cas9, directed by crRNAs, cleaves viral DNA upon reinfection, preventing further infection (Image created using Biorender.com)

(gRNAs) are designed and coupled with bacterial Cas proteins for genome editing purposes [438].

The foundation of CRISPR system is based on a CRISPR array, consisting of three essential components: a leader sequence, repeats, and spacers [950]. Leader sequence is short and non-repetitive promoter like sequence that helps to initiate the process of transcription in a CRISPR array [26]. Spacers are short repetitive sequences that are incorporated from the viral genome and repeats are repetitive sequences that are present next to spacers [916]. Through a process of CRISPR array transcription, CRISPR RNA (crRNA) and tracer RNA (trRNA) are produced. These two RNAs unite to form guide RNA (gRNA) which interacts with the CRISPR-associated (Cas) protein. It possesses the capability to direct the Cas protein to the target sequence and initiate the necessary cleavage. The cleavage induced by CRISPR is then repaired by DNA repair mechanisms [744]. In the domain of drug designing, a synthetic gRNA is synthesized which is then used along with Cas9 proteins to induce a cleavage on the site of interest. In this way, genetic errors are corrected to cure a diverse set of diseases [424].

7.5.2 *Characteristics of Studies and Problem Distribution*

The purpose of this section is to summarise the distribution of AI predictors across 10 different CRISPR tasks. Predictor distribution analysis under individual tasks offers insights into the most active CRISPR tasks. This consolidated distribution provides a centralized platform for researchers to access valuable information about their area of interest.

Table 7.2 illustrates the distribution of predictors across 10 different CRISPR tasks. Among the 50 predictors, 27 are tailored to predict on/off-target activity in CRISPR [70, 218, 257, 298, 390, 462, 537, 579, 625, 712, 759, 760, 766, 828, 997, 1114, 1150, 1152, 1192, 1195, 1202, 1204]. Additionally, approximately 7 predictors are designed to predict acr proteins [222, 285, 383, 406, 590, 1033, 1238]. Furthermore, 4 predictors are specialized for CRISPR arrays [249, 721, 751, 857] and Cas prediction [790, 797, 1025, 1194]. Finally, only a few predictors are tailored for tasks such as CRISPR operons [1148, 1164], gene editing outcomes [27, 606, 642, 908], and aca proteins [1158].

Among 10 distinct CRISPR tasks, the prediction of on/off-target activity and acr proteins emerge as prominent trends in current CRISPR research [222, 257, 383, 462, 590, 625, 760, 997, 1114, 1150, 1195, 1238]. These tasks garner significant attention due to their crucial roles in refining the specificity and controllability of CRISPR-based genome editing. The ability to accurately anticipate undesired effects prior to laboratory experimentation holds immense value, potentially conserving financial and biochemical resources as well as time. Moreover, integrating predictions of on- and off-target effects and activities with acr proteins holds significant promise for optimizing gene therapies, potentially resulting in safe, inert, and non-detrimental outcomes.

Table 7.2 Problem distribution of reviewed papers

Problem	Count	Reference
CRISPR arrays	2	[249, 721]
CRISPR loci	2	[751, 857]
CRISPR systems CrRNA	2	[790, 797]
Acr proteins	6	[222, 285, 383, 590, 1033, 1238]
Aca proteins	1	[1158]
CRISPR operons	2	[1148, 1164]
Cas protein	2	[1157, 1194]
Off target activity	10	[462, 625, 670, 760, 952, 997, 1025, 1150, 1195, 1202, 1204]
On target activity	16	[218, 257, 298, 390, 537, 579, 712, 759, 766, 828, 1114, 1152, 1183, 1192, 1202, 1204]
Editable target region	1	[70]
Gene editing outcome prediction	4	[27, 606, 642, 908]
Acr proteins activity	2	[406, 736]

Cas and CRISPR arrays prediction [249, 721, 790, 797, 857, 1194] are emerging as prominent research focuses within the realm of CRISPR based gene editing. The ongoing discovery of novel and enhanced Cas systems plays a pivotal role in advancing these tasks, promising better precision and efficacy in gene editing. The identification and characterization of new CRISPR arrays within DNA sequences holds immense potential for optimizing gene editing strategies, facilitating targeted modifications with unprecedented accuracy. While fewer papers are dedicated for the prediction of cleavage, gene editing outcomes, CRISPR operons and acr-associated proteins, their significance cannot be understated. Understanding these processes is essential to fine tune gene editing and to regulate and modulate the activity of CRISPR complexes.

7.5.2.1 CRISPR Arrays

Figure 7.6 illustrates that CRISPR arrays are short sequences of repetitive DNA (repeats) interspersed with unique sequences (spacers) derived from viral or plasmid DNA that help bacteria to identify external genomes. Once the external genome is identified, CRISPR RNA (crRNA) acts as an immune mechanism, forming a small molecule with Cas proteins that destroys the external genome. Particularly, CRISPR arrays contain two distinct components i.e., spacers, which are small parts of the external genome incorporated inside the bacterial DNA, and repeats, which are small palindromic sequences that are repeated in the CRISPR array. Both of these components help the crRNA to bind with Cas proteins. Research in this field includes the detection of CRISPR arrays and the discrimination between valid and non-valid arrays [249].

Table 7.3 presents 3 different benchmark datasets developed to identify CRISPR arrays. Mitrofanov et al. [722] gathered archaeal and bacterial CRISPR arrays and generated two different benchmark datasets. On the basis of the these datasets, authors checked the validity of CRISPR arrays. Deshmukh et al. [249] proposed a CRISPR detection method with three stages: detect potential CRISPR arrays, classify repeats, and filter invalid arrays. First, the CRT tool identifies potential arrays from DNA sequences using specific parameters. Next, an LSTM model

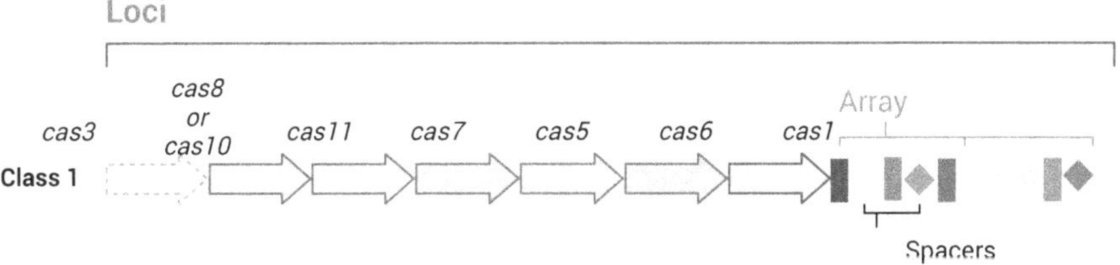

Fig. 7.6 The arrangement of CRISPR loci and CRISPR arrays (Image created using Biorender. com)

Table 7.3 Benchmark datasets for CRISPR arrays prediction

Datasets	Specie	Type	Positive	Negative
CrisprIdentify [722]	Archea	Train	400	400
		Test	550	200
	Bacteria	Train	600	600
		Test	550	200
CRISPRLstm [249]	–	–	11,407	12,000

with a sigmoid activation function scores the repeats. Finally, the method averages these scores to calculate the overall array score, discarding arrays below a certain threshold. The authors assessed the accuracy of classifying short DNA segments as CRISPR repeats using a dataset of 11,407 CRISPR repeats and 12,000 invalid repeats [367, 1046]. They validated the CRISPRLstm pipeline with 309 CRISPR arrays from 60 organisms.

7.5.2.2 CRISPR Loci

The CRISPR locus consists of the CRISPR array and the Cas genes that form an operon as shown in Fig. 7.6. The CRISPR locus is responsible for the complete adaptive immune response in prokaryotes, including spacer sensing, crRNA processing, and foreign DNA interference. The arrays store genetic information from previous infections, while the Cas genes encode proteins necessary for processing and fighting invaders.

Nethery et al. [751] created a benchmark dataset for CRISPR loci subtype identification. First, authors downloaded genomes with previously classified CRISPR loci from the National Center for Biotechnology Information (NCBI) [890]. Authors obtained repeats using MinCED [101], retaining all detectable sequences. The data, comprising 7808 CRISPR loci and 15,669 repeat sequences across 30 subtypes, were used to train the model. Overall, The training set included 12,534 repeats, and the validation set contained 3135 repeats across 30 subtypes. The dataset can be obtained from the following link https://github.com/CRISPRlab/CRISPRclassify.

Russel et al. [857] created another benchmark dataset for CRISPR loci identification by using MinCED v0.4.2 [101]. They included consensus repeats from all arrays located within 1 kbp of a Cas operon, resulting in a total of 5838 subtyped repeat sequences. The benchmark dataset can be downloaded from the following link https://github.com/Russel88/CRISPRCasTyper/tree/master.

7.5.2.3 Cas Proteins

The CRISPR-associated protein has the purpose of cleaving at the DNA target site. Over the years different Cas proteins have been discovered and designing a

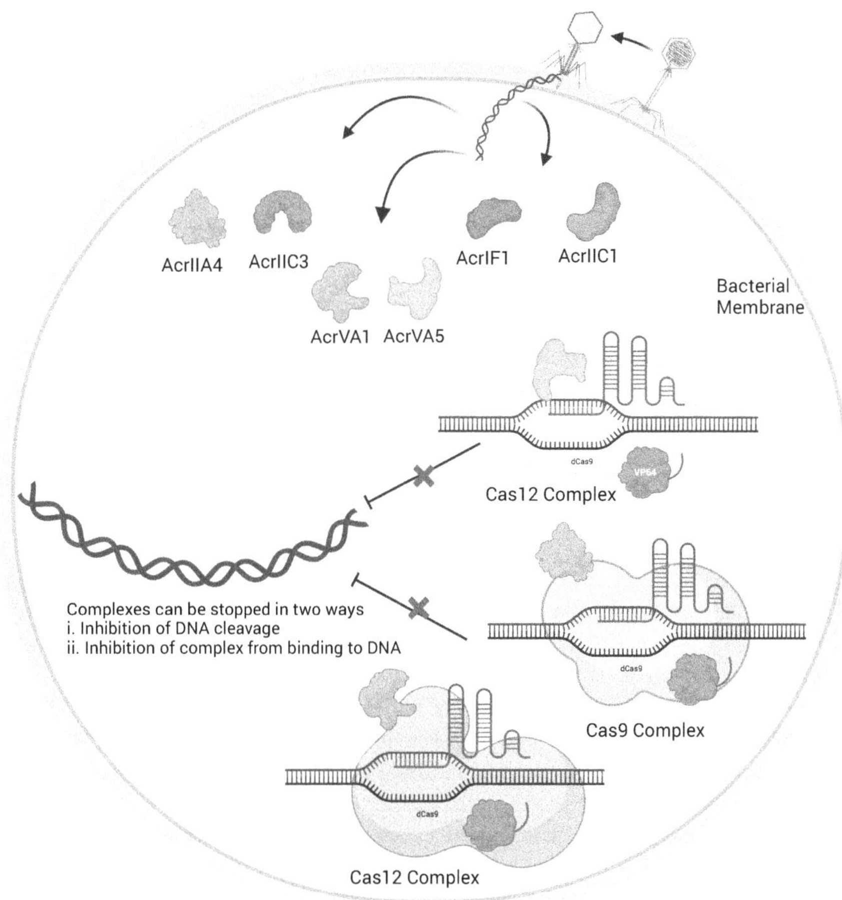

Fig. 7.7 Viral genome contains information about various acr proteins. Once translated, these proteins can interact with Cas complexes and inhibit them from cleaving the viral genome (Image created using Biorender.com)

CRISPR/Cas system with specific Cas proteins aids in the precise performance of the cleavage [1194]. As a result, undesired on-target effects are minimized. A persisting challenge is that the variety and number of available Cas proteins are still not meeting the researchers' needs, hindering the development of CRISPR/Cas editing tools. The large size of the currently known Cas proteins often leads to limitations in the gene editing process, thereby encouraging the continuous search for smaller Cas proteins. In this field ML and DL techniques contribute to the research by predicting whether a protein has the potential to be a Cas protein or not [1194].

Yang et al. [1157] proposed the first benchmark dataset for Cas protein prediction dataset. The authors gathered Cas protein sequences from the UniProt database and applied the CD-HIT tool to yield 155 Cas protein sequences. Authors collected non-Cas protein sequences form Uniprot having no or less similarity with Cas protein sequences. This resulted in 155 non-Cas protein sequences. Building on their work, Zhang et al. [1194] followed similar protocol to collect Cas protein sequences. In addition, Zhang et al. [1194] the non-Cas protein sequences from the work of Yang et al. [1157]. Overall, Zhang et al. [1194] dataset contained 418 Cas and non-Cas protein sequences.

7.5.2.4 Anti-CRISPR Proteins

Anti-CRISPR (acr) proteins act against the CRISPR mechanism. Figure 7.7 shows that acr proteins play a crucial role as a control mechanism for the CRISPR system's activity [1170] and can work in two different ways, i.e., they can prevent the Cas-gRNA complex from binding to target DNA, and they can also block cleavage by deactivating the Cas effector [1170]. With the help of acr proteins in the gene editing process, timing and precision are enhanced, and undesired effects are mitigated. In terms of acr proteins, there are 3 crucial tasks which include predicting acr family classes [590], binary classification of acr proteins [222], and acr-Cas protein interaction prediction.

Anti-CRISPRdb [267] categorizes a variety of acr proteins, which inhibit different subtypes of CRISPR systems. For instance, Type I-F [697] includes 12 identified Acr proteins. These proteins inhibit Type I-F CRISPR-Cas systems found in various bacteria such as *Pseudomonas aeruginosa*. Type I-E [1171] Acr proteins target the I-E subtype, which is another common type of CRISPR-Cas system. Similarly, Type II-A Acr proteins [1209] are used to inhibit Cas9 protein commonly used in gene editing technologies. Type I-C, I-D, III-B, III-I, V-A, VI-A, VI-B, have a varied number of Acr proteins identified that inhibit their respective CRISPR-Cas systems. The sequences of these proteins are updated daily in databases like Anti-CRISPRdb [267], and AcrHub [1063]. Based on these databases and types of acr proteins, multiple acr proteins benchmark datasets have been proposed.

Table 7.4 provides an overview of 6 different benchmark datasets used to train AI acr protein predictors. Li et al. [589] proposed 3 different benchmark acr protein datasets. Authors collected acr proteins from anti-CRISPRdb [267] and PaCRISPR [1061]. In order to test the generalizability of the acr protein predictors in a better way, the authors created three different variants of the datasets based on different train and test configurations. For instance, in AcrNet-1, they chose types I-F, II-C, and I-D as testing samples and used the remaining Acrs as training samples. In AcrNet-2, types I-F, I-E, V-A, I-C, VI-A, VI-B, III-I, III-B, and I-B were selected as testing data. In AcrNet-3, types I-D, II-C, I-E, V-A, I-C, VI-A, VI-B, III-I, III-B, and I-B were chosen as testing data.

Table 7.4 Benchmark Datasets for ACR protein prediction

Data		Positive	Negative	Additional details	Databases used	Link	Year
AcrNet-5-fold [589]	Train	1094	1162		paCRISPR, CRISPRDb	https://acranker.pythonanywhere.com/	2023
AcrNet-1 [589]	Train	884	902	From type I-F, II-C, and I-D in anti-CRISPRdb			
	Test	210	260				
AcrNet-2 [589]	Train	904	902	From type I-F, I-E, V-A, I-C, VI-A, VI-B, III-I, III-B, and I-B in anti-CRISPRdb			
	Test	190	260				
AcrNet-3 [589]	Train	962	902	From type I-D, II-C, I-E, V-A, I-C, VI-A, VI-B, III-I, III-B, and I-B in anti-CRISPRdb			
	Test	132	260				
AcRanker [286]	Train	432	432	12 of the proteins are active against subtype I-F CRISPR Cas systems, four against I-E, and four against II-A	AntiCrisprDb	https://academic.oup.com/nar/article/48/9/4698/5819938	2020
	Test		–	–	–	–	–
PreAcrs [1242]	Train	412	412		Anti-CRISPRDb, AcrDb, AcrCatalog	https://github.com/Lyn-666/anti_CRISPR/tree/main/data	2022
	Test	176	176				
PaCRISPR [1061]	Train	98	902	–	AntiCrisprDb, and literature	https://pacrispr.erc.monash.edu/download.jsp	2020
	Test	26	260				
Gussow et al. [383]		488	488	–	AntiCrisprDb, and literature	–	
Dao et al. [222]	Train	205	902	–	–	–	2023
	Test	26	260	–	–	–	

Etzinger et al. [286] collected acr protein data from the Anti-CRISPRdb [267], ensuring a non-redundant set with a 40% sequence identity threshold using CD-HIT, resulting in 20 verified Acrs for the positive class. This included 12 against subtype I-F, 4 against I-E, 4 four against II-A. They downloaded complete proteomes of source species and filtered out proteins with $\geq$ 40% similarity to known Acrs to form the negative dataset. For independent testing, they used a separate dataset of 20 known Acrs covering various mechanisms and sequences, primarily from the same subtypes as the training set.

Zhu et al. [1242] collected 1378 validated Acrs from Anti-CRISPRdb and 17 new Acrs from NCBI, then used CD-HIT with a 70% identity threshold to filter redundant sequences which resulted in 588 Acrs. These were split into 412 for training and 176 for testing. For negative samples, 1571 non-Acrs were selected from UniProt based on four strict criteria, 412 were used in training while 176 were used in testing. Finally, training dataset had 412 positive and 412 negative samples, and the test dataset had 176 positive and 176 negative samples.

Wang et al. [1061] collected 488 experimentally validated acr proteins from Anti-CRISPRdb and literature. After removing redundant sequences with more than 70% identity, they obtained 98 sequences as positive samples for training. Negative samples were selected based on four criteria: they must not be acrs, must come from phages or bacterial MGEs, must have <40% sequence similarity to each other and the positive samples, and must have lengths between 50 and 350 residues. This resulted in a training dataset of 98 positive and 902 negative samples. For further testing, they collected 26 new acrs with <10% similarity to the training set (except two) and 260 non-acrs using similar criteria, forming an independent dataset with 26 positive and 260 negative samples.

Table 7.5 provides an overview of 2 benchmark datasets for acr-Cas protein interaction prediction. Hasani et al. [406] proposed an acr-mediated CRISPR-Cas inhibition dataset. The dataset comprises 227 pairs of Acr and CRISPR-Cas systems, with 132 pairs showing positive (functional) inhibition and 95 pairs negative (non-functional) inhibition. These sequences are taken from AcrHub [1063], Anti-CRISPRdb [267], and several published works. Each sample includes the Acr and Cas protein sequences, organism identity, CRISPR-Cas system type (I-C, I-E, or I-F), bacterial species/strain, and an inhibition label (1 for positive, 0 for negative). Focused on type I CRISPR-Cas systems, the dataset excludes subtypes I-B and I-D due to insufficient information. It features systems from *Pseudomonas aeruginosa*, P*ectobacterium atrosepticum*, *Escherichia coli*, and *Serratia species*.

Murmu et al. [736] developed a Cas-acr interaction dataset: positive (interacting pairs) and negative (non-interacting pairs). They compiled 192 interacting Acr and Cas protein pairs from the Anti-CRISPRDb [267] and removed 85 redundant pairs. Cas protein sequences were retrieved from protein data bank (PDB) [121], UniProt [204], and GenBank [88]. Negative pairs were generated by shuffling amino acid sequences to create a balanced dataset.

Table 7.5 Benchmark datasets for acr-Cas protein interaction prediction

Type	Positive	Negative	Additional details	Databases	Link	Year
AcrTransAct [406]	132	95	type I-C, I-E, or I-F), and Acr inhibits the CRISPR-Cas system (label 1) or not (label 0)	AcrHub, and antiCRISPRDb	https://github.com/USask-BINFO/AcrTransAct/tree/main/data	2023
AcrCasPPI [736]	107	107	–	PDB, AntiCRISPRDb, and Genbank	https://pypi.org/project/acrcasppi-ml/	2023

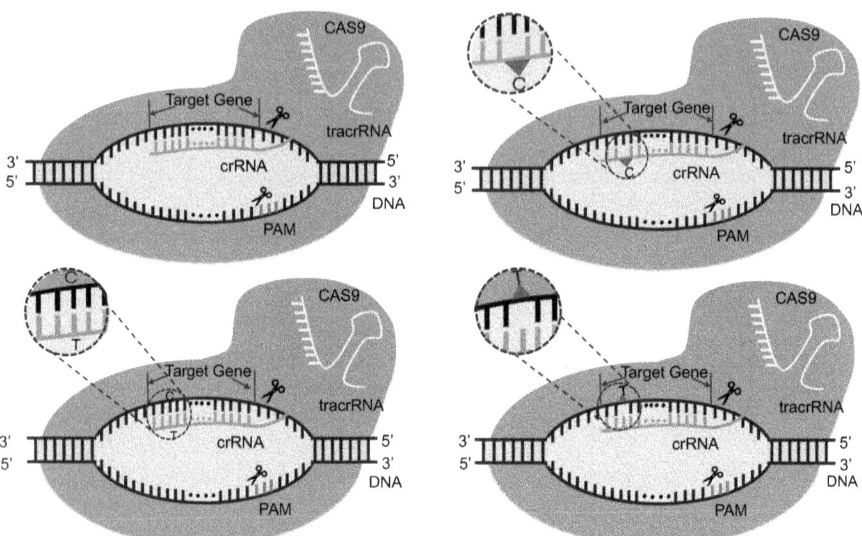

Fig. 7.8 3 different types of off-target effects. (**a**) corresponds to the normal gene editing process. (**b**) refers to the bulge of RNA, (**c**) shows the mismatch case, where the target is not fully recognized and a cleavage is made at the wrong location, and (**d**) a bulge of the DNA

7.5.2.5 Off-target Activity Prediction

In the CRISPR gene editing process, the single guide RNA (sgRNA) directs the Cas9 protein to the precise location for the intended genetic modification. This process is not always executed as desired, as Cas9 may cleave at unintended locations. Such unintended cuts can lead to unstable gene sequences and malfunctions in normal genes [143]. This phenomenon is referred to as off-target effects or off-target activity as shown in Fig. 7.8. These effects are influenced by factors such as the structure and length of sgRNA [143]. In this particular task, AI predictors are trained in two different paradigms i.e., classification: Off-target sites are labeled "1" for unintended edits by CRISPR9, while on-target or non off-target sites are labeled "0" for intended edits, regression: a continuous value represents the likelihood or magnitude of off-target activities at the target genomic location.

Tables 7.6 and 7.7 present 14 different benchmark off-target activity prediction datasets that have been developed to train and evaluate AI predictors. Table 7.6 encompasses 7 different benchmark datasets for off-target activity prediction across six distinct cell types: HEK293T, K562V, U2OS, K562, HAP1, and Human primary T cells. For example, Dhanjal et al. [253] created an off-target activity benchmark dataset using GUIDE-seq [1002], SITE-seq [128], and CIRCLE-seq [1003]. The inactive targets were chosen from *CRISPCut* [252], resulting in highly imbalanced datasets due to the abundance of negative samples. Chuai et al. [191] developed 2 different datasets for off-target activity prediction using two cell types: 293-

Table 7.6 Off target activity datasets

Dataset	Set	Positive	Negative	IR	Cell type(s)	Link	Year
Dhanjal et al. [253]	Train	6337	7040	1.46	HEK293T K562V U2OS	https://web.iitd.ac.in/crispcut/off-targets/	2018
	Test	2877	4010	–		–	
K562 [191]	–	120	20,199	168.32	K562	https://github.com/bm2-lab/DeepCRISPR	2018
HEK239T [191]	–	536	132,378	246.97	HEK239T	https://github.com/bm2-lab/DeepCRISPR	2018
CRISPOR [801]	742	408,260	550.22	HAP1, HEK293T, K562, and U2OS	[a]	2018	
Zhang et al. [1213]	–	26,412	26,412	1	HEK293T	https://github.com/JiazhiHuLab/CNN_predict	2021
CHANGE-seq [559, 1142]		67,476	2,806,151	41.59	Human primary T cells	https://github.com/OrensteinLab/SysEvalOffTarget	2022

[a] https://academic.oup.com/bioinformatics/article/34/17/i757/5093213

Table 7.7 Off target activity datasets

Type	Technique	Total	Validated off-targets	Guide RNAs	With Indel	Cell type(s)	Link
I/1 [1003]	CIRCLE-Seq	584,949	7371	10	Yes		–
I/2 [636]	GUIDE-Seq	213,943	60	6	Yes		https://codeocean.com/capsule/9553651/tree/v1
II/1 [262]	Protein knockout detection	4853	2273	65	No	A375 BV2 HT29	–
II/2 [387]	PCR, Digenome-Seq and HTGTS	10,129	354	19	–		–
II/3 [128]	SITE-Seq	217,733	3767	9	No	HEK293	–
II/4 [1002]	GUIDE-Seq	294,534	52	9	No	HEK293T U2OS	–
II/5 [529]	GUIDE-Seq	95,829	54	5	No	EGFP U2OS	–
II/6 [636]	GUIDE-Seq	383,463	56	22	No	HCT116 HEK293T HL60 Kbm7 K562 U2OS	–

related cell lines (18 sgRNAs) and K562 cells (12 sgRNAs). By utilizing *bowtie2* [558], they identified approximately 160,000 potential off-target loci across the genome for 30 sgRNAs, allowing up to six mismatches. This dataset was also highly unbalanced, with roughly 1 in 250 loci identified as off-targets. Zhang et al. [1213] proposed a balanced off-target activity dataset of Cas9 variants for HEK293T cell lines. Peng et al. [801] created another dataset from 9 different experiments performed on CRISPR, including Targeted PCR [184, 442, 1053], PCR [523, 524], Flanking PCR [835], GUIDE-seq [1002], Digenome-seq [523], HTGTS [318], Multiplex Digenome-seq [524], and CIRCLE-seq [1003], encompassing a total of 76 gRNAs. Lazzarotto et al. [559] recently introduced a new dataset for off-target activity based on in-vitro and in-cellular experiments i.e., CHANGE-seq (110 gRNAs). In this specific dataset, active on-targets with up to six mismatches were experimentally determined, while inactive off-targets were identified using Cas-OFFinder [68] (Fig. 7.8).

Lin et al. [630] classified the off-target effects of CRISPR gRNAs into three categories: (a) sites with base mismatches; (b) sites with missing bases (RNA bulge or insertion); (c) sites with additional bases (DNA bulge or deletion). Instances (b) and (c) are recognized as indel (insertion or deletion) off-target occurrences. Building on the similar idea, 8 different datasets have been proposed which are presented in Table 7.7. For instance, I/1 [1003] and I/2 [636] encompass pairs of gRNA and target DNA sequences exhibiting mismatches and indels. Specifically, I/1 [1003] comprises pairs sourced from 10 distinct gRNAs, among which 7371 active off-targets (430 featuring indels) were empirically affirmed through CIRCLE-seq experimentation. Similarly, I/2 [636] comprises pairs sources from 6 different sgRNA with approximately 60 validated active off targets. Furthermore, utilizing the gRNA sequences, Cas-Offinder [68] a flexible tool designed for identifying potential off-target sites of Cas9 RNA-guided endonucleases was employed to acquire inactive off-target sites in the genome associated with the aforementioned two types.

Six independent gRNA-target pairs based off-target activity datasets do not incorporate mismatches and indels together, but rather focus on only mismatches i.e., II/1 $\cdots$ II/6. Donech et al. [262] provided II/1 which contains 65 gRNAs related 4853 validated off targets with human sequence target CD33, belonging to three different cell lines i.e., A375, BV2, and HT29. Similarly, Haeussler et al. [387] provided II/2 dataset of 19 gRNAs with a total of 350 validated off targets. II/3, proposed by Cameron et al. [128] contains 3767 positive off-target sites from 9 different gRNAs validated by SITE-Seq. Datasets II/4, II/5, and II/6 comprise validated gRNA-target pairs confirmed through GUIDE-Seq, each sourced from distinct research works: Tasi et al. [1002], Listgarten et al. [636], and Kleinstiver et al. [529]. It is noteworthy to mention that Tasi et al. [1002], Listgarten et al. [636], and Kleinstiver et al. [529] solely provided the active off-target sites. Consequently, employing Cas-Offinder [68], all potential off-targeting sites with up to six mismatches in the human genome were identified, and the corresponding datasets were formulated.

From the pool of the studies selected for the review in this paper, multiple datasets have been utilized for off-target activity prediction. Störtz et al. [952] and Daneshpajouh et al. [218] utilized the CrisprSQL dataset [951], a comprehensive collection of 17 base-pair resolved off-target cleavage studies on SpCas9, totaling 25,632 samples. It includes data from various cell lines, primarily U2OS, HEK293, and K562.

Toukifuzzaman et al. [997] utilized sgRNA-DNA pairs of DeepCRIPR study [191]. Imani et al. [462] used the K562 and HEK293T cell lines related Deep-CRISPR dataset [191] for training DL models. On the other hand, Lin et al. [625] trained and assessed their models on both CRISPOR [387] and GUIDE-seq [1002, 1142] datasets.

Neu et al. [760] utilized 7 different off-target activity prediction datasets namely, CIRCLE-seq [1003] (contains mismatch, insertion, and deletion off-target sites), protein knockout detection (II/1) [262], Digenome PDH (II/2) [387], II/3 SITE [128] and GUIDE-seq I, II, III (II/4, II/5, II/6) [529, 636, 1002]. Yang et al. [1150] utilized all the datasets presented in Table 7.7 and K562, HEK293T datasets of Chuai et al. [191]. Toufikuzzaman et al. [997] used the augmented datasets of DeepCrispr [191] with a maximum of six nucleotide mismatches. This specific dataset contains 293-related cell lines (18 sgRNAs) and K562t (12 sgRNAs).

7.5.2.6 On-target Activity Prediction

When Guide RNA along with CRISPR system is directed for a specific DNA sequence, the Cas9 protein induces double-stranded breaks at that specific genomic location. Subsequently, these breaks are repaired by the cell's DNA repair mechanisms such as non-homologous end joining (NHEJ) and homology-directed repair (HDR) [550] as shown in Fig. 7.9. These mechanisms can introduce challenges and potentially cause unwanted effects at the target site, such as insertions and deletions [550]. AI methods are utilized to predict the efficiency of gRNA or on-target activity.

In the era of CRISPR, numerous datasets have emerged to assess on-target activity that stem from various origins such as in vitro experiments, or in vivo studies. This diversity underscores the necessity for novel algorithms to benchmark against these datasets. Considering a similar notion, researchers have recently endeavored to gather disparate on-target activity datasets onto a unified platform. For instance, Haeussler et al. [387] gathered 15 different CRISPR on-target activity datasets. These datasets are subdivided into two main groups on the basis of the origin of the gRNA i..e, from U6 or T7 promoter. Table 7.8 shows samples statistics of CRISPR on-target activity benchmark datasets based on U6 and T7 promoters. The U6 promoter groups include 12 different datasets which include Wang/Xu HL60 [1052, 1129], Donech Mouse-EL4 [263], Koike-Yusa 1 M-ESC [535], Chari 293T [145], Donech A375 [262], Hart Repl2Lib1 HCT116 [402], Gandhi Eelectrop. Ciona [323], Farboud C. elegans [301], Ren Drosophilla [845]. Similarly, the T7 promoter based datasets include Varshney Zebrafish [1015], Gagnon Zebrafish [322], and Morneo-Mateis Zebrafish [727].

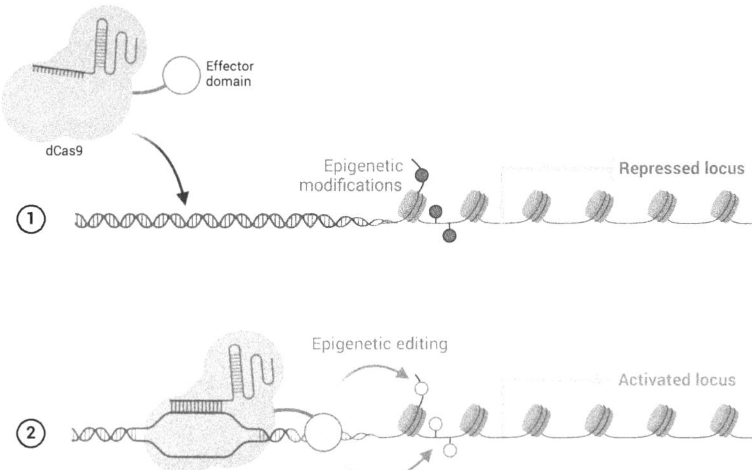

Fig. 7.9 The effects of epigenetic modifications and the potential for reversal using dCas9. The top part shows that epigenetic modifications can cause gene repression. The bottom part demonstrates how dCas9 can be used to perform epigenetic editing to reverse these modifications, leading to gene activation (Image created using Biorender.com)

Table 7.8 On-target activity datasets based on U6 and T7 promoters and bacteria. These datasets are available under following repositories https://github.com/maximilianh/crisporPaper, and https://github.com/VKonstantakos/CRISPRedict

	Dataset	Specie	No. of samples	Year
U6	Chari 293T	HM	1193	2015
	Doech HS	HM	110	
	Doech MM	MM	150	2014
	Doench azd Hg19	HM	431	2016
	Hart HCT116	HM	4199	2016
	Hart HeLALib1	HM	4217	2016
	Hart HeLALib2	HM	3816	2016
	Hart RPE1	HM	4175	2016
	Xu HL60	HM	2057	2015
	Xu KMB7	HM	2057	2015
	Gandhi	CN	72	2016
	Farboud	CE	50	2015
T7	Gagnon	ZB	111	2014
	Moreno-Mateos	ZB	1020	2015
	Varshney	ZB	102	2015
Bacteria	Guo E.Coli	E.Coli	40, 468	2016

Table 7.9 Counts of train and test samples for each CRISPR variant

CRISPR variant	Train	Test	Link	Year
SpCas9	34,713	5415	Clickable Link	2019
SpCas9-NG	.	.		.
VRQR variant	.	.		.
xCas	.	.		.
Sniper-Cas9	.	.		.
eSpCas9(1.1)	.	.		.
SpCas9-HF.1	.	.		.
HypaCas9	.	.		.
evoCas9	34,713	5415		.
HT-Cas9 (kim)	12,832	542		2019
Xiang-gRNA	10,592	.		2021

As the CRISPR field is burgeoning, a steady stream of new datasets continues to emerge regarding CRISPR's on-target activity. For instance, Wang et al. [1042] proposed new on-target activity datasets based on the different Cas proteins i.e., SpCas9-HF1, or High Fidelity Cas9 (ESP), which is a modified version of the Cas9 protein derived from Streptococcus pyogenes (SpCas9)and exhibits higher specificity in targeting DNA sequences. Similarly, SpCas9-HF1 (High Fidelity Cas9) (HF1) is another variant of the Cas9 enzyme from Streptococcus pyogenes (SpCas9). Like SpCas9-HF1 (ESP), this version is designed to improve the specificity of CRISPR genome editing. WT-SpCas9 refers to the wild-type form of the Cas9 enzyme isolated from Streptococcus pyogenes. This unmodified version of Cas9 is the original enzyme used in CRISPR genome editing. While it remains a powerful tool for gene editing, WT-SpCas9 may exhibit higher off-target effects compared to engineered high-fidelity variants such as SpCas9-HF1 (ESP) and SpCas9-HF1 (HF1). Researchers often use WT-SpCas9 alongside modified versions to compare their editing efficiency and specificity.

Kim et al. [521, 525, 1112] proposed more on-target activity datasets across different settings in CRISPR with the same or enhanced variants of Cas9 protein i.e., SpCas9, SpCas9-NG, VRQR variant, xCas, Sniper-Cas9, eSpCas9(1.1), SpCas9-HF.1, HypaCas9, and evoCas9. The sample statistics are provided in Table 7.9.

As the influx of datasets continues to grow, researchers endeavor to establish a consensus by proposing various categorizations of datasets. For instance, Zhang et al. [1198] devised a taxonomy that classifies on-target activity datasets into 3 groups based on size, i.e., small, large, and medium datasets as shown in Table 7.10. This approach aids in organizing and understanding the diverse array of datasets available for analysis and research purposes, facilitating more efficient data utilization and fostering collaboration within the scientific community. Although it is simplistic and organized, authors neglect multiple datasets of Kim et al. [525] and the majority of datasets from Table 7.8 which are initially collected and presented by Haeussler et al. [387].

Like gRNA's role in gene editing, it also exhibits activity in epigenome editing [1152]. This enables the regulation of gene expression without altering the underlying DNA sequence. Yang et al. [1152] gathered these datasets from different

Table 7.10 Samples statistics of protospacer and PAM combinations for on-target activity. Datasets with '*' are reported in earlier tables as well

Types	Dataset	No. of samples	Link	Year
LARGE	SpCas9-HF1 (High Fidelity Cas9) (ESP) [1042]	58,616	https://github.com/izhangcd/DeepHF	2019
	SpCas9-HF1 (High Fidelity Cas9) (HF1) [1042]	56,887	https://github.com/izhangcd/DeepHF	2019
	WT-SpCas9 (Wild-Type Streptococcus pyogenes Cas9) [1042]	55,603	https://github.com/izhangcd/DeepHF	2019
MEDIUM	Sniper*	37,974	https://www.nature.com/articles/s41587-020-0537-9	
	SpCas9*	30,585	https://www.nature.com/articles/s41587-020-0537-9	2018
	xCas9*	37,738	https://www.nature.com/articles/s41587-020-0537-9	2018
SMALL	Hart HCT116*	4239	https://github.com/bm2-lab/DeepCRISPR	2015
	HELA (Hart HeLALib1 + HeLALib2)*	8101	https://github.com/bm2-lab/DeepCRISPR	2015
	Wang/Xu HL60*	2076	https://github.com/bm2-lab/DeepCRISPR	2014

Table 7.11 On-target activity prediction datasets for epigenome editing

Type	Dataset	gRNA	Genes	Source
CRISPRoff editing	CRISPRoff-tiling	111,682	520	[770]
	CRISPRoff-genome	20,221	18,779	[770]
	Endogenous Genes (H2B)	326	1	[770]
	Endogenous Genes (CLTA)	415	1	
	Endogenous Genes (RAB11A)	392	1	
	Endogenous Genes (VIM)	528	1	
CRISPRi editing	CRISPRi-activity score	18,079	1539	[435]
	hCRISPRi-v2	199,523	18,549	[435]
	CRISPRi-genome	107,595	14,361	–
	CRISPRi-K562	111,283	520	–
CRISPRa editing	hCRISPRa-v2	198,756	18,495	–
	CRISPRa-activityscore	2779	236	–

literature sources [435, 770]. The statistics of these 9 datasets are provided in Table 7.11.

Niu et al. [759] created 4 distinct agronomic species datasets i.e., Glycine max, Zea mays, Sorghum bicolor, and Triticum aestivum by gathering sgRNA sequences with high and low on-target activities from Sun et al. [962] which included exper-

Table 7.12 Number of sequences with high and low on-target activities for the four crops. These datasets can be downloaded from http://crispr.hzau.edu.cn/CRISPR-Local/

Crop	Pos. seq.	Neg. seq.
Glycine	135,800	122,880
Zea	643,939	442,190
Sorghum	722,906	837,222
Triticum	581,120	429,900

imentally verified seed sgRNAs with known knockout effects. The initial dataset contained around 15,000 sgRNAs from seed experiments. The authors utilized CD-HIT to remove redundant sequences from positive and negative samples. The sample statistics of these datasets are presented in Table 7.12.

Overall, 27-39 unique CRISPR on-target datasets can be considered to design and benchmark novel on-target activity prediction tools/applications based on the problem setting. Inside each dataset, a sample contains nucleotide PAM sequence along with a numerical on-target activity value.

Over the past three years, 6 different studies have been conducted to enhance the accuracy of on-target activity predictions. Each study utilizes a specific set of datasets to train and evaluate the performance of ML and DL models. For instance, Xiao et al. [1114] trained and assessed the performance of their DL model namely, AttCRISPR on 3 different publicly available datasets from DeepHF [628] namely, WT-SpCas9, eSpCas9(1.1) and SpCas9-HF1. These datasets contain 55,604 (WT-SpCas9), 58,617 (eSpCas9(1.1)), and 56,888 (SpCas9-HF1) sgRNAs with continuous activity values.

Dimauro et al. [257] utilized the datasets gathered by Xu et al. [1140] which are also presented in Table 7.8. Particularly, the authors used 10 out of 15 different datasets namely, Wang/Xu HL60 [1052, 1129], Donech Mouse-EL4 [263], Chari 293T [145], Donech A375 [262], Hart Repl2Lib1 HCT116 [402], Gandhi Eelectrop. Ciona [323], Farboud C. elegans [301], Varshney Zebrafish [1015], Gagnon Zebrafish [322], and Morneo-Mateis Zebrafish [727]. Similarly, Zhang et al. [1192], Rafid et al. [828], Li et al. [579], and Fanaras et al. [298] utilized only 4 different datasets namely, Hart Repl2Lib1 HCT116 [402], Chari 293T [145], Hart Repl2Lib1 HCT116 [402], and Wang/Xu HL60 [1052, 1129]. It is important to mention that Zhang et al. [1192], Li et al. [579], and Fanaras et al. [298] did not utilize the original versions of these datasets. Instead they made use of augmented datasets as done in DeepCrispr study [191]. The researchers expanded these datasets by introducing two mismatches in the PAM-distal region of original sgRNA sequences, a technique that does not affect cleavage efficacy. This process generated approximately 200,000 unique sgRNAs, each assigned the same efficacy labels as the original sequences. The augmented dataset provides a diverse and biologically meaningful set of sgRNAs for training purposes.

Previous studies have shown that PAM-distal region has a high tolerance for sequence mismatches [524, 529]. To be specific, gRNAs with two mismatches in the first two positions from the 5' end has little influence on cleavage efficiency

[262, 263]. Inspired by these studies, Chuai et al. applied a data augmentation procedure by changing each gRNA into a new one with two mismatches in the PAM distal region [191]. Consequently, a 23-nt gRNA sequence can be expanded into 16 gRNAs with identical cleavage efficacy. The augmented dataset was generated from 15,000 gRNAs with known on-target cleavage efficacy. By adopting this data augmentation strategy, they obtained 180,512 non-redundant gRNAs. Each observation in the data contains a 23-nt gRNA sequence and its corresponding cleavage efficiency. In this work, we used this augmented dataset as the benchmark data for model selection and pre-training.

Ham et al. [390] created a new on-target activity prediction dataset recently with a motivation that current models poorly predict SpCas9/sgRNA activity because the underlying datasets are inaccurate and fail to distinguish between cleavage activity and toxicity. To address this, authors utilized a two-plasmid positive selection system to generate high-quality data that accurately measures SpCas9/sgRNA cleavage activity and separates it from toxicity. It is important to mention that the last study related to on-target activity prediction explores the performance of DL predictors on 9 different datasets that are presented in Table 7.10 and discussed earlier [1198] (Table 7.13).

Noshay et al. [766] utilized spCas9 dataset as presented in Table 7.9. In addition, Konstantakos et al. [537] provided a web server of existing tools by training and assessing these models on gRNAs expressed in U6 and T7 promoters. U6 and T7 promoter based datasets are already discussed earlier and presented in Table 7.8.

7.5.2.7 CRISPR Gene Editing Outcome Prediction

Upon locating the target site and inducing a double-stranded break, the cell's DNA repair mechanisms are activated (Fig. 7.9). These repair mechanisms include i.e., Homology-directed repair (HDR), and end joining. In a HDR, cells repair the damage by copying from the sister chromatid, filling in gaps around the break site. Scientists can exploit this by introducing a DNA template similar to the CRISPR cut's surroundings but with a modification. The cell uses this template for repair, resulting in precise, controllable DNA editing at the target location. In addition, the repairs done by end joining i.e., microhomology-mediated end-joining (MMEJ) and non-homologous end-joining (NHEJ), are not controllable as compared to HDR. End joining can be seen as a haphazard attempt to repair the cut in a way that prevents CRISPR from targeting it again. This process results in diverse and heterogeneous insertions and deletions across different cells. While HDR has been the preferred method in genome editing, end joining is often considered undesirable noise, despite being more efficient than HDR. While HDR has been the preferred method in genome editing, end joining is often considered undesirable noise, despite being more efficient than HDR as shown in Fig. 7.10.

Upon completion of the repair process, various disease-related mutations may occur, including insertions, deletions, frameshifts, inversions, translocations, and point mutations. This field encompasses predicting gene editing outcomes [606].

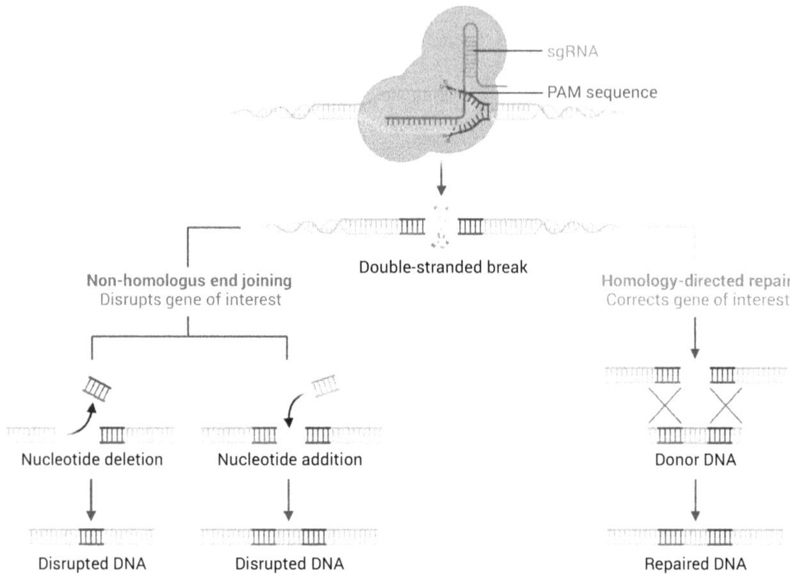

Fig. 7.10 DNA cleavage is repaired by two different pathways i.e., NHEJ and HDJ. NHEJ repairs DNA breaks by directly joining the broken ends, often resulting in small insertions or deletions. HDR uses a homologous sequence as a template to accurately repair DNA breaks which ensures high-fidelity restoration (Image created using Biorender.com)

Table 7.13 Sample statistics of benchmark datasets for gene editing outcome prediction

Name	Cell line	Indels	Deletions	Samples	Link	Year
Apindel [642]	–	–	–	–	https://github.com/MoonLBH/Apindel	2022
Lindel [173]	HEK293T	21	536		https://github.com/shendurelab/Lindel/tree/data_analysis	2019
SPROUT [574]	T cell	9 types statistics of the repair outcomes such as average insertion length.		1603	–	2019
FORECasT [27]	K562, RPE1, iPSC, CHO, HAP1, mESCs	20	420	31,617	https://elixir.ut.ee/forecast/	2019
InDelphi [908]	HEK293, K562, HCT116, mESCs, U2OS	4	149		https://github.com/maxwshen/indelphi-dataprocessinganalysis	2018

This specific task includes multi-class classification with soft labels. For instance, the types of mutations are predicted (MH deletions, MH-less deletions, and 1 bp insertions) with the likelihood of specific mutation.

A handful number of approximately 5 tools for gene editing outcome prediction tools have been developed till now, out of which 3 proposed novel benchmark datasets. For instance, Shen et al. [908] proposed the very first gene editing outcome prediction tool namely, InDelphi. InDelphi managed to predict 90 classes of MH Deletion, 59 classes of Non-MH Deletion, and 4 classes of 1 bp Insertion. Authors created a benchmark dataset of 1095 target sites from mouse and human cells i.e., HEK293, K562, HCT116, mESCs, and U2OS [908]. Similarly, ForeCast generated candidate mutations for each gRNA in synthetic contructs to predict repair outcomes [27]. Overall, it had approximately 440 mutational outcomes and more than 31 thousand samples. SPROUT [574] predicts various statistics related to gene editing outcomes such as the fraction of mutant reads with an insertion/deletion, fraction of total reads with insertion/deletion, average insertion length given an insertion, average deletion length given a deletion, diversity, most likely inserted base pair and finally the edit (mutation) efficiency of the CRISPR outcome [574]. Using convolutional neural networks (CNNs) and neural architecture search (NAS), CROTON [606] automates the prediction of 1 bp insertion and deletion probabilities, as well as deletion and frameshift frequencies, directly from raw sequences without any prior knowledge. CROTON [606] utilized the datasets of ForeCast and SPROUT, where the models were trained on synethic construct dataset from ForeCast and evaluated on endogenous T-cell dataset from SPROUT. Apindel [642] uses ForeCast and Lindel datasets and predicts 557 different labels related to different mutations such as 1bp insertion $\cdots \geq$ 3bp insertions, and 5 kinds of 1bp insertions, 6 kinds of 2bp insertions, $\cdots$ 32 different kinds of 29bp insertions.

7.5.2.8 Acr-associated Proteins

The acr-associated (aca) protein can be described as a defense mechanism of the bacterial cell against acr proteins [1158]. They hinder the acr-protein from blocking the cleavage of the Cas protein. Therefore aca-proteins can be seen and used as a regulatory mechanism for CRISPR gene editing. In this field, researchers try to predict aca proteins and their associated acr-aca operons [1158].

While the genomic locations of acr proteins are diverse, they often coexist near the gene loci of aca-proteins. Their genes oftentimes form an operon with the genes encoding for acr proteins [567]. An operon is a functional unit within the genomic DNA. Identifying these operons contributes to the improvement of gene editing tools [1148].

Although Yang et al. [1158] proposed a framework for the identification of aca proteins and their operons, there is no evidence suggesting the application of AI in this domain or any relevant benchmark datasets [1148].

7.5.2.9 Other Tasks

While tasks like on-target and off-target effects, anti-CRISPR (Acr) proteins, and CRISPR arrays have garnered significant attention, there are other topics within the CRISPR research landscape that remain less explored. These tasks include CRISPR-Cas system identification, and crRNA classification. These topics have been covered in detailed earlier, therefore, hereby we only discuss their relevant datasets.

Padilha et al. [790] constructed CRISPR-Cas systems datasets by collecting Cas protein sequences from classified archaeal and bacterial CRISPR-Cas systems available in public databases such as NCBI. The sequences were clustered using the Markov Cluster Algorithm to identify protein families, and Hidden Markov Model profiles were used to determine the presence of specific proteins within CRISPR-Cas systems. The final dataset consisted of thousands of samples categorized into 17 distinct CRISPR-Cas subtypes, providing a robust foundation for training and evaluating the machine learning models.

Park et al. [797] developed crRNA classification dataset by using CRISPR-Cas systems from the CRISPRCasdb [814] database, with CRISPR arrays labeled by their co-localized Cas system types. The dataset included multiple major classes, each with over 1000 samples.

7.5.2.10 CRISPR Databases for the Development of New Benchmark Datasets

This section provides an overview of databases that can be used to develop novel CRISPR-related benchmark datasets. Additionally, it entails the types and quantities of data available in 17 different databases which can help researchers to identify valuable resources for compiling comprehensive and diverse datasets necessary for effective CRISPR research.

The rapid advancements in CRISPR technology have generated a vast amount of data, leading to the creation of numerous databases. Table 7.14 summarizes the list of public databases categorized based on different CRISPR tasks. These databases encompass data related to various aspects of CRISPR systems, such as CRISPR arrays, acr proteins, operons, Cas proteins, and on/off-target activities. This abundance of data presents a significant opportunity for the development of novel benchmark datasets, which can enhance the performance and accuracy of AI tools designed for CRISPR research.

Databases such as CRISPRBank [98], and CRISPRCasDb [814] provide a wealth of sequences and annotations for CRISPR arrays. These databases include data on hundreds of thousands of CRISPR arrays and spacers from a vast number of bacterial and archaeal genomes. This creates opportunities for training AI predictors to accurately identify and annotate CRISPR arrays in newly sequenced genomes.

Databases such as AcrHub [1063], Anti-Crisprdb [267], AcrDb [450], and AcrCatalog [385] offer extensive data on anti-CRISPR proteins, including their sequences, structures, and functional annotations. These datasets span a wide variety

Table 7.14 A pool of CRISPR related databases

	Database	Data type	URL	Description
CRISPR arrays	CRISPRBank [98]	Arrays, repeats, and spacers in FASTA	Link	CRISPRBank contains analysis of genome from RefSeq 95 July, 2019. Particulalry, CRISPRDetect 2.4 was employed to comprehensively analyze all 151,845 bacterial genomes and 855 archaeal genomes. In total, there are 132,379 CRISPR arrays and 1,992,510 spacers.
	CRISPRCasDb [814]	FASTA	Link	CRISPRCasdb contains comprehensive data on CRISPR-Cas systems including information on 2086 CRISPR arrays and 130,293 spacers, along with details on 19,232 Cas proteins and 7125 associated Cas proteins.
Acr	AcrHub [1063]	XLSX, FASTA	Link	AcrHub offers extensive annotations and functional data for anti-CRISPR associated proteins. It features information on 1800 proteins and their interactions, spanning various species within the bacterial and archaeal domains.
	Anti-Crisprdb [450]	XLSX, CSV, JSON	Link	Anti-CRISPRdb catalogs a wide array of anti-CRISPR proteins with detailed annotations, encompassing sequences and structural data for 1200 proteins. The database covers anti-CRISPR proteins found in numerous bacterial species, providing insights into their diversity and functionalities.
	AcrDb			AcrDB is a comprehensive database providing sequences and structural information of anti-CRISPR proteins. It includes data on 2500 anti-CRISPR proteins across diverse bacterial and archaeal species.
	AcrCatalog [385]	FASTA	Link	AcrCatalog is a specialized database that catalogs anti-CRISPR proteins and their interactions across various CRISPR-Cas systems. It contains sequences, structural information, and functional annotations for approximately 16,919 putative acr proteins. These proteins are associated with specific CRISPR-Cas systems, including Cas-IA to IE, Cas-IIA to IIC, Cas-IIIA to IIID, Cas-IVA, Cas-VA, and Cas-VIA to VIC.

(continued)

Table 7.14 (continued)

	Database	Data type	URL	Description
Aca	UniProt Database [204]	TXT, FASTA, XML, JSON	Link	Universal Protein Resource database is a resource for protein sequence and annotation data
	AcrCatalog [385]	TXT	Link	Anti-CRISPR proteins predicted with ML [384]
	AcrHub [1063]	XLSX, FASTA	Link	AcrHub predicts Anti-CRISPR proteins
Cas	CasPDB [978]	FASTA	Link	CasPDB is an integrated database housing 287 reviewed Cas proteins, 257,745 putative Cas proteins, and 3593 Cas operons from 32,023 bacterial species and 1802 archaeal species. The database comprehensively contains all 3593 putative Cas operons, including 328 operons associated with the type II CRISPR-Cas system.
	CRISPRCasdb [814]	SQL, FASTA	Link	CRISPRCasdb contains CRISPR arrays andcas genes from complete genome sequences
	UniProt Database [204]	TXT, FASTA, XML, JSON	Link	Universal Protein Resource database is a resource for protein sequence and annotation data
	CasPedia [6]	FASTA	Link	CasPedia is an annotated database for Cas proteins from bacteria and archaea, featuring 287 reviewed Cas proteins, 257,745 putative Cas proteins, and 3593 Cas operons from 32,023 bacterial and 1802 archaeal species. It offers free access, a user-friendly interface, and details on all operons, including 328 from the type II CRISPR-Cas system.
On/Off target activity	cripsrSQL	CSV	Link	crisprSQL is a SQL-based database forCRISPR/Cas9 off-target cleavage assays and epigenetically annotated, base-pair resolved cleavage frequency distributions
	Ensembl BioMart	FASTA, GTF,GFF, SQL	Link	Ensembl is a genome browser for vertebrate genomes that supports research in comparative genomics, evolution, sequence variation and transcriptional regulation,BioMart is a data mining tool
	crisprSQL	CSV	Link	crisprSQL is a SQL-based database forCRISPR/Cas9 off-target cleavage assays and epigenetically annotated, base-pair resolved cleavage frequency distributions
	CRISPOR	TSV	Link	Data from CRISPOR paper [387]

of species which provides a comprehensive view of acr proteins diversity. The richness of this data can be utilized to create benchmark datasets for training AI predictors that predict acr proteins based on sequence data. Moreover, these datasets can be used to develop models to map and predict interactions between acr proteins and CRISPR-Cas systems, as well as benchmark tools that annotate the function and efficacy of acr proteins.

The CasPDB [978], CRISPRCasdb [814], UniProt [204], and CasPedia [6] contain extensive data on Cas proteins, including reviewed and putative proteins, as well as comprehensive operon information. This presents opportunities for using these datasets to benchmark AI tools that predict the three-dimensional structures of Cas proteins. Additionally, benchmark datasets can be developed for annotating the function of Cas proteins in various CRISPR-Cas systems, and tools can be created to study the evolutionary relationships between different Cas proteins and operons.

Databases such as crisprSQL [951], Ensembl BioMart [1161] provide valuable data on the on/off-target activity of CRISPR-Cas systems. These datasets include detailed information on off-target cleavage assays and epigenetically annotated cleavage frequency distributions. They offer opportunities to benchmark AI predictors that predict off-target effects of CRISPR-Cas editing and develop datasets that help optimize CRISPR tools for higher specificity and reduced off-target activity. Additionally, these datasets can be used to create benchmarks for assessing the safety and efficacy of CRISPR-based gene editing in various organisms.

In conclusion, the diverse and extensive CRISPR-related databases provide a rich source of data for the development of novel benchmark datasets. These datasets can significantly advance AI tools in CRISPR research, improving the understanding and application of CRISPR technology. By leveraging these opportunities, researchers can enhance the accuracy, functionality, and safety of CRISPR-based applications, driving forward the fields of genetics and biotechnology.

7.5.3 Feature Extraction Methods in AI Driven CRISPR Tasks

This section provides an overview of the most commonly utilized feature extraction methods. First, we categorize these feature extraction methods into groups and then explore their distribution across 10 different CRISPR tasks.

Table 7.15 enlists 75 unique feature extraction methods and their utilization in 10 different CRISPR tasks. These feature extraction methods can be categorized into several groups. For instance, sequence-based feature extraction methods include methods like k-mer [857], nucleotide composition [537], dipeptide composition [1238], position-specific nucleotides [537], one hot encoding [1114], and position-specific scoring matrices [222]. Structural and physicochemical properties-based methods include GC content [751], melting temperature [712], minimum free energy [1183], NetSurfP-3.0 [406] and RaptorX-based [590] secondary structure features, and deletion frequencies [606]. Epigenetic and genomic features-based methods encompass nucleosome positioning [1152], chromatin accessibility [1152],

Table 7.15 Features and feature engineering methods used in the reviewed papers

	Study	Feature engineering
On/Off target activity	[1150, 1204]	K-mer embedding (default embedding layer of keras or Pytorch)
	[766]	GC content, temperature of melting of the DNA duplex, minimum free energy (calculated with ViennaRNA), distance of the target sequence to the closest downstream PAM, location relative to the target gene, HOMO-LUMO energy gap.
	[462]	W2vec embeddings
	[997]	OHE
	[537]	nucleotide composition, position-specific nucleotides, GC content, number of Adenine in the middle, presence of certain motifs
	[712]	Position-independent nucleotides, position-independent dinucleotides, position-specific nucleotides, position-specific dinucleotides, GC content, melting temperature, self-folding energy, Shannon entropy
	[1202]	Basic Sequence features/aligned sequence features of gRNAs and targets, mismatch positions, PAM nucleotides in the target sequences, numbers of mismatches
	[1152]	One hot encoding, nucleosome positioning features, chromatin accessibility, DNA methylation, gene expression
	[298, 579, 1192]	One hot encoding of gRNA and epigenetic information including CTCF binding, H3K4me3, chromatin accessibility, and DNA methylation.
	[1114]	Embedding and one hot encoding
	[828]	sequence-based features, position-independent features, position-specific features, n-gaped di-nucleotides
	[760], [1025]	One hot encoding
	[952]	Guide and target loci, sequence, cell line, assay type cleavage frequency, CRISPRoff score, nucleosome organisation-related features/scores such as GC content, Nucleotide BDM, NuPoP Affinity
	[1183]	GC content, frequency of k-mers, number of poly-T segments, length of the longest poly-T segment, position-dependent k-mer (k=1, 2) instances, melting temperatures, minimum free energy metrics
Acr proteins/Activity	[406]	Pre-trained Evolutionary Scale Modeling (ESM) protein transformer, NetSurfP-3.0 based secondary structure features
	[222]	Dipeptide composition (DPC), Composition-transition-distribution (CTD), Position specific scoring matrices, PSSM-composition, DPC-PSSM, PSSM-AC, RPSSM

(continued)

Table 7.15 (continued)

	Study	Feature engineering
	[590]	RaptorX based structure and solvent accessibility features , Transformer embeddings from ESM-1b, POSSUM, and one hot encoding
	[1238]	Sequence: Amino acid composition (AAC), Pseudo amino acid composition (PAAC), Composition of k-spaced amino acid pair (CKSAAP), dipeptide deviation from expected mean (DDE), dipeptide composition (DPC) Evolutionary: PSSM-composition, DPC-PSSM, PSSM-AC, RPSSM, PSSM-SMITH. Pretrained: LM, SSA, TAPE-Bert, Unirep, W2vec, ESM, ProtTrans
	[383]	Containing Genome is Self-Targeting, Directon Annotated Protein Fraction, Directon Protein Lengths Mean, Directon Size, Protein has HTH-Downstream, Protein Length, Protein Hydrophobicity
	[1033]	PSSM-composition, DPC-PSSM, PSSM-AC, and RPSSM
	[285]	Conjoint-Triad
Edit. out.	[606]	1 bp insertion frequency, 1 bp deletion frequency, deletion frequency, 1 bp frameshift frequency, 2 bp frameshift frequency, total frameshift frequency
	[908]	MH length MH GC frac. Del. length, Del. length
	[27]	–
	[574]	One hot encoding and genomic features
	[642]	GloVe and positional embeddings
C-arrays	[721]	Repeat length, number repeats, repeat similarity, AT richness, average spacer length, spacer similarity, repeat number mismatches, spacer evenness, MFE score, ORF score, tandem protein score, BLAST score known repeats, BLAST score similarly known repeats
	[249]	Randomly initialized Embeddings
C-loci	[857]	k-mer counts (4-mer)
	[751]	Length, GC content, palindromic index, k-mers
Cas proteins	[1157]	Di-peptide composition
	[1194]	Amino Acid Composition (AAC), Adaptive skip dipeptide composition (ASDC), Composition of K-Spaced Amino Acid Pairs (CKSAAP), Dipeptide Deviation from Expected Mean (DDE), Quasi-Sequence Order (QSO), Dipeptide Composition, DPC.

DNA methylation [1152], gene expression, and other epigenetic information. Embeddings-based methods include methods such as GloVe [642], positional embeddings [642], and transformer embeddings from models like ESM-1b [406]. Lastly, miscellaneous methods comprise Shannon entropy [712], cell line characteristics [952], CRISPRoff scores [952], ORF scores [721], tandem protein scores [721], and various repeat and spacer-related representation [721].

A deeper analysis of Table 7.15 reveals that certain feature extraction methods are important in specific CRISPR-tasks because of their ability to capture critical biological information. For on/off-target activity, sequence-based methods such as k-mer, nucleotide composition, and position-specific nucleotides are frequently utilized because they provide detailed insights into the sequence-specific interactions and potential mismatches [298, 579, 1114, 1152, 1183, 1192]. Epigenetic and genomic features, including nucleosome positioning, chromatin accessibility, and DNA methylation, are also crucial as they offer context about genomic accessibility and regulation, influencing CRISPR efficacy [579, 1152]. It is important to mention here that one hot encoding proves to be quite effective in representing gRNA and DNA sequences for on/off-target activity prediction [298, 579, 1114, 1183]. This happens because one hot encoding ensures that all possible sequence variations and mismatches are explicitly represented. This allows ML/DL to accurately learn and differentiate the subtle sequence patterns that influence CRISPR targeting efficacy and off-target effects. By retaining the full granularity of sequence data, one hot encoding helps models identify critical nucleotide positions and motifs that are essential for high-fidelity CRISPR targeting. This precision is particularly important given the potential consequences of off-target effects in CRISPR tasks, making one hot encoding a reliable and effective method for on/off-target activity prediction [298, 579, 1114, 1183].

In acr proteins and their activity prediction, protein-based features like pre-trained models (e.g., ESM), secondary structure features, and dipeptide compositions are vital for understanding protein structure and function, which is key for predicting acr proteins and their activity. Evolutionary features such as position-specific scoring matrices provide insights into conserved sequences and structural stability, enhancing Acr protein prediction accuracy. For gene editing outcomes prediction, structural and physicochemical properties like insertion/deletion frequencies and MH length/GC fraction are essential because they directly influence the types and frequencies of CRISPR edits. Embeddings and learned representations, such as GloVe and positional embeddings, capture contextual and positional information, improving editing outcome predictions. In CRISPR array prediction, sequence-based and miscellaneous features like repeat length, spacer similarity, and AT richness are important as they are specific to the structure and composition of CRISPR arrays, aiding in their identification and classification. For CRISPR Loci, sequence-based and structural features like k-mer counts and palindromic index help characterize the loci where CRISPR systems are integrated, facilitating their identification. In Cas Proteins, protein-based features such as di-peptide composition, amino acid composition, and advanced methods like adaptive skip dipeptide composition (ASDC) and quasi-sequence order (QSO) are crucial for capturing detailed information about protein sequences and structures, essential for predicting Cas protein functions. Overall, sequence-based features and protein-based features are fundamental across multiple tasks, while epigenetic and genomic features are vital for on/off target activity, and structural properties are crucial for acr protein prediction, demonstrating the importance of capturing diverse biological patterns for accurate CRISPR-related predictions.

Though the mentioned and general DL approaches gain more and more popularity, the feature extraction procedure remains a black box for researchers using these approaches in genomics, leading to a lack of interpretability of models [918]. But to truly understand the aspects of CRISPR, interpretability is crucial. Handcrafted features provide a way to interpret and analyze models so that certain aspects of CRISPR can be derived from experiments. Unfortunately, the creation of handcrafted features frequently is a time-consuming and complex task, where many decisions have to be made that influence models' potential outcomes and performance in many ways.

7.5.4 Classifiers and Regressors Utilization in AI-Driven CRISPR Tasks

This section presents insights into distinct types of classifiers and regressors that have been utilized to develop AI-driven applications for 10 distinct CRISPR tasks. It thoroughly examines emerging trends of classifiers and regressors across distinct CRISPR tasks.

Table 7.18 provides an overview of 35 classifiers/regressors that are used to develop AI-driven applications for 10 distinct CRISPR tasks. These classifiers/regressors include: convolutional neural networks (CNNs) [952], recurrent neural networks (RNNs) [997], long short-term memory networks (LSTM) [462], fully convolutional networks (FCN) [590], bidirectional long short-term memory networks (BiLSTM) [1150], neural architecture search for CNNs (NAS-CNN) [606], bidirectional gated recurrent units (BiGRU) [952], gated recurrent units (GRU) [462], bidirectional encoder representations from transformers (BERT) [670], hierarchical neural networks (HNN) [1150], knowledge-infused neural networks (KINN) [1195], attention-based CNN [1204], multilayer perceptrons (MLP) [1114], support vector machines (SVM) [1033], k-nearest neighbors (KNN) [797], logistic regression (LR) [537], random forests (RF) [537], extreme gradient boosting (XGBoost) [537], light gradient boosting machine (LightGBM) [1238], categorical boosting (CatBoost) [1238], extra trees [721], hidden Markov models (HMM) [857], gradient boosting decision trees (GBDT) [1194], classification and regression trees (CART) [790], elastic net logistic regression (ENLOR) [712], iterative random forests (iRF) [766], random intersection trees (RIT) [766], and reinforcement learning in the form of collaborative multi-agent reinforcement learning (CMT-MARL) [70].

These methods can be broadly classified into three different categories i.e., ML, DL, and generic which includes methods from statistics, and reinforcement learning (RL). A deeper analysis of Table 7.18 and Fig. 7.11 reveals that DL-based methods have been utilized the most among the 3 categories. Particularly, CNNs, RNNs, LSTMs, and GRUs have been used commonly. The prime focus of researchers has been on CNNs because of multiple reasons. For instance,

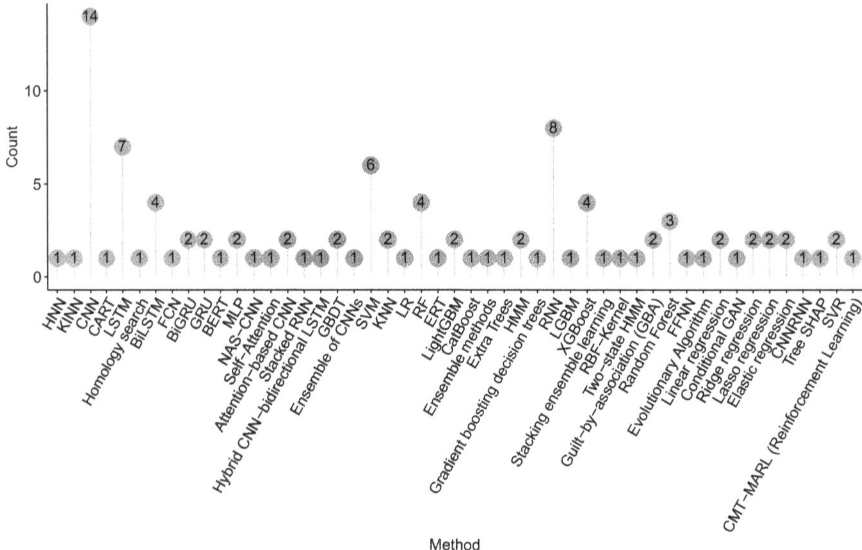

Fig. 7.11 Overall count of classification/regression methods used to develop CRISPR-related applications

CNNs are highly effective in capturing spatial hierarchies in data due to their convolutional layers, making them suitable for various CRISPR-related tasks such as predicting off-target and on-target activities [160]. This is crucial in genome editing applications where understanding and minimizing off-target effects are essential for ensuring precision and safety. In addition to CNNs, methods such as RNNs and their variants (like LSTMs and GRUs) are frequently employed due to their ability to handle sequential data, which is valuable in tasks involving time series or order-dependent biological data such as gene editing outcomes or CRISPR array predictions [291]. Moreover, in some studies, researchers have harnessed the potential of CNNs to extract global features from sequence data and combined this with the contextual learning capabilities of RNNs, LSTMs, and GRUs [1025]. This powerful combination leverages the strengths of both architectures: CNNs excel at identifying spatial patterns and features across the entire sequence, while RNNs, LSTMs, and GRUs are adept at capturing temporal dependencies and sequential relationships. Integration of distinct architectures into one predictor enables researchers to develop more sophisticated predictive pipelines capable of providing more. Despite the advantages of DL methods, they also suffer due to some limitations. First, DL models require an extensive amount of data to train their weights [566]. In addition, a thorough hyperparameter tuning is required to obtain suitable results. Particularly, if training data is not representative of the broader sample population, DL methods can inherit and amplify biases present in the data which leads to faulty and inaccurate predictions [765]. Finally, DL methods are black box, which means that it is challenging to interpret how DL models make their predictions which can be challenging in sequence analysis tasks [1023].

Figure 7.11 reveals that among ML models, SVM, RF, and XGBoost have been commonly utilized [222, 383, 1238]. These models typically perform well under scenarios where features are hand-crafted. Hand-crafted features, derived from domain-specific knowledge, can significantly enhance the performance of ML models by providing relevant and discriminative information [738]. In CRISPR-related studies, carefully designed features that capture the biological intricacies of genomic sequences, protein interactions, and gene editing outcomes can lead to more accurate and reliable predictions. SVMs, with their ability to find optimal decision boundaries, and XGBoost, with its powerful boosting framework, are particularly effective in leveraging these features to achieve high prediction performance. Similar to DL methods, ML methods also have certain limitations. For instance, hand-crafted features are first complex to generate and can increase the dimensionality significantly [156]. Due to this, such models become data sensitive and they fail on noisy samples.

Figure 7.12 illustrates the distribution of AI predictors across various CRISPR tasks. Two distinct patterns emerge from this data: first, the specific methods that have been utilized in multiple CRISPR tasks, and second, the potential opportunities for leveraging existing AI methods to enhance predictive performance across different CRISPR tasks.

Out of 10 CRISPR tasks, 23 unique AI predictors have been utilized in on-target activity. Particularly, DL models such as CNN, RNNs, LSTMs, and ML models such as LR, RF, and XGBoost have been employed commonly as compared to other models. In spite of on-target activity prediction being a regression task, traditional models have not been explored properly such as ridge, lasso, and elastic regression. In terms of off-target activity prediction, RNN and LSTMs have been more commonly used. It is noteworthy to mention here that only for off-target activity prediction [670] the potential of language models is explored. Similarly, for CRISPR arrays, loci, editable gene target identification, cas proteins, and aca have witnessed development of multifarious predictive pipelines but only a few pipelines encompass DL models. In these tasks, CNNs, hybrid models, DL models with attention, and language models potential have not been explored yet.

7.5.5 Experimental Setting and Evaluation Strategies for CRISPR Tasks

In the evaluation of AI predictors for CRISPR tasks, predictors are typically trained and tested in two different experimental settings: cross-validation [790] and independent testing [722]. In cross-validation, first the data is divided into *K* equal subsets. Then the predictor is trained on *K-1* subsets and tested on the remaining subset. This process is repeated k times to ensure each sample participates into

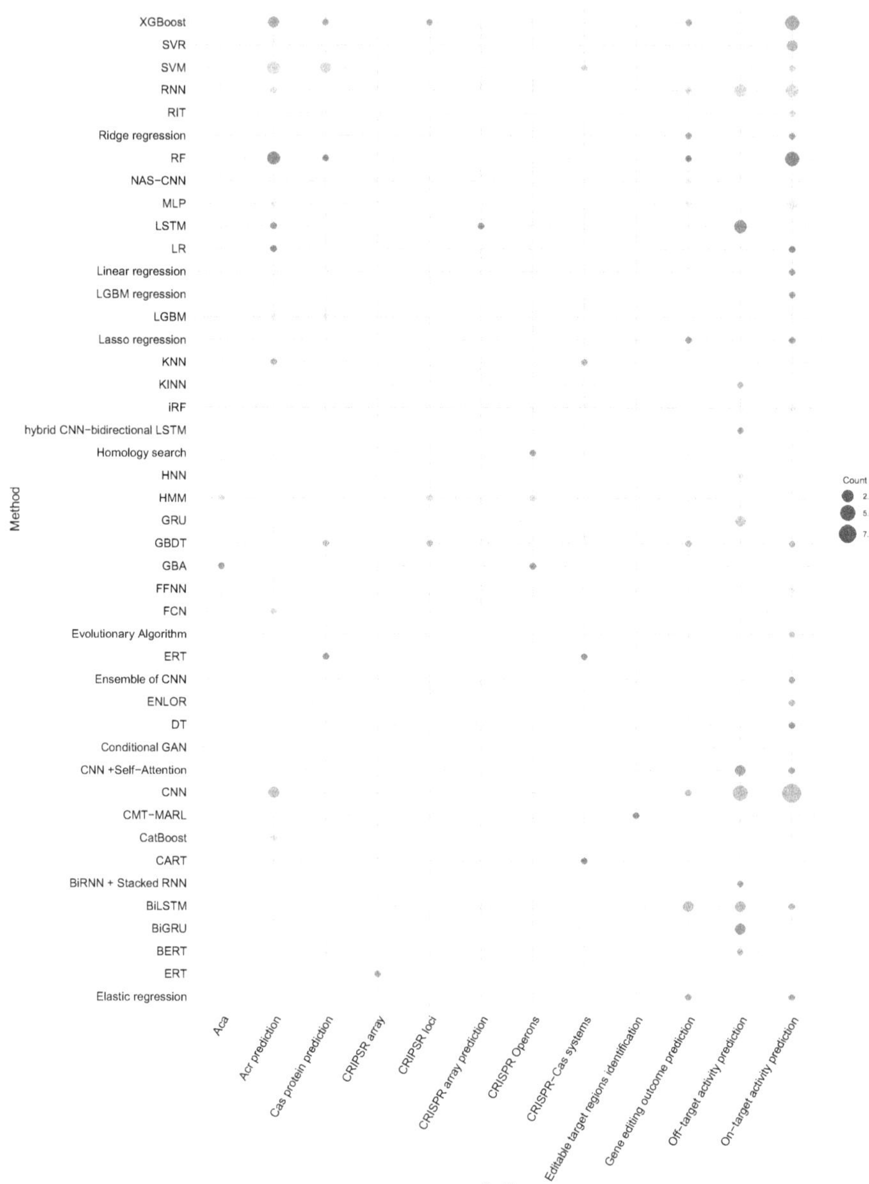

Fig. 7.12 Distribution of AI predictors across different CIRSPR tasks

model training and evaluation [790]. Independent testing, on the other hand, uses a separate dataset that was not involved in the training process which provides an unbiased evaluation of predictor performance. This approach helps to validate the predictor's ability to generalize to new data and also ensures that the observed performance is not due to overfitting on the training data [722].

In the realm of CRISPR tasks, along with these two experimental settings researchers have employed various evaluation metrics to measure predictor effectiveness. Table 7.16 and Fig. 7.13 show that 12 different evaluation measures have been utilized by existing studies. 8 out of 12 different evaluation measures have been used to evaluate classification predictors which include measures like accuracy (ACC) [759], precision [47, 1150], recall [843, 1025], specificity (SP) [222], F1-score [49, 670], area under the ROC curve (AUROC) [462], area under the precision-recall curve (AUPRC) [760], and Matthews correlation coefficient (MCC) [590]. On the other hand, regression-based studies utilized 4 distinct evaluation measures namely, mean squared error (MSE) [952], Pearson correlation coefficient (PCC) [1202], Spearman correlation coefficient (SCC) [1202], and Kendall Tau [642].

For classification tasks, ACC measures the proportion of true results (both true positives and true negatives) among the total number of samples [759]. Precision is the ratio of correctly predicted positive samples to the total predicted positives [1150]. Recall is the ratio of correctly predicted positive observations to all the observations in the actual class [1025]. Specificity is the ratio of true negative predictions to the total number of actual negative instances [222]. F1 score is the weighted average of precision and recall and provides a balance between the two [670]. It is particularly useful when the class distribution is imbalanced. AUROC evaluates the ability of the predictor to distinguish between classes [462], and AUPRC focuses on the performance of the predictor for the positive class [760], especially important in datasets with class imbalance. MCC takes into account true and false positives and negatives, providing a balanced measure even if the classes are of different sizes [590].

For regression tasks, MSE measures the average of the squares of the errors [952]. The Pearson correlation coefficient assesses the linear correlation between predicted and actual values [1202]. The Spearman coefficient measures the rank correlation between predicted and actual values [390, 1202]. Kendall Tau is a statistic used to measure the ordinal association between two measured quantities which is useful to understand the strength and direction of association [642].

Although a plethora of evaluation measures exist for the performance evaluation of AI predictors, it is important to recognize that each evaluation measure has its pros and cons. For instance, metrics like ACC provide a straightforward measure of overall predictor correctness but may not account for class imbalances [474]. Meanwhile, regression metrics like MSE quantify prediction errors but may be sensitive to outliers, while correlation coefficients like Pearson and Spearman assess the strength and direction of relationships but may not capture all nuances of predictive accuracy [507].

Table 7.16 Evaluation metrics used by the reviewed papers

	Study	Evaluation metrics
Targ.	[70]	Average mutual score, hybrid score, microhomology score, vertical score
On-target activity pred.	[1114]	SCC
	[828]	AUROC
	[390]	SCC
	[298]	PC, AUROC
	[579]	AUROC, SCC, PC
	[759]	AUROC, SN, SP, MCC, ACC
	[257]	SCC
	[390]	SCC
	[766]	PC
	[537]	SCC, nDCG, R-Precision, AUC
	[712]	ROC, AUC, Kolmogorov–Smirnov test
	[218]	SCC, MSE
	[1202]	PCC, MSE, Steiger's Test, SCC, ANOVA, Tukey'spost hoc test
	[1152]	MSE, cosine similarity
	[1192]	SCC, AUROC, Kolmogorov–Smirnov test
	[1183]	SCC
	[1204]	SCC, AUROC, PRAUC
Off-target activity pred.	[997]	ACC, Precision, Recall, F1-score, AUROC, AUPRC
	[462]	Validation Loss, ACC, AUPRC, AUROC
	[625]	Recall, Precision, ROC
	[760]	ROC, Precision, Recall, AUPRC
	[1150]	AUPRC, AUROC, Recall, Precision
	[1195]	PCC and AUPRC (for mutations)
	[670]	AUROC, AUPRC, F1 score, and MCC
	[1204]	SCC, AUROC, PRAUC
	[1025]	Recall, Precision, MCC, R-squared
	[952]	MSE, AUPRC
Acr/Activity pred.	[590]	ACC, Precision, Recall, F1-score, MCC
	[406]	F1 score, ACC, AUC, Precision, Recall
	[222]	ACC, SN, SP, MCC, AUC
	[1238]	PRE, SN, SP, F-score, ACC, MCC
	[383]	ROC, AUC, Precision, Recall, ACC
	[1033]	SN, SP, ACC, F-Value, MCC, AUC
	[285]	–
	[736]	Precision, recall, ACC, F1-score, MCC

(continued)

Table 7.16 (continued)

	Study	Evaluation metrics
Mut.	[642]	MSE, AUC, PC, Kendall Tau
	[606]	AUC-ROC, Pearsons Coefficient
	[1202]	
CR. array	[249]	AUC, SN, SP
	[721]	SN, SP, learned evaluation function
	[857]	ACC
	[751]	AUC, Recall, Precision, F-1 score
Cas.	[790]	Adjusted balanced ACC, F-score with macro-averaging, MAE
	[1194]	ACC, SP, SN, MCC
OP.	[1148]	Recall
	[1164]	Recall
ACA	[1158]	Recall
C.-sys.	[790]	ACC, F1-score, SN
	[797]	ACC, F1-score

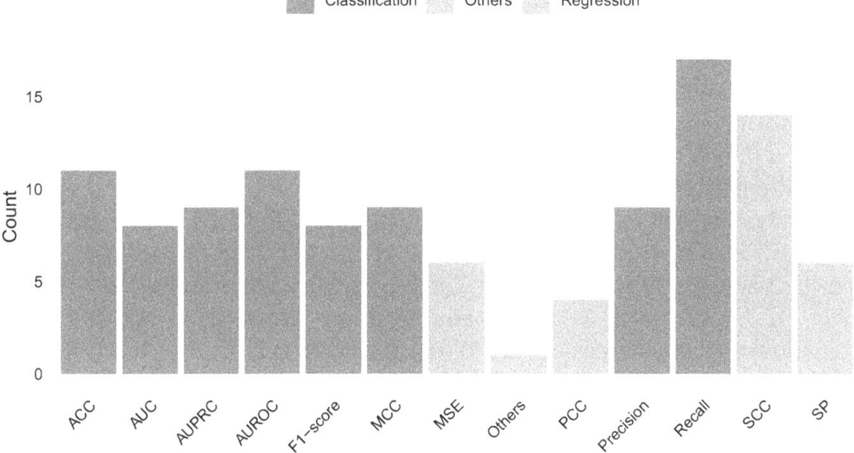

Fig. 7.13 Count of evaluation measures used to assess the predictive performance of classifiers and regressors

It is essential to highlight that some studies did not use a sufficient number of evaluation metrics, which can result in potential issues in the evaluation of AI predictor. For instance, Rafid et al. [828] uses only AUROC, potentially overlooking precision and recall, while Yi et al. [1164] focuses only on recall, which can lead to a high number of false positives if precision is not considered. Niu et al. [760] uses ROC, precision, recall, and AUPRC but omits metrics like the F1-score which provides a balanced view of performance.

Overall, it can be concluded that utilizing limited metrics can lead to an incomplete understanding of predictor performance, especially in imbalanced datasets. Therefore, employing a comprehensive evaluation approach is crucial. Figure 7.13 graphically illustrates the count of evaluation measures used to assess the predictive performance of classifiers and regressors. For classification predictors, combining metrics such as accuracy, precision, recall, F1-score, AUROC, and AUPRC captures various performance aspects, ensuring robustness across scenarios. For regression predictors, using MSE, Pearson, Spearman, and Kendall Tau evaluate both prediction error magnitude and the strength of relationships, offering a holistic view of predictive power and reliability. This comprehensive evaluation strategy is vital for advancing CRISPR research and developing effective gene-editing tools.

7.5.6 Libraries and AI Driven CRISPR Applications Source Codes

This section compiles detailed information on open source predictors and the libraries they leverage in various CRISPR tasks. By providing this comprehensive overview, researchers can build upon existing tools, promote collaboration and advance the development of effective CRISPR prediction predictors.

Table 7.17 encompasses links to the 45 open source code repositories and respective libraries utilized by them. Among the 50 CRISPR-related studies, 40 have provided publicly accessible source code. Among these studies, 22 have utilized Python libraries such as TensorFlow and Keras [257, 390, 462, 537, 579, 606, 642, 670, 721, 759, 828, 1025, 1114, 1150, 1152, 1183, 1192, 1195, 1204, 1204]. Additionally, 15 studies have employed PyTorch [285, 406, 590, 625, 952, 997, 1238]. Other ML libraries such as Scikit-learn, XGBoost, ViennaRNA, and BioPython have also been commonly integrated to design predictors [257, 390, 462, 537, 606, 625, 721, 759, 828, 1025, 1114, 1150, 1152, 1195, 1204].

A detailed analysis of these open source codes reveals that the majority of these tools have been developed using well-established libraries, promoting a standardized approach while also fostering innovation. This integration of well-established libraries contributes to the robustness and effectiveness of CRISPR prediction models within the research community.

The selection of a specific library for CRISPR tasks is inherently subjective and depends on factors such as the preferred development platform, the choice of prediction models, and the specific research questions at hand. Therefore, recommendations are made based on the variety of models and evaluation measures each library offers. For Python, TensorFlow and Keras are highly recommended due to their extensive support for DL models and user-friendly APIs [462, 1150, 1195]. PyTorch is also favored for its flexibility in model development and dynamic computational graphs [625, 952, 997]. Additionally, libraries such as Scikit-learn and XGBoost are valuable for more traditional ML approaches due to their

Table 7.17 Source links and libraries

Problem	Study	Source code	Libraries
Editable target region ident.	[70]	Link	
Off-target activity prediction	[997]	Link	pytorch
	[462]	Link	tensorflow, keras, sklearn, gensim
	[625]	Link	torch, sklearn
	[760]		
	[1150]	Link	tensorflow, keras, sklearn
	[1195]	Link	tensorflow, sklearn
	[670]	Link	tensorflow, keras, sklearn
	[1204]	Link	tensorflow, keras
	[1025]	Link	tensorflow, keras
	[952]	Link	sklearn, pytorch
On-target activity prediction	[1114]	Link	tensorflow, keras, sklearn, biopython, viennarna
	[298]	–	–
	[218]	–	–
	[1202]	–	–
	[1192]	Link	sklearn, keras, tensorflow
	[712]	Link	viennarna
	[1202]	Link	
	[1152]	Link	tensorflow
	[537]	Link	sklearn, xgboost
	[766]	Link1Link2	sklearn
	[1204]	Link	keras, tensorflow
	[390]	Link	keras, sklearn, tensorflow, biopython
	[257]	Link	tensorflow, sklearn
	[828]	Link	sklearn
	[579]	Link	sklearn, keras, tensorflow
	[759]	Link	sklearn, keras
	[1183]	Link	viennarna, biopython
Acr	[406]	Link	bio, biolib, biopython, pytorch, sklearn, transformers
	[590]	Link	torch, sklearn
	[1238]	Link	sklearn, lightgbm, xgboost, catboost
	[383]	Link	sklearn
	[285]	Link	sklearn, xgboost
	[222]	Link	–
CRIPSR array	[721]	Link	keras, sklearn, biopython, viennarna, hmmer, blast
	[857]	Link1Link2	sklearn, biopython, xgboost
Mutations	[642]	Link	mittens, keras, pytorch, tensorflow
	[606]	Link	tensorflow, keras

(continued)

Table 7.17 (continued)

Problem	Study	Source code	Libraries
Cas protein prediction	[1025]	Link	tensorflow, keras, sklearn
	[790]	Link	sklearn, hmmer,
	[797]	Link	–
	[1194]	Link	keras, tensorflow, sklearn, biopython, prodigal, blast, hmmer, viennarna, xgboost
	[952]	Link	torch, tensorflow, sklearn, xgboost
Operons	[1148]	Link	biopython
	[1164]	Link	CRISPRCasFinder, psiblast+, blastn (NCIB), blastp (NCIB)
Aca	[1158]	Link	VIBRANT, cctyper, diamond, hmmer, prodigal

comprehensive suite of algorithms and ease of integration into various workflows [285, 537]. Ultimately, selecting the right library aligned with individual research needs not only streamlines the development process but also enhances the overall reliability and effectiveness of CRISPR prediction models.

7.5.7 Performance Values of AI-Predictors in CRISPR

This section presents the predictive performance values of 37 AI predictors across 10 different CRISPR tasks, evaluated on 77 benchmark datasets. In-depth analyses of these predictors using evaluation measures such as precision, recall, F1-score, SCC, PCC, and AUC, offer insights into the strengths and weaknesses of various feature extraction and classification methods specific to different CRISPR tasks. This comprehensive analysis aids in selecting the most suitable classifiers and feature extraction methods, optimizing experimental design. Additionally, it identifies tasks for improvement, promoting innovation in AI predictor development and facilitating cross-disciplinary research.

Table 7.18 presents performance values of 37 predictors across 8 different CRISPR tasks namely, i.e., on/off-target activity prediction, CRISPR array identification, Cas proteins prediction, acr and aca proteins identification, acr proteins activity prediction, and gene editing outcome prediction. It encompasses predictive performance values in terms of 13 distinct evaluation measures ACC, SN, SP, AUCROC, FP, F1, MCC, AUPRC, Cohen's Kappa, R^2, MSE, SCC, and PCC. Two different types of trends can be observed here i.e., which feature extraction and AI method performs well in a single task and secondly which specific set of feature extraction and AI method performs well across multiple CRISPR tasks.

For CRISPR array identification, the predictor by [249] demonstrates superior performance. This approach uses randomly initialized embeddings for representation learning and an LSTM classifier, achieving high predictive performance: ACC:

Table 7.18 Performance values of 39 different predictors across 76 different benchmark datasets related to 10 different CRISPR tasks

	Name	Author, year	Dataset	Rep. learning	Classifier	Performance
B. Cls.	CRISPR array	[249]	[249]	Randomly initialized embeddings	LSTM	Fivefold: ACC: 94.58, SN:91.99, SP:97.17, AUCROC:98.72
		[721]				
MC Cls.	CRISPR loci	[857]	[857]	k-mer counts (4-mer)	gradient boosting decision trees	IND: med ACC: 98.6, FP: 28 (0.4)
		[751]	[751]	Length, GC content, palindromic index, k-mers	multivariate logistic regression, XGBoost, OVA XGBoost	OVA (one-vs-all) XGBoost: F1: 0.97
B.Cls	Cas protein	[1157]	[1157]	Di-peptide composition	SVM	ind: SN: 83.71, SP: 86.77, ACC: 84.84, MCC: 70.0, AUCROC: 89.45
		[1194]	[1194]	AAC, ASDC, CKSAAP, DDE, QSO, DPC, PSSM, AATP, Pse-PSSM, TTri-gram-PSSM, CTD, CTDC, CTDT Transition, UniRep	Stacked ML: Baseline Classifiers (LGBM, RF, ERT, GBDT, XGBoost), Meta Classifier: SVM	Cas300: ind: ACC: 97.28, MCC: 0.944, SN: 97.71, SP: 96.31
						Cas300: ind: ACC: 94.07, MCC: 0.866, SN: 96.61, SP: 91.52
	Anti-CRISPR proteins	[590]	AcrNet-1	RaptorX based structure and solvent accessibility features , Transformer embeddings from ESM-1b, POSSUM, and one hot encoding	CNNs and FCNs	fivefold: ACC: 79.79, P: 83.63, SN: 68.10, F1: 75.05, MCC: 59.24

(continued)

Table 7.18 (continued)

	Name	Author, year	Dataset	Rep. learning	Classifier	Performance
			Gussow			fivefold: ACC: 95.43, P: 97.62, SN: 93.60, F1: 95.53, MCC: 91.01
			AcrNet-2			fivefold: SP: 95.77, ACC: 89.42, P: 48.95, SN: 76.00, F1: 63.27, MCC: 52.39
			AcrNet-3			fivefold: SP: 95.38, ACC: 77.36, P: 31.06, SN: 73.72, F1: 44.32, MCC: 36.55
			Combined (1,2,3)			fivefold: ACC: 94.42, P: 94.71, SN: 94.09, F1: 94.18, MCC: 88.83
		[222]	AcrPred	DPC, CTD, PSSM, PSSM-composition, DPC-PSSM, PSSM-AC, RPSSM	ML ensemble	fivefold: SN: 92.3, SP: 87.7, ACC: 88.1, AUC: 95.2
		[1238]	PreAcrs	PSSM-AC, RPSSM and SSA	LR ensemble of SVM, KNN, MLP, LR, RF, XGBoost, LightGBM, CatBoost	IND: P: 98.6, SN: 79.5, SP: 98.9, F1: 88.1, ACC: 89.2, MCC: 79.9, AUC: 97.2, AUPRC: 97.6
		[383]	[383]	self-targeting genomes, annotated protein fractions in directions, protein lengths, presence of HTH domains downstream, directon size, and protein hydrophobicity	Extra Trees	IND: AUC-ROC: 83.0, 15-fold: 93.0
		[1033]	[1033]	PSSM-Composition, DPC-PSSM, PSSM-AC, and RPSSM	SVM	IND: SN: 90.9, SP: 85.6, ACC: 88.2, F-value: 88.3, MCC: 76.5
		[285]	[285]	AAC, Grouped Dimer and Trimer Frequency Counts	XGBoost	–
				QSO, DPC.		

B.Cls.	Off-target Activity Prediction	[997]	[191]	OHE	LSTM (Best)	IND: ACC:99.7, P: 73.4, SN: 61.1, F1: 66.7, AUCROC; 99.0, AUPRC; 72.11
		[462]	[191]	W2vec embedding	BiLSTM	K562: IND: ACC: 99.40, AUPRC:86.67, AUROC:99.61
						HEK293T: IND: ACC: 99.40, AUPRC:66.20 AUROC:99.21
		[760]	CIRCLE-seq [1003]	OHE	LSTM	fivefold: AUCROC: 97.6, AUORC: 48.0
			PKD (II/1) [262]			–
			Digenome PDH (II/2) [387]			–
			II/3 SITE			–
			II/4 [1002]			Train (II-4), Test (II-5): IND: AUROC: 99.1, AUPRC: 31.9,
			II/5 [529]			Train (CIRCLE, PCR, Digenome, SITE, and II-4), Test (II-5): (AUROC = 99.3, AUPRC = 29.7),"Train (CIRCLE, PKD, PDH, GUIDE-I), Test: (AUROC = 98.9, AUPRC = 25.4)"
						Train (PKD, PDH, SITE, GUIDE-I), Test: (AUROC = 99.1, AUPRC = 31.9)

(continued)

Table 7.18 (continued)

	Name	Author, year	Dataset	Rep. learning	Classifier	Performance
						Train (CIRCLE), Test: (AUROC = 99.3, AUPRC = 17.3)
						Train (CIRCLE, PKD, PDH, SITE, GUIDE-I), Test (AUROC = 99.1, AUPRC = 31.2)
						Train (SITE), Test: (AUROC = 99.1, AUPRC = 25.01)
						Train (PKD, PDH, GUIDE-I), Test: (AUROC = 99.2, AUPRC = 26.5)
						Train (CIRCLE, SITE), Test: (AUROC = 99.4, AUPRC = 22.0),"Train (CIRCLE, PKD, PDH, SITE, GUIDE-I), Test: (AUROC = 99.3, AUPRC = 13.1)"
			II/6 [636]			Train (PKD, PDH, SITE, GUIDE-I), Test: (AUROC = 99.8, AUPRC = 18.4)
						Train (CIRCLE, PKD, PDH, SITE, GUIDE-I), Test: (AUROC = 99.92, AUPRC = 14.3)
						Train (PKD, PDH, GUIDE-I), Test: (AUROC = 99.94, AUPRC = 15.0)
						Train (CIRCLE, PKD, PDH, SITE, GUIDE-I), Test: (AUROC = 99.6, AUPRC = 11.9)

		[1150]	I/1 [1003]	Embedding	BiLSTM	fivefold: AUPRC: 58.58, AUCROC: 98.74
			I/2 [636]			–
			II/1 [262]			Train (I/1,II/5, I/2) AUCROC:87.31; AUPRC: 53.21
			II/2 [387]			Train (I/1,II/5, I/2) AUCROC:87.31; AUPRC: 53.21
			II/3 [128]			AUPRC:79.6
			II/4 [1002]			–
			II/5 [529]			–
			II/6 [636]			–
			K562 [191]			AUCROC: 99.79 AUPRC: 80.49
			HEK293T [191]			AUCROC:98.79, AUPRC:78.39
		[670]	K562 [191]	Positional encoding and OHE	BERT	AUROC: 99.9, PRAUC: 97.6, F1: 88.9, MCC: 88.6
			HEK293T [191]			AUROC: 97.0, PRAUC: 52.2, F1: 33.9, MCC: 40.4
			II/4 [1002]			IND: Train (HEK293t, K562, II5 (combined)), AUROC: 99.8, PRAUC: 63.0, F1: 48.0, MCC: 53.2
			II/5 [529]			AUROC:0.998, PRAUC: 0.444, F1 score: 0.333, MCC: 0.344
			II/6 [636]			AUCROC: 99.7, AUPRC: 44.4

(continued)

Table 7.18 (continued)

	Name	Author, year	Dataset	Rep. learning	Classifier	Performance
			I/1 [1003]			AUROC: 98.7, PRAUC: 76.4, F1: 64.6, MCC: 65.5
			I/2 [636]			AUROC: 99.8, PRAUC: 64.1, F1: 56.4, MCC: 59.6
		[1204]	K562 [191]	Embedding	CNN + attention	AUROC: 99.4, PRAUC: 81.6
			HEK293T [191]			IND: AUROC: 97.3, PRAUC: 79.0
		[1025]	I/1 [1003] (negative data doesnot have Cas-cleavage)	OHE	CNN+BiLSTM	IND: P: 91.0 SN: 87.0 F1: 89.0, MCC: 80.0, Cohen's Kappa: 0.77, R: 0.71
		[952]	CrisprSQL [951]	OHE, and nucleosome+epigenetic features	CNN+BiGRU	AUCROC: 99.5, AUPRC: 78.2
		[1195]	II [636]	OHE	CNN	AUPRC: 26.2
			II/5 [529]			AUPRC: 36.4
			II [636]			AUPRC: 32.4
		[625]	II/4 [1002]	OHE	CNN	fivefold: AUCROC:98.0, AUPRC:32.0
			II/2 [387]			fivefold: AUC-ROC: 98.0, AUPRC:42.0
	Gene editing outcome	[642]	Lindel	Glove and Positional encoding	BiLSTM + Attention	IND: MSE: 0.000164

			Forecast			Deletion frequency (AUC: 0.91, PCC: 0.53) 1 bp Insertion frequency (AUC: 0.94, PCC: 0.86) 1 bp Deletion frequency (AUC: 0.83, PCC: 0.70) 1 bp Frameshift frequency (AUC: 0.77, PCC: 0.43) 2 bp Frameshift frequency (AUC: 0.73, PCC: 0.46) Frameshift frequency (AUC: 0.69, PCC: 0.26)
			CROTON			Deletion frequency (AUC: 84.6, PCC: 0.7), 1 bp Insertion frequency (AUC: 88.1, PCC: 76.9)
Regression	On-target Activity	[1114]	SpCas9-HF (ESP) [1042]	OHE + Embedding	CNN, RNN+Attention	IND: (SCC: 0.867; Mean: 3.71×10^{-3})
			SpCas9-HF1 [1042]			IND: (SCC: 0.867; Mean: 2.65×10^{-3})
			WT-SpCas9 [1042]		IND: (SCC: 0.872; Mean: 2.55×10^{-3})	
		[257]	Chari	OHE	CNN	SCC: 0.49
			Wang/Xu			SCC: 0.69
			Doench Mouse			SCC: 0.51
			Doench Human			SCC: 0.23
			Hart HCT116			SCC: 0.55
			Moreno-Mateos			SCC: 0.19

(continued)

Table 7.18 (continued)

	Name	Author, year	Dataset	Rep. learning	Classifier	Performance
			Gandhi			SCC: 0.36
			Farboud			SCC: 0.60
			Varshney			SCC: 0.35
			Gagnon			SCC: 0.35
Classification		[828]	Hart HCT116	Position Independent Features (PIF), Position Specific Features (PSF), and n-Gapped Di-nucleotides (nGD).	SVM	AUCROC: 87.9
			Chari 293T			AUCROC: 44.4
			Hart HeLA			AUCROC: 79.7
			Xu HL60			AUCROC: 75.9
AUC-ROC: 0.9817			Chari 293T			AUC-ROC: 99.62
			Hart HeLA			AUC-ROC: 97.23
			Xu HL60			AUC-ROC: 98.42
Regression		[298]	Hart HCT116			PCC: 0.6696
			Chari 293T			PCC: 0.7417
			Hart HeLA			PCC: 0.6247
			Xu HL60			AUCROC: 75.9 PCC: 0.5913

Regression		[579]	Hart HCT116	Sequence and epigenetic features based OHE	CNN	fivefold: SCC: 0.6548
			Chari 293T			fivefold: SCC: 0.7352
			Hart HeLA			fivefold: SCC: 0.6397
			Xu HL60			fivefold: SCC:0.5473
Classification		[579]	Hart HCT116			fivefold: AUCROC: 97.32
			Chari 293T			fivefold: AUCROC: 99.65
			Hart HeLA			fivefold: AUCROC: 97.14
			Xu HL60			fivefold: AUCROC: 97.06
		[759]	Glycine	OHE	CNN	ACC: 82.43, AUCROC: 85.29, SE: 99.83, SP: 64.25, MCC: 67.91
			Zea			ACC: 81.26, AUCROC: 80.94, SE: 81.83, SP: 81.66, MCC: 71.56
			Sorghum			ACC: 78.25, AUCROC: 83.06, SE: 79.00, SP: 77.50, MCC: 55.27
			Triticum			ACC: 87.49, AUCROC: 78.95, SE: 96.47, SP: 77.52, MCC: 68.26
		[1204]	SpCas9-HF (ESP) [1042]	Embedding	CNN+Attention	10-fold: SCC: 0.850
			SpCas9-HF1 [1042]			10-fold: SCC: 0.853
			WT-SpCas9 [1042]			10-fold: SCC: 0.848

(continued)

Table 7.18 (continued)

	Name	Author, year	Dataset	Rep. learning	Classifier	Performance
			Sniper-Cas9 [525]			10-fold: SCC: 0.931
			xCas9 [525]			10-fold: SCC: 0.864
Regression		[766]	Hart HCT116, Chari 293T, Hart HeLA, Xu HL60,	raw values, one-hot encoding, quantum chemical properties (QCT), and k-mers	iRF	R2: 0.229489979, PCC: 0.486193
			Combined: Hart HCT116, Chari 293T, Hart HeLA, Xu HL60, and Donech HM			R2: 0.211671332, PCC: 0.4964907
			Donech HM			R2: 0.389120714, PCC: 0.6525512
			Combined: Guo, Hart HCT116, Chari 293T, Hart HeLA, Xu HL60, and Donech HM			R2: 0.486194, PCC: 0.6972761, [E.coli 0.504], [H.sapien 0.491]
			Guo E.coli			R2: 0.249, PCC: 0.5019173

Regression		[1202]	Hart HeLa-Lib1	OHE	BiLSTM	10-Fold: SCC: 0.438
			Hart HCT116-Lib1			10-Fold: SCC: 0.479
			Hart RPE			10-Fold: SCC: 0.375
			Hart HeLa-Lib2			10-Fold: SCC: 0.493
			Doench A375			10-Fold: SCC: 0.471
			Xu HL60			10-Fold: SCC: 0.622
			Xu KMB7			10-Fold: SCC: 0.644
			Chari 293T			10-Fold: SCC: 0.52
			Doench MM			10-Fold: SCC: 0.645
			Doench MOLM13			10-Fold: SCC: 0.705
			Endo-293T			10-Fold: SCC: 0.652
			Endo-H1			10-Fold: SCC: 0.468
			Endo-K562			10-Fold: SCC: 0.503
		[1152]	CRISPRoff-tiling	OHE	CNN	AUC: 87.9, SCC: 0.58-0-60
			CRISPRoff-genome			AUC: 68.7
			CRISPRi-activityscore			AUC: 71.6
			CRISPRi-K562			AUC. 83.3
			hCRISPRa-V2			AUC: 71.6

(continued)

Table 7.18 (continued)

	Name	Author, year	Dataset	Rep. learning	Classifier	Performance
			CRISPRi-genome			AUC: 61.9
			hCRISPRi-V2			AUC: 60.9
Regression		[1192]	Combined: Hart HCT116, Chari 293T, Hart HeLA, Xu HL60	OHE of sequence and epigenetic features	CNN-SVR	10-Fold: HCT116 (SCC: 0.719, AUCROC: 0.933) HEK293T (SCC: 0.807, AUCROC: 0.983) HELA (SCC: 0.699, AUCROC: 0.933) HL60 (SCC: 0.589, AUCROC: 0.934)
			Hart HCT116			leave one cell out: SCC: 0.719, AUCROC: 93.3
			Chari 293T			leave one cell out: SCC: 0.807, AUCROC: 98.3
			Hart HeLA			leave one cell out: SCC: 0.699, AUCROC: 93.3
			Xu HL60			leave one cell out: SCC: 0.589, AUCROC: 93.4
		[1183]	Train: Combined (HTCas9 Kim, Xiang-gRNA)	GC, k-mer frequencies (k=1, 2, 3), poly-T segments features, melting temperatures , and minimum free energy	LightGBM regression	–
			Chari 293T			SCC: 0.466

			Doench Hs			SCC: 0.704
			Doench Mm			SCC: 0.603
			Doench azd-hg19			SCC: 0.413
			Hart HCT1162Lib1			SCC: 0.479
			Hart HeLaLib1			SCC: 0.45
			Hart HeLaLib2			SCC: 0.503
			Hart RPE1			SCC: 0.355
			Xu HL60			SCC: 0.604
			Xu KBM7			SCC: 0.622
			Gagnon			SCC: 0.29
			Moreno-Mateos			SCC: 0.181
			Varshney			SCC: 0.327
	Cas. sys.	[790]	–	HMM profiles of cassettes	SVM, HMM, CART, ERT	IND: ACC: 98.56, 50-fold: CART Mean F1: 0.97, SVM Mean F1: 0.98,ERT Mean F1: 0.99
	CrRNA	[797]	[797]	–	KNN	F1: 89.0, and ACC: 92.3

94.58, SN: 91.99, SP: 97.17, and AUCROC: 98.72 in a fivefold cross-validation setting. The use of LSTM classifiers is particularly effective here due to their ability to capture long-range dependencies in sequence data, which is crucial for identifying complex patterns in CRISPR arrays.

In the domain of CRISPR loci classification, the predictor by [857] performs exceptionally well with a k-mer counts (4-mer) feature extraction method and a gradient boosting decision trees classifier, yielding a median accuracy (ACC) of 98.6 and a false positive count (FP) of 28 (0.4). K-mer based features effectively capture sequence composition, while gradient boosting classifiers leverage these features to distinguish between different loci types. Additionally, the predictor by [751], which utilizes features such as length, GC content, palindromic index, and k-mers, combined with multivariate logistic regression, XGBoost, and OVA XGBoost classifiers, achieves an F1-score of 0.97. This combination is effective because it integrates both sequence and structural information, enhancing predictive accuracy.

For Cas protein prediction, two approaches stand out. The predictor developed by [1157] uses di-peptide composition for feature extraction and an SVM classifier, achieving metrics of SN: 83.71, SP: 86.77, ACC: 84.84, MCC: 0.70, and AUCROC: 0.8945. Di-peptide composition captures essential biochemical properties of proteins, while SVM classifiers effectively separate classes in high-dimensional spaces. Another significant predictor is by [1194], which employs a wide range of features including AAC, ASDC, CKSAAP, DDE, QSO, DPC, PSSM, AATP, Pse-PSSM, TTri-gram-PSSM, CTD, CTDC, CTDT Transition, and UniRep. This method uses a stacked ML approach with baseline classifiers such as LGBM, RF, ERT, GBDT, and XGBoost, and a meta classifier SVM. The performance on the Cas300 dataset is outstanding with ACC: 97.28, MCC: 0.944, SN: 97.71, and SP: 96.31. The combination of diverse peptide features and stacked ML models is effective because it captures various aspects of protein sequences, enhancing prediction robustness.

In the area of acr protein prediction, the work by [590] shows that using RaptorX-based structure and solvent accessibility features, transformer embeddings from ESM-1b, POSSUM, and one-hot encoding, processed through CNNs and FCNs, achieves strong results. For instance, AcrNet-1 achieves ACC: 0.7979, P: 0.8363, SN: 0.6810, F1: 0.7505, and MCC: 0.5924, while the combined dataset achieves ACC: 0.9442, P: 0.9471, SN: 0.9409, F1: 0.9418, and MCC: 0.8883. These methods work well because they combine structural, sequence, and embedding features, providing a comprehensive representation of proteins. Similarly, the predictor by [222], which uses PSSM-based features and an ML ensemble classifier, achieves high metrics with fivefold cross-validation, including SN: 0.923, SP: 0.877, ACC: 0.881, and AUC: 0.952. The success of PSSM-based features and ensemble classifiers lies in their ability to capture evolutionary information and aggregate multiple models' strengths.

In the domain of gene editing outcome prediction, the study by [642] uses GloVe and positional encoding with BiLSTM and attention mechanisms, achieving an MSE of 0.000164 and high AUC and PCC values for various editing outcomes. Word embeddings like GloVe capture contextual meaning, while attention-based

BiLSTM models are adept at handling dependencies and variations in gene editing outcomes.

The prediction of on/off-target activity presents unique challenges due to significant variability in datasets and predictive models. Each study often uses different datasets, making direct comparisons difficult and limiting the generalizability of findings. For instance, [997] utilizes the [191] dataset OHE and LSTM classifiers, achieving high performance metrics such as ACC: 0.997 and AUCROC: 0.990. While the use of publicly available datasets like [191] enhances reproducibility, differences in cell types and experimental conditions between studies still pose challenges for direct comparison. In another example, [462] utilizes word2vec embeddings and BiLSTM on the K562 cell line dataset, achieving ACC: 99.40 and AUPRC: 86.67. While embeddings capture semantic information effectively, the results are highly specific to the K562 dataset, complicating generalizability. The study by [760] uses multiple datasets, including CIRCLE-seq [1003], PKD (II/1) [262], and others, with LSTM and OHE, yielding an AUROC of 0.976. The use of diverse datasets aims to improve generalizability but introduces variability in experimental conditions, making uniform assessment challenging. Similarly, [1150] employs several datasets, such as I/1 [1003] and II/2 [387], with BiLSTM and embedding, achieving an AUPRC of 58.58 and AUCROC of 98.74. The inclusion of diverse datasets enhances robustness but complicates performance evaluation due to varying dataset characteristics.

Finally, here we make some recommendations related to CRISPR tasks and the use of feature extraction methods and AI predictors. Feature extraction methods such as k-mer counts, PSSM, structural features from RaptorX, OHE, and random and transformer embeddings consistently show high performance across various CRISPR tasks. These methods should be prioritized because they capture essential biological and sequence-specific information. Ensemble methods (e.g., gradient boosting, XGBoost), LSTM/BiLSTM, and attention-based neural networks prove effective due to their ability to handle complex patterns and integrate diverse features.

7.5.8 Discussion

The integration of AI has markedly enhanced the efficiency and accuracy of CRISPR systems, particularly in the identification of target sites, prediction of off-target effects, and optimization of gene editing outcomes. Our analysis reveals that while AI predictors have been developed for 10 different CRISPR tasks, there is a notable emphasis on the prediction of on/off-target activities and acr proteins. These tasks are critical due to their direct impact on the specificity and safety of CRISPR-based genome editing. The ability to predict and minimize off-target effects is paramount in ensuring the precision and efficacy of CRISPR interventions [191, 1003]. Despite these advancements, there remains a substantial need for further innovation to address the complexities inherent in genetic diseases and the variability in individual genetic makeup.

In terms of 10 different CRISPR tasks, 75 distinct benchmark datasets have been developed i.e., CRISRP arrays: 2, CRISPR loci:2, Cas-proteins: 2, acr proteins: 9, acr proteins activity: 2, off-target activity: 15, on-target activity: 39, and gene editing outcomes: 4. In the current research landscape of AI in CRISPR tasks, the majority of studies rely on public datasets rather than proprietary in-house data. This trend ensures fair performance comparisons between new predictors and existing models. Despite the vast array of available datasets, only a few are commonly utilized. This heterogeneity among different studies for a single application can result in models that perform well on certain datasets but poorly on others, limiting their applicability in real-world scenarios. For instance, off-target activity prediction models like DeepCRISPR has been evaluated on 2 cell lines datasets, such as K562 and HEK293T [191]. Other off-target activity prediction models, such as RCrispr, have been evaluated on datasets like CIRCLE-seq [1003] and PKD [262]. MisIndel used datasets like I/1 [1003], I/2 [636], and II/3 [128]. Models like piCRISPR [951] and hybrid multitask [1003] have also limited cohort of datasets such as CrisprSQL and I/1. On-target activity prediction models have been evaluated similarly on limited number of datasets. For example, AttCrispr, was evaluated on datasets like SpCas9-HF1 and WT-SpCas9 [1042], and CRISPRPred was evaluated on Hart HCT116 and Chari 293T [828]. GanOnTarget used datasets such as Hart HCT116, Chari 293T, Hart HeLA, and Xu HL60 [298]. Models like CnnXg [579] and quantum [766] have utilized a few datasets including Hart HCT116, Chari 293T, Hart HeLA, and Xu HL60. Similar trends exist for other CRISPR tasks where benchmarking is not conducted properly across a broad cohort of datasets, leading to inconsistent performance comparisons and the development of less powerful predictors.

To address these challenges, it is recommended to develop and use standardized, publicly available benchmark datasets and establish consistent evaluation protocols. This would enable more reliable comparisons and enhance reproducibility. By addressing this issue, the field can move towards more robust and generalizable predictive models for CRISPR tasks, advancing both research and clinical applications.

In developing AI predictors for CRISPR tasks, the selection of feature extraction methods and classifiers or regressors should be done carefully, as these choices can significantly impact the model's performance and interpretability. Inappropriate feature extraction methods may fail to capture crucial genomic information, while suboptimal classifiers or regressors can lead to poor prediction accuracy and generalization issues. In terms of CRISPR tasks, methods such as k-mer counts, OHE, and advanced embedding techniques have proven effective in capturing the complex details of DNA, RNA, and protein sequences. However, the potential of 29 different types of embedding methods has yet not been explored in CRISPR tasks, such as DANE [615], DeepWalk [711, 748, 1071, 1240], ELMo [479, 1146, 1193], FastText [562, 1000], GATNE [1175], GEMSEC [123], GraRep [52, 179, 1076], MetaGraph2Vec [276], HAKE [1069], HIN2Vec [1087], HOPE [1075, 1234], Laplacian eigen maps [20], LINE [957, 1058, 1071], Locally linear embedding [20], Mashup [1028, 1101], Node2Vec [20, 607, 1028], OPA2Vec [769], Random Watcher-Walker (RW2) [686], RotatE [131, 816, 1069], RWR [990], SDNE [793, 1071], SocDim [1076, 1136], Struc2Vec [190, 1040], SVD [607, 1237], Topo2Vec [691], TransE [1019], and Graph2vec [746]. Moreover, the effectiveness of 19 distinct language models in CRISPR tasks remains untested. i.e., ALBERT [859, 1159, 1186], AlphaFold [313, 440, 445, 899, 1078], AlphaFold2 [677, 851], BERT [576, 623, 723, 859, 899, 1186], BigBird [212], ELECTRA [35, 1186], ESM-1 [440, 707, 1133], ESM-2 [676, 851, 1133, 1146], GPT [443, 497, 920, 998], Graph Transformer Network [581], Heterogeneous Graph Transformer [1250], IgFold [708], LongFormer [212], RoBERTa [576, 859], T5 [292, 409, 1133, 1179], Transformer [212, 841, 948, 1070], Transformer-XL [196], ULMFiT [704], Vision Transformer [479], and XLNet [1186].

The evaluation of AI-driven CRISPR models has predominantly relied on a range of evaluation measures, including accuracy, precision, recall, and the area under the ROC curve. These metrics provide a comprehensive assessment of model performance, but the variability in evaluation strategies across studies underscores the necessity for standardized usage of evaluation measures. Such standardization would facilitate the comparison of different models and accelerate the development of accurate and better CRISPR applications [760, 1114].

In summary, the integration of AI with CRISPR technology holds immense promise for advancing genetic research and therapy. To fully harness this potential, future research must focus on developing more interpretable AI predictors, standardizing evaluation metrics, and creating comprehensive benchmarking datasets. By addressing these challenges, researchers can enhance the precision, safety, and effectiveness of CRISPR-based interventions, paving the way for groundbreaking advancements in genetic medicine.

References

1. Abbas, Z., Rehman, M., & Chong, K. (2022). TC-6mA-Pred: Prediction of DNA N6-methyladenine sites using CNN with transformer. *2022 IEEE international conference on bioinformatics and biomedicine (BIBM)* (pp. 2506–2510).
2. Abbasi, A., Asim, M., Trygg, J., Dengel, A., & Ahmed, S. (2023). Deep learning architectures for the prediction of YY1-mediated chromatin loops. *International symposium on bioinformatics research and applications* (pp. 72–84).
3. Abbasi, A. F., Asim, M. N., & Dengel, A. (2025). Transitioning from wet lab to artificial intelligence: A systematic review of AI predictors in CRISPR. *Journal of Translational Medicine, 23*(1), 153.
4. Abdullah, S., Kusuma, W., & Wijaya, S. (2022). Prediksi interaksi protein-protein berbasis sekuens protein menggunakan fitur autocorrelation dan machine learning. *Jurnal Teknologi Dan Sistem Komputer*, *10*, 1–11.
5. Abrams, Z., Johnson, T., Huang, K., Payne, P., & Coombes, K. (2019). A protocol to evaluate RNA sequencing normalization methods. *BMC Bioinformatics*, *20*, 1–7.
6. Adler, B., Trinidad, M., Bellieny-Rabelo, D., Zhang, E., Karp, H., Skopintsev, P., Thornton, B., Weissman, R., Yoon, P., Chen, L., Others. (2024). CasPEDIA database: A functional classification system for class 2 CRISPR-Cas enzymes. *Nucleic Acids Research*, *52*, D590–D596.
7. Adnan, M., Islam, W., Waheed, A., Hussain, Q., Shen, L., Wang, J., & Liu, G. (2023). SNARE protein Snc1 is essential for vesicle trafficking, membrane fusion and protein secretion in fungi. *Cells*, *12*, 1547.
8. Aggarwal, S., Gupta, S., Gupta, D., Gulzar, Y., Juneja, S., Alwan, A., & Nauman, A. (2023). An artificial intelligence-based stacked ensemble approach for prediction of protein subcellular localization in confocal microscopy images. *Sustainability*, *15*, 1695.
9. Aghajanbaglo, S., Moosavi, S., Rahgozar, M., & Rahimi, A. (2017). Predicting protein-protein interactions based on rotation of proteins in 3D-space. Preprint. ArXiv:1712.09332.
10. Aguilera-Mendoza, L., Ayala-Ruano, S., Martinez-Rios, F., Chavez, E., Garcıa-Jacas, C., Brizuela, C., & Marrero-Ponce, Y. (2023). StarPep toolbox: An open-source software to assist chemical space analysis of bioactive peptides and their functions using complex networks. *Bioinformatics*, *39*, btad506.
11. Ahmad, S., & Sarai, A. (2004). Moment-based prediction of DNA-binding proteins. *Journal of Molecular Biology*, *341*, 65–71.
12. Ai, H. (2022). Gsca-sdbe: A gene selection method for breast cancer classification based on gsea and analyzing differences in performance metrics. *Plos One*, *17*, Article e0263171.

M. Nabeel Asim et al., *Artificial Intelligence for Molecular Biology*,
https://doi.org/10.1007/978-3-031-90450-9

13. Aida, M. (1988). An ab initio molecular orbital study on the sequence-dependency of DNA conformation: An evaluation of intra-and inter-strand stacking interaction energy. *Journal of Theoretical Biology*, *130*, 327–335.
14. Akalın, F., & Yumuşak, N. (2023). Classification of exon and intron regions on dna sequences with hybrid use of sbert and anfis approaches. *Politeknik Dergisi*, 1–1.
15. Akbar, S., Hayat, M., Iqbal, M., & Jan, M. (2017). iACP-GAEnsC: Evolutionary genetic algorithm based ensemble classification of anticancer peptides by utilizing hybrid feature space. *Artificial Intelligence in Medicine*, *79*, 62–70.
16. Akiyama, M., & Sakakibara, Y. (2022). Informative RNA base embedding for RNA structural alignment and clustering by deep representation learning. *NAR Genomics and Bioinformatics*, *4*, lqac012.
17. Akiyama, M., Sato, K., & Sakakibara, Y. (2018). A max-margin training of RNA secondary structure prediction integrated with the thermodynamic model. *Journal of Bioinformatics and Computational Biology*, *16*, 1840025.
18. Akiyama, M., Sakakibara, Y., & Sato, K. (2022). Direct inference of base-pairing probabilities with neural networks improves prediction of RNA secondary structures with pseudoknots. *Genes*, *13*, 2155.
19. Akmal, M., Rasool, N., & Khan, Y. (2017). Prediction of N-linked glycosylation sites using position relative features and statistical moments. *PloS One*, *12*, Article e0181966.
20. Al Taweraqi, N., & King, R. (2022). Improved prediction of gene expression through integrating cell signalling models with machine learning. *BMC Bioinformatics*, *23*, 323.
21. Alanis-Lobato, G., Andrade-Navarro, M., & Schaefer, M. (2016). HIPPIE v2. 0: Enhancing meaningfulness and reliability of protein–protein interaction networks. *Nucleic Acids Research*, gkw985.
22. Alberts, B., Johnson, A., Lewis, J., Raff, M., Roberts, K., & Walter, P. (2002). The shape and structure of proteins. *Molecular Biology of the Cell* (4th ed.). Garland Science.
23. Alexeev, D., Lipanov, A., & Skuratovskii, Y. I. (1987). Poly (dA) poly (dT) is a B-type double helix with a distinctively narrow minor groove. *Nature*, *325*, 821–823.
24. Alharbi, W., & Rashid, M. (2022). A review of deep learning applications in human genomics using next-generation sequencing data. *Human Genomics*, *16*, 1–20.
25. Ali, A., Zafar, M., Farooq, Z., Ahmed, S., Ijaz, A., Anwar, Z., Abbas, H., Tariq, M., Tariq, H., Mustafa, M., Others. (2023). Breakthrough in CRISPR/Cas system: Current and future directions and challenges. *Biotechnology Journal*, *18*, 2200642.
26. Alkhnbashi, O., Shah, S., Garrett, R., Saunders, S., Costa, F., & Backofen, R. (2016). Characterizing leader sequences of CRISPR loci. *Bioinformatics*, *32*, i576–i585.
27. Allen, F., Crepaldi, L., Alsinet, C., Strong, A., Kleshchevnikov, V., De Angeli, P., Pálenıková, P., Khodak, A., Kiselev, V., Kosicki, M., Others. (2019). Predicting the mutations generated by repair of Cas9-induced double-strand breaks. *Nature Biotechnology*, *37*, 64–72.
28. Allison, L. (2021). *Fundamental molecular biology*. John Wiley & Sons.
29. Alotaibi, F., Attique, M., & Khan, Y. (2021). AntiFlamPred: An anti-inflammatory peptide predictor for drug selection strategies. *Computers, Materials and Continua*, *69*.
30. Al-Saggaf, U., Usman, M., Naseem, I., Moinuddin, M., Jiman, A., Alsaggaf, M., Alshoubaki, H., & Khan, S. (2021). ECM-LSE: Prediction of extracellular matrix proteins using deep latent space encoding of K-spaced amino acid pairs. *Frontiers in Bioengineering and Biotechnology*, *9*, Article 752658.
31. Amdursky, N., Marchak, D., Sepunaru, L., Pecht, I., Sheves, M., & Cahen, D. (2014). Electronic transport via proteins. *Advanced Materials*, *26*, 7142–7161.
32. Amin, N., McGrath, A., & Chen, Y. (2019). Evaluation of deep learning in non-coding RNA classification. *Nature Machine Intelligence*, *1*, 246.
33. Amin, A., Awais, M., Sahai, S., Hussain, W., & Rasool, N. (2021). idrp-pseaac: Identification of DNA replication proteins using general PSEAAC and position dependent features. *International Journal of Peptide Research and Therapeutics*, *27*, 1315–1329.
34. Ammari, M., Gresham, C., McCarthy, F., & Nanduri, B. (2016). HPIDB 2.0: A curated database for host–pathogen interactions. *Database*, *2016*, baw103.

35. An, W., Guo, Y., Bian, Y., Ma, H., Yang, J., Li, C., & Huang, J. (2022). MoDNA: Motif-oriented pre-training for DNA language model. *Proceedings of the 13th ACM international conference on bioinformatics, computational biology and health informatics* (pp. 1–5).
36. An, J., & Weng, X. (2022). Collectively encoding protein properties enriches protein language models. *BMC Bioinformatics*, *23*, 467.
37. Anaya, J., Sidhom, J., Cummings, C., Baras, A., Others. (2020). *Aggregation tool for genomic concepts (ATGC): A deep learning framework for sparse genomic measures*. BioRxiv.
38. Anders, G., Mackowiak, S., Jens, M., Maaskola, J., Kuntzagk, A., Rajewsky, N., Landthaler, M., & Dieterich, C. (2012). doRiNA: A database of RNA interactions in post-transcriptional regulation. *Nucleic Acids Research*, *40*, D180–D186.
39. Anderson, T., & Wheeler, T. (2024). An FPGA-based hardware accelerator supporting sensitive sequence homology filtering with profile hidden Markov models. *BMC Bioinformatics*, *25*, 247.
40. Ansorge, W. (2009). Next-generation DNA sequencing techniques. *New Biotechnology*, *25*, 195–203.
41. Arakawa, K., & Tomita, M. (2007). The GC skew index: A measure of genomic compositional asymmetry and the degree of replicational selection. *Evolutionary Bioinformatics*, *3*, 117693430700300000.
42. Arango-Argoty, G., Heath, L., Pruden, A., Vikesland, P., & Zhang, L. (2021). MetaMLP: A fast word embedding based classifier to profile target gene databases in metagenomic samples. *Journal of Computational Biology*, *28*, 1063–1074.
43. Ashraf, M., Khan, Y., Shoaib, B., Khan, M., Khan, F., Whangbo, T., Others. (2021). BLact-Pred: A predictor developed for identification of beta-lactamases using statistical moments and PseAAC via 5-step rule. *Computational Intelligence and Neuroscience*, *2021*.
44. Asim, M., Ibrahim, M., Malik, M., Dengel, A., & Ahmed, S. (2021). ChrSLoc-Net: Machine learning-based prediction of channelrhodopsins proteins within plasma membrane. *2021 IEEE EMBS international conference on biomedical and health informatics (BHI)* (pp. 1–4).
45. Asim, M., Ibrahim, M., Malik, M., Dengel, A., & Ahmed, S. (2020). Enhancer-DSNet: A supervisedly prepared enriched sequence representation for the identification of enhancers and their strength. *International conference on neural information processing* (pp. 38–48).
46. Asim, M., Ibrahim, M., Zehe, C., Cloarec, O., Sjogren, R., Trygg, J., Dengel, A., & Ahmed, S. (2021). L2S-MirLoc: A lightweight two stage MiRNA sub-cellular localization prediction framework. *2021 International joint conference on neural networks (IJCNN)* (pp. 1–8).
47. Asim, M., Khan, M., Malik, M., Dengel, A., & Ahmed, S. (2019). A robust hybrid approach for textual document classification. *2019 International conference on document analysis and recognition (ICDAR)* (pp. 1390–1396).
48. Asim, M., Malik, M., Dengel, A., & Ahmed, S. (2020). K-mer neural embedding performance analysis using amino acid codons. *2020 international joint conference on neural networks (IJCNN)* (pp. 1–8).
49. Asim, M., Wasim, M., Khan, M., Mahmood, N., & Mahmood, W. (2019). The use of ontology in retrieval: a study on textual, multilingual, and multimedia retrieval. *IEEE Access*, *7*, 21662–21686.
50. Asim, M., Malik, M., Zehe, C., Trygg, J., Dengel, A., & Ahmed, S. (2020). A robust and precise convnet for small non-coding rna classification (rpc-snrc). *IEEE Access*, *9*, 19379–19390.
51. Asim, M., Ghani, M., Ibrahim, M., Mahmood, W., Dengel, A., & Ahmed, S. (2021). Benchmarking performance of machine and deep learning-based methodologies for Urdu text document classification. *Neural Computing and Applications*, *33*, 5437–5469.
52. Asim, M., Ibrahim, M., Malik, M., Zehe, C., Cloarec, O., Trygg, J., Dengel, A., & Ahmed, S. (2022). EL-RMLocNet: An explainable LSTM network for RNA-associated multi-compartment localization prediction. *Computational and Structural Biotechnology Journal*, *20*, 3986–4002.

53. Asim, M., Ibrahim, M., Imran Malik, M., Dengel, A., & Ahmed, S. (2022). Circ-LocNet: A computational framework for circular RNA sub-cellular localization prediction. *International Journal of Molecular Sciences*, *23*, 8221.
54. Asim, M., Ibrahim, M., Zehe, C., Trygg, J., Dengel, A., & Ahmed, S. (2022). Bot-net: A lightweight bag of tricks-based neural network for efficient lncrna-mirna interaction prediction. *Interdisciplinary Sciences: Computational Life Sciences*, *14*, 841–862.
55. Asim, M., Ibrahim, M., Malik, M., Dengel, A., & Ahmed, S. (2022). LGCA-VHPPI: A local-global residue context aware viral-host protein-protein interaction predictor. *Plos One*, *17*, Article e0270275.
56. Asim, M., Fazeel, A., Ibrahim, M., Dengel, A., & Ahmed, S. (2022). MP-VHPPI: Meta predictor for viral host protein-protein interaction prediction in multiple hosts and viruses. *Frontiers in Medicine*, *9*, 1025887.
57. Asim, M. N., Ibrahim, M. A., Zaib, A., & Dengel, A. (2025). DNA sequence analysis landscape: A comprehensive review of DNA sequence analysis task types, databases, datasets, word embedding methods, and language models. *Frontiers in Medicine, 12*, 1503229.
58. Asim, M. N., Ibrahim, M. A., Asif, T., & Dengel, A. (2025). RNA sequence analysis landscape: A comprehensive review of task types, databases, datasets, word embedding methods, and language models. *Heliyon, 11*(2).
59. Asim, M. N., Asif, T., Hassan, F., & Dengel, A. (2025). Protein sequence analysis landscape: A systematic review of task types, databases, datasets, word embeddings methods, and language models. *Database, 2025*, baaf027.
60. Asim, M. N., Asif, T., Mehmood, F., & Dengel, A. (2025). Peptide classification landscape: An in-depth systematic literature review on peptide types, databases, datasets, predictors architectures and performance. *Computers in Biology and Medicine, 188*, 109821.
61. Attique, M., Farooq, M., Khelifi, A., & Abid, A. (2020). Prediction of therapeutic peptides using machine learning: Computational models, datasets, and feature encodings. *IEEE Access*, *8*, 148570–148594.
62. Auron, P., Webb, A., Rosenwasser, L., Mucci, S., Rich, A., Wolff, S., & Dinarello, C. (1984). Nucleotide sequence of human monocyte interleukin 1 precursor cDNA. *Proceedings of the National Academy of Sciences*, *81*, 7907–7911.
63. Aurora, R., Donlin, M., Cannon, N., Tavis, J., Others. (2009). Genome-wide hepatitis C virus amino acid covariance networks can predict response to antiviral therapy in humans. *The Journal of Clinical Investigation*, *119*, 225–236.
64. Ayyadevara, V. (2018). Random forest. *Pro machine learning algorithms: A hands-on approach to implementing algorithms in Python and R* (pp. 105–116). Springer.
65. Azodi, C., Pardo, J., VanBuren, R., Los Campos, G., & Shiu, S. (2020). Transcriptome-based prediction of complex traits in maize. *The Plant Cell*, *32*, 139–151.
66. Azouri, D., Abadi, S., Mansour, Y., Mayrose, I., & Pupko, T. (2021). Harnessing machine learning to guide phylogenetic-tree search algorithms. *Nature Communications*, *12*, 1983.
67. Bachmayr-Heyda, A., Reiner, A., Auer, K., Sukhbaatar, N., Aust, S., Bachleitner-Hofmann, T., Mesteri, I., Grunt, T., Zeillinger, R., & Pils, D. (2015). Correlation of circular RNA abundance with proliferation-exemplified with colorectal and ovarian cancer, idiopathic lung fibrosis and normal human tissues. *Scientific Reports*, *5*, 1–10.
68. Bae, S., Park, J., & Kim, J. (2014). Cas-OFFinder: A fast and versatile algorithm that searches for potential off-target sites of Cas9 RNA-guided endonucleases. *Bioinformatics*, *30*, 1473–1475.
69. Bagaev, D., Vroomans, R., Samir, J., Stervbo, U., Rius, C., Dolton, G., Greenshields-Watson, A., Attaf, M., Egorov, E., Zvyagin, I., Others. (2020). VDJdb in 2019: Database extension, new analysis infrastructure and a T-cell receptor motif compendium. *Nucleic Acids Research*, *48*, D1057–D1062.
70. Baidya, S., Choudhury, S., & De, R. (2022). A novel CRISPR-MultiTargeter Multi-agent Reinforcement learning (CMT-MARL) algorithm to identify editable target regions using a Hybrid scoring from multiple similar sequences. *Applied Intelligence*, *53*, 9562–9579.

71. Balogh, O., Benczik, B., Horváth, A., Pétervári, M., Csermely, P., Ferdinandy, P., & Ágg, B. (2022). Efficient link prediction in the protein-protein interaction network using topological information in a generative adversarial network machine learning model. *BMC Bioinformatics*, *23*, 78.
72. Banerjee, J., & Sen, C. (2013). MicroRNAs in skin and wound healing. *MicroRNA Protocols*, 343–356.
73. Barker, W., George, D., & Hunt, L. (1990). [3] *Protein Sequence Database*. Elsevier.
74. Barman, R., Mukhopadhyay, A., Maulik, U., & Das, S. (2019). Identification of infectious disease-associated host genes using machine learning techniques. *BMC Bioinformatics*, *20*, 1–12.
75. Barrangou, R., & Doudna, J. (2016). Applications of CRISPR technologies in research and beyond. *Nature Biotechnology*, *34*, 933–941.
76. Barrett, T., Wilhite, S., Ledoux, P., Evangelista, C., Kim, I., Tomashevsky, M., Marshall, K., Phillippy, K., Sherman, P., Holko, M., Others. (2012). NCBI GEO: Archive for functional genomics data sets—update. *Nucleic Acids Research*, *41*, D991–D995.
77. Barukab, O., Khan, Y., Khan, S., Chou, K., Others. (2022). DNAPred_Prot: Identification of DNA-binding proteins using composition-and position-based features. *Applied Bionics and Biomechanics*, *2022*.
78. Barukab, O., Ali, F., Alghamdi, W., Bassam, Y., & Khan, S. (2022). DBP-CNN: Deep learning-based prediction of DNA-binding proteins by coupling discrete cosine transform with two-dimensional convolutional neural network. *Expert Systems with Applications*, *197*, Article 116729.
79. Bateman, A., Agrawal, S., Birney, E., Bruford, E., Bujnicki, J., Cochrane, G., Cole, J., Dinger, M., Enright, A., Gardner, P., & Others RNAcentral: A vision for an international database of RNA sequences. *Rna*, *17*, 1941–1946 (2011)
80. Beal, P., & Dervan, P. (1991). Second structural motif for recognition of DNA by oligonucleotide-directed triple-helix formation. *Science*, *251*, 1360–1363.
81. Bechinger, B., & Gorr, S. (2017). Antimicrobial peptides: mechanisms of action and resistance. *Journal of Dental Research*, *96*, 254–260.
82. Beknazarov, N., Jin, S., & Poptsova, M. (2020). Deep learning approach for predicting functional Z-DNA regions using omics data. *Scientific Reports*, *10*, 19134.
83. Belkin, M., & Niyogi, P. (2001). Laplacian eigenmaps and spectral techniques for embedding and clustering. *Advances in Neural Information Processing Systems*, *14*. https://proceedings.neurips.cc/paper_files/paper/2001/file/f106b7f99d2cb30c3db1c3cc0fde9ccb-Paper.pdf.
84. Beltagy, I., Peters, M., & Cohan, A. (2020). Longformer: The long-document transformer. Preprint. ArXiv:2004.05150.
85. Ben Chorin, A., Masrati, G., Kessel, A., Narunsky, A., Sprinzak, J., Lahav, S., Ashkenazy, H., & Ben-Tal, N. (2020). ConSurf-DB: An accessible repository for the evolutionary conservation patterns of the majority of PDB proteins. *Protein Science*, *29*, 258–267.
86. Bengio, Y., Simard, P., & Frasconi, P. (1994). Learning long-term dependencies with gradient descent is difficult. *IEEE Transactions on Neural Networks*, *5*, 157–166.
87. Benson, D., Karsch-Mizrachi, I., Lipman, D., Ostell, J., Rapp, B., & Wheeler, D. (2000). GenBank. *Nucleic Acids Research*, *28*, 15–18.
88. Benson, D., Cavanaugh, M., Clark, K., Karsch-Mizrachi, I., Lipman, D., Ostell, J., & Sayers, E. (2012). GenBank. *Nucleic Acids Research*, *41*, D36–D42.
89. Benson, D., Cavanaugh, M., Clark, K., Karsch-Mizrachi, I., Lipman, D., Ostell, J., & Sayers, E. (2017). GenBank. *Nucleic Acids Research*, *45*, D37.
90. Berman, H., Westbrook, J., Feng, Z., Gilliland, G., Bhat, T., Weissig, H., Shindyalov, I., & Bourne, P. (2000). The protein data bank. *Nucleic Acids Research*, *28*, 235–242.
91. Betsch, C., Schmid, P., Heinemeier, D., Korn, L., Holtmann, C., & Böhm, R. (2018). Beyond confidence: Development of a measure assessing the 5C psychological antecedents of vaccination. *PloS One*, *13*, Article e0208601.

92. Bhardwaj, K., Banyal, S., & Sharma, D. (2019). Artificial intelligence based diagnostics, therapeutics and applications in biomedical engineering and bioinformatics. *Internet of things in biomedical engineering* (pp. 161–187). Elsevier.
93. Bhasin, M., & Raghava, G. (2004). Classification of nuclear receptors based on amino acid composition and dipeptide composition. *Journal of Biological Chemistry*, *279*, 23262–23266.
94. Bhat, A., Nisar, S., & Al, S. (2022). Integration of CRISPR/Cas9 with artificial intelligence for improved cancer therapeutics. *Journal of Translational Medicine*, *20*, 534.
95. Bhattamisra, S., Banerjee, P., Gupta, P., Mayuren, J., Patra, S., & Candasamy, M. (2023). Artificial intelligence in pharmaceutical and healthcare research. *Big Data and Cognitive Computing*, *7*, 10.
96. Bi, Y., Xiang, D., Ge, Z., Li, F., Jia, C., & Song, J. (2020). An interpretable prediction model for identifying N7-methylguanosine sites based on XGBoost and SHAP. *Molecular Therapy-Nucleic Acids*, *22*, 362–372.
97. Bin, L. (2019). BioSeq-Analysis: A platform for DNA, RNA and protein sequence analysis based on machine learning approaches. *Briefings in Bioinformatics*, *20*, 1280–1294.
98. Biswas, A., Staals, R., Morales, S., Fineran, P., & Brown, C. (2016). CRISPRDetect: A flexible algorithm to define CRISPR arrays. *BMC Genomics*, *17*, 1–14.
99. Blake, R., & Delcourt, S. (1996). Thermodynamic effects of formamide on DNA stability. *Nucleic Acids Research*, *24*, 2095–2103.
100. Blanco-Gonzalez, A., Cabezon, A., Seco-Gonzalez, A., Conde-Torres, D., Antelo-Riveiro, P., Pineiro, A., & Garcia-Fandino, R. (2023). The role of AI in drug discovery: Challenges, opportunities, and strategies. *Pharmaceuticals*, *16*, 891.
101. Bland, C., Ramsey, T., Sabree, F., Lowe, M., Brown, K., Kyrpides, N., & Hugenholtz, P. (2007). CRISPR recognition tool (CRT): A tool for automatic detection of clustered regularly interspaced palindromic repeats. *BMC Bioinformatics*, *8*, 1–8.
102. Blohm, P., Frishman, G., Smialowski, P., Goebels, F., Wachinger, B., Ruepp, A., & Frishman, D. (2014). Negatome 2.0: A database of non-interacting proteins derived by literature mining, manual annotation and protein structure analysis. *Nucleic Acids Research*, *42*, D396–D400.
103. Blum, M., Chang, H., Chuguransky, S., Grego, T., Kandasaamy, S., Mitchell, A., Nuka, G., Paysan-Lafosse, T., Qureshi, M., Raj, S., Others. (2021). The InterPro protein families and domains database: 20 years on. *Nucleic Acids Research*, *49*, D344–D354.
104. Bojar, D., & Lisacek, F. (2022). Glycoinformatics in the artificial intelligence era. *Chemical Reviews*, *122*, 15971–15988.
105. Boland, C. (2017). Non-coding RNA: It's not junk. *Digestive Diseases And Sciences*, *62*, 1107.
106. Bonidia, R., Domingues, D., Sanches, D., & Carvalho, A. (2022). MathFeature: Feature extraction package for DNA, RNA and protein sequences based on mathematical descriptors. *Briefings in Bioinformatics*, *23*, bbab434.
107. Bonidia, R., Sampaio, L., Domingues, D., Paschoal, A., Lopes, F., Carvalho, A., & Sanches, D. (2021). Feature extraction approaches for biological sequences: A comparative study of mathematical features. *Briefings in Bioinformatics*, *22*, bbab011.
108. Bonidia, R., Sampaio, L., Domingues, D., Paschoal, A., Lopes, F., Leon Ferreira, A., Sanches, D., Others. (2020). Feature extraction approaches for biological sequences: A comparative study of mathematical models. *BioRxiv*.
109. Bonidia, R., Sanches, D., & Carvalho, A. (2020). Mathfeature: Feature extraction package for biological sequences based on mathematical descriptors. *BioRxiv*.
110. Boschiero, C., Dai, X., Lundquist, P., Roy, S., Bang, T., Zhang, S., Zhuang, Z., Torres-Jerez, I., Udvardi, M., Scheible, W., Others. (2020). MtSSPdb: The Medicago truncatula small secreted peptide database. *Plant Physiology*, *183*, 399–413.
111. Boutet, E., Lieberherr, D., Tognolli, M., Schneider, M., & Bairoch, A. (2007). UniProtKB/Swiss-Prot: The manually annotated section of the UniProt KnowledgeBase. *Plant Bioinformatics: Methods and Protocols*, 89–112.

112. Bouvrie, J. (2006). *Notes on convolutional neural networks*. Massachusetts Institute of Technology.
113. Bowman, L., Sheridan, S., Wham, B., & Wright, S. (2023). *FASTA/FASTQ data curation primer*. Data Curation Network.
114. Brachman, P. (2003). Infectious diseases–past, present, and future. *International Journal of Epidemiology, 32*, 684–686.
115. Brandes, N., Ofer, D., Peleg, Y., Rappoport, N., & Linial, M. (2022). ProteinBERT: A universal deep-learning model of protein sequence and function. *Bioinformatics, 38*, 2102–2110.
116. Breda, A., Valadares, N., Souza, O., & Garratt, R. (2007). Protein structure, modelling and applications. *Bioinformatics in tropical disease research: A practical and case-study approach* [Internet].
117. Breslauer, K., Frank, R., Blöcker, H., & Marky, L. (1986). Predicting DNA duplex stability from the base sequence. *Proceedings of the National Academy of Sciences, 83*, 3746–3750.
118. Brukner, I., Sanchez, R., Suck, D., & Pongor, S. (1995). Sequence-dependent bending propensity of DNA as revealed by DNase I: Parameters for trinucleotides. *The EMBO Journal, 14*, 1812–1818.
119. Brunak, S., Danchin, A., Hattori, M., Nakamura, H., Shinozaki, K., Matise, T., & Preuss, D. (2002). Nucleotide sequence database policies. *Science, 298*, 1333–1334.
120. Buniello, A., MacArthur, J., Cerezo, M., Harris, L., Hayhurst, J., Malangone, C., McMahon, A., Morales, J., Mountjoy, E., Sollis, E., Others. (2019). The NHGRI-EBI GWAS Catalog of published genome-wide association studies, targeted arrays and summary statistics 2019. *Nucleic Acids Research, 47*, D1005–D1012.
121. Burley, S., Berman, H., Kleywegt, G., Markley, J., Nakamura, H., & Velankar, S. (2017). Protein Data Bank (PDB): The single global macromolecular structure archive. *Protein Crystallography: Methods and Protocols*, 627–641.
122. Busch, A., Richter, A., & Backofen, R. (2008). IntaRNA: Efficient prediction of bacterial sRNA targets incorporating target site accessibility and seed regions. *Bioinformatics, 24*, 2849–2856.
123. Cai, K., & Zhu, Y. (2022). A method for identifying essential proteins based on deep convolutional neural network architecture with particle swarm optimization. *2022 Asia conference on advanced robotics, automation, and control engineering (ARACE)* (pp. 7–12).
124. Cai, Y., Liu, X., Xu, X., & Chou, K. (2002). Support vector machines for predicting membrane protein types by incorporating quasi-sequence-order effect. *Internet Electronic Journal of Molecular Design, 1*, 219–226.
125. Cai, C., Han, L., Ji, Z., Chen, X., & Chen, Y. (2003). SVM-Prot: Web-based support vector machine software for functional classification of a protein from its primary sequence. *Nucleic Acids Research, 31*, 3692–3697.
126. Cai, C., Han, L., Ji, Z., & Chen, Y. (2004). Enzyme family classification by support vector machines. *Proteins: Structure, Function, and Bioinformatics, 55*, 66–76.
127. Calvin, C. (2005). Descartes: A new generation system for neutronic calculations. International topical meeting on mathematics and computation, super-computing, reactor physics and nuclear and biological applications. SFEN.
128. Cameron, P., Fuller, C., Donohoue, P., Jones, B., Thompson, M., Carter, M., Gradia, S., Vidal, B., Garner, E., Slorach, E., Others. (2017). Mapping the genomic landscape of CRISPR–Cas9 cleavage. *Nature Methods, 14*, 600–606.
129. Canese, K., & Weis, S. (2013). PubMed: The bibliographic database. *The NCBI handbook* (Vol. 2). National Center for Biotechnology Information (US).
130. Cannone, J., Subramanian, S., Schnare, M., Collett, J., D'Souza, L., Du, Y., Feng, B., Lin, N., Madabusi, L., Müller, K., Others. (2002). The comparative RNA web (CRW) site: An online database of comparative sequence and structure information for ribosomal, intron, and other RNAs. *BMC Bioinformatics, 3*, 1–31.
131. Cao, W., Chen, Y., Yang, J., Xue, F., Yu, Z., Feng, J., Wu, Z., Gong, J., & Niu, X. (2023). Metapath-aggregated multilevel graph embedding for miRNA–disease association

prediction. *2023 IEEE international conference on bioinformatics and biomedicine (BIBM)* (pp. 468–473).

132. Cao, Y., Fanning, S., Proos, S., Jordan, K., & Srikumar, S. (2017). A review on the applications of next generation sequencing technologies as applied to food-related microbiome studies. *Frontiers in Microbiology*, 1829.
133. Cao, S., Lu, W., & Xu, Q. (2015). Grarep: Learning graph representations with global structural information. *Proceedings of the 24th ACM international on conference on information and knowledge management* (pp. 891–900).
134. Cao, Z., & Gao, G. (2022). Multi-omics single-cell data integration and regulatory inference with graph-linked embedding. *Nature Biotechnology*, *40*, 1458–1466.
135. Carson, S., Miller, H., Srougi, M., & Witherow, D. (2019). *Molecular biology techniques: A classroom laboratory manual*. Academic Press.
136. Chaffey, N. Alberts, B., Johnson, A., Lewis, J., Raff, M., Roberts, K., & Walter, P. (2003). *Molecular biology of the cell* (4th ed.). Oxford University Press.
137. Chakraborty, A., Mitra, S., Bhattacharjee, M., De, D., & Pal, A. (2023). Determining human-coronavirus protein-protein interaction using machine intelligence. *Medicine in Novel Technology and Devices*, *18*, Article 100228.
138. Chalikian, T., Völker, J., Plum, G., & Breslauer, K. (1999). A more unified picture for the thermodynamics of nucleic acid duplex melting: A characterization by calorimetric and volumetric techniques. *Proceedings of the National Academy of Sciences*, *96*, 7853–7858.
139. Chan, P., & Lowe, T. (2009). GtRNAdb: A database of transfer RNA genes detected in genomic sequence. *Nucleic Acids Research*, *37*, D93–D97.
140. Chan, W., Zhang, H., Yang, J., Brender, J., Hur, J., Özgür, A., & Zhang, Y. (2015). GLASS: A comprehensive database for experimentally validated GPCR-ligand associations. *Bioinformatics*, *31*, 3035–3042.
141. Chan, G., Koch, C., & Connors, L. (2017). Blood proteomic profiling in inherited (ATTRm) and acquired (ATTRwt) forms of transthyretin-associated cardiac amyloidosis. *Journal of Proteome Research*, *16*, 1659–1668.
142. Chang, A., Jeske, L., Ulbrich, S., Hofmann, J., Koblitz, J., Schomburg, I., Neumann-Schaal, M., Jahn, D., & Schomburg, D. (2021). BRENDA, the ELIXIR core data resource in 2021: New developments and updates. *Nucleic Acids Research*, *49*, D498–D508.
143. Chao, R., & Fei, J. (2023). Off-target effects of CRISPR/Cas9 and their solutions. *Highlights in Science, Engineering and Technology*, *45*, 296–301.
144. Chao, C., Tsai, Y., Lee, W., Yeh, H., & Chiang, C. (2022). Deep learning-assisted repurposing of plant compounds for treating vascular calcification: An in silico study with experimental validation. *Oxidative Medicine and Cellular Longevity*, *2022*, 4378413.
145. Chari, R., Mali, P., Moosburner, M., & Church, G. (2015). Unraveling CRISPR-Cas9 genome engineering parameters via a library-on-library approach. *Nature Methods*, *12*, 823–826.
146. Charoenkwan, P., Nantasenamat, C., Hasan, M., Moni, M., Lio', P., & Shoombuatong, W. (2021). iBitter-fuse: A novel sequence-based bitter peptide predictor by fusing multi-view features. *International Journal of Molecular Sciences*, *22*, 8958.
147. Charoenkwan, P., Nantasenamat, C., Hasan, M., Moni, M., Manavalan, B., & Shoombuatong, W. (2021). UMPred-FRL: A new approach for accurate prediction of umami peptides using feature representation learning. *International Journal of Molecular Sciences*, *22*, 13124.
148. Charoenkwan, P., Ahmed, S., Nantasenamat, C., Quinn, J., Moni, M., Lio', P., & Shoombuatong, W. (2022). AMYPred-FRL is a novel approach for accurate prediction of amyloid proteins by using feature representation learning. *Scientific Reports*, *12*, 7697.
149. Chavda, V., Vihol, D., Patel, A., Redwan, E., & Uversky, V. (2023). Introduction to Bioinformatics, AI, and ML for Pharmaceuticals. *Bioinformatics tools for pharmaceutical drug product development* (pp. 1–18). Wiley Online Library.
150. Chen, J., Lin, J., Hu, Y., Ye, M., Yao, L., Wu, L., Zhang, W., Wang, M., Deng, T., Guo, F., Others. (2023). RNADisease v4. 0: An updated resource of RNA-associated diseases,

providing RNA-disease analysis, enrichment and prediction. *Nucleic Acids Research*, *51*, D1397–D1404.

151. Chen, Z., Liu, X., Zhao, P., Li, C., Wang, Y., Li, F., Akutsu, T., Bain, C., Gasser, R., Li, J., Others. (2022). iFeatureOmega: An integrative platform for engineering, visualization and analysis of features from molecular sequences, structural and ligand data sets. *Nucleic Acids Research*, *50*, W434–W447.
152. Chen, W., McKenna, A., Schreiber, J., Haeussler, M., Yin, Y., Agarwal, V., Noble, W., & Shendure, J. (2019). Massively parallel profiling and predictive modeling of the outcomes of CRISPR/Cas9-mediated double-strand break repair. *Nucleic Acids Research*, *47*, 7989–8003.
153. Chen, C., Wang, J., Pan, D., Wang, X., Xu, Y., Yan, J., Wang, L., Yang, X., Yang, M., & Liu, G. (2023). Applications of multi-omics analysis in human diseases. *Medcomm*, *4*.
154. Chen, M., Zhang, W., Gou, Y., Xu, D., Wei, Y., Liu, D., Han, C., Huang, X., Li, C., Ning, W., Others. (2023). GPS 6.0: An updated server for prediction of kinase-specific phosphorylation sites in proteins. *Nucleic Acids Research*, gkad383.
155. Chen, Z., Zhao, P., Li, F., Leier, A., Marquez-Lago, T., Wang, Y., Webb, G., Smith, A., Daly, R., Chou, K., Others. (2018). iFeature: A python package and web server for features extraction and selection from protein and peptide sequences. *Bioinformatics*, *34*, 2499–2502.
156. Chen, Z., Zhao, P., Li, C., Li, F., Xiang, D., Chen, Y., Akutsu, T., Daly, R., Webb, G., Zhao, Q., Others. (2021). iLearnPlus: A comprehensive and automated machine-learning platform for nucleic acid and protein sequence analysis, prediction and visualization. *Nucleic Acids Research*, *49*, e60–e60.
157. Chen, Z., Zhao, P., Li, F., Marquez-Lago, T., Leier, A., Revote, J., Zhu, Y., Powell, D., Akutsu, T., Webb, G., Others. (2020) iLearn: An integrated platform and meta-learner for feature engineering, machine-learning analysis and modeling of DNA, RNA and protein sequence data. *Briefings in Bioinformatics*, *21*, 1047–1057.
158. Chen, K., Zhou, Y., Ding, M., Wang, Y., Ren, Z., & Yang, Y. (2023). Self-supervised learning on millions of pre-mRNA sequences improves sequence-based RNA splicing prediction. *BioRxiv*.
159. Chen, Z., Zhou, Y., Song, J., & Zhang, Z. (2013). hCKSAAP_UbSite: Improved prediction of human ubiquitination sites by exploiting amino acid pattern and properties. *Biochimica Et Biophysica Acta (BBA)-Proteins and Proteomics*, *1834*, 1461–1467.
160. Chen, X. (2021). Genomics with deep learning. *Deep Learning for Biomedical Applications*, 181–227.
161. Chen, D., & Li, Y. (2022). PredMHC: An effective predictor of major histocompatibility complex using mixed features. *Frontiers in Genetics*, *13*, Article 875112.
162. Chen, H., Xue, Y., Huang, N., Yao, X., & Sun, Z. (2006). MeMo: A web tool for prediction of protein methylation modifications. *Nucleic Acids Research*, *34*, W249–W253.
163. Chen, K., Kurgan, L., & Ruan, J. (2007). Prediction of flexible/rigid regions from protein sequences using k-spaced amino acid pairs. *BMC Structural Biology*, *7*, 1–13.
164. Chen, Z., Chen, Y., Wang, X., Wang, C., Yan, R., & Zhang, Z. (2011). Prediction of ubiquitination sites by using the composition of k-spaced amino acid pairs. *PloS One*, *6*, Article e22930.
165. Chen, W., Feng, P., & Lin, H. (2012). Prediction of replication origins by calculating dna structural properties. *Febs Letters*, *586*, 934–938.
166. Chen, G., Wang, Z., Wang, D., Qiu, C., Liu, M., Chen, X., Zhang, Q., Yan, G., & Cui, Q. (2012). LncRNADisease: A database for long-non-coding RNA-associated diseases. *Nucleic Acids Research*, *41*, D983–D986.
167. Chen, Y., Chen, Z., Gong, Y., & Ying, G. (2012). SUMOhydro: A novel method for the prediction of sumoylation sites based on hydrophobic properties. *PloS One*, *7*, Article e39195.
168. Chen, W., Feng, P., Lin, H., & Chou, K. (2013). iRSpot-PseDNC: Identify recombination spots with pseudo dinucleotide composition. *Nucleic Acids Research*, *41*, e68–e68.

169. Chen, W., Feng, P., Deng, E., Lin, H., & Chou, K. (2014). iTIS-PseTNC: A sequence-based predictor for identifying translation initiation site in human genes using pseudo trinucleotide composition. *Analytical Biochemistry*, *462*, 76–83.
170. Chen, W., Tran, H., Liang, Z., Lin, H., & Zhang, L. (2015). Identification and analysis of the N6-methyladenosine in the Saccharomyces cerevisiae transcriptome. *Scientific Reports*, *5*, 1–8.
171. Chen, X., Han, P., Zhou, T., Guo, X., Song, X., & Li, Y. (2016). circRNADb: A comprehensive database for human circular RNAs with protein-coding annotations. *Scientific Reports*, *6*, 34985.
172. Chen, W., Yang, H., Feng, P., Ding, H., & Lin, H. (2017). iDNA4mC: Identifying DNA N4-methylcytosine sites based on nucleotide chemical properties. *Bioinformatics*, *33*, 3518–3523.
173. Chen, W., McKenna, A., Schreiber, J., Haeussler, M., Yin, Y., Agarwal, V., Noble, W., & Shendure, J. (2019). Massively parallel profiling and predictive modeling of the outcomes of CRISPR/Cas9-mediated double-strand break repair. *Nucleic Acids Research*, *47*, 7989–8003.
174. Chen, W., Song, X., Lv, H., & Lin, H. (2019). iRNA-m2G: Identifying N2-methylguanosine sites based on sequence-derived information. *Molecular Therapy-Nucleic Acids*, *18*, 253–258.
175. Chen, W., Chen, G., Zhao, L., & Chen, C. (2021). Predicting drug-target interactions with deep-embedding learning of graphs and sequences. *The Journal of Physical Chemistry A*, *125*, 5633–5642.
176. Chen, J., Cheong, H., & Siu, S. (2021). xDeep-AcPEP: Deep learning method for anticancer peptide activity prediction based on convolutional neural network and multitask learning. *Journal of Chemical Information and Modeling*, *61*, 3789–3803.
177. Chen, Z., Li, X., Yang, M., Zhang, H., & Xu, X. (2022). Optimization of deep learning models for the prediction of gene mutations using unsupervised clustering. *The Journal of Pathology Clinical Research*, *9*, 3–17.
178. Chen, X., Zhang, Q., Li, B., Lu, C., Yang, S., Long, J., He, B., Chen, H., & Huang, J. (2022). BBPpredict: A web service for identifying blood-brain barrier penetrating peptides. *Frontiers in Genetics*, *13*, Article 845747.
179. Chen, Z., Zhao, B., Li, J., Guo, Z., & You, Z. (2023). GraphCPIs: A novel graph-based computational model for potential compound-protein interactions. *Molecular Therapy-Nucleic Acids*, *32*, 721–728.
180. Chen, L., Hu, Z., Rong, Y., & Lou, B. (2024). Deep2Pep: A deep learning method in multi-label classification of bioactive peptide. *Computational Biology and Chemistry*, *109*, Article 108021.
181. Cherednichenko, O., & Poptsova, M. (2024). Generative models for prediction of Non-B DNA structures. *BioRxiv*.
182. Cherry, J., Adler, C., Ball, C., Chervitz, S., Dwight, S., Hester, E., Jia, Y., Juvik, G., Roe, T., Schroeder, M., Others. (1998). SGD Saccharomyces genome database. *Nucleic Acids Research*, *26*, 73–79.
183. Cho, K., Van Merriënboer, B., Gulcehre, C., Bahdanau, D., Bougares, F., Schwenk, H., & Bengio, Y. (2014). Learning phrase representations using RNN encoder-decoder for statistical machine translation. Preprint. ArXiv:1406.1078.
184. Cho, S., Kim, S., Kim, Y., Kweon, J., Kim, H., Bae, S., & Kim, J. (2014). Analysis of off-target effects of CRISPR/Cas-derived RNA-guided endonucleases and nickases. *Genome Research*, *24*, 132–141.
185. Chong, S., & Peyton, P. (2012). A meta-analysis of the accuracy and precision of the ultrasonic cardiac output monitor (USCOM). *Anaesthesia*, *67*, 1266–1271.
186. Choo, K., Tan, T., & Ranganathan, S. (2005). SPdb-a signal peptide database. *BMC Bioinformatics*, *6*, 1–8.
187. Chou, K. (2000). Prediction of protein subcellular locations by incorporating quasi-sequence-order effect. *Biochemical and Biophysical Research Communications*, *278*, 477–483.

188. Chou, K. (2001). Prediction of protein cellular attributes using pseudo-amino acid composition. *Proteins: Structure, Function, and Bioinformatics*, *43*, 246–255.
189. Chou, K. (2005). Using amphiphilic pseudo amino acid composition to predict enzyme subfamily classes. *Bioinformatics*, *21*, 10–19.
190. Chu, X., Guan, B., Dai, L., Liu, J., Li, F., & Shang, J. (2023). Network embedding framework for driver gene discovery by combining functional and structural information. *BMC Genomics*, *24*, 426.
191. Chuai, G., Ma, H., Yan, J., Chen, M., Hong, N., Xue, D., Zhou, C., Zhu, C., Chen, K., Duan, B., Others. (2018). DeepCRISPR: Optimized CRISPR guide RNA design by deep learning. *Genome Biology*, *19*, 1–18.
192. Chung, J., Gulcehre, C., Cho, K., & Bengio, Y. (2014). Empirical evaluation of gated recurrent neural networks on sequence modeling. Preprint. ArXiv:1412.3555.
193. Chung, C., Chang, Y., Hsu, Y., Chen, S., Wu, L., Horng, J., & Lee, T. (2020). Incorporating hybrid models into lysine malonylation sites prediction on mammalian and plant proteins. *Scientific Reports*, *10*, 10541.
194. Clark, K., Luong, M., Le, Q., & Manning, C. (2020). Electra: Pre-training text encoders as discriminators rather than generators. Preprint. ArXiv:2003.10555.
195. Clauwaert, J., Menschaert, G., & Waegeman, W. (2021). Explainability in transformer models for functional genomics. *Briefings in Bioinformatics*, *22*, bbab060.
196. Clauwaert, J., & Waegeman, W. (2020). Novel transformer networks for improved sequence labeling in genomics. *IEEE/ACM Transactions on Computational Biology and Bioinformatics*, *19*, 97–106.
197. Clemons, Jr., W., May, J., Wimberly, B., McCutcheon, J., Capel, M., & Ramakrishnan, V. (1999). Structure of a bacterial 30S ribosomal subunit at 5.5 Åresolution. *Nature*, *400*, 833–840.
198. Clough, E., & Barrett, T. (2016). The gene expression omnibus database. *Statistical Genomics: Methods and Protocols* (pp. 93–110).
199. Coenye, T., Gevers, D., Peer, Y., Vandamme, P., & Swings, J. (2005). Towards a prokaryotic genomic taxonomy. *Fems Microbiology Reviews*, *29*, 147–167.
200. Commichaux, S. (2023). Database size positively correlates with the loss of species-level taxonomic resolution for the 16s rrna and other prokaryotic marker genes. *PLOS Computational Biology, 20*(8), e1012343.
201. Conrad, T., Genzel, M., Cvetkovic, N., Wulkow, N., Leichtle, A., Vybiral, J., Kutyniok, G., & Schütte, C. (2017). Sparse Proteomics Analysis-a compressed sensing-based approach for feature selection and classification of high-dimensional proteomics mass spectrometry data. *BMC Bioinformatics*, *18*, 1–20.
202. Consortium, G. (2004). The gene ontology (GO) database and informatics resource. *Nucleic Acids Research*, *32*, D258–D261.
203. Consortium, E. (2011). A user's guide to the encyclopedia of DNA elements (ENCODE). *PLoS Biology*, *9*, Article e1001046.
204. Consortium, U. (2015). UniProt: A hub for protein information. *Nucleic Acids Research*, *43*, D204–D212.
205. Cravedi, K. (2008). GenBank celebrates 25 years of service with two-day conference. Leading scientists will discuss the DNA database at April 7–8 meeting. *NIH News*.
206. Crick, F. (1970). Central dogma of molecular biology. *Nature*, *227*, 561–563.
207. Cueto, M. (2023). *Global health essentials* (pp. 421–424). The World Health Organization.
208. Cui, T., Dou, Y., Tan, P., Ni, Z., Liu, T., Wang, D., Huang, Y., Cai, K., Zhao, X., Xu, D., Others. (2022). RNALocate v2. 0: An updated resource for RNA subcellular localization with increased coverage and annotation. *Nucleic Acids Research*, *50*, D333–D339.
209. Cui, D., Li, W., Wu, J., Xie, J., & Wu, Y. (2021). Advancсs in multi omics applications in hbv-associated hepatocellular carcinoma. *Frontiers in Medicine*, *8*.
210. Cursons, J., Pillman, K., Scheer, K., Gregory, P., Foroutan, M., Hediyeh-Zadeh, S., Toubia, J., Crampin, E., Goodall, G., Bracken, C., Others. (2018). Combinatorial targeting by microRNAs co-ordinates post-transcriptional control of EMT. *Cell Systems*, *7*, 77–91.

211. Dağliyan, O., Üney-Yüksektepe, F., Kavaklı, İ, & Türkay, M. (2011). Optimization based tumor classification from microarray gene expression data. *Plos One*, *6*, Article e14579.
212. Dai, Z., & Deng, F. (2023). LncPNdeep: A long non-coding RNA classifier based on large language model with peptide and nucleotide embedding. *BioRxiv*.
213. Dai, W., Chang, Q., Peng, W., Zhong, J., & Li, Y. (2020). Network embedding the protein-protein interaction network for human essential genes identification. *Genes*, *11*, 153.
214. Dalla-Torre, H., Gonzalez, L., Mendoza-Revilla, J., Carranza, N., Grzywaczewski, A., Oteri, F., Dallago, C., Trop, E., Almeida, B., Sirelkhatim, H., Others. (2023). The nucleotide transformer: Building and evaluating robust foundation models for human genomics. *BioRxiv*.
215. Damane, B., Mkhize-Kwitshana, Z., Kgokolo, M., Luvhengo, T., & Dlamini, Z. (2023). Applying artificial intelligence prediction tools for advancing precision oncology in immunotherapy: Future perspectives in personalized care. *Artificial Intelligence and Precision Oncology: Bridging Cancer Research and Clinical Decision Support* (pp. 239–258). Springer.
216. Damodaran, S. (2008). Amino acids, peptides and proteins. *Fennema's Food Chemistry*, *4*, 425–439.
217. Danaee, P., Rouches, M., Wiley, M., Deng, D., Huang, L., & Hendrix, D. (2018). bpRNA: Large-scale automated annotation and analysis of RNA secondary structure. *Nucleic Acids Research*, *46*, 5381–5394.
218. Daneshpajouh, A., Fowler, M., & Wiese, K. (2023). Navitas/Optimus: A novel computational tool for enhanced CRISPR/Cas genome editing. *Proceedings of the Canadian conference on artificial intelligence*.
219. Dang, T., & Vu, T. (2024). xCAPT5: Protein-protein interaction prediction using deep and wide multi-kernel pooling convolutional neural networks with protein language model. *BMC Bioinformatics*, *25*, 106.
220. Daniel A. G. (2023). *N. Genome Gov*. https://www.genome.gov/genetics-glossary/Mitosis, Accessed on March 26, 2023.
221. Daniel D., & Lee, H. (2000). Algorithms for non-negative matrix factorization . *Neural Computation, 13*, 1–7.
222. Dao, F., Liu, M., Su, W., Lv, H., Zhang, Z., Lin, H., & Liu, L. (2022). AcrPred: A hybrid optimization with enumerated machine learning algorithm to predict anti-CRISPR proteins. *International Journal of Biological Macromolecules*, *228*, 706–714.
223. Dao, F., Lv, H., Zhang, D., Zhang, Z., Liu, L., & Lin, H. (2021). DeepYY1: A deep learning approach to identify YY1-mediated chromatin loops. *Briefings in Bioinformatics*, *22*, bbaa356.
224. D'Argenio, V. (2018). The high-throughput analyses era: Are we ready for the data struggle? *High-throughput*, *7*, 8.
225. Das, L., Kumar, A., Nanda, S., & Das, J. (2019). Improved protein coding region prediction using dipole moment based SVD algorithm. *2019 5th international conference on signal processing, computing and control (ISPCC)* (pp. 90–95).
226. Das, L., Nanda, S., & Das, J. (2017). A novel DNA mapping scheme for improved exon prediction using digital filters. *2017 2nd international conference on man and machine interfacing (MAMI)* (pp. 1–6).
227. Das, J., & Barman, S. (2017). DSP based entropy estimation for identification and classification of Homo sapiens cancer genes. *Microsystem Technologies*, *23*, 4145–4154.
228. Das, L., Nanda, S., & Das, J. (2019). An integrated approach for identification of exon locations using recursive Gauss Newton tuned adaptive Kaiser window. *Genomics*, *111*, 284–296.
229. Das, L., Kumar, A., Das, J., & Nanda, K. (2019). Modified gabor wavelet transform in prediction of cancerous genes. *International Journal of Engineering and Advanced Technology*, *19*, 902–907.
230. Das, D., Jaiswal, M., Khan, F., Ahamad, S., & Kumar, S. (2020). PlantPepDB: A manually curated plant peptide database. *Scientific Reports*, *10*, 2194.

231. Das, L., Das, J., Mohapatra, S., & Nanda, S. (2021). DNA numerical encoding schemes for exon prediction: A recent history. *Nucleosides, Nucleotides and Nucleic Acids*, *40*, 985–1017.
232. Dasgupta, S. (2013). Experiments with random projection. Preprint. ArXiv:1301.3849.
233. Davis, C., Hitz, B., Sloan, C., Chan, E., Davidson, J., Gabdank, I., Hilton, J., Jain, K., Baymuradov, U., Narayanan, A., Others. (2018). The encyclopedia of DNA elements (ENCODE): Data portal update. *Nucleic Acids Research*, *46*, D794–D801.
234. Datta, S., Asim, M., Dengel, A., & Ahmed, S. (2023). NTpred: A robust and precise machine learning framework for in-silico identification of Tyrosine nitration sites in protein sequences. *BioRxiv*.
235. Davis, A., Grondin, C., Johnson, R., Sciaky, D., Wiegers, J., Wiegers, T., & Mattingly, C. (2021). Comparative toxicogenomics database (CTD): Update 2021. *Nucleic Acids Research*, *49*, D1138–D1143.
236. Davis, A., Wiegers, T., Johnson, R., Sciaky, D., Wiegers, J., & Mattingly, C. (2023). Comparative toxicogenomics database (CTD): Update 2023. *Nucleic Acids Research*, *51*, D1257–D1262.
237. Dawson, J. (2021). Tertiary structure protein stability. *BIOC* 2580: Introduction to biochemistry*. Open Library Publishing Platform.
238. Dayhoff, M., & Eck, R. (1972). *Atlas of protein sequence and structure*. National Biomedical Research Foundation.
239. Debie, E., & Shafi, K. (2019). Implications of the curse of dimensionality for supervised learning classifier systems: Theoretical and empirical analyses. *Pattern Analysis and Applications*, *22*, 519–536.
240. Dee, W. (2022). LMPred: Predicting antimicrobial peptides using pre-trained language models and deep learning. *Bioinformatics Advances*, *2*, vbac021.
241. Del Vecchio, F., Mastroiaco, V., Di Marco, A., Compagnoni, C., Capece, D., Zazzeroni, F., Capalbo, C., Alesse, E., & Tessitore, A. (2017). Next-generation sequencing: Recent applications to the analysis of colorectal cancer. *Journal of Translational Medicine*, *15*, 1–19.
242. Demirci, S., Leonard, A., Haro-Mora, J., Uchida, N., & Tisdale, J. (2019). CRISPR/Cas9 for sickle cell disease: Applications, future possibilities, and challenges. *Cell Biology and Translational Medicine, Volume 5: Stem Cells: Translational Science To Therapy* (pp. 37–52). Springer.
243. Deng, F., Yang, K., & Zheng, G. (2020). Period family of clock genes as novel predictors of survival in human cancer: A systematic review and meta-analysis. *Disease Markers*, *2020*, 1–9.
244. Deng, L., Lin, W., Wang, J., & Zhang, J. (2020). DeepciRGO: Functional prediction of circular RNAs through hierarchical deep neural networks using heterogeneous network features. *BMC Bioinformatics*, *21*, 1–18.
245. Deng, L., Jiang, Y., Hu, X., Zheng, R., Huang, Z., & Zhang, J. (2023). ABLNCPP: Attention mechanism-based bidirectional long short-term memory for noncoding RNA coding potential prediction. *Journal of Chemical Information and Modeling*, *63*, 3955–3966.
246. Deng, H., Ding, M., Wang, Y., Li, W., Liu, G., & Tang, Y. (2023). ACP-MLC: A two-level prediction engine for identification of anticancer peptides and multi-label classification of their functional types. *Computers in Biology and Medicine*, *158*, Article 106844.
247. Derat, E., & Kamerlin, S. (2022). Computational advances in protein engineering and enzyme design. *The Journal of Physical Chemistry B*, *126*, 2449–2451.
248. Desai, D., & Kulkarni-Kale, U. (2014). T-cell epitope prediction methods: An overview. *Immunoinformatics*, 333–364.
249. Deshmukh, S., Heller, P., & Khuri, N. (2019). ICMLA - A long-short term memory network for detecting CRISPR arrays. *2019 18th IEEE international conference on machine learning and applications (ICMLA)* (pp. 619–624).
250. Devlin, J., Chang, M., Lee, K., & Toutanova, K. (2018). Bert: Pre-training of deep bidirectional transformers for language understanding. Preprint. ArXiv:1810.04805.

251. Dey, L., & Mukhopadhyay, A. (2019). A classification-based approach to prediction of dengue virus and human protein-protein interactions using amino acid composition and conjoint triad features. *2019 IEEE region 10 symposium (TENSYMP)* (pp. 373–378).
252. Dhanjal, J., Radhakrishnan, N., & Sundar, D. (2019). CRISPcut: A novel tool for designing optimal sgRNAs for CRISPR/Cas9 based experiments in human cells. *Genomics*, *111*, 560–566.
253. Dhanjal, J., Dammalapati, S., Pal, S., & Sundar, D. (2020). Evaluation of off-targets predicted by sgRNA design tools. *Genomics*, *112*, 3609–3614.
254. Di Luca, M., Maccari, G., Maisetta, G., & Batoni, G. (2015). BaAMPs: The database of biofilm-active antimicrobial peptides. *Biofouling*, *31*, 193–199.
255. Dias, R., & Torkamani, A. (2019). Artificial intelligence in clinical and genomic diagnostics. *Genome Medicine*, *11*, 1–12.
256. Diella, F., Cameron, S., Gemünd, C., Linding, R., Via, A., Kuster, B., Sicheritz-Pontén, T., Blom, N., & Gibson, T. (2004). Phospho. ELM: A database of experimentally verified phosphorylation sites in eukaryotic proteins. *BMC Bioinformatics*, *5*, 1–5.
257. Dimauro, G., Colagrande, P., Carlucci, R., Ventura, M., Bevilacqua, V., & Caivano, D. (2019). Crisprlearner: A deep learning-based system to predict crispr/cas9 sgrna on-target cleavage efficiency. *Electronics*, *8*, 1478.
258. Dimauro, G., Barletta, V., Catacchio, C., Colizzi, L., Maglietta, R., & Ventura, M. (2022). A systematic mapping study on machine learning techniques for the prediction of CRISPR/Cas9 sgRNA target cleavage. *Computational and Structural Biotechnology Journal*, *20*, 5813–5823.
259. Ding, F., Sharma, S., Chalasani, P., Demidov, V., Broude, N., & Dokholyan, N. (2008). Ab initio rna folding by discrete molecular dynamics: From structure prediction to folding mechanisms. *Rna*, *14*, 1164–1173.
260. Do, D., Le, T., & Le, N. (2021). Using deep neural networks and biological subwords to detect protein S-sulfenylation sites. *Briefings in Bioinformatics*, *22*, bbaa128.
261. Do, D., & Le, N. (2019). A sequence-based approach for identifying recombination spots in Saccharomyces cerevisiae by using hyper-parameter optimization in FastText and support vector machine. *Chemometrics and Intelligent Laboratory Systems*, *194*, Article 103855.
262. Doench, J., Fusi, N., Sullender, M., Hegde, M., Vaimberg, E., Donovan, K., Smith, I., Tothova, Z., Wilen, C., Orchard, R., Others. (2016). Optimized sgRNA design to maximize activity and minimize off-target effects of CRISPR-Cas9. *Nature Biotechnology*, *34*, 184–191.
263. Doench, J., Hartenian, E., Graham, D., Tothova, Z., Hegde, M., Smith, I., Sullender, M., Ebert, B., Xavier, R., & Root, D. (2014). Rational design of highly active sgRNAs for CRISPR-Cas9-mediated gene inactivation. *Nature Biotechnology*, *32*, 1262–1267.
264. Doherty, E., & Doudna, J. (2001). Ribozyme structures and mechanisms. *Annual Review of Biophysics and Biomolecular Structure*, *30*, 457–475.
265. Dong, Q., Zhou, S., & Guan, J. (2009). A new taxonomy-based protein fold recognition approach based on autocross-covariance transformation. *Bioinformatics*, *25*, 2655–2662.
266. Dong, C., Yuan, Y., Zhang, F., Hua, H., Ye, Y., Labena, A., Lin, H., Chen, W., & Guo, F. (2016). Combining pseudo dinucleotide composition with the Z curve method to improve the accuracy of predicting DNA elements: A case study in recombination spots. *Molecular BioSystems*, *12*, 2893–2900.
267. Dong, C., Hao, G., Hua, H., Liu, S., Labena, A., Chai, G., Huang, J., Rao, N., & Guo, F. (2018). Anti-CRISPRdb: A comprehensive online resource for anti-CRISPR proteins. *Nucleic Acids Research*, *46*, D393–D398.
268. Dosovitskiy, A., Beyer, L., Kolesnikov, A., Weissenborn, D., Zhai, X., Unterthiner, T., Dehghani, M., Minderer, M., Heigold, G., Gelly, S., Others. (2020). An image is worth 16x16 words: Transformers for image recognition at scale. Preprint. ArXiv:2010.11929.
269. Doudna, J., & Charpentier, E. (2014). The new frontier of genome engineering with CRISPR-Cas9. *Science*, *346*, 1258096.

270. Dragovich, B., & Mišić, N. (2019). p-Adic hierarchical properties of the genetic code. *Biosystems*, *185*, Article 104017.
271. Dreos, R., Ambrosini, G., Cavin Périer, R., & Bucher, P. (2013). EPD and EPDnew, high-quality promoter resources in the next-generation sequencing era. *Nucleic Acids Research*, *41*, D157–D164.
272. Du, M., Zhang, C., Wang, H., Liu, S., Wei, W., & Guo, F. (2018). The GC content as a main factor shaping the amino acid usage during bacterial evolution process. *Frontiers in Microbiology*, *9*, 2948.
273. Du, Z., Zhong, X., Wang, F., & Uversky, V. (2022). Inference of gene regulatory networks based on the light gradient boosting machine. *Computational Biology and Chemistry*, *101*, Article 107769.
274. Du, W., Zhao, L., Wu, R., Huang, B., Liu, S., Liu, Y., Huang, H., & Shi, G. (2024). Predicting drug-Protein interaction with deep learning framework for molecular graphs and sequences: Potential candidates against SAR-CoV-2. *Plos One*, *19*, Article e0299696.
275. Duan, C., Zang, Z., Xu, Y., He, H., Liu, Z., Song, Z., Zheng, J., & Li, S. (2024). FGBERT: Function-driven pre-trained gene language model for metagenomics. Preprint. ArXiv:2402.16901.
276. Duan, T., Kuang, Z., Wang, J., & Ma, Z. (2021). GBDTLRL2D predicts LncRNA-disease associations using MetaGraph2Vec and K-means based on heterogeneous network. *Frontiers in Cell and Developmental Biology*, *9*, Article 753027.
277. Dubchak, I., Muchnik, I., Holbrook, S., & Kim, S. (1995). Prediction of protein folding class using global description of amino acid sequence. *Proceedings of the National Academy of Sciences*, *92*, 8700–8704.
278. Dubchak, I., Muchnik, I., Mayor, C., Dralyuk, I., & Kim, S. (1999). Recognition of a protein fold in the context of the SCOP classification. *Proteins: Structure, Function, and Bioinformatics*, *35*, 401–407.
279. Dudekula, D., Panda, A., Grammatikakis, I., De, S., Abdelmohsen, K., & Gorospe, M. (2016). CircInteractome: A web tool for exploring circular RNAs and their interacting proteins and microRNAs. *RNA Biology*, *13*, 34–42.
280. Dunbar, J., Krawczyk, K., Leem, J., Baker, T., Fuchs, A., Georges, G., Shi, J., & Deane, C. (2014). SAbDab: The structural antibody database. *Nucleic Acids Research*, *42*, D1140–D1146.
281. Durland, J., & Ahmadian-Moghadam, H. (2022). Genetics, mutagenesis. *StatPearls* [Internet].
282. Eakin, R. (1971). Gregor mendel (1822–1884). *JSTOR*.
283. Edgar, R., Domrachev, M., & Lash, A. (2002). Gene expression omnibus: NCBI gene expression and hybridization array data repository. *Nucleic Acids Research*, *30*, 207–210.
284. Edwards, R., McNair, K., Faust, K., Raes, J., & Dutilh, B. (2015). Computational approaches to predict bacteriophage-host relationships. *Fems Microbiology Reviews*, *40*, 258–272.
285. Eitzinger, S., Asif, A., Watters, K., Iavarone, A., Knott, G., Doudna, J., & Amir Afsar Minhas, F. (2020). Machine learning predicts new anti-CRISPR proteins. *Nucleic Acids Research*, *48*, 4698–4708.
286. Eitzinger, S., Asif, A., Watters, K., Iavarone, A., Knott, G., Doudna, J., & Minhas, F. (2020). Machine learning predicts new anti-CRISPR proteins. *Nucleic Acids Research*, *48*, 4698–4708.
287. Ekimler, S., & Sahin, K. (2014). Computational methods for microrna target prediction. *Genes*, *5*, 671–683.
288. El Allali, A., & Rose, J. MGC: Gene calling in metagenomic sequences. *8th international symposium on bioinformatics research and applications*.
289. El Hassan, M., & Calladine, C. (1996). Propeller-twisting of base-pairs and the conformational mobility of dinucleotide steps in DNA. *Journal of Molecular Biology*, *259*, 95–103.
290. El-Badawy, I., Gasser, S., Aziz, A., & Khedr, M. (2015). On the use of pseudo-EIIP mapping scheme for identifying exons locations in DNA sequences. *2015 IEEE international conference on signal and image processing applications (ICSIPA)* (pp. 244–247).

291. Elhassani, M., Maisonnasse, L., Olgiati, A., Jerome, R., Rehali, M., Duroux, P., Giudicelli, V., & Kossida, S. (2022). Deep Learning concepts for genomics: An overview. *EMBnet. Journal*, *27*, Article e990.
292. Elnaggar, A., Essam, H., Salah-Eldin, W., Moustafa, W., Elkerdawy, M., Rochereau, C., & Rost, B. (2023). Ankh: Optimized protein language model unlocks general-purpose modelling. Preprint. ArXiv2301.06568. https://doi.org/10.48550.
293. Eric Green Genome Gov. (2023). https://www.genome.gov/genetics-glossary/Shotgun-Sequencing, Accessed on March 26, 2023.
294. Espe, S. (2018). MalaCards: The human disease database. *Journal of the Medical Library Association: JMLA*, *106*, 140.
295. Fan, K., & Zhang, Y. (2020). Pseudo2GO: A graph-based deep learning method for pseudogene function prediction by borrowing information from coding genes. *Frontiers in Genetics*, *11*, Article 538028.
296. Fan, L., Sun, J., Zhou, M., Zhou, J., Lao, X., Zheng, H., & Xu, H. (2016). DRAMP: A comprehensive data repository of antimicrobial peptides. *Scientific Reports*, *6*, 24482.
297. Fan, X., Lin, B., Hu, J., & Guo, Z. (2023). I-DNAN6mA: Accurate identification of DNA N6-Methyladenine sites using the base-pairing map and deep learning. *Journal of Chemical Information and Modeling*, *63*, 1076–1086.
298. Fanaras, K., Antoniadis, C., & Massoud, Y. (2023). Enhanced sgRNA on-target cleavage efficacy prediction using conditional GANs. *2023 IEEE international symposium on circuits and systems (ISCAS)*.
299. Fang, S., Zhang, L., Guo, J., Niu, Y., Wu, Y., Li, H., Zhao, L., Li, X., Teng, X., Sun, X., Others. (2018). NONCODEV5: A comprehensive annotation database for long non-coding RNAs. *Nucleic Acids Research*, *46*, D308–D314 .
300. Fang, G., Zeng, F., Li, X., & Yao, L. (2021). Word2vec based deep learning network for DNA N4-methylcytosine sites identification. *Procedia Computer Science*, *187*, 270–277.
301. Farboud, B., & Meyer, B. (2015). Dramatic enhancement of genome editing by CRISPR/Cas9 through improved guide RNA design. *Genetics*, *199*, 959–971.
302. Fazeel, A., Agha, A., Dengel, A., & Ahmed, S. (2023). NP-BERT: A two-staged BERT based nucleosome positioning prediction architecture for multiple species. *Bioinformatics*, 175–187.
303. Federhen, S. (2012). The NCBI taxonomy database. *Nucleic Acids Research*, *40*, D136–D143.
304. Feng, J., Zeng, A., Chen, Y., Payne, P., & Li, F. (2020). Signaling interaction link prediction using deep graph neural networks integrating protein-protein interactions and omics data. *BioRxiv*.
305. Feng, Z., & Zhang, C. (2000). Prediction of membrane protein types based on the hydrophobic index of amino acids. *Journal of Protein Chemistry*, *19*, 269–275.
306. Fishman, V., Kuratov, Y., Petrov, M., Shmelev, A., Shepelin, D., Chekanov, N., Kardymon, O., & Burtsev, M. (2023). GENA-LM: A family of open-source foundational DNA language models for long sequences. *BioRxiv*.
307. Flotho, A., & Melchior, F. (2013). Sumoylation: A regulatory protein modification in health and disease. *Annual Review of Biochemistry*, *82*, 357–385.
308. Forghani, M., & Khani, R. (2017). A multivariate clustering of AAindex database for protein numerical representation. *2017 3rd Iranian conference on intelligent systems and signal processing (ICSPIS)* (pp. 1–4).
309. Forman, G., Others. (2003). An extensive empirical study of feature selection metrics for text classification. *Journal of Machine Learning Research*, *3*, 1289–1305.
310. Formenti, G., Abueg, L., Brajuka, A., Brajuka, N., Gallardo-Alba, C., Giani, A., Fedrigo, O., & Jarvis, E. (2022). Gfastats: Conversion, evaluation and manipulation of genome sequences using assembly graphs. *Bioinformatics*, *38*, 4214–4216.
311. Fornes, O., Castro-Mondragon, J., Khan, A., Lee, R., Zhang, X., Richmond, P., Modi, B., Correard, S., Gheorghe, M., Baranašić, D., Others. (2020). JASPAR 2020: Update of the

open-access database of transcription factor binding profiles. *Nucleic Acids Research*, *48*, D87–D92.
312. Fox, N., Brenner, S., & Chandonia, J. (2014). SCOPe: Structural classification of proteins–extended, integrating SCOP and ASTRAL data and classification of new structures. *Nucleic Acids Research*, *42*, D304–D309.
313. Franke, J., Runge, F., Koeksal, R., Backofen, R., & Hutter, F. (2024). RNAformer: A simple yet effective deep learning model for RNA secondary structure prediction. *BioRxiv*.
314. Frankish, A., Diekhans, M., Ferreira, A., Johnson, R., Jungreis, I., Loveland, J., Mudge, J., Sisu, C., Wright, J., Armstrong, J., Others. (2019). GENCODE reference annotation for the human and mouse genomes. *Nucleic Acids Research*, *47*, D766–D773.
315. Frankish, A., Diekhans, M., Jungreis, I., Lagarde, J., Loveland, J., Mudge, J., Sisu, C., Wright, J., Armstrong, J., Barnes, I., Others. (2021). GENCODE 2021. *Nucleic Acids Research*, *49*, D916–D923.
316. Frank-Kamenetskii, M., & Mirkin, S. (1995). Triplex DNA structures. *Annual Review Of Biochemistry*, *64*, 65–95.
317. Franzén, O., Gan, L., & Björkegren, J. (2019). PanglaoDB: A web server for exploration of mouse and human single-cell RNA sequencing data. *Database*, *2019*, baz046.
318. Frock, R., Hu, J., Meyers, R., Ho, Y., Kii, E., & Alt, F. (2015). Genome-wide detection of DNA double-stranded breaks induced by engineered nucleases. *Nature Biotechnology*, *33*, 179–186.
319. Fu, X., Ke, L., Cai, L., Chen, X., Ren, X., & Gao, M. (2019). Improved prediction of cell-penetrating peptides via effective orchestrating amino acid composition feature representation. *IEEE Access*, *7*, 163547–163555.
320. Fu, L., Cao, Y., Wu, J., Peng, Q., Nie, Q., & Xie, X. (2022). UFold: Fast and accurate RNA secondary structure prediction with deep learning. *Nucleic Acids Research*, *50*, e14–e14.
321. Gabriel, L., Becker, F., Hoff, K., & Stanke, M. (2024). Tiberius: End-to-end deep learning with an HMM for gene prediction. *BioRxiv*.
322. Gagnon, J., Valen, E., Thyme, S., Huang, P., Ahkmetova, L., Pauli, A., Montague, T., Zimmerman, S., Richter, C., & Schier, A. (2014). Efficient mutagenesis by Cas9 protein-mediated oligonucleotide insertion and large-scale assessment of single-guide RNAs. *PloS One*, *9*, Article e98186.
323. Gandhi, S., Christiaen, L., & Stolfi, A. (2016). Rational design and whole-genome predictions of single guide RNAs for efficient CRISPR/Cas9-mediated genome editing in Ciona. *BioRxiv*, 041632.
324. Gao, T., & Qian, J. (2020). EnhancerAtlas 2.0: An updated resource with enhancer annotation in 586 tissue/cell types across nine species. *Nucleic Acids Research*, *48*, D58–D64.
325. Gao, S., Ding, B., & Lou, W. (2020). Microrna-dependent modulation of genes contributes to esr1's effect on erα positive breast cancer. *Frontiers in Oncology*, *10*.
326. Gao, Z., Liu, Q., Zeng, W., Jiang, R., & Wong, W. (2023). EpiGePT: A Pretrained Transformer model for epigenomics. *BioRxiv*.
327. Gao, Y., Shang, S., Guo, S., Li, X., Zhou, H., Liu, H., Sun, Y., Wang, J., Wang, P., Zhi, H., Others. (2021). Lnc2Cancer 3.0: An updated resource for experimentally supported lncRNA/circRNA cancer associations and web tools based on RNA-seq and scRNA-seq data. *Nucleic Acids Research*, *49*, D1251–D1258.
328. Gao, F., & Zhang, C. (2004). Comparison of various algorithms for recognizing short coding sequences of human genes. *Bioinformatics*, *20*, 673–681.
329. Gao, F., Luo, H., & Zhang, C. (2012). DeOri: A database of eukaryotic DNA replication origins. *Bioinformatics*, *28*, 1551–1552.
330. Garcia-Herrero, S., Simon, B., & Garcia-Planells, J. (2020). The reproductive journey in the genomic era: From preconception to childhood. *Genes*, *11*, 1521.
331. Garg, P., & Sharma, S. (2020). Sensitivity enhancement of dwt based algorithm for detection of cpg islands in dna sequences. *Procedia Computer Science*, *167*, 1829–1838.

332. Gata, W., & Bayhaqy, A. (2020). Analysis sentiment about islamophobia when christchurch attack on social media. *Telkomnika (Telecommunication Computing Electronics And Control)*, *18*, 1819.
333. Gautam, A., Chaudhary, K., Singh, S., Joshi, A., Anand, P., Tuknait, A., Mathur, D., Varshney, G., & Raghava, G. (2014). Hemolytik: a database of experimentally determined hemolytic and non-hemolytic peptides. *Nucleic Acids Research*, *42*, D444–D449.
334. Gauthier, J., Vincent, A., Charette, S., & Derome, N. (2019). A brief history of bioinformatics. *Briefings in Bioinformatics*, *20*, 1981–1996.
335. Ge, R., Feng, G., Jing, X., Zhang, R., Wang, P., & Wu, Q. (2020). Enacp: An ensemble learning model for identification of anticancer peptides. *Frontiers in Genetics*, *11*, Article 553906.
336. Geary, R. (1954). The contiguity ratio and statistical mapping. *The Incorporated Statistician*, *5*, 115–146.
337. Geer, L., Marchler-Bauer, A., Geer, R., Han, L., He, J., He, S., Liu, C., Shi, W., & Bryant, S. (2010). The NCBI biosystems database. *Nucleic Acids Research*, *38*, D492–D496.
338. Geng, Q., Yang, R., & Zhang, L. (2022). A deep learning framework for enhancer prediction using word embedding and sequence generation. *Biophysical Chemistry*, *286*, Article 106822.
339. Ghandi, M., Huang, F., Jané-Valbuena, J., Kryukov, G., Lo, C., McDonald III, E., Barretina, J., Gelfand, E., Bielski, C., Li, H., Others. (2019). Next-generation characterization of the cancer cell line encyclopedia. *Nature*, *569*, 503–508.
340. Ghandi, M., Mohammad-Noori, M., & Beer, M. (2014). Robust k-mer frequency estimation using gapped k-mers. *Journal of Mathematical Biology*, *69*, 469–500.
341. Ghosh, A., Singh, T., Singla, V., Bagga, R., Srinivasan, R., & Khandelwal, N. (2019). Dti histogram parameters correlate with the extent of myoinvasion and tumor type in endometrial carcinoma: A preliminary analysis. *Acta Radiologica*, *61*, 675–684.
342. Ghosh, N., Santoni, D., Saha, I., & Felici, G. (2024). Predicting transcription factor binding sites with deep learning. *International Journal of Molecular Sciences*, *25*, 4990.
343. Ghulam, A., Ali, F., Sikander, R., Ahmad, A., Ahmed, A., & Patil, S. (2022). ACP-2DCNN: Deep learning-based model for improving prediction of anticancer peptides using two-dimensional convolutional neural network. *Chemometrics and Intelligent Laboratory Systems*, *226*, Article 104589.
344. Gialitsis, N., Giannakopoulos, G., & Athanasouli, M. (2020). Evaluation of distributed DNA representations on the classification of conserved non-coding elements. *11th Hellenic conference on artificial intelligence* (pp. 41–47).
345. Gillani, M., & Pollastri, G. (2024). SCLpred-ECL: Subcellular localization prediction by deep N-to-1 convolutional neural networks. *International Journal of Molecular Sciences*, *25*, 5440.
346. Giri, N., & Cheng, J. . De. (2024). Novo atomic protein structure modeling for cryoEM density maps using 3D transformer and HMM. *Nature Communications*, *15*, 5511.
347. Giudice, G., Sánchez-Cabo, F., Torroja, C., & Lara-Pezzi, E. (2016). ATtRACT—a database of RNA-binding proteins and associated motifs. *Database*, *2016*, baw035.
348. Glažar, P., Papavasileiou, P., & Rajewsky, N. (2014). circBase: A database for circular RNAs. *Rna*, *20*, 1666–1670.
349. Goeman, J. (2008). Autocorrelated logistic ridge regression for prediction based on proteomics spectra. *Statistical Applications in Genetics and Molecular Biology*, *7*.
350. Gogoladze, G., Grigolava, M., Vishnepolsky, B., Chubinidze, M., Duroux, P., Lefranc, M., & Pirtskhalava, M. (2014). DBAASP: Database of antimicrobial activity and structure of peptides. *FEMS Microbiology Letters*, *357*, 63–68.
351. Goldtzvik, Y., Sen, N., Lam, S., & Orengo, C. (2023). Protein diversification through post-translational modifications, alternative splicing, and gene duplication. *Current Opinion in Structural Biology*, *81*, Article 102640.
352. Golzadeh, A., Kamandi, A., & Rahami, H. (2023). An attributed network embedding method to predict missing links in protein-protein interaction networks. *Journal of Algorithms and Computation*, *55*, 79–99.

353. Gong, Y., Liao, B., Peng, D., & Zou, Q. (2021). Accurate prediction and key feature recognition of immunoglobulin. *Applied Sciences*, *11*, 6894.
354. Goñi, J., Pérez, A., Torrents, D., & Orozco, M. (2007). Determining promoter location based on DNA structure first-principles calculations. *Genome Biology*, *8*, R263.
355. Goodfellow, I., Bengio, Y., & Courville, A. (2016). *Deep learning*. MIT Press.
356. Goodsell, D., & Dickerson, R. (1994). Bending and curvature calculations in B-DNA. *Nucleic Acids Research*, *22*, 5497.
357. Goodwin, S., McPherson, J., & McCombie, W. (2016). Coming of age: Ten years of next-generation sequencing technologies. *Nature Reviews Genetics*, *17*, 333–351.
358. Gorin, A., Zhurkin, V., & Wilma, K. (1995). B-DNA twisting correlates with base-pair morphology. *Journal of Molecular Biology*, *247*, 34–48.
359. Gotoh, O., & Tagashira, Y. (1981). Stabilities of nearest-neighbor doublets in double-helical DNA determined by fitting calculated melting profiles to observed profiles. *Biopolymers: Original Research on Biomolecules*, *20*, 1033–1042.
360. Govindan, G., & Nair, A. (2011). Composition, Transition and Distribution (CTD)—a dynamic feature for predictions based on hierarchical structure of cellular sorting. *2011 annual IEEE India conference* (pp. 1–6).
361. Grabowski, P., & Rappsilber, J. (2019). A primer on data analytics in functional genomics: How to move from data to insight? *Trends in Biochemical Sciences*, *44*, 21–32.
362. Grantham, R. (1974). Amino acid difference formula to help explain protein evolution. *Science*, *185*, 862–864.
363. Gregoriadis, G., Jain, S., Papaioannou, I., & Laing, P. (2005). Improving the therapeutic efficacy of peptides and proteins: A role for polysialic acids. *International Journal of Pharmaceutics*, *300*, 125–130.
364. Griffiths-Jones, S. (2006). miRBase: The microRNA sequence database. *MicroRNA Protocols* (pp. 129–138). Springer.
365. Griffiths-Jones, S., Bateman, A., Marshall, M., Khanna, A., & Eddy, S. (2003). Rfam: An RNA family database. *Nucleic Acids Research*, *31*, 439–441.
366. Grigoriev, A. (1998). Analyzing genomes with cumulative skew diagrams. *Nucleic Acids Research*, *26*, 2286–2290.
367. Grissa, I., Vergnaud, G., & Pourcel, C. (2007). The CRISPRdb database and tools to display CRISPRs and to generate dictionaries of spacers and repeats. *BMC Bioinformatics*, *8*, 1–10.
368. Gromiha, M., & Ponnuswamy, P. (1996). Hydrophobic distribution and spatial arrangement of amino acid residues in membrane proteins. *International Journal of Peptide and Protein Research*, *48*, 452–460.
369. Grover, A., & Leskovec, J. (2016). node2vec: Scalable feature learning for networks. *Proceedings of the 22nd ACM SIGKDD international conference on knowledge discovery and data mining* (pp. 855–864).
370. Gu, X., Chen, Z., & Wang, D. (2020). Prediction of G protein-coupled receptors with CTDC extraction and MRMD2. 0 dimension-reduction methods. *Frontiers in Bioengineering and Biotechnology*, *8*, 635.
371. Gu, J. Liu t., Wang X., Wang G., Cai J., & Chen T. (2018). *Recent advances in convolutional neural networks pattern recognition*, *77*, 354–377.
372. Gu, F., Chen, H., & Ni, J. (2008). Protein structural class prediction based on an improved statistical strategy. *BMC Bioinformatics*, *9*, 1–9.
373. Gui, Y., Wang, R., Wei, Y., & Wang, X. (2019). DNN-PPI: A large-scale prediction of protein-protein interactions based on deep neural networks. *Journal of Biological Systems*, *27*, 1–18.
374. Guo, L., Wang, L., You, Z., Yu, C., Hu, M., Zhao, B., & Li, Y. (2024). Likelihood-based feature representation learning combined with neighborhood information for predicting circRNA–miRNA associations. *Briefings in Bioinformatics*, *25*, bbae020.
375. Guo, Y., Yu, L., Wen, Z., & Li, M. (2008). Using support vector machine combined with auto covariance to predict protein-protein interactions from protein sequences. *Nucleic Acids Research*, *36*, 3025–3030.

376. Guo, Y., Zhou, D., Nie, R., Ruan, X., & Li, W. (2020). DeepANF: A deep attentive neural framework with distributed representation for chromatin accessibility prediction. *Neurocomputing*, *379*, 305–318.
377. Guo, W., Liu, X., Ma, Y., & Zhang, R. (2021). iRspot-DCC: Recombination hot/cold spots identification based on dinucleotide-based correlation coefficient and convolutional neural network. *Journal of Intelligent & Fuzzy Systems*, *41*, 1309–1317.
378. Gupta, S., & Shankar, R. (2023) miWords: Transformer-based composite deep learning for highly accurate discovery of pre-miRNA regions across plant genomes. *Briefings in Bioinformatics*, *24*, bbad088.
379. Gupta, S., & Shankar, R. (2023). miWords: Transformer-based composite deep learning for highly accurate discovery of pre-miRNA regions across plant genomes. *Briefings in Bioinformatics*, *24*, bbad088.
380. Gupta, M., Gouda, G., Donde, R., Goswami, P., Rajesh, N., Pati, P., Rathore, S., Vadde, R., & Behera, L. (2021). Rna structure prediction. *Bioinformatics in Rice Research: Theories and Techniques*, (pp. 209–237). Springer.
381. Gupta, S., Dennis, J., Thurman, R., Kingston, R., Stamatoyannopoulos, J., & Noble, W. (2008). Predicting human nucleosome occupancy from primary sequence. *PLoS Computational Biology*, *4*, Article e1000134.
382. Gurumayum, S., Jiang, P., Hao, X., Campos, T., Young, N., Korhonen, P., Gasser, R., Bork, P., Zhao, X., He, L., Others. (2021). OGEE v3: Online GEne essentiality database with increased coverage of organisms and human cell lines. *Nucleic Acids Research*, *49*, D998–D1003.
383. Gussow, A., Park, A., Borges, A., Shmakov, S., Makarova, K., Wolf, Y., Bondy-Denomy, J., & Koonin, E. (2020). Machine-learning approach expands the repertoire of anti-CRISPR protein families. *Nature Communications*, *11*, 3784–3784.
384. Gussow, A., Shmakov, S., Makarova, K., Wolf, Y., Bondy-Denomy, J., & Koonin, E. (2020). Vast diversity of anti-CRISPR proteins predicted with a machine-learning approach. *BioRxiv*.
385. Gussow, A., Shmakov, S., Makarova, K., Wolf, Y., Bondy-Denomy, J., & Koonin, E. (2020). Vast diversity of anti-CRISPR proteins predicted with a machine-learning approach. *BioRxiv*.
386. Guyon, I., & Elisseeff, A. (2003). An introduction to variable and feature selection. *Journal of Machine Learning Research*, *3*, 1157–1182.
387. Haeussler, M., Schönig, K., Eckert, H., Eschstruth, A., Mianné, J., Renaud, J., Schneider-Maunoury, S., Shkumatava, A., Teboul, L., Kent, J., Others. (2016). Evaluation of off-target and on-target scoring algorithms and integration into the guide RNA selection tool CRISPOR. *Genome Biology*, *17*, 1–12.
388. Hagerman, P. (1990). Sequence-directed curvature of DNA. *Annual Review of Biochemistry*, *59*, 755–781.
389. Hakim, A., Erwin, A., Eng, K., Galinium, M., & Muliady, W. (2014). Automated document classification for news article in Bahasa Indonesia based on term frequency inverse document frequency (TF-IDF) approach. *2014 6th international conference on information technology and electrical engineering (ICITEE)* (pp. 1–4).
390. Ham, D., Browne, T., Banglorewala, P., Wilson, T., Michael, R., Gloor, G., & Edgell, D. (2023). A generalizable Cas9/sgRNA prediction model using machine transfer learning with small high-quality datasets. *Nature Communications*, *14*, 5514.
391. Hamosh, A., Scott, A., Amberger, J., Valle, D., & McKusick, V. (2000). Online Mendelian inheritance in man (OMIM). *Human Mutation*, *15*, 57–61.
392. Han, S., Liang, Y., Li, Y., & Du, W. (2016). Long noncoding RNA identification: comparing machine learning based tools for long noncoding transcripts discrimination. *BioMed Research International*, *2016*.
393. Han, Y., & Zhang, S. (2023). ncRPI-LGAT: Prediction of ncRNA-protein interactions with line graph attention network framework. *Computational and Structural Biotechnology Journal*, *21*, 2286–2295.

394. Han, Y., & Zhang, S. (2023). ncRPI-LGAT: Prediction of ncRNA-protein interactions with line graph attention network framework. *Computational and Structural Biotechnology Journal, 21*, 2286–2295.
395. Han, L., Cai, C., Lo, S., Chung, M., & Chen, Y. (2004). Prediction of RNA-binding proteins from primary sequence by a support vector machine approach. *Rna, 10*, 355–368.
396. Han, G., Li, Q., & Li, Y. (2021). Comparative analysis and prediction of nucleosome positioning using integrative feature representation and machine learning algorithms. *BMC Bioinformatics, 22*, 129.
397. Hansen, P. (1990). Truncated singular value decomposition solutions to discrete ill-posed problems with ill-determined numerical rank. *SIAM Journal on Scientific and Statistical Computing, 11*, 503–518.
398. Haraty, R., Dimishkieh, M., & Masud, M. (2015). An enhanced k-means clustering algorithm for pattern discovery in healthcare data. *International Journal of Distributed Sensor Networks, 11*, Article 615740.
399. Harbeck, N., & Thomssen, C. (2010). A new look at node-negative breast cancer. *The Oncologist, 15*, 29–38.
400. Harrow, J., Denoeud, F., Frankish, A., Reymond, A., Chen, C., Chrast, J., Lagarde, J., Gilbert, J., Storey, R., Swarbreck, D., Others. (2006). GENCODE: Producing a reference annotation for ENCODE. *Genome Biology, 7*, 1–9.
401. Harrow, J., Frankish, A., Gonzalez, J., Tapanari, E., Diekhans, M., Kokocinski, F., Aken, B., Barrell, D., Zadissa, A., Searle, S., Others. (2012). GENCODE: The reference human genome annotation for The ENCODE Project. *Genome Research, 22*, 1760–1774.
402. Hart, T., Chandrashekhar, M., Aregger, M., Steinhart, Z., Brown, K., MacLeod, G., Mis, M., Zimmermann, M., Fradet-Turcotte, A., Sun, S., Others. (2015). High-resolution CRISPR screens reveal fitness genes and genotype-specific cancer liabilities. *Cell, 163*, 1515–1526.
403. Hartmann, B., Malfoy, B., & Lavery, R. (1989). Theoretical prediction of base sequence effects in DNA: Experimental reactivity of Z-DNA and BZ transition enthalpies. *Journal of Molecular Biology, 207*, 433–444.
404. Hasan, M., & Kurata, H. (2018). GPSuc: Global prediction of generic and species-specific succinylation sites by aggregating multiple sequence features. *PloS One, 13*, Article e0200283.
405. Hasan, M., Zhou, Y., Lu, X., Li, J., Song, J., & Zhang, Z. (2015). Computational identification of protein pupylation sites by using profile-based composition of k-spaced amino acid pairs. *PloS One, 10*, Article e0129635.
406. Hasani, M., Trost, C., Timmerman, N., & Jin, L. (2023). AcrTransAct: Pre-trained protein transformer models for the detection of type I Anti-CRISPR activities. *14th ACM international conference on bioinformatics, computational biology and health informatics*.
407. Hasanzadeh, A., Hamblin, M., Kiani, J., Noori, H., Hardie, J., Karimi, M., & Shafiee, H. (2022). Could artificial intelligence revolutionize the development of nanovectors for gene therapy and mRNA vaccines? *Nano Today, 47*.
408. Haseeb, A., Bashir, M., & Wali, A. (2023). BERTDom: Protein domain boundary prediction using BERT. *Computing and Informatics, 42*, 667–689.
409. Haselbeck, F., John, M., Zhang, Y., Pirnay, J., Fuenzalida-Werner, J., Costa, R., & Grimm, D. (2023). Superior protein thermophilicity prediction with protein language model embeddings. *NAR Genomics and Bioinformatics, 5*, lqad087.
410. Hayat, M., Tahir, M., Alarfaj, F., Alturki, R., & Gazzawe, F. (2022). NLP-BCH-Ens: NLP-based intelligent computational model for discrimination of malaria parasite. *Computers in Biology and Medicine, 149*, Article 105962.
411. He, W., & Jia, C. (2017). EnhancerPred2. 0: Predicting enhancers and their strength based on position-specific trinucleotide propensity and electron ion interaction potential feature selection. *Molecular Biosystems, 13*, 767–774.
412. He, S., Gao, B., Sabnis, R., & Sun, Q. (2023). RNAdegformer: Accurate prediction of mRNA degradation at nucleotide resolution with deep learning. *Briefings in Bioinformatics, 24*, bbac581.

413. He, W., Jia, C., & Zou, Q. (2019). 4mCPred: Machine learning methods for DNA N4-methylcytosine sites prediction. *Bioinformatics*, *35*, 593–601.
414. Heinzinger, M., Weissenow, K., Sanchez, J., Henkel, A., Steinegger, M., & Rost, B. (2023). Prostt5: Bilingual language model for protein sequence and structure. *BioRxiv*.
415. Hellberg, S., Sjoestroem, M., Skagerberg, B., & Wold, S. (1987). Peptide quantitative structure-activity relationships, a multivariate approach. *Journal of Medicinal Chemistry*, *30*, 1126–1135.
416. Helmy, M., Eldaydamony, E., Mekky, N., Elmogy, M., & Soliman, H. (2022). Predicting Parkinson disease related genes based on PyFeat and gradient boosted decision tree. *Scientific Reports*, *12*, 10004.
417. Henry, R. (2012). Next-generation sequencing for understanding and accelerating crop domestication. *Briefings in Functional Genomics*, *11*, 51–56.
418. Hermjakob, H., Montecchi-Palazzi, L., Lewington, C., Mudali, S., Kerrien, S., Orchard, S., Vingron, M., Roechert, B., Roepstorff, P., Valencia, A., Others. (2004). IntAct: An open source molecular interaction database. *Nucleic Acids Research*, *32*, D452–D455.
419. Herrera-Bravo, J., Belén, L., Farias, J., & Beltrán, J. (2021). TAP 1.0: A robust immunoinformatic tool for the prediction of tumor T-cell antigens based on AAindex properties. *Computational Biology and Chemistry*, *91*, 107452.
420. Heston, W. (1948). Genetics of cancer. *Advances in Genetics*, *2*, 99–125.
421. Hie, B., & Yang, K. (2022). Adaptive machine learning for protein engineering. *Current Opinion in Structural Biology*, *72*, 145–152.
422. Higgs, P. (2000). RNA secondary structure: Physical and computational aspects. *Quarterly Reviews of Biophysics*, *33*, 199–253.
423. Hinton, G., Srivastava, N., Krizhevsky, A., Sutskever, I., & Salakhutdinov, R. (2012). Improving neural networks by preventing co-adaptation of feature detectors. Preprint. ArXiv:1207.0580.
424. Hirakawa, M., Krishnakumar, R., Timlin, J., Carney, J., & Butler, K. (2020). Gene editing and CRISPR in the clinic: Current and future perspectives. *Bioscience Reports*, *40*, BSR20200127.
425. Ho Thanh Lam, L., Le, N., Van Tuan, L., Tran Ban, H., Nguyen Khanh Hung, T., Nguyen, N., Huu Dang, L., & Le, N. (2020). Machine learning model for identifying antioxidant proteins using features calculated from primary sequences. *Biology*, *9*, 325.
426. Ho, Q., Phan, V., Ou, Y., Others. (2020). Use Chou's 5-steps rule with different word embedding types to boost performance of electron transport protein prediction model. *IEEE/ACM Transactions on Computational Biology and Bioinformatics*, *19*, 1235–1244.
427. Ho, P., Ellison, M., Quigley, G., & Rich, A. (1986). A computer aided thermodynamic approach for predicting the formation of Z-DNA in naturally occurring sequences. *The EMBO Journal*, *5*, 2737–2744.
428. Hoang, T., Yin, C., & Yau, S. (2016). Numerical encoding of DNA sequences by chaos game representation with application in similarity comparison. *Genomics*, *108*, 134–142.
429. Hochreiter, S. (1998). The vanishing gradient problem during learning recurrent neural nets and problem solutions. *International Journal of Uncertainty, Fuzziness and Knowledge-Based Systems*, *6*, 107–116.
430. Hol, W., & Verlinde, C. (2006). Non-communicable diseases. *Insulin*, *106*, 107.
431. Holden, T., Subramaniam, R., Sullivan, R., Cheung, E., Schneider, C., Tremberger, G., Jr., Flamholz, A., Lieberman, D., & Cheung, T. (2007). ATCG nucleotide fluctuation of deinococcus radiodurans radiation genes. *Instruments, Methods, and Missions for Astrobiology X*, *6694*, 402–411.
432. Holdt, L., Kohlmaier, A., & Teupser, D. (2018). Circular RNAs as therapeutic agents and targets. *Frontiers in Physiology*, *9*, 1262.
433. Holley, R., Apgar, J., Everett, G., Madison, J., Marquisee, M., Merrill, S., & Penswick, J., & Zamir, A. (1965). Structure of a ribonucleic acid. *Science*, *147*, 1462–1465
434. Hong, Z., Zeng, X., Wei, L., & Liu, X. (2020). Identifying enhancer-promoter interactions with neural network based on pre-trained DNA vectors and attention mechanism. *Bioinformatics*, *36*, 1037–1043.

435. Horlbeck, M., Gilbert, L., Villalta, J., Adamson, B., Pak, R., Chen, Y., Fields, A., Park, C., Corn, J., Kampmann, M., Others. (2016). Compact and highly active next-generation libraries for CRISPR-mediated gene repression and activation. *Elife*, *5*, Article e19760.
436. Hornbeck, P., Kornhauser, J., Tkachev, S., Zhang, B., Skrzypek, E., Murray, B., Latham, V., & Sullivan, M. (2012). PhosphoSitePlus: A comprehensive resource for investigating the structure and function of experimentally determined post-translational modifications in man and mouse. *Nucleic Acids Research*, *40*, D261–D270.
437. Horne, D. (1988). Prediction of protein helix content from an autocorrelation analysis of sequence hydrophobicities. *Biopolymers: Original Research on Biomolecules*, *27*, 451–477.
438. Horvath, P., & Barrangou, R. (2010). CRISPR/Cas, the immune system of bacteria and archaea. *Science*, *327*, 167–170.
439. Hotelling, H. (1933). Analysis of a complex of statistical variables into principal components. *Journal of Educational Psychology*, *24*, 417.
440. Hou, X., Wang, Y., Bu, D., Wang, Y., & Sun, S. (2023). EMNGly: Predicting N-linked glycosylation sites using the language models for feature extraction. *Bioinformatics*, *39*, btad650.
441. Howard, J., & Ruder, S. (2018). Universal language model fine-tuning for text classification. Preprint. ArXiv:1801.06146.
442. Hsu, P., Scott, D., Weinstein, J., Ran, F., Konermann, S., Agarwala, V., Li, Y., Fine, E., Wu, X., Shalem, O., Others. (2013). DNA targeting specificity of RNA-guided Cas9 nucleases. *Nature Biotechnology*, *31*, 827–832.
443. Hu, M., Alkhairy, S., Lee, I., Pillich, R., Fong, D., Smith, K., Bachelder, R., Ideker, T., & Pratt, D. (2023). Evaluation of large language models for discovery of gene set function. *ArXiv*.
444. Hu, B., Zheng, L., Long, C., Song, M., Li, T., Yang, L., & Zuo, Y. (2019). EmExplorer: A database for exploring time activation of gene expression in mammalian embryos. *Open Biology*, *9*, Article 190054.
445. Hu, M., Yuan, F., Yang, K., Ju, F., Su, J., Wang, H., Yang, F., & Ding, Q. (2022). Exploring evolution-aware & -free protein language models as protein function predictors. *Advances in Neural Information Processing Systems*, *35*, 38873–38884.
446. Hu, E., Lan, X., Liu, Z., Gao, J., & Niu, D. (2022). A positive correlation between GC content and growth temperature in prokaryotes. *BMC Genomics*, *23*, 110.
447. Huang, N., Shoichet, B., & Irwin, J. (2006). Benchmarking sets for molecular docking. *Journal of Medicinal Chemistry*, *49*, 6789–6801.
448. Huang, L., Zhang, H., Deng, D., Zhao, K., Liu, K., Hendrix, D., & Mathews, D. (2019). LinearFold: Linear-time approximate RNA folding by 5'-to-3'dynamic programming and beam search. *Bioinformatics*, *35*, i295–i304.
449. Huang, Q., Zhou, W., Guo, F., Xu, L., & Zhang, L. (2021). 6mA-Pred: Identifying DNA N6-methyladenine sites based on deep learning. *PeerJ*, *9*, Article e10813.
450. Huang, L., Yang, B., Yi, H., Asif, A., Wang, J., Lithgow, T., Zhang, H., Minhas, F., & Yin, Y. (2021). AcrDB: A database of anti-CRISPR operons in prokaryotes and viruses. *Nucleic Acids Research*, *49*, D622–D629.
451. Hubbard, T., Barker, D., Birney, E., Cameron, G., Chen, Y., Clark, L., Cox, T., Cuff, J., Curwen, V., Down, T., Others. (2002). The Ensembl genome database project. *Nucleic Acids Research*, *30*, 38–41.
452. Hubert, B. (2022). SkewDB, a comprehensive database of GC and 10 other skews for over 30,000 chromosomes and plasmids. *Scientific Data*, *9*, 92.
453. Hull, R., Ramagaga, S., Nkosi, N., Marina, R., Kabahuma, R., & Dlamini, Z. (2023). Epigenetics analysis using artificial intelligence in the era of precision oncology. *Artificial intelligence and precision oncology: Bridging cancer research and clinical decision support* (pp. 117–137). Springer.
454. Hulo, N., Bairoch, A., Bulliard, V., Cerutti, L., De Castro, E., Langendijk-Genevaux, P., Pagni, M., & Sigrist, C. (2006). The PROSITE database. *Nucleic Acids Research*, *34*, D227–D230.

455. Hulsen, T., Huynen, M., Vlieg, J., & Groenen, P. (2006). Untitled. *Genome Biology*, *7*, R31.
456. Huntley, R., Sawford, T., Mutowo-Meullenet, P., Shypitsyna, A., Bonilla, C., Martin, M., & O'Donovan, C. (2015). The GOA database: Gene ontology annotation updates for 2015. *Nucleic Acids Research*, *43*, D1057–D1063.
457. Huson, D., & Zeng, W. (2023). MR-DNA: Flexible 5mC-Methylation-Site recognition in DNA sequences using token classification. *bio Rxiv*.
458. Ibrahim, A., Al-Turjman, F., & Al, Z. (2022). Futuristic CRISPR-based biosensing in the cloud and internet of things era: an overview. *Multimedia Tools and Applications*, *81*, 35143–35171.
459. Idhaya, T., Suruliandi, A., & Raja, S. (2023). Stacked framework of machine learning classifiers for protein family prediction using protein characteristics. *Current Science*, *125*, 508.
460. IDT Integrated DNA Technologies. (2023). CRISPR-Cas9: What are the pros and cons? https://eu.idtdna.com/pages/education/decoded/article/crispr-cas9-what-are-the-pros-and-cons.
461. Ikehara, K. (2014). [GADV]-protein world hypothesis on the origin of life. *Origins of Life and Evolution of Biospheres*, *44*, 299–302.
462. Imani, A., Valiant, J., & Gunawan, A. (2023). Deep learning-based approach on sgRNA off-target prediction in CRISPR/Cas9. *2023 international conference on computer science, information technology and engineering (ICCoSITE)*.
463. Imker, H. (2018). 25 years of molecular biology databases: A study of proliferation, impact, and maintenance. *Frontiers in Research Metrics and Analytics*, *3*, 18.
464. Landrum, M., Chitipiralla, S., Brown, G., Chen, C., Gu, B., Hart, J., Hoffman, D., Jang, W., Kaur, K., Liu, C., Others. (2020). ClinVar: Improvements to accessing data. *Nucleic Acids Research*, *48*, D835–D844.
465. Inayat, N., Khan, M., Iqbal, N., Khan, S., Raza, M., Khan, D., Khan, A., & Wei, D. (2021). iEnhancer-DHF: Identification of enhancers and their strengths using optimize deep neural network with multiple features extraction methods. *IEEE Access*, *9*, 40783–40796.
466. Sillitoe, I., Bordin, N., Dawson, N., Waman, V., Ashford, P., Scholes, H., Pang, C., Woodridge, L., Rauer, C., Sen, N., Others. (2021). CATH: Increased structural coverage of functional space. *Nucleic Acids Research*, *49*, D266–D273.
467. Inzamam-Ul-Hossain, M., & Islam, M. (2023). Identification of essential protein using chemical reaction optimization and machine learning technique. *IEEE/ACM transactions on computational biology and bioinformatics*.
468. Ioffe, S., & Szegedy, C. (2015). Batch normalization: Accelerating deep network training by reducing internal covariate shift. Preprint. ArXiv:1502.03167.
469. Ishaq, M., Abid, A., Farooq, M., Manzoor, M., Farooq, U., Abid, K., & Helou, M. (2023). Advances in database systems education: Methods, tools, curricula, and way forward. *Education and Information Technologies*, *28*, 2681–2725.
470. Ivanov, V., Minchenkova, L., Chernov, B., McPhie, P., Ryu, S., Garges, S., Barber, A., Zhurkin, V., & Adhya, S. (1995). CRP-DNA complexes: Inducing theA-likeForm in the binding sites with an extended central spacer. *Journal of Molecular Biology*, *245*, 228–240.
471. Jackson, S., McKenzie, R., Fagerlund, R., Kieper, S., Fineran, P., & Brouns, S. (2017). CRISPR-Cas: Adapting to change. *Science*, *356*, eaal5056.
472. Jagodnik, K., Shvili, Y., & Bartal, A. (2023). HetIG-PreDiG: A heterogeneous integrated graph model for predicting human disease genes based on gene expression. *Plos One*, *18*, Article e0280839.
473. Jan, A., Hayat, M., Wedyan, M., Alturki, R., Gazzawe, F., Ali, H., & Alarfaj, F. (2022). Target-AMP: Computational prediction of antimicrobial peptides by coupling sequential information with evolutionary profile. *Computers In Biology And Medicine*, *151*, Article 106311.
474. Japkowicz, N., & Shah, M. (2015). Performance evaluation in machine learning. *Machine Learning in Radiation Oncology: Theory and Applications* (pp. 41–56).

475. Jasiak, A., Koczkowska, M., Stukan, M., Wydra, D., Biernat, W., Izycka-Swieszewska, E., Buczkowski, K., Eccles, M., Walker, L., Wasag, B., Others. (2023). Analysis of BRCA1 and BRCA2 alternative splicing in predisposition to ovarian cancer. *Experimental and Molecular Pathology*, *130*, Article 104856.
476. Jaworska, N., & Chupetlovska-Anastasova, A. (2009). A review of multidimensional scaling (MDS) and its utility in various psychological domains. *Tutorials in Quantitative Methods for Psychology*, *5*, 1–10.
477. Jenuth, J. (1999). The NCBI: Publicly available tools and resources on the web. *Bioinformatics Methods and Protocols*, 301–312.
478. Jeong, Y., Gerhäuser, C., Sauter, G., Schlomm, T., Rohr, K., & Lutsik, P. (2023). MethylBERT: A transformer-based model for read-level DNA methylation pattern identification and tumour deconvolution. *BioRxiv*.
479. Jha, K., Saha, S., & Karmakar, S. (2023). Prediction of protein-protein interactions using vision transformer and language model. *IEEE/ACM Transactions on Computational Biology and Bioinformatics*, *20*, 3215–3225.
480. Jhong, J., Chi, Y., Li, W., Lin, T., Huang, K., & Lee, T. (2019). dbAMP: An integrated resource for exploring antimicrobial peptides with functional activities and physicochemical properties on transcriptome and proteome data. *Nucleic Acids Research*, *47*, D285–D297.
481. Ji, Y., Zhou, Z., Liu, H., & Davuluri, R. (2021). DNABERT: Pre-trained Bidirectional Encoder Representations from Transformers model for DNA-language in genome. *Bioinformatics*, *37*, 2112–2120.
482. Jia, C., & He, W. (2016). EnhancerPred: A predictor for discovering enhancers based on the combination and selection of multiple features. *Scientific Reports*, *6*, 38741.
483. Jiang, Q., Wang, J., Wu, X., Ma, R., Zhang, T., Jin, S., Han, Z., Tan, R., Peng, J., Liu, G., Others. (2015). LncRNA2Target: A database for differentially expressed genes after lncRNA knockdown or overexpression. *Nucleic Acids Research*, *43*, D193–D196.
484. Jiang, Q., Wang, Y., Hao, Y., Juan, L., Teng, M., Zhang, X., Li, M., Wang, G., & Liu, Y. (2009). miR2Disease: A manually curated database for microRNA deregulation in human disease. *Nucleic Acids Research*, *37*, D98–D104.
485. Jiang, P., Ning, W., Shi, Y., Liu, C., Mo, S., Zhou, H., Liu, K., & Guo, Y. (2021). FSL-Kla: A few-shot learning-based multi-feature hybrid system for lactylation site prediction. *Computational and Structural Biotechnology Journal*, *19*, 4497–4509.
486. Jiang, J., Song, B., Meng, J., & Zhou, J. (2024). Tissue-specific RNA methylation prediction from gene expression data using sparse regression models. *Computers in Biology and Medicine*, *169*, Article 107892.
487. Jin, J., Lu, P., Xu, Y., Li, Z., Yu, S., Liu, J., Wang, H., Chua, N., & Cao, P. (2021). PLncDB V2. 0: A comprehensive encyclopedia of plant long noncoding RNAs. *Nucleic Acids Research*, *49*, D1489–D1495.
488. Jin, J., Yu, Y., Wang, R., Zeng, X., Pang, C., Jiang, Y., Li, Z., Dai, Y., Su, R., Zou, Q., Others. (2022). iDNA-ABF: Multi-scale deep biological language learning model for the interpretable prediction of DNA methylations. *Genome Biology*, *23*, 219.
489. John, A. C., & Mona, S. (2007). Predicting functionally important residues from sequence conservation. *Bioinformatics, 23*(15), 1875–1882.
490. Jonsson, J., Eriksson, L., Hellberg, S., Sjöström, M., & Wold, S. (1989). Multivariate parametrization of 55 coded and non-coded amino acids. *Quantitative Structure-Activity Relationships*, *8*, 204–209.
491. Jordan, K., He, F., Soto, M., Akhunova, A., & Akhunov, E. (2020). Differential chromatin accessibility landscape reveals structural and functional features of the allopolyploid wheat chromosomes. *Genome Biology*, *21*.
492. Ju, Z., & Wang, S. (2020). Prediction of lysine formylation sites using the composition of k-spaced amino acid pairs via Chou's 5-steps rule and general pseudo components. *Genomics*, *112*, 859–866.
493. Jumper, J., Evans, R., Pritzel, A., Green, T., Figurnov, M., Ronneberger, O., Tunyasuvunakool, K., Bates, R., Žıdek, A., Potapenko, A., Others. (2021). Highly accurate protein structure prediction with AlphaFold. *Nature*, *596*, 583–589.

494. Justyna, M., Antczak, M., & Szachniuk, M. (2023). Machine learning for RNA 2D structure prediction benchmarked on experimental data. *Briefings in Bioinformatics*, *24*, bbad153.
495. Kabanga, E., Yun, S., Van Messem, A., & De Neve, W. (2024). Impact of U2-type introns on splice site prediction in Arabidopsis thaliana using deep learning. *BioRxiv*.
496. Kabir, A., Bhattarai, M., Rasmussen, K., Shehu, A., Bishop, A., Alexandrov, B., & Usheva, A. (2024). Advancing transcription factor binding site prediction using DNA breathing dynamics and sequence transformers via cross attention. *BioRxiv*.
497. Kafkas, Ş., Abdelhakim, M., Althagafi, A., Toonsi, S., Alghamdi, M., Schofield, P., & Hoehndorf, R. (2023). The application of large language models to the phenotype-based prioritization of causative genes in rare disease patients. *MedRxiv*.
498. Källberg, M., Wang, H., Wang, S., Peng, J., Wang, Z., Lu, H., & Xu, J. (2012). Template-based protein structure modeling using the RaptorX web server. *Nature Protocols*, *7*, 1511–1522.
499. Kalra, K., Others. (2023). *Predicting the genotype to phenotype relationship in plants using machine learning and deep learning*. University of Saskatchewan.
500. Kaminuma, E., Kosuge, T., Kodama, Y., Aono, H., Mashima, J., Gojobori, T., Sugawara, H., Ogasawara, O., Takagi, T., Okubo, K., Others. (2010). DDBJ progress report. *Nucleic Acids Research*, *39*, D22–D27.
501. Kandaswamy, K., Pugalenthi, G., Hartmann, E., Kalies, K., Möller, S., Suganthan, P., & Martinetz, T. (2010). SPRED: A machine learning approach for the identification of classical and non-classical secretory proteins in mammalian genomes. *Biochemical and Biophysical Research Communications*, *391*, 1306–1311.
502. Kanehisa, M. (2002). The KEGG database. *'In Silico' simulation of biological processes: Novartis foundation symposium* (Vol. 247, pp. 91–103).
503. Kang, X., Dong, F., Shi, C., Liu, S., Sun, J., Chen, J., Li, H., Xu, H., Lao, X., & Zheng, H. (2019). DRAMP 2.0, an updated data repository of antimicrobial peptides. *Scientific Data*, *6*, 148.
504. Kang, Y., Wang, X., Xie, C., Zhang, H., & Xie, W. (2023). BBLN: A bilateral-branch learning network for unknown protein-protein interaction prediction. *Computers in Biology and Medicine*, *167*, Article 107588.
505. Kanz, C., Aldebert, P., Althorpe, N., Baker, W., Baldwin, A., Bates, K., Browne, P., Broek, A., Castro, M., Cochrane, G., Others. (2005). The EMBL nucleotide sequence database. *Nucleic Acids Research*, *33*, D29–D33.
506. Karczewski, K., & Francioli, L. (2017). The genome aggregation database (gnomAD). *MacArthur Lab* (pp. 1–10).
507. Karunasingha, D. (2022). Root mean square error or mean absolute error? Use their ratio as well. *Information Sciences*, *585*, 609–629.
508. Karve, T., & Cheema, A. (2011). Small changes huge impact: The role of protein posttranslational modifications in cellular homeostasis and disease. *Journal of Amino Acids*, *2011*, Article 207691.
509. Katuwawala, A., Zhao, B., & Kurgan, L. (2022). DisoLipPred: Accurate prediction of disordered lipid-binding residues in protein sequences with deep recurrent networks and transfer learning. *Bioinformatics*, *38*, 115–124.
510. Kaur, I., Sandhu, A., & Kumar, Y. (2022). Artificial intelligence techniques for predictive modeling of vector-borne diseases and its pathogens: A systematic review. *Archives of Computational Methods in Engineering*, *29*, 3741–3771.
511. Kawashima, S., & Kanehisa, M. (2000). AAindex: amino acid index database. *Nucleic Acids Research*, *28*, 374–374.
512. Kawashima, S., Pokarowski, P., Pokarowska, M., Kolinski, A., Katayama, T., & Kanehisa, M. (2007). AAindex: amino acid index database, progress report 2008. *Nucleic Acids Research*, *36*, D202–D205.
513. Kenneth Jr, M., LeGrand, S., & Others The protein folding problem and tertiary structure prediction. (Springer Science & Business Media,2012)

514. Kha, Q., Ho, Q., & Le, N. (2022). Identifying SNARE proteins using an alignment-free method based on multiscan convolutional neural network and PSSM profiles. *Journal of Chemical Information and Modeling*, *62*, 4820–4826.
515. Khanal, J., Tayara, H., & Chong, K. (2020). Identifying enhancers and their strength by the integration of word embedding and convolution neural network. *IEEE Access*, *8*, 58369–58376.
516. Khanal, J., Tayara, H., Zou, Q., & Chong, K. (2021). Identifying DNA N4-methylcytosine sites in the rosaceae genome with a deep learning model relying on distributed feature representation. *Computational and Structural Biotechnology Journal*, *19*, 1612–1619.
517. Khatun, M., Hasan, M., & Kurata, H. (2019). PreAIP: Computational prediction of anti-inflammatory peptides by integrating multiple complementary features. *Frontiers in Genetics*, *10*, 129.
518. Khojasteh, H., Pirgazi, J., & Ghanbari Sorkhi, A. (2023). Improving prediction of drug-target interactions based on fusing multiple features with data balancing and feature selection techniques. *Plos One*, *18*, Article e0288173.
519. Khoshandam, M., Soltaninejad, H., & Hamidieh, A. (2023). CRISPR and Artificial intelligence to improve precision medicine: Future perspectives and potential limitations. *Multimedia Tools and Applications, 42*, 1–6.
520. Kim, S., Chen, J., Cheng, T., Gindulyte, A., He, J., He, S., Li, Q., Shoemaker, B., Thiessen, P., Yu, B., Others. (2023). PubChem 2023 update. *Nucleic Acids Research*, *51*, D1373–D1380.
521. Kim, H., Kim, Y., Lee, S., Min, S., Bae, J., Choi, J., Park, J., Jung, D., Yoon, S., & Kim, H. (2019). SpCas9 activity prediction by DeepSpCas9, a deep learning–based model with high generalization performance. *Science Advances*, *5*, eaax9249.
522. Kim, S., Mollaei, P., Antony, A., Magar, R., & Barati Farimani, A. (2024). GPCR-BERT: Interpreting sequential design of G protein-coupled receptors using protein language models. *Journal of Chemical Information and Modeling*, *64*, 1134–1144.
523. Kim, D., Bae, S., Park, J., Kim, E., Kim, S., Yu, H., Hwang, J., Kim, J., & Kim, J. (2015). Digenome-seq: genome-wide profiling of CRISPR-Cas9 off-target effects in human cells. *Nature Methods*, *12*, 237–243.
524. Kim, D., Kim, S., Kim, S., Park, J., & Kim, J. (2016). Genome-wide target specificities of CRISPR-Cas9 nucleases revealed by multiplex Digenome-seq. *Genome Research*, *26*, 406–415.
525. Kim, N., Kim, H., Lee, S., Seo, J., Choi, J., Park, J., Min, S., Yoon, S., Cho, S., & Kim, H. (2020). Prediction of the sequence-specific cleavage activity of Cas9 variants. *Nature Biotechnology*, *38*, 1328–1336.
526. Kim, D., Lee, J., Cho, C., Kim, E., Bhattacharya, D., & Yoon, H. (2022). Group II intron and repeat-rich red algal mitochondrial genomes demonstrate the dynamic recent history of autocatalytic RNAs. *BMC Biology*, *20*, 2.
527. King, R., Rotter, J., & Motulsky, A. (2002). *The genetic basis of common diseases*. Oxford University Press.
528. Kitsiranuwat, S., Suratanee, A., & Plaimas, K. (2021). Multi-data aspects of protein similarity with a learning technique to identify drug-disease associations. *Applied Sciences*, *11*, 2914.
529. Kleinstiver, B., Pattanayak, V., Prew, M., Tsai, S., Nguyen, N., Zheng, Z., & Joung, J. (2016). High-fidelity CRISPR-Cas9 nucleases with no detectable genome-wide off-target effects. *Nature*, *529*, 490–495.
530. Knops, J., Bradley, K., & Wedin, D. (2002). Mechanisms of plant species impacts on ecosystem nitrogen cycling. *Ecology Letters*, *5*, 454–466.
531. Kodama, Y., Mashima, J., Kaminuma, E., Gojobori, T., Ogasawara, O., Takagi, T., Okubo, K., & Nakamura, Y. (2012). The DNA Data Bank of Japan launches a new resource, the DDBJ Omics Archive of functional genomics experiments. *Nucleic Acids Research*, *40*, D38–D42.

532. Kodama, Y., Shumway, M., & Leinonen, R. (2012). The sequence read archive: Explosive growth of sequencing data. *Nucleic Acids Research, 40*, D54–D56.
533. Kohavi, R., & John, G. (1997). Wrappers for feature subset selection. *Artificial Intelligence, 97*, 273–324.
534. Kohli, R., & Zhang, Y. (2013). Tet enzymes, tdg and the dynamics of dna demethylation. *Nature, 502*, 472–479.
535. Koike-Yusa, H., Li, Y., Tan, E., Velasco-Herrera, M., & Yusa, K. (2014). Genome-wide recessive genetic screening in mammalian cells with a lentiviral CRISPR-guide RNA library. *Nature Biotechnology, 32*, 267–273.
536. Kong, M., Zhang, Y., Xu, D., Chen, W., & Dehmer, M. (2020). FCTP-WSRC: Protein-protein interactions prediction via weighted sparse representation based classification. *Frontiers in Genetics, 11*, 18.
537. Konstantakos, V., Nentidis, A., Krithara, A., & Paliouras, G. (2022). *CRISPRedict: The case for simple and interpretable efficiency prediction for CRISPR-Cas9 gene editing*. Preprints.
538. Konstantakos, V., Nentidis, A., Krithara, A., & Paliouras, G. (2022). CRISPR- Cas9 gRNA efficiency prediction: an overview of predictive tools and the role of deep learning. *Nucleic Acids Research, 50*, 3616–3637.
539. Koo, H., Wu, H., & Crothers, D. (1986). DNA bending at adenine · thymine tracts. *Nature, 320*, 501–506.
540. Kovaltsuk, A., Leem, J., Kelm, S., Snowden, J., Deane, C., & Krawczyk, K. (2018). Observed antibody space: A resource for data mining next-generation sequencing of antibody repertoires. *The Journal of Immunology, 201*, 2502–2509.
541. Koyama, T., Tsumura, H., Matsumoto, S., Okita, R., Kojima, R., & Okuno, Y. (2024). ChemGLaM: Chemical genomics language models for compound-protein interaction prediction. *BioRxiv*.
542. Kubota, M., Tran, C., & Spitale, R. (2015). Progress and challenges for chemical probing of RNA structure inside living cells. *Nature Chemical Biology, 11*, 933–941.
543. Kukurba, K., & Montgomery, S. (2015). RNA sequencing and analysis. *Cold Spring Harbor Protocols, 2015*, pdb-top084970
544. Kulakovskiy, I., Vorontsov, I., Yevshin, I., Sharipov, R., Fedorova, A., Rumynskiy, E., Medvedeva, Y., Magana-Mora, A., Bajic, V., Papatsenko, D., Others. (2018). HOCOMOCO: Towards a complete collection of transcription factor binding models for human and mouse via large-scale ChIP-Seq analysis. *Nucleic Acids Research, 46*, D252–D259.
545. Kumar, H., & Kim, P. (2024). Artificial intelligence in fusion protein three-dimensional structure prediction: Review and perspective. *Clinical and Translational Medicine, 14*, Article e1789.
546. Kumar, R., Chaudhary, K., Sharma, M., Nagpal, G., Chauhan, J., Singh, S., Gautam, A., & Raghava, G. (2015). AHTPDB: A comprehensive platform for analysis and presentation of antihypertensive peptides. *Nucleic Acids Research, 43*, D956–D962.
547. Kurgan, L. (2023). Machine learning in bioinformatics of protein sequences: Algorithms, databases and resources for modern protein bioinformatics. World Scientific.
548. Kuru, H., Tepeli, Y., & Taştan, Ö. (2022). GEGE: Predicting gene essentiality with graph embeddings. *Düzce Üniversitesi Bilim Ve Teknoloji Dergisi, 10*, 1567–1577.
549. Kyte, J. (2006). *Structure in protein chemistry*. Garland Science.
550. Lackner, M., Helmbrecht, N., Pääbo, S., & Riesenberg, S. (2023). Detection of unintended on-target effects in CRISPR genome editing by DNA donors carrying diagnostic substitutions. *Nucleic Acids Research, 51*, e26–e26.
551. Lai, M. (2013). Common applications of next-generation sequencing technologies in genomic research. *Translational Cancer Research*, 2(1)
552. Lal, T., Chapelle, O., Weston, J., & Elisseeff, A. (2006). Embedded methods. *Feature Extraction*, 137–165.
553. Lalović, D., & Veljković, V. (1990). The global average DNA base composition of coding regions may be determined by the electron-ion interaction potential. *Biosystems, 23*, 311–316.

554. Lan, W., Zhu, M., Chen, Q., Chen, B., Liu, J., Li, M., & Chen, Y. (2020). CircR2Cancer: A manually curated database of associations between circRNAs and cancers. *Database, 2020*, baaa085.
555. Landrum, M., Lee, J., Benson, M., Brown, G., Chao, C., Chitipiralla, S., Gu, B., Hart, J., Hoffman, D., Hoover, J., Others. (2016). ClinVar: Public archive of interpretations of clinically relevant variants. *Nucleic Acids Research*, *44*, D862–D868.
556. Lane, A., Chaires, J., Gray, R., & Trent, J. (2008). Stability and kinetics of G-quadruplex structures. *Nucleic Acids Research*, *36*, 5482–5515.
557. Langfelder, P., & Horvath, S. (2008). Wgcna: An r package for weighted correlation network analysis. *BMC Bioinformatics*, *9*.
558. Langmead, B., & Salzberg, S. (2012). Fast gapped-read alignment with Bowtie 2. *Nature Methods*, *9*, 357–359.
559. Lazzarotto, C., Malinin, N., Li, Y., Zhang, R., Yang, Y., Lee, G., Cowley, E., He, Y., Lan, X., Jividen, K., Others. CHANGE-seq reveals genetic and epigenetic effects on CRISPR–Cas9 genome-wide activity. *Nature Biotechnology*, *38*, 1317–1327 (2020)
560. Le, N., Ho, Q., Nguyen, T., & Ou, Y. (2021). A transformer architecture based on BERT and 2D convolutional neural network to identify DNA enhancers from sequence information. *Briefings in Bioinformatics*, *22*, bbab005.
561. Le, N., & Huynh, T. (2019). Identifying SNAREs by incorporating deep learning architecture and amino acid embedding representation. *Frontiers in Physiology*, *10*, 1501.
562. Le, N., Yapp, E., Nagasundaram, N., & Yeh, H. (2019). Classifying promoters by interpreting the hidden information of DNA sequences via deep learning and combination of continuous fasttext N-grams. *Frontiers in Bioengineering and Biotechnology*, *7*, 305.
563. Le, N., Yapp, E., Ho, Q., Nagasundaram, N., Ou, Y., & Yeh, H. (2019). iEnhancer-5Step: Identifying enhancers using hidden information of DNA sequences via Chou's 5-step rule and word embedding. *Analytical Biochemistry*, *571*, 53–61.
564. Le, N., Do, D., Hung, T., Lam, L., Huynh, T., & Nguyen, N. (2020). A computational framework based on ensemble deep neural networks for essential genes identification. *International Journal of Molecular Sciences*, *21*, 9070.
565. LeCun, Y., Bottou, L., Orr, G., & Müller, K. (2012). Efficient backprop. *Neural Networks: Tricks of the Trade*, 9–48
566. LeCun, Y., Bengio, Y., & Hinton, G. (2015). Deep learning. *Nature*, *521*, 436–444.
567. Lee, S., Birkholz, N., Fineran, P., & Park, H. (2022). Molecular basis of anti-CRISPR operon repression by Aca10. *Nucleic Acids Research*, *50*, 8919–8928
568. Lee, C., Gallagher, P., & Tu, Z. (2016). Generalizing pooling functions in convolutional neural networks: Mixed, gated, and tree. *Artificial Intelligence and Statistics*, 464–472.
569. Lee, M. (2023). Deep learning in CRISPR-Cas systems: A review of recent studies. *Frontiers in Bioengineering and Biotechnology, 11,* 1–14.
570. Lee, T. (1998). Independent component analysis. *Independent Component Analysis*, 27–66.
571. Lee, D., Karchin, R., & Beer, M. (2011). Discriminative prediction of mammalian enhancers from DNA sequence. *Genome Research*, *21*, 2167–2180.
572. Lee, T., Chen, S., Hung, H., & Ou, Y. (2011). Incorporating distant sequence features and radial basis function networks to identify ubiquitin conjugation sites. *PloS One*, *6*, Article e17331.
573. Leem, J., Oliveira, S., Krawczyk, K., & Deane, C. (2018). STCRDab: The structural T-cell receptor database. *Nucleic Acids Research*, *46*, D406–D412.
574. Leenay, R., Aghazadeh, A., Hiatt, J., Tse, D., Roth, T., Apathy, R., Shifrut, E., Hultquist, J., Krogan, N., Wu, Z., Others. (2019). Large dataset enables prediction of repair after CRISPR–Cas9 editing in primary T cells. *Nature Biotechnology*, *37*, 1034–1037.
575. Leinonen, R., Sugawara, H., Shumway, M., & Collaboration, I. (2010). The sequence read archive. *Nucleic Acids Research*, *39*, D19–D21.
576. Lennox, M., Robertson, N., & Devereux, B. (2021). Modelling drug-target binding affinity using a BERT based graph neural network. *2021 43rd annual international conference of the IEEE engineering in medicine & biology society (EMBC)* (pp. 4348–4353).

577. Amberger, J., Bocchini, C., Scott, A., & Hamosh, A. (2019). OMIM.org.: Leveraging knowledge across phenotype-gene relationships. *Nucleic Acids Research*, *47*, D1038–D1043.
578. Lewis, J., & Sankey, O. (1995). Geometry and energetics of DNA basepairs and triplets from first principles quantum molecular relaxations. *Biophysical Journal*, *69*, 1068–1076.
579. Li, B., Ai, D., & Liu, X. (2022). CNN-XG: A hybrid framework for sgRNA on-target prediction. *Biomolecules*, *12*, 409–409. https://lens.org/141-724-023-233-250.
580. Li, Z., Gao, E., Zhou, J., Han, W., Xu, X., & Gao, X. (2023). Applications of deep learning in understanding gene regulation. *Cell Reports Methods*, *3*.
581. Li, Y., Guo, Z., Wang, K., Gao, X., & Wang, G. (2023). End-to-end interpretable disease–gene association prediction. *Briefings in Bioinformatics*, *24*, bbad118.
582. Li, Y., Guo, Z., Wang, K., Gao, X., & Wang, G. (2023). End-to-end interpretable disease–gene association prediction. *Briefings in Bioinformatics*, *24*, bbad118.
583. Li, P., Hastie, T., & Church, K. (2006). Very sparse random projections. *Proceedings of the 12th ACM SIGKDD international conference on knowledge discovery and data mining* (pp. 287–296).
584. Li, Z., Jin, J., Wang, Y., Long, W., Ding, Y., Hu, H., & Wei, L. (2023). ExamPle: Explainable deep learning framework for the prediction of plant small secreted peptides. *Bioinformatics*, *39*, btad108.
585. Li, Z., Li, S., Luo, M., Jhong, J., Li, W., Yao, L., Pang, Y., Wang, Z., Wang, R., Ma, R., Others. (2022). dbPTM in 2022: An updated database for exploring regulatory networks and functional associations of protein post-translational modifications. *Nucleic Acids Research*, *50*, D471–D479.
586. Li, J., Liu, S., Zhou, H., Qu, L., & Yang, J. (2014). starBase v2. 0: Decoding miRNA-ceRNA, miRNA-ncRNA and protein–RNA interaction networks from large-scale CLIP-Seq data. *Nucleic Acids Research*, *42*, D92–D97.
587. Li, S., Moayedpour, S., Li, R., Bailey, M., Riahi, S., Kogler-Anele, L., Miladi, M., Miner, J., Zheng, D., Wang, J., Others. (2023). CodonBERT: Large language models for mRNA design and optimization. *BioRxiv*.
588. Li, Y., Qiu, C., Tu, J., Geng, B., Yang, J., Jiang, T., & Cui, Q. (2014). HMDD v2. 0: A database for experimentally supported human microRNA and disease associations. *Nucleic Acids Research*, *42*, D1070–D1074.
589. Li, Y., Wei, Y., Xu, S., Tan, Q., Zong, L., Wang, J., Wang, Y., Chen, J., Hong, L., & Li, Y. (2023). AcrNET: Predicting anti-CRISPR with deep learning. *Bioinformatics*, *39*, btad259.
590. Li, Y., Wei, Y., Xu, S., Tan, Q., Zong, L., Wang, J., Wang, Y., Chen, J., Hong, L., & Li, Y. (2023). AcrNET: Predicting anti-CRISPR with deep learning. *Bioinformatics (Oxford, England)*, *39*.
591. Li, J., Wu, Z., Lin, W., Luo, J., Zhang, J., Chen, Q., & Chen, J. (2023). iEnhancer-ELM: Improve enhancer identification by extracting position-related multiscale contextual information based on enhancer language models. *Bioinformatics Advances*, *3*, vbad043.
592. Li, H., Wu, B., Sun, M., Zhu, Z., Chen, K., & Ge, H. (2024). Cross-domain contrastive graph neural network for LncRNA-protein interaction prediction. *Knowledge-Based Systems*, 111901.
593. Li, Q., Xu, L., Li, Q., Zhang, L., Others. (2020). Identification and classification of enhancers using dimension reduction technique and recurrent neural network. *Computational and Mathematical Methods in Medicine*, *2020*.
594. Li, J., Zou, Q., & Yuan, L. (2023). A review from biological mapping to computation-based subcellular localization. *Molecular Therapy-Nucleic Acids*, *32*, 507–521.
595. Li, M., Zhou, Y., Luo, Q., & Li, Z. (2010). The 3D structures of G-Quadruplexes of HIV-1 integrase inhibitors: Molecular dynamics simulations in aqueous solution and in the gas phase. *Journal of Molecular Modeling*, *16*, 645–657.
596. Li, W., Zhong, Z., Zhu, P., Deng, E., Ding, H., Chen, W., & Lin, H. (2014). Sequence analysis of origins of replication in the Saccharomyces cerevisiae genomes. *Frontiers in Microbiology*, *5*, 574.

597. Li, P., Chen, S., Chen, H., Mo, X., Li, T., Shao, Y., Xiao, B., & Guo, J. (2015). Using circular RNA as a novel type of biomarker in the screening of gastric cancer. *Clinica Chimica Acta, 444*, 132–136.
598. Li, W., Cowley, A., Uludag, M., Gur, T., McWilliam, H., Squizzato, S., Park, Y., Buso, N., & Lopez, R. (2015). The EMBL-EBI bioinformatics web and programmatic tools framework. *Nucleic Acids Research, 43*, W580–W584.
599. Li, Q., Zhang, C., Chen, H., Xue, J., Guo, X., Liang, M., & Chen, M. (2018). BioPepDB: An integrated data platform for food-derived bioactive peptides. *International Journal of Food Sciences and Nutrition, 69*, 963–968.
600. Li, Z., Tian, D., Wang, B., Wang, J., Wang, S., Chen, H., Xu, X., Wang, C., & He, N. (2019). Microbes drive global soil nitrogen mineralization and availability. *Global Change Biology, 25*, 1078–1088.
601. Li, Y., Huang, C., Ding, L., Li, Z., Pan, Y., & Gao, X. (2019). Deep learning in bioinformatics: Introduction, application, and perspective in the big data era. *Methods, 166*, 4–21.
602. Li, Q., Zhou, W., Wang, D., Wang, S., & Li, Q. (2020). Prediction of anticancer peptides using a low-dimensional feature model. *Frontiers in Bioengineering and Biotechnology, 8*, 892.
603. Li, F., Chen, J., Ge, Z., Wen, Y., Yue, Y., Hayashida, M., Baggag, A., Bensmail, H., & Song, J. (2021). Computational prediction and interpretation of both general and specific types of promoters in Escherichia coli by exploiting a stacked ensemble-learning framework. *Briefings in Bioinformatics, 22*, 2126–2140.
604. Li, S., Li, L., Meng, X., Sun, P., Liu, Y., Song, Y., Zhang, S., Jiang, C., Cai, J., & Zhao, Z. (2021). DREAM: A database of experimentally supported protein-coding RNAs and drug associations in human cancer. *Molecular Cancer, 20*, 1–6.
605. Li, Z., Jiang, H., Kong, L., Chen, Y., Lang, K., Fan, X., Zhang, L., & Pian, C. (2021). Deep6ma: A deep learning framework for exploring similar patterns in dna n6-methyladenine sites across different species. *Plos Computational Biology, 17*, Article e1008767.
606. Li, V., Zhang, Z., & Troyanskaya, O. (2021). CROTON: An automated and variant-aware deep learning framework for predicting CRISPR/Cas9 editing outcomes. *Bioinformatics, 37*, i342–i348.
607. Li, J., Li, J., Kong, M., Wang, D., Fu, K., & Shi, J. (2021). SVDNVLDA: Predicting lncRNA-disease associations by Singular Value Decomposition and node2vec. *BMC Bioinformatics, 22*, 1–18.
608. Li, J., He, S., Guo, F., & Zou, Q. (2021). HSM6AP: A high-precision predictor for the Homo sapiens N6-methyladenosine (mundefined 6 A) based on multiple weights and feature stitching. *RNA Biology, 18*, 1882–1892.
609. Li, H., Pang, Y., & Liu, B. (2021). BioSeq-BLM: A platform for analyzing DNA, RNA and protein sequences based on biological language models. *Nucleic Acids Research, 49*, e129–e129.
610. Li, F., Chen, J., Ge, Z., Wen, Y., Yue, Y., Hayashida, M., Baggag, A., Bensmail, H., & Song, J. (2021). Computational prediction and interpretation of both general and specific types of promoters in Escherichia coli by exploiting a stacked ensemble-learning framework. *Briefings in Bioinformatics, 22*, 2126–2140.
611. Li, F., Guo, X., Xiang, D., Pitt, M., Bainomugisa, A., & Coin, L. (2022). Computational analysis and prediction of PE_PGRS proteins using machine learning. *Computational and Structural Biotechnology Journal, 20*, 662–674.
612. Li, M., Fan, Y., Zhang, Y., & Lv, Z. (2022). Using sequence similarity based on CKSNP features and a graph neural network model to identify miRNA-disease associations. *Genes, 13*, 1759.
613. Li, Q., Zhang, L., Xu, L., Zou, Q., Wu, J., & Li, Q. (2022). Identification and classification of promoters using the attention mechanism based on long short-term memory. *Frontiers of Computer Science, 16*, Article 164348.

614. Li, W., Guo, Y., Wang, B., & Yang, B. (2023). Learning spatiotemporal embedding with gated convolutional recurrent networks for translation initiation site prediction. *Pattern Recognition, 136*, Article 109234.
615. Li, W., Liu, W., Guo, Y., Wang, B., & Qing, H. (2023). Deep contextual representation learning for identifying essential proteins via integrating multisource protein features. *Chinese Journal of Electronics, 32*, 868–881.
616. Li, J., Chiu, T., & Rohs, R. (2024). Predicting DNA structure using a deep learning method. *Nature Communications, 15*, 1243.
617. Liang, Z., Ye, H., Ma, J., Wei, Z., Wang, Y., Zhang, Y., Huang, D., Song, B., Meng, J., Rigden, D., Others. (2024). m6A-Atlas v2. 0: Updated resources for unraveling the N 6-methyladenosine (m6A) epitranscriptome among multiple species. *Nucleic Acids Research, 52*, D194–D202.
618. Liang, Y., Wu, Y., Zhang, Z., Liu, N., Peng, J., & Tang, J. (2022). Hyb4mC: A hybrid DNA2vec-based model for DNA N4-methylcytosine sites prediction. *BMC Bioinformatics, 23*, 258.
619. Liao, J., Yang, B., Zhang, Y., Wang, X., Ye, Y., Peng, J., Yang, Z., He, J., Zhang, Y., Hu, K., Others. (2020). EuRBPDB: A comprehensive resource for annotation, functional and oncological investigation of eukaryotic RNA binding proteins (RBPs). *Nucleic Acids Research, 48*, D307–D313.
620. Liao, M., Zhao, J., Tian, J., & Zheng, C. (2022). iEnhancer-DCLA: Using the original sequence to identify enhancers and their strength based on a deep learning framework. *Bmc Bioinformatics, 23*, 480.
621. Liberzon, A., Subramanian, A., Pinchback, R., Thorvaldsdóttir, H., Tamayo, P., & Mesirov, J. (2011). Molecular signatures database (MSigDB) 3.0. *Bioinformatics, 27*, 1739–1740.
622. Licata, L., Briganti, L., Peluso, D., Perfetto, L., Iannuccelli, M., Galeota, E., Sacco, F., Palma, A., Nardozza, A., Santonico, E., Others. (2012). MINT, the molecular interaction database: 2012 update. *Nucleic Acids Research, 40*, D857–D861.
623. Ligeti, B., Szepesi-Nagy, I., Bodnár, B., Ligeti-Nagy, N., & Juhász, J. (2024). ProkBERT family: Genomic language models for microbiome applications. *Frontiers in Microbiology, 14*, 1331233.
624. Limongelli, I., Marini, S., & Bellazzi, R. (2015). PaPI: Pseudo amino acid composition to score human protein-coding variants. *BMC Bioinformatics, 16*, 1–14.
625. Lin, J., Chen, X., & Wong, K. (2022). An artificial intelligence approach for gene editing off-target quantification: Convolutional self-attention neural network designs and considerations. *Statistics in Biosciences*. https://lens.org/008-841-938-882-196.
626. Lin, M., Chen, Q., & Yan, S. (2013). Network in network. Preprint. ArXiv:1312.4400.
627. Lin, Z., & Pan, X. (2001). Accurate prediction of protein secondary structural content. *Journal of Protein Chemistry, 20*, 217–220.
628. Lin, J., & Wong, K. (2018). Off-target predictions in CRISPR-Cas9 gene editing using deep learning. *Bioinformatics, 34*, i656–i663.
629. Lin, K., May, A., & Taylor, W. (2002). Amino acid encoding schemes from protein structure alignments: Multi-dimensional vectors to describe residue types. *Journal of Theoretical Biology, 216*, 361–365.
630. Lin, Y., Cradick, T., Brown, M., Deshmukh, H., Ranjan, P., Sarode, N., Wile, B., Vertino, P., Stewart, F., & Bao, G. (2014). CRISPR/Cas9 systems have off-target activity with insertions or deletions between target DNA and guide RNA sequences. *Nucleic Acids Research, 42*, 7473–7485.
631. Lin, W., Shen, P., Liu, H., Cho, Y., Hsu, M., Lin, I., Chen, F., Yang, J., Ma, W., & Cheng, W. (2021). LipidSig: A web-based tool for lipidomic data analysis. *Nucleic Acids Research, 49*, W336–W345.
632. Linder, J., Srivastava, D., Yuan, H., Agarwal, V., & Kelley, D. (2023). Predicting RNA-seq coverage from DNA sequence as a unifying model of gene regulation. *BioRxiv*.
633. Lino, C., Harper, J., Carney, J., & Timlin, J. (2018). Delivering CRISPR: A review of the challenges and approaches. *Drug Delivery, 25*, 1234–1257.

634. Lisio, L., Gómez-López, G., Sánchez-Beato, M., Gómez-Abad, C., Rodriguez, M., Villuendas, R., Ferreira, B., Carro, A., Rico, D., Mollejo, M., Martínez, M., Menárguez, J., Díaz-Alderete, A., Gil, J., Cigudosa, J., Pisano, D., Piris, M., & Martínez, N. (2010). Mantle cell lymphoma: Transcriptional regulation by micrornas. *Leukemia*, *24*, 1335–1342.
635. Lisser, S., & Margalit, H. (1994). Determination of common structural features in Escherichia coli promoters by computer analysis. *European Journal of Biochemistry*, *223*, 823–830.
636. Listgarten, J., Weinstein, M., Kleinstiver, B., Sousa, A., Joung, J., Crawford, J., Gao, K., Hoang, L., Elibol, M., Doench, J., Others. (2018). Prediction of off-target activities for the end-to-end design of CRISPR guide RNAs. *Nature Biomedical Engineering*, *2*, 38–47.
637. Liu, Y., & Tian, B. (2024). Protein–DNA binding sites prediction based on pre-trained protein language model and contrastive learning. *Briefings in Bioinformatics*, *25*, bbad488.
638. Liu, B., Gao, X., & Zhang, H. (2019). BioSeq-Analysis2. 0: An updated platform for analyzing DNA, RNA and protein sequences at sequence level and residue level based on machine learning approaches. *Nucleic Acids Research*, *47*, e127–e127.
639. Liu, H., Li, D., & Wu, H. (2023). Lnclocator-imb: An imbalance-tolerant ensemble deep learning framework for predicting long non-coding RNA subcellular localization. *IEEE Journal of Biomedical and Health Informatics*.
640. Liu, Y., Ott, M., Goyal, N., Du, J., Joshi, M., Chen, D., Levy, O., Lewis, M., Zettlemoyer, L., & Stoyanov, V. (2019). Roberta: A robustly optimized bert pretraining approach. Preprint. ArXiv:1907.11692.
641. Liu, D., Tang, Y., Fan, C., Chen, Z., & Deng, L. (2016). PredRBR: Accurate prediction of RNA-binding residues in proteins using gradient tree boosting. *2016 IEEE international conference on bioinformatics and biomedicine (BIBM)* (pp. 47–52).
642. Liu, X., Wang, S., & Ai, D. (2022). Predicting CRISPR/Cas9 repair outcomes by attention-based deep learning framework. *Cells*, *11*, 1847–1847.
643. Liu, S., Xu, X., Yang, Z., Zhao, X., Liu, S., & Zhang, W. (2021). Epihc: Improving enhancer-promoter interaction prediction by using hybrid features and communicative learning. *IEEE/Acm Transactions on Computational Biology and Bioinformatics*, 1–1.
644. Liu, Y., Yan, X., Li, J., Ren, X., Wu, Q., Wang, G., Chen, Y., & Zhu, X. (2023). miRNA-disease association prediction based on heterogeneous graph transformer with multi-view similarity and random auto-encoder. *2023 IEEE international conference on bioinformatics and biomedicine (BIBM)* (pp. 885–888).
645. Liu, N., Zhang, Z., Wu, Y., Wang, Y., & Liang, Y. (2023). CRBSP: Prediction of CircRNA-RBP binding sites based on multimodal intermediate fusion. *IEEE/ACM Transactions on Computational Biology and Bioinformatics*.
646. Liu, B. (2016). iEnhancer-PsedeKNC: Identification of enhancers and @articlebgroups based on Pseudo degenerate kmer nucleotide composition. *Neurocomputing*, *217*, 46–52.
647. Liu, T., Lin, Y., Wen, X., Jorissen, R., & Gilson, M. (2007). BindingDB: A web-accessible database of experimentally determined protein-ligand binding affinities. *Nucleic Acids Research*, *35*, D198–D201.
648. Liu, X., Zhao, L., & Dong, Q. (2011). Protein remote homology detection based on auto-cross covariance transformation. *Computers in Biology and Medicine*, *41*, 640–647.
649. Liu, B., Xu, J., Lan, X., Xu, R., Zhou, J., Wang, X., & Chou, K. (2014). iDNA-Protl dis: Identifying DNA-binding proteins by incorporating amino acid distance-pairs and reduced alphabet profile into the general pseudo amino acid composition. *PloS One*, *9*, Article e106691.
650. Liu, B., Liu, F., Wang, X., Chen, J., Fang, L., & Chou, K. (2015). Pse-in-One: A web server for generating various modes of pseudo components of DNA, RNA, and protein sequences. *Nucleic Acids Research*, *43*, W65 W71.
651. Liu, B., Liu, F., Fang, L., Wang, X., & Chou, K. (2015). repDNA: A Python package to generate various modes of feature vectors for DNA sequences by incorporating user-defined physicochemical properties and sequence-order effects. *Bioinformatics*, *31*, 1307–1309.

652. Liu, B., Liu, Y., Jin, X., Wang, X., & Liu, B. (2016). iRSpot-DACC: A computational predictor for recombination hot/cold spots identification based on dinucleotide-based auto-cross covariance. *Scientific Reports*, *6*, 33483.
653. Liu, B., Fang, L., Long, R., Lan, X., & Chou, K. (2016). iEnhancer-2L: A two-layer predictor for identifying enhancers and their strength by pseudo k-tuple nucleotide composition. *Bioinformatics*, *32*, 362–369.
654. Liu, B., Wang, S., Dong, Q., Li, S., & Liu, X. (2016). Identification of DNA-binding proteins by combining auto-cross covariance transformation and ensemble learning. *IEEE Transactions on Nanobioscience*, *15*, 328–334.
655. Liu, B., Li, K., Huang, D., & Chou, K. (2018). iEnhancer-EL: Identifying enhancers and their strength with ensemble learning approach. *Bioinformatics*, *34*, 3835–3842.
656. Liu, B., Li, K., Huang, D., & Chou, K. (2018). iEnhancer-EL: Identifying enhancers and their strength with ensemble learning approach. *Bioinformatics*, *34*, 3835–3842.
657. Liu, B., Zheng, D., Jin, Q., Chen, L., & Yang, J. (2019). VFDB 2019: A comparative pathogenomic platform with an interactive web interface. *Nucleic Acids Research*, *47*, D687–D692.
658. Liu, M., Wang, Q., Shen, J., Yang, B., & Ding, X. (2019). Circbank: A comprehensive database for circRNA with standard nomenclature. *RNA Biology*, *16*, 899–905.
659. Liu, H., Guan, J., Li, H., Bao, Z., Wang, Q., Luo, X., & Xue, H. (2020). Predicting the disease genes of multiple sclerosis based on network representation learning. *Frontiers in Genetics*, *11*, 328.
660. Liu, S., Cui, C., Chen, H., & Liu, T. (2022). Ensemble learning-based feature selection for phosphorylation site detection. *Frontiers in Genetics*, *13*, Article 984068.
661. Liu, Y., Shen, Y., Wang, H., Zhang, Y., & Zhu, X. (2022). m5Cpred-XS: A new method for predicting RNA m5C sites based on XGBoost and SHAP. *Frontiers in Genetics*, *13*, Article 853258.
662. Liu, X., Zhang, H., Zeng, Y., Zhu, X., Zhu, L., & Fu, J. (2024). DRANetSplicer: A splice site prediction model based on deep residual attention networks. *Genes*, *15*, 404.
663. Liwo, A., Wawak, R., Scheraga, H., Pincus, M., & Rackovsky, S. (1993). Calculation of protein backbone geometry from -carbon coordinates based on peptide-group dipole alignment. *Protein Science*, *2*, 1697–1714.
664. Löchel, H., & Heider, D. (2021). Chaos game representation and its applications in bioinformatics. *Computational and Structural Biotechnology Journal*, *19*, 6263–6271.
665. Lovell, D., Müller, W., Taylor, J., Zwart, A., & Helliwell, C. (2011). Proportions, percentages, ppm: Do the molecular biosciences treat compositional data right? *Compositional Data Analysis: Theory and Applications*, 191–207.
666. Lu, P., Yang, P., & Liao, Y. (2023). Deep learning framework for predicting essential proteins with temporal convolutional networks. *Journal of Shanghai Jiaotong University (Science)*, 1–11.
667. Lu, C., Zhang, L., Zeng, M., Lan, W., & Wang, J. (2022). Identifying disease-associated circRNAs based on edge-weighted graph attention and heterogeneous graph neural network. *BioRxiv*.
668. Luisa, B. (2012). *Cellular energy metabolism and its regulation*. Elsevier.
669. Lumbanraja, F., Nguyen, N., Phan, D., Faisal, M., Abipihi, B., Purnama, B., Delimayanti, M., Kubo, M., & Satou, K. (2018). Improved protein phosphorylation site prediction by a new combination of feature set and feature selection. *Journal of Biomedical Science and Engineering*, *11*, 144–157.
670. Luo, Z., Wang, R., Sun, Y., Liu, J., Chen, Z., & Zhang, Y. (2024). Interpretable feature extraction and dimensionality reduction in ESM2 for protein localization prediction. *Briefings in Bioinformatics*, *25*, bbad534.
671. Luo, J., Zhao, K., Chen, J., Yang, C., Qu, F., Yan, K., Zhang, Y., & Liu, B. (2023). Discovery of novel multi-functional peptides by using protein language models and graph-based deep learning. *BioRxiv*.

672. Luo, X., Huang, Y., Li, H., Luo, Y., Zuo, Z., Ren, J., & Xie, Y. (2022). SPENCER: A comprehensive database for small peptides encoded by noncoding RNAs in cancer patients. *Nucleic Acids Research, 50,* D1373–D1381.
673. Luo, H., Shan, W., Chen, C., Ding, P., & Luo, L. (2023). Improving language model of human genome for DNA-protein binding prediction based on task-specific pre-training. *Interdisciplinary Sciences: Computational Life Sciences, 15,* 32–43.
674. Lv, Z., Ding, H., Wang, L., & Zou, Q. (2021). A convolutional neural network using dinucleotide one-hot encoder for identifying DNA N6-methyladenine sites in the rice genome. *Neurocomputing, 422,* 214–221.
675. Lyngsø, R., Zuker, M., & Pedersen, C. (1999). Internal loops in RNA secondary structure prediction. *Proceedings of the third annual international conference on computational molecular biology* (pp. 260–267).
676. Ma, J., Song, J., Young, N., Chang, B., Korhonen, P., Campos, T., Liu, H., & Gasser, R. (2024). 'Bingo'—a large language model-and graph neural network-based workflow for the prediction of essential genes from protein data. *Briefings in Bioinformatics, 25,* bbad472.
677. Ma, J., Zhao, Z., Li, T., Liu, Y., Ma, J., & Zhang, R. (2024). GraphsformerCPI: Graph transformer for compound–protein interaction prediction. *Interdisciplinary Sciences: Computational Life Sciences,* 1–17.
678. Ma, X. (2024). *Deep5hmc: Predicting genome-wide 5-hydroxymethylcytosine landscape via a multimodal deep learning model.* Oxford University Press.
679. Ma, W., Zhang, L., Zeng, P., Huang, C., Li, J., Geng, B., Yang, J., Kong, W., Zhou, X., & Cui, Q. (2017). An analysis of human microbe-disease associations. *Briefings in Bioinformatics, 18,* 85–97.
680. Ma, Z., Davis, S., & Ho, Y. (2022). Flexible copula model for integrating correlated multi-omics data from single-cell experiments. *Biometrics, 79,* 1559–1572.
681. Maaten, L., & Hinton, G. (2008). Visualizing data using t-SNE. *Journal of Machine Learning Research, 9.*
682. Macaya, R., Schultze, P., Smith, F., Roe, J., & Feigon, J. (1993). Thrombin-binding DNA aptamer forms a unimolecular quadruplex structure in solution. *Proceedings of the National Academy of Sciences, 90,* 3745–3749.
683. Maccari, G., Robinson, J., Ballingall, K., Guethlein, L., Grimholt, U., Kaufman, J., Ho, C., De Groot, N., Flicek, P., Bontrop, R., Others. (2017). IPD-MHC 2.0: An improved inter-species database for the study of the major histocompatibility complex. *Nucleic Acids Research, 45,* D860–D864.
684. Macdonald, J., Smalley, S., Benedetti, J., Hundahl, S., Estes, N., Stemmermann, G., Haller, D., Ajani, J., Gunderson, L., Jessup, J., Others. (2001). Chemoradiotherapy after surgery compared with surgery alone for adenocarcinoma of the stomach or gastroesophageal junction. *New England Journal of Medicine, 345,* 725–730.
685. Madan, S., Demina, V., Stapf, M., Ernst, O., & Fröhlich, H. (2022). Accurate prediction of virus-host protein-protein interactions via a Siamese neural network using deep protein sequence embeddings. *Patterns, 3.*
686. Madeddu, L., Stilo, G., & Velardi, P. (2019). Network-based methods for disease-gene prediction. Preprint. ArXiv:1902.10117.
687. Madeira, F., Pearce, M., Tivey, A., Basutkar, P., Lee, J., Edbali, O., Madhusoodanan, N., Kolesnikov, A., & Lopez, R. (2022). Search and sequence analysis tools services from EMBL-EBI in 2022. *Nucleic Acids Research, 50,* W276–W279.
688. Mailman, M., Feolo, M., Jin, Y., Kimura, M., Tryka, K., Bagoutdinov, R., Hao, L., Kiang, A., Paschall, J., Phan, L., Others. (2007). The NCBI dbGaP database of genotypes and phenotypes. *Nature Genetics, 39,* 1181–1186.
689. Malebary, S., Rehman, M., & Khan, Y. (2019). iCrotoK-PseAAC: Identify lysine crotonylation sites by blending position relative statistical features according to the Chou's 5-step rule. *PloS One, 14,* Article e0223993.

690. Malesios, C., Chatzipanagiotou, M., Demiris, N., Kantartzis, A., Chatzilazarou, G., Chatzinikolaou, S., & Kostoulas, P. (2020). A quantitative analysis of the spatial and temporal evolution patterns of the bluetongue virus outbreak in the island of Lesvos, Greece, in 2014. *Transboundary and Emerging Diseases*, *67*, 2073–2085.
691. Mallick, K., Bandyopadhyay, S., Chakraborty, S., Choudhuri, R., & Bose, S. (2019). Topo2vec: A novel node embedding generation based on network topology for link prediction. *IEEE Transactions on Computational Social Systems*, *6*, 1306–1317.
692. Manavalan, B., Basith, S., Shin, T., & Lee, G. (2021). Computational prediction of species-specific yeast DNA replication origin via iterative feature representation. *Briefings in Bioinformatics*, *22*, bbaa304.
693. Mandic, D., & Chambers, J. (2001). *Recurrent neural networks for prediction: Learning algorithms, architectures and stability*. Wiley.
694. Mandl, K., & Manrai, A. (2019). Potential excessive testing at scale: Biomarkers, genomics, and machine learning. *Jama*, *321*, 739–740.
695. Manso, T., Folch, G., Giudicelli, V., Jabado-Michaloud, J., Kushwaha, A., Nguefack Ngoune, V., Georga, M., Papadaki, A., Debbagh, C., Pegorier, P., Others. (2022). IMGT® databases, related tools and web resources through three main axes of research and development. *Nucleic Acids Research*, *50*, D1262–D1272.
696. Marini, J., Levene, S., Crothers, D., & Englund, P. (1982). Bent helical structure in kinetoplast DNA. *Proceedings of the National Academy of Sciences*, *79*, 7664–7668.
697. Marino, N., Pinilla-Redondo, R., Csörgő, B., & Bondy-Denomy, J. (2020). Anti-CRISPR protein applications: Natural brakes for CRISPR-Cas technologies. *Nature Methods*, *17*, 471–479.
698. Marquet, C., Heinzinger, M., Olenyi, T., Dallago, C., Erckert, K., Bernhofer, M., Nechaev, D., & Rost, B. (2022). Embeddings from protein language models predict conservation and variant effects. *Human Genetics*, *141*, 1629–1647.
699. Mathews, D., Turner, D., & Zuker, M. (2007). RNA secondary structure prediction. *Current Protocols in Nucleic Acid Chemistry*, *28*, 11–12.
700. Matsui, M., & Corey, D. (2017). Non-coding RNAs as drug targets. *Nature Reviews Drug Discovery*, *16*, 167.
701. Matsuo, Y., Komiya, S., Yasumizu, Y., Yasuoka, Y., Mizushima, K., Takagi, T., Kryukov, K., Fukuda, A., Morimoto, Y., Naito, Y., Okada, H., Bono, H., Nakagawa, S., & Hirota, K. (2021). Full-length 16s rrna gene amplicon analysis of human gut microbiota using minion™ nanopore sequencing confers species-level resolution. *BMC Microbiology*, *21*.
702. Mbodi, L., Mathebela, P., & Dlamini, Z. (2023). Association of metabolomics with AI in precision oncology: Emerging perspectives for more effective cancer care. *Artificial intelligence and precision oncology: Bridging cancer research and clinical decision support* (pp. 139–156). Springer.
703. McArthur, A., Waglechner, N., Nizam, F., Yan, A., Azad, M., Baylay, A., Bhullar, K., Canova, M., De Pascale, G., Ejim, L., Others. (2013). The comprehensive antibiotic resistance database. *Antimicrobial Agents and Chemotherapy*, *57*, 3348–3357.
704. Mehmood, F., Arshad, S., & Shoaib, M. (2024). ADH-Enhancer: An attention-based deep hybrid framework for enhancer identification and strength prediction. *Briefings in Bioinformatics*, *25*, bbae030.
705. Mehmood, F., Ghani, M., Asim, M., Shahzadi, R., Mehmood, A., & Mahmood, W. (2021). MPF-Net: A computational multi-regional solar power forecasting framework. *Renewable and Sustainable Energy Reviews*, *151*, Article 111559.
706. Mehmood, F., Arshad, S., & Shoaib, M. (2023). RPPSP: A robust and precise protein solubility predictor by utilizing novel protein sequence encoder. *IEEE Access*, *11*, 59397–59416.
707. Meier, J., Rao, R., Verkuil, R., Liu, J., Sercu, T., & Rives, A. (2021). Language models enable zero-shot prediction of the effects of mutations on protein function. *Advances in Neural Information Processing Systems*, *34*, 29287–29303.

708. Melnyk, I., Chenthamarakshan, V., Chen, P., Das, P., Dhurandhar, A., Padhi, I., & Das, D. (2023). Reprogramming pretrained language models for antibody sequence infilling. *International conference on machine learning* (pp. 24398–24419).
709. Melnyk, I., Chenthamarakshan, V., Chen, P., Das, P., Dhurandhar, A., Padhi, I., & Das, D. (2023). Reprogramming pretrained language models for antibody sequence infilling. *International conference on machine learning* (pp. 24398–24419).
710. Meng, L., Chen, X., Cheng, K., Chen, N., Zheng, Z., Wang, F., Sun, H., & Wong, K. (2024). TransPTM: A transformer-based model for non-histone acetylation site prediction. *Briefings in Bioinformatics*, *25*, bbae219.
711. Meng, X., Xiang, J., Zheng, R., Wu, F., & Li, M. (2021). DPCMNE: Detecting protein complexes from protein-protein interaction networks via multi-level network embedding. *IEEE/ACM Transactions on Computational Biology and Bioinformatics*, *19*, 1592–1602.
712. Menon, A., Jang-Sohn & Nam, J. (2020). CGD: Comprehensive guide designer for CRISPR-Cas systems. *Computational and Structural Biotechnology Journal*, *18*, 814–820.
713. Meyer, J. (2021). Deep learning neural network tools for proteomics. *Cell Reports Methods*, 100003.
714. Miao, Y., Liu, W., Zhang, Q., & Guo, A. (2018). lncRNASNP2: An updated database of functional SNPs and mutations in human and mouse lncRNAs. *Nucleic Acids Research*, *46*, D276–D280.
715. Mikolov, T., Chen, K., Corrado, G., & Dean, J. (2013). Efficient estimation of word representations in vector space. Preprint. ArXiv:1301.3781.
716. Mikolov, T., Grave, E., Bojanowski, P., Puhrsch, C., & Joulin, A. (2017). Advances in pre-training distributed word representations. Preprint. ArXiv:1712.09405.
717. Min, X., Ye, C., Liu, X., & Zeng, X. (2021). Predicting enhancer-promoter interactions by deep learning and matching heuristic. *Briefings in Bioinformatics*, *22*, bbaa254.
718. Min, X., Zeng, W., Chen, S., Chen, N., Chen, T., & Jiang, R. (2017). Predicting enhancers with deep convolutional neural networks. *BMC Bioinformatics*, *18*, 478.
719. Minnoye, L., Marinov, G., Krausgruber, T., Pan, L., Marand, A., Secchia, S., Greenleaf, W., Furlong, E., Zhao, K., Schmitz, R., Bock, C., & Aerts, S. (2021). Chromatin accessibility profiling methods. *Nature Reviews Methods Primers*, *1*.
720. Mirdita, M., Von Den Driesch, L., Galiez, C., Martin, M., Söding, J., & Steinegger, M. (2017). Uniclust databases of clustered and deeply annotated protein sequences and alignments. *Nucleic Acids Research*, *45*, D170–D176.
721. Mitrofanov, A., Alkhnbashi, O., Shmakov, S., Makarova, K., Koonin, E., & Backofen, R. (2020). CRISPRidentify: Identification of CRISPR arrays using machine learning approach. *Nucleic Acids Research*, *49*, e20–e20.
722. Mitrofanov, A., Alkhnbashi, O., Shmakov, S., Makarova, K., Koonin, E., & Backofen, R. (2021). CRISPRidentify: Identification of CRISPR arrays using machine learning approach. *Nucleic Acids Research*, *49*, e20–e20.
723. Mo, S., Fu, X., Hong, C., Chen, Y., Zheng, Y., Tang, X., Shen, Z., Xing, E., & Lan, Y. (2021). Multi-modal self-supervised pre-training for regulatory genome across cell types. Preprint. ArXiv:2110.05231.
724. Mock, F., Kretschmer, F., Kriese, A., Böcker, S., & Marz, M. (2022). Taxonomic classification of DNA sequences beyond sequence similarity using deep neural networks. *Proceedings of the National Academy of Sciences*, *119*, e2122636119.
725. Mohr, S., Hu, Y., Ewen-Campen, B., Housden, B., Viswanatha, R., & Perrimon, N. (2016). CRISPR guide RNA design for research applications. *The FEBS Journal*, *283*, 3232–3238.
726. Montange, R., & Batey, R. (2008). Riboswitches: Emerging themes in RNA structure and function. *Annual Review of Biophysics*, *37*, 117–133.
727. Moreno-Mateos, M., Vejnar, C., Beaudoin, J., Fernandez, J., Mis, E., Khokha, M., & Giraldez, A. (2015). CRISPRscan: Designing highly efficient sgRNAs for CRISPR-Cas9 targeting in vivo. *Nature Methods*, *12*, 982–988.
728. Morris, R., Black, K., & Stollar, E. (2022). Uncovering protein function: from classification to complexes. *Essays in Biochemistry*, *66*, 255–285.

729. Motmaen, A., Dauparas, J., Baek, M., Abedi, M., Baker, D., & Bradley, P. (2023). Peptide-binding specificity prediction using fine-tuned protein structure prediction networks. *Proceedings of the National Academy of Sciences*, *120*, e2216697120.
730. Mu, Z., Tan, Y., Zhang, B., Liu, J., & Shi, Y. (2022). Ab initio predictions for 3D structure and stability of single-and double-stranded DNAs in ion solutions. *PLOS Computational Biology*, *18*, Article e1010501.
731. Muhammod, R., Ahmed, S., Md Farid, D., Shatabda, S., Sharma, A., & Dehzangi, A. (2019). PyFeat: A python-based effective feature generation tool for DNA, RNA and protein sequences. *Bioinformatics*, *35*, 3831–3833.
732. Mukundan, V., Do, N., & Phan, A. (2011). HIV-1 integrase inhibitor T30177 forms a stacked dimeric G-quadruplex structure containing bulges. *Nucleic Acids Research*, *39*, 8984–8991.
733. Thumuluri, V., Almagro Armenteros, J., Johansen, A., Nielsen, H., & Winther, O. (2022). DeepLoc 2.0: Multi-label subcellular localization prediction using protein language models. *Nucleic Acids Research*, *50*, W228–W234.
734. Munteanu, M., Vlahovicek, K., Parthasarathy, S., Simon, I., & Pongor, S. (1998). Rod models of DNA: Sequence-dependent anisotropic elastic modelling of local bending phenomena. *Trends in Biochemical Sciences*, *23*, 341–347.
735. Murad, T., Ali, S., Chourasia, P., & Patterson, M. (2023). Advancing protein-DNA binding site prediction: Integrating sequence models and machine learning classifiers. *BioRxiv*.
736. Murmu, S., Chaurasia, H., Guha Majumdar, S., Rao, A., Rai, A., & Archak, S. (2023). Prediction of protein–protein interactions between anti-CRISPR and CRISPR-Cas using machine learning technique. *Journal of Plant Biochemistry and Biotechnology*, *32*, 818–830.
737. Mysinger, M., Carchia, M., Irwin, J., & Shoichet, B. (2012). Directory of useful decoys, enhanced (DUD-E): Better ligands and decoys for better benchmarking. *Journal of Medicinal Chemistry*, *55*, 6582–6594.
738. Nabeel Asim, M., Ali Ibrahim, M., Fazeel, A., Dengel, A., & Ahmed, S. (2023). DNA-MP: A generalized DNA modifications predictor for multiple species based on powerful sequence encoding method. *Briefings in Bioinformatics*, *24*, bbac546.
739. Naeem, S., Mabrouk, M., Eldosoky, M., & Sayed, A. (2020). Moment invariants for cancer classification based on electron-ion interaction pseudo potentials (EIIP). *Network Modeling Analysis in Health Informatics and Bioinformatics*, *9*, 1–5.
740. Nair, A., & Sreenadhan, S. (2006). A coding measure scheme employing electron-ion interaction pseudopotential (EIIP). *Bioinformation*, *1*, 197.
741. Nair, P., & Vihinen, M. (2013). V ari B ench: A benchmark database for variations. *Human Mutation*, *34*, 42–49.
742. Nakamura, Y., Collaboration, I., Cochrane, G., Collaboration, I., Karsch-Mizrachi, I., & Collaboration, I. (2012). The international nucleotide sequence database collaboration. *Nucleic Acids Research*, *41*, D21–D24.
743. Nallapareddy, V., Bordin, N., Sillitoe, I., Heinzinger, M., Littmann, M., Waman, V., Sen, N., Rost, B., & Orengo, C. (2023). CATHe: Detection of remote homologues for CATH superfamilies using embeddings from protein language models. *Bioinformatics*, *39*, btad029.
744. Nambiar, T., Baudrier, L., Billon, P., & Ciccia, A. (2022). CRISPR-based genome editing through the lens of DNA repair. *Molecular Cell*, *82*, 348–388.
745. Narayanan, A., Chandramohan, M., Venkatesan, R., Chen, L., Liu, Y., & Jaiswal, S. (2017). graph2vec: Learning distributed representations of graphs. Preprint. ArXiv:1707.05005.
746. Narayanan, S., Ramachandran, A., Aakur, S., & Bagavathi, A. (2020). Genome sequence classification for animal diagnostics with graph representations and deep neural networks. Preprint. ArXiv:2007.12791.
747. Narayanan, H., Dingfelder, F., Butté, A., Lorenzen, N., Sokolov, M., & Arosio, P. (2021). Machine learning for biologics: Opportunities for protein engineering, developability, and formulation. *Trends in Pharmacological Sciences*, *42*, 151–165.
748. Nasiri, E., Berahmand, K., Rostami, M., & Dabiri, M. (2021). A novel link prediction algorithm for protein-protein interaction networks by attributed graph embedding. *Computers in Biology and Medicine*, *137*, Article 104772.

749. Nasko, D., Koren, S., Phillippy, A., & Treangen, T. (2018). RefSeq database growth influences the accuracy of k-mer-based lowest common ancestor species identification. *Genome Biology*, *19*, 1–10.
750. Navarez, A., & Roxas, R. (2022). An evaluation of multitask transfer learning methods in identifying 6mA and 5mC methylation sites of rice and maize. Available at SSRN:4178244.
751. Nethery, M., Korvink, M., Makarova, K., Wolf, Y., Koonin, E., & Barrangou, R. (2021). CRISPRclassify: Repeat-based classification of CRISPR loci. *The CRISPR Journal*, *4*, 558–574.
752. Narayanan, B. C., Westbrook, J., Ghosh, S., Petrov, A., Sweeney, B., Zirbel, C., Leontis, N., & Berman, H. (2014). The nucleic acid database: New features and capabilities. *Nucleic Acids Research*, *42*, D114–D122.
753. Nguyen, E., Poli, M., Faizi, M., Thomas, A., Wornow, M., Birch-Sykes, C., Massaroli, S., Patel, A., Rabideau, C., Bengio, Y., Others. (2024). Hyenadna: Long-range genomic sequence modeling at single nucleotide resolution. *Advances in Neural Information Processing Systems*, *36*.
754. Nguyen-Vo, T., Nguyen, Q., Do, T., Nguyen, T., Rahardja, S., & Nguyen, B. (2019). iPseU-NCP: Identifying RNA pseudouridine sites using random forest and NCP-encoded features. *BMC Genomics*, *20*, 1–11.
755. Nguyen-Vo, T., Trinh, Q., Nguyen, L., Nguyen-Hoang, P., Rahardja, S., & Nguyen, B. (2023). i4mC-GRU: Identifying DNA N4-Methylcytosine sites in mouse genomes using bidirectional gated recurrent unit and sequence-embedded features. *Computational and Structural Biotechnology Journal*, *21*, 3045–3053.
756. Ni, Y., Fan, L., Wang, M., Zhang, N., Zuo, Y., & Liao, M. (2022). EPI-mind: Identifying enhancer-promoter interactions based on transformer mechanism. *Interdisciplinary Sciences: Computational Life Sciences*, *14*, 786–794.
757. Nieduszynski, C., Hiraga, S., Ak, P., Benham, C., & Donaldson, A. (2006). Oridb: A dna replication origin database. *Nucleic Acids Research*, *35*, D40–D46.
758. Nilamyani, A., Auliah, F., Moni, M., Shoombuatong, W., Hasan, M., & Kurata, H. (2021). PredNTS: Improved and robust prediction of nitrotyrosine sites by integrating multiple sequence features. *International Journal of Molecular Sciences*, *22*, 2704.
759. Niu, M., Lin, Y., & Zou, Q. (2021). sgRNACNN: Identifying sgRNA on-target activity in four crops using ensembles of convolutional neural networks. *Plant Molecular Biology*, *105*, 483–495.
760. Niu, R., Peng, J., Zhang, Z., & Shang, X. (2021). R-CRISPR: A deep learning network to predict off-target activities with mismatch, insertion and deletion in CRISPR-Cas9 System. *Genes*, *12*, 1878.
761. Niu, M., Wang, C., Chen, Y., Zou, Q., Qi, R., & Xu, L. (2024). CircRNA identification and feature interpretability analysis. *BMC Biology*, *22*, 44.
762. Noble, W., Kuehn, S., Thurman, R., Yu, M., & Stamatoyannopoulos, J. (2005). Predicting the in vivo signature of human gene regulatory sequences. *Bioinformatics*, *21*, i338–i343.
763. Noguchi, S., Arakawa, T., Fukuda, S., Furuno, M., Hasegawa, A., Hori, F., Ishikawa-Kato, S., Kaida, K., Kaiho, A., Kanamori-Katayama, M., Others. (2017). FANTOM5 CAGE profiles of human and mouse samples. *Scientific Data*, *4*, 1–10.
764. Noller, H. (1984). Structure of ribosomal RNA. *Annual Review of Biochemistry*, *53*, 119–162.
765. Norori, N., Hu, Q., Aellen, F., Faraci, F., & Tzovara, A. (2021). Addressing bias in big data and AI for health care: A call for open science. *Patterns*, *2*.
766. Noshay, J., Walker, T., Alexander, W., Klingeman, D., Romero, J., Walker, A., Prates, E., Eckert, C., Irle, S., Kainer, D., & Jacobson, D. (2023). Quantum biological insights into CRISPR-Cas9 sgRNA efficiency from explainable AI driven feature engineering. *Nucleic Acids Research*, *51*, 10147–10161.
767. Novakovsky, G., Dexter, N., Libbrecht, M., Wasserman, W., & Mostafavi, S. (2023). Obtaining genetics insights from deep learning via explainable artificial intelligence. *Nature Reviews Genetics*, *24*, 125–137.

768. Novković, M., Simunić, J., Bojović, V., Tossi, A., & Juretić, D. (2012). DADP: The database of anuran defense peptides. *Bioinformatics*, *28*, 1406–1407.
769. Nunes, S., Sousa, R., & Pesquita, C. (2023). Multi-domain knowledge graph embeddings for gene-disease association prediction. *Journal of Biomedical Semantics*, *14*, 11.
770. Nuñez, J., Chen, J., Pommier, G., Cogan, J., Replogle, J., Adriaens, C., Ramadoss, G., Shi, Q., Hung, K., Samelson, A., Others. (2021). Genome-wide programmable transcriptional memory by CRISPR-based epigenome editing. *Cell*, *184*, 2503–2519.
771. Nusinow, D., Szpyt, J., Ghandi, M., Rose, C., McDonald, E., Kalocsay, M., Jané-Valbuena, J., Gelfand, E., Schweppe, D., Jedrychowski, M., Others. (2020). Quantitative proteomics of the cancer cell line encyclopedia. *Cell*, *180*, 387–402.
772. Nwankpa, C., Ijomah, W., Gachagan, A., & Marshall, S. (2018). Activation functions: Comparison of trends in practice and research for deep learning. Preprint. ArXiv:1811.03378.
773. O'Brien, A., Burgio, G., & Bauer, D. (2021). Domain-specific introduction to machine learning terminology, pitfalls and opportunities in CRISPR-based gene editing. *Briefings in Bioinformatics*, *22*, 208–314.
774. O'Connor, C., Adams, J., & Fairman, J. (2010). Essentials of cell biology. *Cambridge, MA: NPG Education*, *1*, 5.
775. Ogasawara, O., Mashima, J., Kodama, Y., Kaminuma, E., Nakamura, Y., Okubo, K., & Takagi, T. (2012). DDBJ new system and service refactoring. *Nucleic Acids Research*, *41*, D25–D29.
776. O'Leary, N., Wright, M., Brister, J., Ciufo, S., Haddad, D., McVeigh, R., Rajput, B., Robbertse, B., Smith-White, B., Ako-Adjei, D., Others. (2016). Reference sequence (RefSeq) database at NCBI: Current status, taxonomic expansion, and functional annotation. *Nucleic Acids Research*, *44*, D733–D745.
777. Oliveira, E., Santana, K., Josino, L., Lima, A., & Sales Júnior, C. (2021). Predicting cell-penetrating peptides using machine learning algorithms and navigating in their chemical space. *Scientific Reports*, *11*, 7628.
778. Olson, W., Gorin, A., Lu, X., Hock, L., & Zhurkin, V. (1998). DNA sequence-dependent deformability deduced from protein-DNA crystal complexes. *Proceedings of the National Academy of Sciences*, *95*, 11163–11168.
779. Omar, N., Wong, Y., Li, X., Chong, Y., Abdullah, M., & Lee, N. (2017). Enhancer prediction in proboscis monkey genome: A comparative study. *Journal of Telecommunication, Electronic and Computer Engineering (JTEC)*, *9*, 175–179.
780. Ong, S., Lin, H., Chen, Y., Li, Z., & Cao, Z. (2007). Efficacy of different protein descriptors in predicting protein functional families. *BMC Bioinformatics*, *8*, 1–14.
781. Orlov, Y., & Anashkina, A. (2021). Life: Computational genomics applications in life sciences. *Life*, *11*, 1211.
782. Ornstein, R., Rein, R., Breen, D., & Macelroy, R. (1978). An optimized potential function for the calculation of nucleic acid interaction energies I. Base stacking. *Biopolymers: Original Research on Biomolecules*, *17*, 2341–2360.
783. Osguthorpe, D. (2000). Ab initio protein folding. *Current Opinion in Structural Biology*, *10*, 146–152.
784. Osseni, M., Tossou, P., Laviolette, F., & Corbeil, J. (2022). MOT: A multi-omics transformer for multiclass classification tumour types predictions. *BioRxiv*.
785. Ostrovsky-Berman, M., Frankel, B., Polak, P., & Yaari, G. (2021). Immune2vec: Embedding B/T cell receptor sequences in N using natural language processing. *Frontiers in Immunology*, *12*, Article 680687.
786. Ou, M., Cui, P., Pei, J., Zhang, Z., & Zhu, W. (2016). Asymmetric transitivity preserving graph embedding. *Proceedings of the 22nd ACM SIGKDD international conference on knowledge discovery and data mining* (pp. 1105–1114).
787. Oubounyt, M., Louadi, Z., Tayara, H., & Chong, K. (2018). Deep learning models based on distributed feature representations for alternative splicing prediction. *IEEE Access*, *6*, 58826–58834.

788. Ough, M., Lewis, A., Bey, E., Gao, J., Ritchie, J., Bornmann, W., Boothman, D., Oberley, L., & Cullen, J. (2005). Efficacy of beta-lapachone in pancreatic cancer treatment: Exploiting the novel, therapeutic target NQO1. *Cancer Biology & Therapy*, *4*, 102–109.
789. Oughtred, R., Rust, J., Chang, C., Breitkreutz, B., Stark, C., Willems, A., Boucher, L., Leung, G., Kolas, N., Zhang, F., Others. (2021). The BioGRID database: A comprehensive biomedical resource of curated protein, genetic, and chemical interactions. *Protein Science*, *30*, 187–200.
790. Padilha, V., Alkhnbashi, O., Shah, S., Carvalho, A., & Backofen, R. (2020). CRISPRcasIdentifier: Machine learning for accurate identification and classification of CRISPR-Cas systems. *GigaScience*, *9*.
791. Paigen, K., & Petkov, P. (2010). Mammalian recombination hot spots: properties, control and evolution. *Nature Reviews Genetics*, *11*, 221–233.
792. Zhang, W., Wang, L., Liu, K., Wei, X., Yang, K., Du, W., Wang, S., Guo, N., Ma, C., Luo, L., Others. (2020). PIRD: Pan immune repertoire database. *Bioinformatics*, *36*, 897–903.
793. Pan, J., You, W., Lu, X., Wang, S., You, Z., & Sun, Y. (2023). GSPHI: A novel deep learning model for predicting phage-host interactions via multiple biological information. *Computational and Structural Biotechnology Journal*, *21*, 3404–3413.
794. Pándy-Szekeres, G., Caroli, J., Mamyrbekov, A., Kermani, A., Keserű, G., Kooistra, A., & Gloriam, D. (2023). GPCRdb in 2023: State-specific structure models using AlphaFold2 and new ligand resources. *Nucleic Acids Research*, *51*, D395–D402.
795. Pang, Y., & Liu, B. (2024). DisoFLAG: Accurate prediction of protein intrinsic disorder and its functions using graph-based interaction protein language model. *BMC Biology*, *22*, 3.
796. Panwar, B., Omenn, G., & Guan, Y. (2017). miRmine: A database of human miRNA expression profiles. *Bioinformatics*, *33*, 1554–1560.
797. Park, H., Won, J., Park, Y., Anzaku, E., Vankerschaver, J., Messem, A., Neve, W., & Shim, H. (2023). CRISPR-Cas-Docker: Web-based in silico docking and machine learning-based classification of crRNAs with Cas proteins. *BMC Bioinformatics*, *24*, 167.
798. Park, C., Took, C., & Seong, J. (2018). Machine learning in biomedical engineering. *Biomedical Engineering Letters*, *8*, 1–3.
799. Pascanu, R., Mikolov, T., & Bengio, Y. (2013). On the difficulty of training recurrent neural networks. *International conference on machine learning* (pp. 1310–1318).
800. Patel, R., Guo, Y., Alhudhaif, A., Alenezi, F., Althubiti, S., & Polat, K. (2022). Graph-based link prediction between human phenotypes and genes. *Mathematical Problems in Engineering*, *2022*, 7111647.
801. Peng, H., Zheng, Y., Zhao, Z., Liu, T., & Li, J. (2018). Recognition of CRISPR/Cas9 off-target sites through ensemble learning of uneven mismatch distributions. *Bioinformatics*, *34*, i757–i765.
802. Pennington, J., Socher, R., & Manning, C. (2014). Glove: Global vectors for word representation. *Proceedings of the 2014 conference on empirical methods in natural language processing (EMNLP)* (pp. 1532–1543).
803. Périer, R., Praz, V., Junier, T., Bonnard, C., & Bucher, P. (2000). The eukaryotic promoter database (EPD). *Nucleic Acids Research*, *28*, 302–303.
804. Pernot, M., Vanderesse, R., Frochot, C., Guillemin, F., & Barberi-Heyob, M. (2011). Stability of peptides and therapeutic success in cancer. *Expert Opinion On Drug Metabolism & Toxicology*, *7*, 793–802.
805. Perozzi, B., Al-Rfou, R., & Skiena, S. (2014). Deepwalk: Online learning of social representations. *Proceedings of the 20th ACM SIGKDD international conference on knowledge discovery and data mining* (pp. 701–710).
806. Peters, M., Neumann, M., Iyyer, M., Gardner, M., Clark, C., Lee, K., & Zettlemoyer, L. (2018). Deep contextualized word representations. CoRR, abs/1802.05365. http://arxiv.org/abs/1802.05365.
807. Phan, A. (2010). Human telomeric G-quadruplex: Structures of DNA and RNA sequences. *The FEBS Journal*, *277*, 1107–1117.

808. Piñero, J., Bravo, À., Queralt-Rosinach, N., Gutiérrez-Sacristán, A., Deu-Pons, J., Centeno, E., Garcıa-Garcıa, J., Sanz, F., & Furlong, L. (2016). DisGeNET: A comprehensive platform integrating information on human disease-associated genes and variants. *Nucleic Acids Research*, gkw943.
809. Piñero, J., Queralt-Rosinach, N., Bravo, A., Deu-Pons, J., Bauer-Mehren, A., Baron, M., Sanz, F., & Furlong, L. (2015). DisGeNET: A discovery platform for the dynamical exploration of human diseases and their genes. *Database*, *2015*, bav028.
810. Pio, G., Ceci, M., Prisciandaro, F., & Malerba, D. (2020). Exploiting causality in gene network reconstruction based on graph embedding. *Machine Learning*, *109*, 1231–1279.
811. Pio, G., Mignone, P., Magazzù, G., Zampieri, G., Ceci, M., & Angione, C. (2022). Integrating genome-scale metabolic modelling and transfer learning for human gene regulatory network reconstruction. *Bioinformatics*, *38*, 487–493.
812. Piovesan, D., Del Conte, A., Clementel, D., Monzon, A., Bevilacqua, M., Aspromonte, M., Iserte, J., Orti, F., Marino-Buslje, C., & Tosatto, S. (2023). MobiDB: 10 years of intrinsically disordered proteins. *Nucleic Acids Research*, *51*, D438–D444.
813. Pirogova, E., Others. (2001). Development of new computational amino acid parameters for protein structure/function analysis within the resonant recognition model. *2001 conference proceedings of the 23rd annual international conference of the IEEE engineering in medicine and biology society* (Vol. 3, pp. 2890–2893).
814. Pourcel, C., Touchon, M., Villeriot, N., Vernadet, J., Couvin, D., Toffano-Nioche, C., & Vergnaud, G. (2020). CRISPRCasdb a successor of CRISPRdb containing CRISPR arrays and cas genes from complete genome sequences, and tools to download and query lists of repeats and spacers. *Nucleic Acids Research*, *48*, D535–D544.
815. Prabarna Ganguly, P. (2023). Genome Gov. https://www.genome.gov/genetics-glossary/Transcription, Accessed on March 26, 2023.
816. Prabhakar, V., & Liu, K. (2022). Unsupervised co-optimization of a graph neural network and a knowledge graph embedding model to prioritize causal genes for Alzheimer's Disease. *MedRxiv*.
817. Preciat Gonzalez, G., El Assal, L., Noronha, A., Thiele, I., Haraldsdóttir, H., & Fleming, R. (2017). Comparative evaluation of atom mapping algorithms for balanced metabolic reactions: Application to Recon 3D. *Journal of Cheminformatics*, *9*, 1–15.
818. Pruitt, K., Brown, G., Tatusova, T., & Maglott, D. (2012). The reference sequence (RefSeq) database. *The NCBI Handbook*, *2*.
819. Puglisi, R. (2022). Protein mutations and stability, a link with disease: the case study of frataxin. *Biomedicines*, *10*, 425.
820. Pujar, S., O'Leary, N., Farrell, C., Loveland, J., Mudge, J., Wallin, C., Girón, C., Diekhans, M., Barnes, I., Bennett, R., Others. (2018). Consensus coding sequence (CCDS) database: A standardized set of human and mouse protein-coding regions supported by expert curation. *Nucleic Acids Research*, *46*, D221–D228.
821. Qiu, Y., Guo, D., Zhao, P., & Zou, Q. (2024). scMNMF: A novel method for single-cell multi-omics clustering based on matrix factorization. *Briefings in Bioinformatics*, *25*, bbae228.
822. Qiu, J., Nie, W., Ding, H., Dai, J., Wei, Y., Li, D., Zhang, Y., Xie, J., Tian, X., Wu, N., Others. (2024). PB-LKS: A python package for predicting phage–bacteria interaction through local K-mer strategy. *Briefings in Bioinformatics*, *25*, bbae010.
823. Qiu, Y. (2024). Scmnmf: A novel method for single-cell multi-omics clustering based on matrix factorization. *Briefings in Bioinformatics*, *25*.
824. Qu, K., Wei, L., Yu, J., & Wang, C. (2019). Identifying plant pentatricopeptide repeat coding gene/protein using mixed feature extraction methods. *Frontiers in Plant Science*, *9*, 1961.
825. Quaglia, F., Mészáros, B., Salladini, E., Hatos, A., Pancsa, R., Chemes, L., Pajkos, M., Lazar, T., Peña-Dıaz, S., Santos, J., Others. (2022). DisProt in 2022: Improved quality and accessibility of protein intrinsic disorder annotation. *Nucleic Acids Research*, *50*, D480–D487.

826. Quazi, S. (2022). Artificial intelligence and machine learning in precision and genomic medicine. *Medical Oncology*, *39*, 120.
827. Radivojac, P., Vacic, V., Haynes, C., Cocklin, R., Mohan, A., Heyen, J., Goebl, M., & Iakoucheva, L. (2010). Identification, analysis, and prediction of protein ubiquitination sites. *Proteins: Structure, Function, and Bioinformatics*, *78*, 365–380.
828. Rafid, A., Toufikuzzaman, Rahman, M., & Rahman, M. (2020). CRISPRpred(SEQ): A sequence-based method for sgRNA on target activity prediction using traditional machine learning. *BMC Bioinformatics*, *21*, 1–13.
829. Raghunathan, G., Miles, T., & Sasisekharan, V. (1993). Molecular-structure of a dna triple helix. *Biophysical Journal*, *64*, A11–A11.
830. Ramachandran, P., Zoph, B., & Le, Q. (2017). *Swish: A self-gated activation function* (Vol. 7). Preprint. ArXiv:1710.05941.
831. Ramachandran, P., & Antoniou, A. (2008). Identification of hot-spot locations in proteins using digital filters. *IEEE Journal of Selected Topics in Signal Processing*, *2*, 378–389.
832. Ramachandran, P., Lu, W., & Antoniou, A. (2012). Filter-based methodology for the location of hot spots in proteins and exons in DNA. *IEEE Transactions on Biomedical Engineering*, *59*, 1598–1609.
833. Ramazi, S., & Zahiri, J. (2021). Post-translational modifications in proteins: resources, tools and prediction methods. *Database*, *2021*, baab012.
834. Ramos, J., Others. (2003). Using tf-idf to determine word relevance in document queries. *Proceedings of the first instructional conference on machine learning* (Vol. 242, pp. 29–48).
835. Ran, F., Cong, L., Yan, W., Scott, D., Gootenberg, J., Kriz, A., Zetsche, B., Shalem, O., Wu, X., Makarova, K., Others. (2015). In vivo genome editing using Staphylococcus aureus Cas9. *Nature*, *520*, 186–191.
836. Ranzato, M., Huang, F., Boureau, Y., & LeCun, Y. (2007). Unsupervised learning of invariant feature hierarchies with applications to object recognition. *2007 IEEE conference on computer vision and pattern recognition* (pp. 1–8).
837. Rath, A., Davidson, A., & Deber, C. (2005). The structure of "unstructured" regions in peptides and proteins: role of the polyproline II helix in protein folding and recognition. *Peptide Science: Original Research on Biomolecules*, *80*, 179–185.
838. Rawat, W., & Wang, Z. (2017). Deep convolutional neural networks for image classification: A comprehensive review. *Neural Computation*, *29*, 2352–2449.
839. Ray, S., Lall, S., & Bandyopadhyay, S. (2022). A deep integrated framework for predicting SARS-CoV2-human protein-protein interaction. *IEEE Transactions on Emerging Topics in Computational Intelligence*, *6*, 1463–1472.
840. Raza, A., Uddin, J., Almuhaimeed, A., Akbar, S., Zou, Q., & Ahmad, A. (2023). AIPs-SnTCN: Predicting anti-inflammatory peptides using fastText and transformer encoder-based hybrid word embedding with self-normalized temporal convolutional networks. *Journal of Chemical Information and Modeling*, *63*, 6537–6554.
841. Reddy, A., Herschl, M., Geng, X., Kolli, S., Lu, A., Kumar, A., Hsu, P., Levine, S., & Ioannidis, N. (2023). Strategies for effectively modelling promoter-driven gene expression using transfer learning. *BioRxiv*.
842. Rehman, I., Kerndt, C., & Botelho, S. (2017). *Biochemistry: Tertiary protein structure*. StatPearls Publishing.
843. Rehman, A., Javed, K., Babri, H., & Asim, M. (2018). Selection of the most relevant terms based on a max-min ratio metric for text classification. *Expert Systems with Applications*, *114*, 78–96.
844. Rein, D., Ternes, P., Demin, R., Gierke, J., Helgason, T., & Schön, C. (2019). Artificial intelligence identified peptides modulate inflammation in healthy adults. *Food and Function*, *10*, 6030–6041.
845. Ren, X., Yang, Z., Xu, J., Sun, J., Mao, D., Hu, Y., Yang, S., Qiao, H., Wang, X., Hu, Q., Others. (2014). Enhanced specificity and efficiency of the CRISPR/Cas9 system with optimized sgRNA parameters in Drosophila. *Cell Reports*, *9*, 1151–1162.

846. Ren, R., Yin, C., & S.-T. Yau, S. (2022). kmer2vec: A novel method for comparing DNA sequences by word2vec embedding. *Journal of Computational Biology*, *29*, 1001–1021.
847. Reynaud, E., Others. (2010). Protein misfolding and degenerative diseases. *Nature Education*, *3*, 28.
848. Richardson, L., Allen, B., Baldi, G., Beracochea, M., Bileschi, M., Burdett, T., Burgin, J., Caballero-Pérez, J., Cochrane, G., Colwell, L., Others. (2023). MGnify: The microbiome sequence data analysis resource in 2023. *Nucleic Acids Research*, *51*, D753–D759.
849. Rizqiana, A., Faisal, M., & Lumbanraja, F. (2021). Implementation protein sequence segmentation in AAC and DC as protein descriptors for improving a classification performance of acetylation prediction. *Journal of Physics: Conference Series*, *1751*, Article 012031.
850. Robinson, A. (1993). Sickle cell disease, the thalassemias, Tay-Sachs disease: Protein to gene. *CMAJ: Canadian Medical Association Journal*, *148*, 1481.
851. Roche, R., Moussad, B., Shuvo, M., Tarafder, S., & Bhattacharya, D. (2024). EquiPNAS: Improved protein-nucleic acid binding site prediction using protein-language-model-informed equivariant deep graph neural networks. *Nucleic Acids Research*, *52*, e27–e27.
852. Rophina, M., Sharma, D., Poojary, M., & Scaria, V. (2020). Circad: A comprehensive manually curated resource of circular RNA associated with diseases. *Database*, *2020*, baaa019.
853. Ross, D., Cho, J., Zhang, R., Hines, K., & Xu, L. (2020). LiPydomics: A Python package for comprehensive prediction of lipid collision cross sections and retention times and analysis of ion mobility-mass spectrometry-based lipidomics data. *Analytical Chemistry*, *92*, 14967–14975.
854. Roweis, S., & Saul, L. (2000). Nonlinear dimensionality reduction by locally linear embedding. *Science*, *290*, 2323–2326.
855. Roy, S., Martinez, D., Platero, H., Lane, T., & Werner-Washburne, M. (2009). Exploiting amino acid composition for predicting protein-protein interactions. *PloS One*, *4*, Article e7813.
856. Rummel, R. (1988). *Applied factor analysis*. Northwestern University Press.
857. Russel, J., Pinilla-Redondo, R., Mayo-Muñoz, D., Shah, S., & Sørensen, S. (2020). CRISPRCasTyper: Automated identification, annotation, and classification of CRISPR-Cas loci. *The CRISPR Journal*, *3*, 462–469.
858. Ryabov, A. (1991). The biochemical reactions of organometallics with enzymes and proteins. *Angewandte Chemie International Edition in English*, *30*, 931–941.
859. Saadat, M., Behjati, A., Zare-Mirakabad, F., & Gharaghani, S. (2021). Drug-target binding affinity prediction using transformers.
860. Safran, M., Dalah, I., Alexander, J., Rosen, N., Iny Stein, T., Shmoish, M., Nativ, N., Bahir, I., Doniger, T., Krug, H., Others. (2010). GeneCards Version 3: The human gene integrator. *Database*, *2010*, baq020.
861. Saha, S., Halder, R., & Uddin, M. (2023). Particle swarm optimization-assisted multilayer ensemble model to predict DNA 4mC sites. *Informatics in Medicine Unlocked*, *42*, Article 101374.
862. Saha, S., Chatterjee, P., Basu, S., & Nasipuri, M. (2024). EPI-SF: Essential protein identification in protein interaction networks using sequence features. *PeerJ*, *12*, Article e17010.
863. Sahu, S., Loaiza, C., & Kaundal, R. (2020). Plant-mSubP: A computational framework for the prediction of single-and multi-target protein subcellular localization using integrated machine-learning approaches. *AoB Plants*, *12*, plz068.
864. Sahu, S., & Panda, G. (2010). Efficient localization of hot spots in proteins using a novel S-transform based filtering approach. *IEEE/ACM Transactions on Computational Biology and Bioinformatics*, *8*, 1235–1246.
865. Saidi, R., Maddouri, M., & Mephu Nguifo, E. (2010). Protein sequences classification by means of feature extraction with substitution matrices. *BMC Bioinformatics*, *11*, 1–13.

866. Saier, Jr., M., Reddy, V., Moreno-Hagelsieb, G., Hendargo, K., Zhang, Y., Iddamsetty, V., Lam, K., Tian, N., Russum, S., Wang, J., Others. (2021). The transporter classification database (TCDB): 2021 update. *Nucleic Acids Research*, *49*, D461–D467.
867. Salgado, H., Santos, A., Garza-Ramos, U., Helden, J., Dıaz, E., & Collado-Vides, J. (1999). RegulonDB (version 2.0): A database on transcriptional regulation in Escherichia coli. *Nucleic Acids Research*, *27*, 59–60.
868. San Biagio, M., Martelli, S., Crocco, M., Cristani, M., & Murino, V. (2013). Encoding classes of unaligned objects using structural similarity cross-covariance tensors. *Progress in pattern recognition, image analysis, computer vision, and applications: 18th iberoamerican congress, CIARP 2013, Havana, Cuba, November 20–23, 2013, proceedings, part I 18* (pp. 133–140).
869. Sandberg, M., Eriksson, L., Jonsson, J., Sjöström, M., & Wold, S. (1998). New chemical descriptors relevant for the design of biologically active peptides. A multivariate characterization of 87 amino acids. *Journal of Medicinal Chemistry*, *41*, 2481–2491.
870. Sanger, F., & Tuppy, H. (1951). The amino-acid sequence in the phenylalanyl chain of insulin. 1. The identification of lower peptides from partial hydrolysates. *Biochemical Journal*, *49*, 463.
871. Sanger, F., & Tuppy, H. (1951). The amino-acid sequence in the phenylalanyl chain of insulin. 2. The investigation of peptides from enzymic hydrolysates. *Biochemical Journal*, *49*, 481.
872. Sanger, F. (1949). The terminal peptides of insulin. *Biochemical Journal*, *45*, 563.
873. SantaLucia, J., Allawi, H., & Seneviratne, P. (1996). Improved nearest-neighbor parameters for predicting DNA duplex stability. *Biochemistry*, *35*, 3555–3562.
874. Santos, P., Fang, Z., Mason, S., Setúbal, J., & Dixon, R. (2012). Distribution of nitrogen fixation and nitrogenase-like sequences amongst microbial genomes. *BMC Genomics*, *13*.
875. Sarah A. (2023). Bates Genome Gov. https://www.genome.gov/genetics-glossary/Chromosome, Accessed on March 26, 2023.
876. Sarah A. (2023). Bates Genome Gov. https://www.genome.gov/genetics-glossary/Deoxyribonucleic-Acid, Accessed on March 26, 2023.
877. Sarai, A., Mazur, J., Nussinov, R., & Jernigan, R. (1989). Sequence dependence of DNA conformational flexibility. *Biochemistry*, *28*, 7842–7849.
878. Saravanan, V., & Gautham, N. (2015). Harnessing computational biology for exact linear B-cell epitope prediction: A novel amino acid composition-based feature descriptor. *Omics: A Journal of Integrative Biology*, *19*, 648–658.
879. Satchwell, S., Drew, H., & Travers, A. (1986). Sequence periodicities in chicken nucleosome core DNA. *Journal of Molecular Biology*, *191*, 659–675.
880. Sato, K., & Hamada, M. (2023). Recent trends in RNA informatics: A review of machine learning and deep learning for RNA secondary structure prediction and RNA drug discovery. *Briefings in Bioinformatics*, bbad186.
881. Sato, K., & Kato, Y. (2022). Prediction of RNA secondary structure including pseudoknots for long sequences. *Briefings in Bioinformatics*, *23*, bbab395.
882. Sato, K., Akiyama, M., & Sakakibara, Y. (2021). RNA secondary structure prediction using deep learning with thermodynamic integration. *Nature Communications*, *12*, 941.
883. Saul, L., & Roweis, S. (2000). An introduction to locally linear embedding. Unpublished. Available At: http://www.Cs.Toronto.Edu/undefined~Roweis/lle/publications.html.
884. Saxonov, S., Daizadeh, I., Fedorov, A., & Gilbert, W. (2000). EID: The Exon-Intron Database–an exhaustive database of protein-coding intron-containing genes. *Nucleic Acids Research*, *28*, 185–190.
885. Schapke, J., Tavares, A., & Recamonde-Mendoza, M. (2021). EPGAT: Gene essentiality prediction with graph attention networks. *IEEE/ACM Transactions on Computational Biology and Bioinformatics*, *19*, 1615–1626.
886. Scherer, D., Müller, A., & Behnke, S. (2010). Evaluation of pooling operations in convolutional architectures for object recognition. *International conference on artificial neural networks* (pp. 92–101).

887. Schmidt-Barbo, P., Kalweit, G., Naouar, M., Paschold, L., Willscher, E., Schultheiß, C., Märkl, B., Dirnhofer, S., Tzankov, A., Binder, M., Others. (2024). Detection of disease-specific signatures in B cell repertoires of lymphomas using machine learning. *PLOS Computational Biology*, *20*, Article e1011570.
888. Schneider, G., & Wrede, P. (1994). The rational design of amino acid sequences by artificial neural networks and simulated molecular evolution: De novo design of an idealized leader peptidase cleavage site. *Biophysical Journal*, *66*, 335–344.
889. Schneider, H., Raiol, T., Brigido, M., Walter, M., & Stadler, P. (2017). A support vector machine based method to distinguish long non-coding RNAs from protein coding transcripts. *BMC Genomics*, *18*, 1–14.
890. Schoch, C., Ciufo, S., Domrachev, M., Hotton, C., Kannan, S., Khovanskaya, R., Leipe, D., Mcveigh, R., O'Neill, K., Robbertse, B., Others. (2020). NCBI Taxonomy: A comprehensive update on curation, resources and tools. *Database*, *2020*, baaa062.
891. Schölkopf, B., Smola, A., & Müller, K. (1997). Kernel principal component analysis. *International conference on artificial neural networks* (pp. 583–588).
892. Searle, S., Frankish, A., Bignell, A., Aken, B., Derrien, T., Diekhans, M., Harte, R., Howald, C., Kokocinski, F., Lin, M., Others. (2010). The GENCODE human gene set. *Genome Biology*, *11*, 1–1.
893. Segura-Campos, M., Chel-Guerrero, L., Betancur-Ancona, D., & Hernandez-Escalante, V. (2011). Bioavailability of bioactive peptides. *Food Reviews International*, *27*, 213–226.
894. Selby, D., Spriestersbach, K., Iwashita, Y., Bappert, D., Warrier, A., Mukherjee, S., Asim, M., Kise, K., & Vollmer, S. (2024). Quantitative knowledge retrieval from large language models. Preprint. ArXiv:2402.07770.
895. Selvaraj, C., Chandra, I., & Singh, S. (2021). Artificial intelligence and machine learning approaches for drug design: Challenges and opportunities for the pharmaceutical industries. *Molecular Diversity*, 1–21.
896. Sethupathy, P., Corda, B., & Hatzigeorgiou, A. (2006). TarBase: A comprehensive database of experimentally supported animal microRNA targets. *Rna*, *12*, 192–197.
897. Sha, M., & Rahamathulla, M. (2024). Splice site recognition-deciphering Exon-Intron transitions for genetic insights using Enhanced integrated Block-Level gated LSTM model. *Gene*, *915*, Article 148429.
898. Shah, S., & Ross, A. (2006). Generating synthetic irises by feature agglomeration. *2006 International conference on image processing* (pp. 317–320).
899. Shah, S., & Ou, Y. (2023). Disto-TRP: An approach for identifying transient receptor potential (TRP) channels using structural information generated by AlphaFold. *Gene*, *871*, Article 147435.
900. Shamseer, L., Moher, D., Clarke, M., Ghersi, D., Liberati, A., Petticrew, M., Shekelle, P., & Stewart, L. (2015). Preferred reporting items for systematic review and meta-analysis protocols (PRISMA-P) 2015: Elaboration and explanation. *Bmj*, *349*.
901. Shao, D., Huang, L., Wang, Y., He, K., Cui, X., Wang, Y., Ma, Q., & Cui, J. (2022). DeepSec: A deep learning framework for secreted protein discovery in human body fluids. *Bioinformatics*, *38*, 228–235.
902. Sharma, K., Marucci, L., & Abdallah, Z. (2024). FluxGAT: Integrating flux sampling with graph neural networks for unbiased gene essentiality classification. Preprint. ArXiv:2403.18666.
903. Sharma, R., Shrivastava, S., Kumar Singh, S., Kumar, A., Saxena, S., & Kumar Singh, R. (2021). Deep-ABPpred: Identifying antibacterial peptides in protein sequences using bidirectional LSTM with word2vec. *Briefings in Bioinformatics*, *22*, bbab065.
904. Sharma, S., Murmu, S., Das, R., Tilgam, J., Saakre, M., & Paul, K. (2023). A review on bioinformatics advances in CRISPR-Cas technology. *Journal of Plant Biochemistry and Biotechnology*, *32*, 791–807.
905. Shen, X., & Li, X. (2024). Reformer: Deep learning model for characterizing protein-RNA interactions from sequence at single-base resolution. *BioRxiv*.

906. Shen, J., Zhang, J., Luo, X., Zhu, W., Yu, K., Chen, K., Li, Y., & Jiang, H. (2007). Predicting protein-protein interactions based only on sequences information. *Proceedings of the National Academy of Sciences*, *104*, 4337–4341.
907. Shen, Z., Bao, W., & Huang, D. (2018). Recurrent neural network for predicting transcription factor binding sites. *Scientific Reports*, *8*, 15270.
908. Shen, M., Arbab, M., Hsu, J., Worstell, D., Culbertson, S., Krabbe, O., Cassa, C., Liu, D., Gifford, D., & Sherwood, R. (2018). Predictable and precise template-free CRISPR editing of pathogenic variants. *Nature*, *563*, 646–651.
909. Sheng, N., Wang, Y., Huang, L., Gao, L., Cao, Y., Xie, X., & Fu, Y. (2023). Multi-task prediction-based graph contrastive learning for inferring the relationship among lncRNAs, miRNAs and diseases. *Briefings in Bioinformatics*, *24*, bbad276.
910. Sherkatghanad, Z., Abdar, M., Charlier, J., & Makarenkov, V. (2023). Using traditional machine learning and deep learning methods for on- and off-target prediction in CRISPR/Cas9: A review. *Briefings in Bioinformatics*, *3*.
911. Shi, L., & Chen, B. (2021). LSHvec: A vector representation of DNA sequences using locality sensitive hashing and FastText word embeddings. *Proceedings of the 12th ACM conference on bioinformatics, computational biology, and health informatics* (pp. 1–10).
912. Shi, Y. (2023). CRISPR/Cas system in human genetic diseases. *Highlights in Science, Engineering and Technology*, *74*, 78–85.
913. Shi, W., Zhan, C., Ignatov, A., Manjasetty, B., Marinkovic, N., Sullivan, M., Huang, R., & Chance, M. (2005). Metalloproteomics: High-throughput structural and functional annotation of proteins in structural genomics. *Structure*, *13*, 1473–1486.
914. Shi, H., Li, S., & Su, X. (2022). Plant6mA: A predictor for predicting N6-methyladenine sites with lightweight structure in plant genomes. *Methods*, *204*, 126–131.
915. Shimada, K., Muhlich, J., & Mitchison, T. (2019). A tool for browsing the cancer dependency map reveals functional connections between genes and helps predict the efficacy and selectivity of candidate cancer drugs. *BioRxiv*.
916. Shmakov, S., Sitnik, V., Makarova, K., Wolf, Y., Severinov, K., & Koonin, E. (2017). The CRISPR spacer space is dominated by sequences from species-specific mobilomes. *MBio*, *8*, 10–1128.
917. Shokralla, S., Spall, J., Gibson, J., & Hajibabaei, M. (2012). Next-generation sequencing technologies for environmental DNA research. *Molecular Ecology*, *21*, 1794–1805.
918. Shrikumar, A., Greenside, P., & Kundaje, A. (2017). Learning important features through propagating activation differences. *International conference on machine learning* (pp. 3145–3153).
919. Shujaat, M., Jin, J., Tayara, H., & Chong, K. (2022). iProm-phage: A two-layer model to identify phage promoters and their types using a convolutional neural network. *Frontiers in Microbiology*, *13*.
920. Shulgina, Y., Trinidad, M., Langeberg, C., Nisonoff, H., Chithrananda, S., Skopintsev, P., Nissley, A., Patel, J., Boger, R., Shi, H., Others. (2024). RNA language models predict mutations that improve RNA function. *BioRxiv*.
921. Shun-xian, Z., Xuan, Z., Wang, L., Ping, P., & Pei, T. (2018). A novel model for predicting associations between diseases and lncrna-mirna pairs based on a newly constructed bipartite network. *Computational and Mathematical Methods in Medicine*, *2018*, 1–11.
922. Siegl, G., & Frösner, G. (1978). Characterization and classification of virus particles associated with hepatitis AI Size, density, and sedimentation. *Journal of Virology*, *26*, 40–47.
923. Sigrist, C., Cerutti, L., De Castro, E., Langendijk-Genevaux, P., Bulliard, V., Bairoch, A., & Hulo, N. (2010). PROSITE, a protein domain database for functional characterization and annotation. *Nucleic Acids Research*, *38*, D161–D166.
924. Sikander, R., Wang, Y., Ghulam, A., & Wu, X. (2021). Identification of enzymes-specific protein domain based on DDE, and convolutional neural network. *Frontiers in Genetics*, 2253

925. Sikander, R., Ghulam, A., & Ali, F. (2022). XGB-DrugPred: Computational prediction of druggable proteins using eXtreme gradient boosting and optimized features set. *Scientific Reports*, *12*, 1–9.
926. Sinden, R. (1994). *DNA structure and function*. Gulf Professional Publishing.
927. Singh, P., FNU, K., & Encarnação, T. (2023). Genetic modification: A gateway to stimulate the industrial production of biofuels. *Marine organisms: A solution to environmental pollution? Uses in bioremediation and in biorefinery* (pp. 237–260).
928. Singh, V., Shrivastava, S., Kumar Singh, S., Kumar, A., & Saxena, S. (2022). Accelerating the discovery of antifungal peptides using deep temporal convolutional networks. *Briefings in Bioinformatics*, *23*, bbac008.
929. Singh, V., & Singh, S. (2023). A separable temporal convolutional networks based deep learning technique for discovering antiviral medicines. *Scientific Reports*, *13*, 13722.
930. Singh, S., Chaudhary, K., Dhanda, S., Bhalla, S., Usmani, S., Gautam, A., Tuknait, A., Agrawal, P., Mathur, D., & Raghava, G. (2016). SATPdb: A database of structurally annotated therapeutic peptides. *Nucleic Acids Research*, *44*, D1119–D1126.
931. Singh, S., Yang, Y., Póczos, B., & Ma, J. (2019). Predicting enhancer-promoter interaction from genomic sequence with deep neural networks. *Quantitative Biology*, *7*, 122–137.
932. Sivolob, A., & Khrapunov, S. (1995). Translational positioning of nucleosomes on DNA: The role of sequence-dependent isotropic DNA bending stiffness. *Journal of Molecular Biology*, *247*, 918–931.
933. Skrylnik, A. Hidden Markov Models for detection of intrinsically disordered regions of proteins.
934. Skutkova, H., Maderankova, D., Sedlar, K., Jugas, R., & Vitek, M. (2019). A degeneration-reducing criterion for optimal digital mapping of genetic codes. *Computational and Structural Biotechnology Journal*, *17*, 406–414.
935. Smaili, F., Gao, X., & Hoehndorf, R. (2019). OPA2Vec: Combining formal and informal content of biomedical ontologies to improve similarity-based prediction. *Bioinformatics*, *35*, 2133–2140.
936. Sneath, P. (1966). Relations between chemical structure and biological activity in peptides. *Journal of Theoretical Biology*, *12*, 157–195.
937. Sokal, R., & Thomson, B. (2006). Population structure inferred by local spatial autocorrelation: an example from an Amerindian tribal population. *American Journal of Physical Anthropology: The Official Publication Of The American Association Of Physical Anthropologists*, *129*, 121–131.
938. Soldner, F., & Jaenisch, R. (2015). Dissecting risk haplotypes in sporadic alzheimer's disease. *Cell Stem Cell*, *16*, 341–342.
939. Sonawane, A., Platig, J., Fagny, M., Chen, C., Paulson, J., Lopes-Ramos, C., DeMeo, D., Quackenbush, J., Glass, K., & Kuijjer, M. (2017). Understanding tissue-specific gene regulation. *Cell Reports, 21*(4), 1077–1088.
940. Sondka, Z., Dhir, N., Carvalho-Silva, D., Jupe, S., Madhumita, McLaren, K., Starkey, M., Ward, S., Wilding, J., Ahmed, M., Others. (2024). COSMIC: A curated database of somatic variants and clinical data for cancer. *Nucleic Acids Research*, *52*, D1210–D1217.
941. Song, C., Diao, J., Brunger, A., & Quake, S. (2016). Simultaneous single-molecule epigenetic imaging of dna methylation and hydroxymethylation. *Proceedings of the National Academy of Sciences*, *113*, 4338–4343.
942. Souza, N. (2012). The ENCODE project. *Nature Methods*, *9*, 1046–1046.
943. Spänig, S., Mohsen, S., Hattab, G., Hauschild, A., & Heider, D. (2021). A large-scale comparative study on peptide encodings for biomedical classification. *NAR Genomics and Bioinformatics*, *3*, lqab039.
944. Sprous, D., Zacharias, W., Wood, Z., & Harvey, S. (1995). Dehydrating agents sharply reduce curvature in DNAs containing A tracts. *Nucleic Acids Research*, *23*, 1816–1821.
945. Srivastava, N., Hinton, G., Krizhevsky, A., Sutskever, I., & Salakhutdinov, R. (2014). Dropout: A simple way to prevent neural networks from overfitting. *The Journal of Machine Learning Research*, *15*, 1929–1958.

946. Ssnger, F., Air, G., BarreH, B., Brown, N., Couison, A., Fiddes, III, J., Hutchison, III, C., Sloeombe, P. M., & Smith, M. (1977). Nucleotide sequence bacteriophageundefined X174 DNA. *Nature, 265*.
947. Stanley, D., Watson-Haigh, N., Cowled, C., & Moore, R. (2013). Genetic architecture of gene expression in the chicken. *BMC Genomics, 14*, 13.
948. Stanojević, D., Li, Z., Foo, R., & Šikić, M. (2022). Rockfish: A transformer-based model for accurate 5-methylcytosine prediction from nanopore sequencing. *BioRxiv*.
949. Stenson, P., Ball, E., Mort, M., Phillips, A., Shiel, J., Thomas, N., Abeysinghe, S., Krawczak, M., & Cooper, D. (2003). Human gene mutation database (HGMD®): 2003 update. *Human Mutation, 21*, 577–581.
950. Sternberg, S., Richter, H., Charpentier, E., & Qimron, U. (2016). Adaptation in CRISPR-Cas systems. *Molecular Cell, 61*, 797–808.
951. Störtz, F., & Minary, P. (2021). crisprSQL: A novel database platform for CRISPR/Cas off-target cleavage assays. *Nucleic Acids Research, 49*, D855–D861.
952. Störtz, F., Mak, J., & Minary, P. (2023). piCRISPR: Physically informed deep learning models for CRISPR/Cas9 off-target cleavage prediction. *Artificial Intelligence in the Life Sciences, 3*, 100075–100075.
953. Stricker, M., Asim, M., Dengel, A., & Ahmed, S. (2022). CircNet: An encoder–decoder-based convolution neural network (CNN) for circular RNA identification. *Neural Computing and Applications*, 1–12
954. Strokach, A., Lu, T., & Kim, P. (2021). ELASPIC2 (EL2): Combining contextualized language models and graph neural networks to predict effects of mutations. *Journal of Molecular Biology, 433*, Article 166810.
955. Stupp, D., Sharon, E., Bloch, I., Zitnik, M., Zuk, O., & Tabach, Y. (2021). Co-evolution based machine-learning for predicting functional interactions between human genes. *Nature Communications, 12*, 6454.
956. Su, W., Liu, M., Yang, Y., Wang, J., Li, S., Lv, H., Dao, F., Yang, H., & Lin, H. (2021). PPD: A manually curated database for experimentally verified prokaryotic promoters. *Journal of Molecular Biology, 433*, Article 166860.
957. Su, X., Hu, L., You, Z., Hu, P., & Zhao, B. (2022). Multi-view heterogeneous molecular network representation learning for protein-protein interaction prediction. *BMC Bioinformatics, 23*, 234.
958. Sugimoto, N., Nakano, S., Yoneyama, M., & Honda, K. (1996). Improved thermodynamic parameters and helix initiation factor to predict stability of DNA duplexes. *Nucleic Acids Research, 24*, 4501–4505.
959. Suleman, M., Alkhalifah, T., Alturise, F., & Khan, Y. (2022). DHU-Pred: Accurate prediction of dihydrouridine sites using position and composition variant features on diverse classifiers. *PeerJ, 10*, Article e14104.
960. Sun, Z., Yang, C., Huang, L., Mo, Z., Zhang, K., Fan, W., Wang, K., Wu, F., Wang, J., Meng, F., Others. (2024). <? mode longmeta?> circRNADisease v2. 0: An updated resource for high-quality experimentally supported circRNA-disease associations. *Nucleic Acids Research, 52*, D1193–D1200.
961. Sun, W., Li, J., Liu, S., Wu, J., Zhou, H., Qu, L., & Yang, J. (2016). RMBase: A resource for decoding the landscape of RNA modifications from high-throughput sequencing data. *Nucleic Acids Research, 44*, D259–D265.
962. Sun, J., Liu, H., Liu, J., Cheng, S., Peng, Y., Zhang, Q., Yan, J., Liu, H., & Chen, L. (2019). CRISPR-Local: A local single-guide RNA (sgRNA) design tool for non-reference plant genomes. *Bioinformatics, 35*, 2501–2503.
963. Sun, J., Yang, H., Yao, J., Ding, H., Han, S., Wu, C., & Tang, H. (2020). Prediction of cyclin protein using two-step feature selection technique. *IEEE Access, 8*, 109535–109542.
964. Sun, M., Hu, H., Pang, W., & Zhou, Y. (2023). Acp-bc: A model for accurate identification of anticancer peptides based on fusion features of bidirectional long short-term memory and chemically derived information. *International Journal of Molecular Sciences, 24*, 15447.

965. Suo, S., Qiu, J., Shi, S., Sun, X., Huang, S., Chen, X., & Liang, R. (2012). Position-specific analysis and prediction for protein lysine acetylation based on multiple features. *PloS One*, *7*, Article e49108.
966. Sutskever, I., Martens, J., & Hinton, G. (2011). Generating text with recurrent neural networks. *ICML*.
967. Suzuki, Y., Yamashita, R., Nakai, K., & Sugano, S. (2002). DBTSS: DataBase of human transcriptional start sites and full-length cDNAs. *Nucleic Acids Research*, *30*, 328–331.
968. Svoboda, P., & Cara, A. (2006). Hairpin RNA: A secondary structure of primary importance. *Cellular and Molecular Life Sciences CMLS*, *63*, 901–908.
969. Sweeney, B., Hoksza, D., Nawrocki, E., Ribas, C., Madeira, F., Cannone, J., Gutell, R., Maddala, A., Meade, C., Williams, L., Others. (2021). R2DT is a framework for predicting and visualising RNA secondary structure using templates. *Nature Communications*, *12*, 3494.
970. Szabat, M., Lorent, D., Czapik, T., Tomaszewska, M., Kierzek, E., & Kierzek, R. (2020). RNA secondary structure as a first step for rational design of the oligonucleotides towards inhibition of influenza a virus replication. *Pathogens*, *9*, 925.
971. Szcześniak, M., Bryzghalov, O., Ciomborowska-Basheer, J., & Makałowska, I. (2019). CANTATAdb 2.0: Expanding the collection of plant long noncoding RNAs. *Plant Long Non-Coding RNAs: Methods And Protocols*, 415–429.
972. Szklarczyk, D., Kirsch, R., Koutrouli, M., Nastou, K., Mehryary, F., Hachilif, R., Gable, A., Fang, T., Doncheva, N., Pyysalo, S., Others. (2023). The STRING database in 2023: protein–protein association networks and functional enrichment analyses for any sequenced genome of interest. *Nucleic Acids Research*, *51*, D638–D646.
973. Szklarczyk, D., Santos, A., Von Mering, C., Jensen, L., Bork, P., & Kuhn, M. (2016). STITCH 5: Augmenting protein-chemical interaction networks with tissue and affinity data. *Nucleic Acids Research*, *44*, D380–D384.
974. Tahir, M., Hayat, M., Gul, S., & Chong, K. (2020). An intelligent computational model for prediction of promoters and their strength via natural language processing. *Chemometrics and Intelligent Laboratory Systems*, *202*, Article 104034.
975. Talukder, A., Saadat, S., Li, X., & Hu, H. (2019). EPIP: A novel approach for condition-specific enhancer-promoter interaction prediction. *Bioinformatics*, *35*, 3877–3883.
976. Tanaka, T., & Kikuchi, Y. (2001). Origin of the cloverleaf shape of transfer RNA-the double-hairpin model: Implication for the role of tRNA intron and the long extra loop. *Viva Origino*, *29*, 134–142.
977. Tang, L., & Liu, H. (2009). Relational learning via latent social dimensions. *Proceedings of the 15th ACM SIGKDD international conference on knowledge discovery and data mining* (pp. 817–826).
978. Tang, Z., Chen, S., Chen, A., He, B., Zhou, Y., Chai, G., Guo, F., & Huang, J. (2019). CasPDB: An integrated and annotated database for Cas proteins from bacteria and archaea. *Database*, *2019*, baz093.
979. Tang, J., Qu, M., Wang, M., Zhang, M., Yan, J., & Mei, Q. (2015). Line: Large-scale information network embedding. *Proceedings of the 24th international conference on world wide web* (pp. 1067–1077).
980. Tang, L., Wu, J., Li, C., Jiang, H., Xu, M., Du, M., Yin, Z., Mei, H., & Hu, Y. (2020). Characterization of immune dysfunction and identification of prognostic immune-related risk factors in acute myeloid leukemia. *Clinical Cancer Research*, *26*, 1763–1772.
981. Tao, H., Li, H., Xu, K., Hong, H., Jiang, S., Du, G., Wang, J., Sun, Y., Huang, X., Ding, Y., Others. (2021). Computational methods for the prediction of chromatin interaction and organization using sequence and epigenomic profiles. *Briefings in Bioinformatics*, *22*, bbaa405.
982. Tate, J., Bamford, S., Jubb, H., Sondka, Z., Beare, D., Bindal, N., Boutselakis, H., Cole, C., Creatore, C., Dawson, E., Others. (2019). COSMIC: The catalogue of somatic mutations in cancer. *Nucleic Acids Research*, *47*, D941–D947.

983. Teng, X., Chen, X., Xue, H., Tang, Y., Zhang, P., Kang, Q., Hao, Y., Chen, R., Zhao, Y., & He, S. (2020). NPInter v4. 0: An integrated database of ncRNA interactions. *Nucleic Acids Research*, *48*, D160–D165.
984. Terzian, P., Olo Ndela, E., Galiez, C., Lossouarn, J., Pérez Bucio, R., Mom, R., Toussaint, A., Petit, M., & Enault, F. (2021). PHROG: Families of prokaryotic virus proteins clustered using remote homology. *NAR Genomics and Bioinformatics*, *3*, lqab067.
985. Terziyski, Z., Terziyska, M., Deseva, I., Hadzhikoleva, S., Krastanov, A., Mihaylova, D., & Hadzhikolev, E. (2023). PepLab platform: Database and software tools for analysis of food-derived bioactive peptides. *Applied Sciences*, *13*, 961.
986. Teufel, F., Almagro Armenteros, J., Johansen, A., Gıslason, M., Pihl, S., Tsirigos, K., Winther, O., Brunak, S., Heijne, G., & Nielsen, H. (2022). SignalP 6.0 predicts all five types of signal peptides using protein language models. *Nature Biotechnology*, *40*, 1023–1025.
987. Thakur, M., Buniello, A., Brooksbank, C., Gurwitz, K., Hall, M., Hartley, M., Hulcoop, D., Leach, A., Marques, D., Martin, M., Others. (2024). EMBL's European bioinformatics institute (EMBL-EBI) in 2023. *Nucleic Acids Research*, *52*, D10–D17.
988. Thomas, S., Karnik, S., Barai, R., Jayaraman, V., & Idicula-Thomas, S. (2010). CAMP: A useful resource for research on antimicrobial peptides. *Nucleic Acids Research*, *38*, D774–D780.
989. Tian, X., Chen, Z., Nie, S., Shi, T., Yan, X., Bao, Y., Li, Z., Ma, H., Jia, K., Zhao, W., Others. (2024). Plant-LncPipe: A computational pipeline providing significant improvement in plant lncRNA identification. *Horticulture Research*, uhae041.
990. Tian, Z., Han, C., Xu, L., Teng, Z., & Song, W. (2024). MGCNSS: miRNA–disease association prediction with multi-layer graph convolution and distance-based negative sample selection strategy. *Briefings in Bioinformatics*, *25*, bbae168.
991. Tickotsky, N., Sagiv, T., Prilusky, J., Shifrut, E., & Friedman, N. (2017). McPAS-TCR: A manually curated catalogue of pathology-associated T cell receptor sequences. *Bioinformatics*, *33*, 2924–2929.
992. Timasheff, S. (2002). Protein-solvent preferential interactions, protein hydration, and the modulation of biochemical reactions by solvent components. *Proceedings of the National Academy of Sciences*, *99*, 9721–9726.
993. Tinoco, Jr. I., & Bustamante, C. (1999). How RNA folds. *Journal of Molecular Biology*, *293*, 271–281.
994. Tiwari, P., Ko, T., Dubey, R., Chouhan, M., Tsai, L., Singh, N., Chaubey, K., Dayal, D., Chiang, C., & Kumar, S. (2023). CRISPR/Cas9 as a therapeutic tool for triple negative breast cancer: From bench to clinics. *Frontiers in Molecular Biosciences, 10,* 1–20.
995. Tomii, K., & Kanehisa, M. (1996). Analysis of amino acid indices and mutation matrices for sequence comparison and structure prediction of proteins. *Protein Engineering, Design and Selection*, *9*, 27–36.
996. Tong, X., & Liu, S. (2019). CPPred: Coding potential prediction based on the global description of RNA sequence. *Nucleic Acids Research*, *47*, e43–e43.
997. Toufikuzzaman, M., Samee, M., & Rahman, M. (2023). CRISPR-DIPOFF: An interpretable deep learning approach for CRISPR Cas-9 off-target prediction. *Briefings in Bioinformatics, 25*(2), bbad530.
998. Toufiq, M., Rinchai, D., Bettacchioli, E., Kabeer, B., Khan, T., Subba, B., White, O., Yurieva, M., George, J., Jourde-Chiche, N., Others. (2023). Harnessing large language models (LLMs) for candidate gene prioritization and selection. *Journal of Translational Medicine*, *21*, 728.
999. Trabelsi, A., Chaabane, M., & Ben-Hur, A. (2019). Comprehensive evaluation of deep learning architectures for prediction of DNA/RNA sequence binding specificities. *Bioinformatics*, *35*, i269–i277.
1000. Tran, T., Pham, D., Ou, Y., Others. (2021). An extensive examination of discovering 5-Methylcytosine sites in genome-wide DNA promoters using machine learning based approaches. *IEEE/ACM Transactions On Computational Biology And Bioinformatics*, *19*, 87–94.

1001. Travers, A., & Muskhelishvili, G. (2015). DNA structure and function. *The FEBS Journal*, *282*, 2279–2295.
1002. Tsai, S., Zheng, Z., Nguyen, N., Liebers, M., Topkar, V., Thapar, V., Wyvekens, N., Khayter, C., Iafrate, A., Le, L., Others. (2015). GUIDE-seq enables genome-wide profiling of off-target cleavage by CRISPR-Cas nucleases. *Nature Biotechnology*, *33*, 187–197.
1003. Tsai, S., Nguyen, N., Malagon-Lopez, J., Topkar, V., Aryee, M., & Joung, J. (2017). CIRCLE-seq: A highly sensitive in vitro screen for genome-wide CRISPR-Cas9 nuclease off-targets. *Nature Methods*, *14*, 607–614.
1004. Tsukiyama, S., Hasan, M., Deng, H., & Kurata, H. (2022). BERT6mA: Prediction of DNA N6-methyladenine site using deep learning-based approaches. *Briefings in Bioinformatics*, *23*, bbac053.
1005. Tu, T., Krishna, G., & Aghazadeh, A. (2023). ProtiGeno: A prokaryotic short gene finder using protein language models. Preprint. ArXiv:2307.10343.
1006. Tu, G., Wang, X., Xia, R., & Song, B. (2024). m6A-TCPred: A web server to predict tissue-conserved human m6A sites using machine learning approach. *BMC Bioinformatics*, *25*, 127.
1007. Tung, C., & Ho, S. (2008). Computational identification of ubiquitylation sites from protein sequences. *BMC Bioinformatics*, *9*, 1–15.
1008. Tzavella, K., Diaz, A., Olsen, C., & Vranken, W. (2023). Combining evolution and protein language models for an interpretable cancer driver mutation prediction with D2Deep. *BioRxiv*.
1009. Ungar, D., & Hughson, F. (2003). SNARE protein structure and function. *Annual Review of Cell and Developmental Biology*, *19*, 493–517.
1010. Usman, M., Khan, S., & Lee, J. (2020). Afp-lse: Antifreeze proteins prediction using latent space encoding of composition of k-spaced amino acid pairs. *Scientific Reports*, *10*, 7197.
1011. Vadapalli, S., Abdelhalim, H., Zeeshan, S., & Ahmed, Z. (2022). Artificial intelligence and machine learning approaches using gene expression and variant data for personalized medicine. *Briefings in Bioinformatics*, *23*, bbac191.
1012. Vanhaeren, T., Divina, F., Garcıa-Torres, M., Gómez-Vela, F., Vanhoof, W., & Martınez-Garcıa, P. (2020). A comparative study of supervised machine learning algorithms for the prediction of long-range chromatin interactions. *Genes*, *11*, 985.
1013. Varadi, M., Anyango, S., Deshpande, M., Nair, S., Natassia, C., Yordanova, G., Yuan, D., Stroe, O., Wood, G., Laydon, A., Others. (2022). AlphaFold protein structure database: Massively expanding the structural coverage of protein-sequence space with high-accuracy models. *Nucleic Acids Research*, *50*, D439–D444.
1014. Varadi, M., De Baets, G., Vranken, W., Tompa, P., & Pancsa, R. (2018). AmyPro: A database of proteins with validated amyloidogenic regions. *Nucleic Acids Research*, *46*, D387–D392.
1015. Varshney, G., Pei, W., LaFave, M., Idol, J., Xu, L., Gallardo, V., Carrington, B., Bishop, K., Jones, M., Li, M., Others. (2015). High-throughput gene targeting and phenotyping in zebrafish using CRISPR/Cas9. *Genome Research*, *25*, 1030–1042.
1016. Venkatesh, B., & Anuradha, J. (2019). A review of feature selection and its methods. *Cybernetics and Information Technologies*, *19*, 3–26.
1017. Verma, B., & Parkinson, J. (2024). HiTaxon: A hierarchical ensemble framework for taxonomic classification of short reads. *Bioinformatics Advances*, *4*, vbae016.
1018. Vijayakrishnan, R. (2009). Structure-based drug design and modern medicine. *Journal of Postgraduate Medicine*, *55*, 301–304.
1019. Vilela, J., Asif, M., Marques, A., Santos, J., Rasga, C., Vicente, A., & Martiniano, H. (2023). Biomedical knowledge graph embeddings for personalized medicine: Predicting disease-gene associations. *Expert Systems*, *40*, Article e13181.
1020. Vita, R., Mahajan, S., Overton, J., Dhanda, S., Martini, S., Cantrell, J., Wheeler, D., Sette, A., & Peters, B. (2019). The immune epitope database (IEDB): 2018 update. *Nucleic Acids Research*, *47*, D339–D343.
1021. Vlahoviček, K., Kajan, L., & Pongor, S. (2003). DNA analysis servers: Plot. it, bend. it, model. it and IS. *Nucleic Acids Research*, *31*, 3686–3687.

1022. Volders, P., Helsens, K., Wang, X., Menten, B., Martens, L., Gevaert, K., Vandesompele, J., & Mestdagh, P. (2013). LNCipedia: a database for annotated human lncRNA transcript sequences and structures. *Nucleic Acids Research*, *41*, D246–D251.
1023. Von Eschenbach, W. (2021). Transparency and the black box problem: Why we do not trust AI. *Philosophy and Technology*, *34*, 1607–1622.
1024. Vora, D., Kalakoti, Y., & Sundar, D. (2022). Computational methods and deep learning for elucidating protein interaction networks. *Computational biology and machine learning for metabolic engineering and synthetic biology* (pp. 285–323). Springer.
1025. Vora, D., Yadav, S., & Sundar, D. (2023). Hybrid multitask learning reveals sequence features driving specificity in the CRISPR/Cas9 system. *Biomolecules*, *13*, 641–641.
1026. Vora, L., Gholap, A., Jetha, K., Thakur, R., Solanki, H., & Chavda, V. (2023). Artificial intelligence in pharmaceutical technology and drug delivery design. *Pharmaceutics*, *15*, 1916.
1027. Wadman, M. (2008). James Watson's genome sequenced at high speed. *Nature*, *452*, 788–789.
1028. Wan, C., Cozzetto, D., Fa, R., & Jones, D. (2019). Using deep maxout neural networks to improve the accuracy of function prediction from protein interaction networks. *PloS One*, *14*, Article e0209958.
1029. Wang, M., Ali, H., Xu, Y., Xie, J., & Xu, S. (2024). BiPSTP: Sequence feature encoding method for identifying different RNA modifications with bidirectional position-specific trinucleotides propensities. *Journal of Biological Chemistry*, *300*.
1030. Wang, Y., Chen, Z., Pan, Z., Huang, S., Liu, J., Xia, W., Zhang, H., Zheng, M., Li, H., Hou, T., Others. (2023). RNAincoder: A deep learning-based encoder for RNA and RNA-associated interaction. *Nucleic Acids Research*, gkad404.
1031. Wang, Z., Combs, S., Brand, R., Calvo, M., Xu, P., Price, G., Golovach, N., Salawu, E., Wise, C., Ponnapalli, S., Others. (2022). Lm-gvp: An extensible sequence and structure informed deep learning framework for protein property prediction. *Scientific Reports*, *12*, 6832.
1032. Wang, D., Cui, P., & Zhu, W. (2016). Structural deep network embedding. *Proceedings of the 22nd ACM SIGKDD international conference on knowledge discovery and data mining* (pp. 1225–1234).
1033. Wang, J., Dai, W., Li, J., Xie, R., Dunstan, R., Stubenrauch, C., Zhang, Y., & Lithgow, T. (2020). PaCRISPR: A server for predicting and visualizing anti-CRISPR proteins. *Nucleic Acids Research*, *48*, W348–W357.
1034. Wang, X., Ding, Z., Wang, R., & Lin, X. (2023). Deepro-Glu: Combination of convolutional neural network and Bi-LSTM models using ProtBert and handcrafted features to identify lysine glutarylation sites. *Briefings in Bioinformatics*, *24*, bbac631.
1035. Wang, X., Gao, X., Wang, G., & Li, D. (2023). miProBERT: Identification of microRNA promoters based on the pre-trained model BERT. *Briefings in Bioinformatics*, *24*, bbad093.
1036. Wang, C., He, Z., Jia, R., Pan, S., Coin, L., Song, J., & Li, F. (2024). PLANNER: A multi-scale deep language model for the origins of replication site prediction. *IEEE Journal of Biomedical and Health Informatics, 28,* 2445–2454.
1037. Wang, T., Wu, D., Coates, A., & Ng, A. (2012). End-to-end text recognition with convolutional neural networks. *Proceedings of the 21st international conference on pattern recognition (ICPR2012)* (pp. 3304–3308).
1038. Wang, C., Wu, J., Xu, L., & Zou, Q. (2020). NonClasGP-Pred: Robust and efficient prediction of non-classically secreted proteins by integrating subset-specific optimal models of imbalanced data. *Microbial Genomics*, *6*.
1039. Wang, X., Yang, K., Jia, T., Gu, F., Wang, C., Xu, K., Shu, Z., Xia, J., Zhu, Q., & Zhou, X. (2024). KDGene: Knowledge graph completion for disease gene prediction using interactional tensor decomposition. *Briefings in Bioinformatics*, *25*, bbae161.
1040. Wang, X., Yu, C., You, Z., Qiao, Y., Li, Z., Huang, W., Zhou, J., & Jin, H. (2023). KS-CMI: A circRNA-miRNA interaction prediction method based on the signed graph neural network and denoising autoencoder. *Iscience*, *26*.

1041. Wang, Y., Zhang, S., Li, F., Zhou, Y., Zhang, Y., Wang, Z., Zhang, R., Zhu, J., Ren, Y., Tan, Y., Others. (2020). Therapeutic target database 2020: Enriched resource for facilitating research and early development of targeted therapeutics. *Nucleic Acids Research*, *48*, D1031–D1041.
1042. Wang, D., Zhang, C., Wang, B., Li, B., Wang, Q., Liu, D., Wang, H., Zhou, Y., Shi, L., Lan, F., Others. (2019). Optimized CRISPR guide RNA design for two high-fidelity Cas9 variants by deep learning. *Nature Communications*, *10*, 4284.
1043. Wang, Y. (2024). EnhancerBD identifing sequence feature. *BioRxiv*. https://www.biorxiv.org/content/early/2024/03/11/2024.03.05.583459.
1044. Wang, X. (2023). *Next-generation sequencing data analysis*. CRC Press.
1045. Wang, H., & Hu, X. (2015). Accurate prediction of nuclear receptors with conjoint triad feature. *BMC Bioinformatics*, *16*, 1–13.
1046. Wang, K., & Liang, C. (2017). CRF: Detection of CRISPR arrays using random forest. *PeerJ*, *5*, Article e3219.
1047. Wang, Y., & Patel, D. (1992). Guanine residues in d (T2AG3) and d (T2G4) form parallel-stranded potassium cation stabilized G-quadruplexes with anti glycosidic torsion angles in solution. *Biochemistry*, *31*, 8112–8119.
1048. Wang, H., & Wu, P. (2018). Prediction of RNA-protein interactions using conjoint triad feature and chaos game representation. *Bioengineered*, *9*, 242–251.
1049. Wang, J., Ma, Q., Shasha, D., & Wu, C. (2001). New techniques for extracting features from protein sequences. *IBM Systems Journal*, *40*, 426–441.
1050. Wang, R., Fang, X., Lu, Y., & Wang, S. (2004). The PDBbind database: Collection of binding affinities for protein- ligand complexes with known three-dimensional structures. *Journal of Medicinal Chemistry*, *47*, 2977–2980.
1051. Wang, G., Li, X., & Wang, Z. (2009). APD2: The updated antimicrobial peptide database and its application in peptide design. *Nucleic Acids Research*, *37*, D933–D937.
1052. Wang, T., Wei, J., Sabatini, D., & Lander, E. (2014). Genetic screens in human cells using the CRISPR-Cas9 system. *Science*, *343*, 80–84.
1053. Wang, X., Wang, Y., Wu, X., Wang, J., Wang, Y., Qiu, Z., Chang, T., Huang, H., Lin, R., & Yee, J. (2015). Unbiased detection of off-target cleavage by CRISPR-Cas9 and TALENs using integrase-defective lentiviral vectors. *Nature Biotechnology*, *33*, 175–178.
1054. Wang, G., Li, X., & Wang, Z. (2016). APD3: The antimicrobial peptide database as a tool for research and education. *Nucleic Acids Research*, *44*, D1087–D1093.
1055. Wang, Y., Yao, H., & Zhao, S. (2016). Auto-encoder based dimensionality reduction. *Neurocomputing*, *184*, 232–242.
1056. Wang, J., Zhang, L., Jia, L., Ren, Y., & Yu, G. (2017). Protein-protein interactions prediction using a novel local conjoint triad descriptor of amino acid sequences. *International Journal of Molecular Sciences*, *18*, 2373.
1057. Wang, W., Han, C., Sun, Y., Chen, T., & Chen, Y. (2019). Noncoding RNAs in cancer therapy resistance and targeted drug development. *Journal of Hematology and Oncology*, *12*, 55.
1058. Wang, J., Zhang, J., Cai, Y., & Deng, L. (2019). Deepmir2go: Inferring functions of human micrornas using a deep multi-label classification model. *International Journal of Molecular Sciences*, *20*, 6046.
1059. Wang, J., Zhang, P., Lu, Y., Li, Y., Zheng, Y., Kan, Y., Chen, R., & He, S. (2019). piRBase: A comprehensive database of piRNA sequences. *Nucleic Acids Research*, *47*, D175–D180.
1060. Wang, J., Zhang, X., Cheng, L., & Luo, Y. (2020). An overview and metanalysis of machine and deep learning-based CRISPR gRNA design tools. *RNA Biology*, *17*, 13–22.
1061. Wang, J., Dai, W., Li, J., Xie, R., Dunstan, R., Stubenrauch, C., Zhang, Y., & Lithgow, T. (2020). PaCRISPR: A server for predicting and visualizing anti-CRISPR proteins. *Nucleic Acids Research*, *48*, W348–W357.
1062. Wang, M., Cui, X., Li, S., Yang, X., Ma, A., Zhang, Y., & Yu, B. (2020). DeepMal: Accurate prediction of protein malonylation sites by deep neural networks. *Chemometrics and Intelligent Laboratory Systems*, *207*, Article 104175.

1063. Wang, J., Dai, W., Li, J., Li, Q., Xie, R., Zhang, Y., Stubenrauch, C., & Lithgow, T. (2021). AcrHub: An integrative hub for investigating, predicting and mapping anti-CRISPR proteins. *Nucleic Acids Research*, *49*, D630–D638.
1064. Wang, N., Zeng, M., Li, Y., Wu, F., & Li, M. (2021). Essential protein prediction based on node2vec and XGBoost. *Journal of Computational Biology*, *28*, 687–700.
1065. Wang, Y., Wang, P., Guo, Y., Huang, S., Chen, Y., & Xu, L. (2021). prPred: A predictor to identify plant resistance proteins by incorporating k-spaced amino acid (group) pairs. *Frontiers in Bioengineering and Biotechnology*, *8*, Article 645520.
1066. Wang, X., Cao, T., Jia, C., Tian, X., & Wang, Y. (2021). Quantitative prediction model for affinity of drug-target interactions based on molecular vibrations and overall system of ligand-receptor. *BMC Bioinformatics*, *22*, 1–18.
1067. Wang, Y., Hou, Z., Yang, Y., Wong, K., & Li, X. (2022). Genome-wide identification and characterization of DNA enhancers with a stacked multivariate fusion framework. *PLOS Computational Biology*, *18*, Article e1010779.
1068. Wang, C., Zheng, C., Wang, H., Zhang, L., Liu, Z., & Xu, P. (2022). The state of the art of PROTAC technologies for drug discovery. *European Journal of Medicinal Chemistry*, *235*, Article 114290.
1069. Wang, H., Zheng, H., & Chen, D. (2022). TANGO: A GO-term embedding based method for protein semantic similarity prediction. *IEEE/ACM Transactions on Computational Biology and Bioinformatics*, *20*, 694–706.
1070. Wang, G., Zhang, X., Pan, Z., Rodrıguez Patón, A., Wang, S., Song, T., & Gu, Y. (2022). Multi-transdti: transformer for drug-target interaction prediction based on simple universal dictionaries with multi-view strategy. *Biomolecules*, *12*, 644.
1071. Wang, L., Wu, M., Wu, Y., Zhang, X., Li, S., He, M., Zhang, F., Wang, Y., & Li, J. (2022). Prediction of the disease causal genes based on heterogeneous network and multi-feature combination method. *Computational Biology and Chemistry*, *97*, Article 107639.
1072. Wang, Z., Xiang, S., Zhou, C., & Xu, Q. (2023). DeepMethylation: A deep learning based framework with GloVe and Transformer encoder for DNA methylation prediction. *PeerJ*, *11*, Article e16125.
1073. Wang, J., Zhou, H., Wang, Y., Xu, M., Yu, Y., Wang, J., & Liu, Y. (2023). Prediction of submitochondrial proteins localization based on gene ontology. *Computers in Biology And Medicine*, *167*, Article 107589.
1074. Wang, R., Zhou, Z., Wu, X., Jiang, X., Zhuo, L., Liu, M., Li, H., Fu, X., & Yao, X. (2023). An effective plant small secretory peptide recognition model based on feature correction strategy. *Journal of Chemical Information and Modeling*, *64*, 2798–2806.
1075. Wang, Z., Gu, Y., Zheng, S., Yang, L., & Li, J. (2023). MGREL: A multi-graph representation learning-based ensemble learning method for gene-disease association prediction. *Computers in Biology and Medicine*, *155*, Article 106642.
1076. Wang, Y., Tai, S., Zhang, S., Sheng, N., & Xie, X. (2023). PromGER: Promoter prediction based on graph embedding and ensemble learning for eukaryotic sequence. *Genes*, *14*, 1441.
1077. Wang, C., Wang, Y., Ding, P., Li, S., Yu, X., & Yu, B. (2024). ML-FGAT: Identification of multi-label protein subcellular localization by interpretable graph attention networks and feature-generative adversarial networks. *Computers in Biology and Medicine*, *170*, Article 107944.
1078. Wang, J., Chen, S., Yuan, Q., Chen, J., Li, D., Wang, L., & Yang, Y. (2024). Predicting the effects of mutations on protein solubility using graph convolution network and protein language model representation. *Journal of Computational Chemistry*, *45*, 436–445.
1079. Wang, T., Chen, H., Li, N., Zhang, B., & Min, H. (2024). Aqueous humor proteomics analyzed by bioinformatics and machine learning in PDR cases versus controls. *Clinical Proteomics*, *21*, 36.
1080. Wasim, M., Asim, M., Ghani, M., Rehman, Z., Rho, S., & Mehmood, I. (2019). Lexical paraphrasing and pseudo relevance feedback for biomedical document retrieval. *Multimedia Tools and Applications*, *78*, 29681–29712.

1081. Watkins, L., & Maier, S. (2005). Immune regulation of central nervous system functions: from sickness responses to pathological pain. *Journal of Internal Medicine*, *257*, 139–155.
1082. Watson, J., & Crick, F. (1953). The structure of DNA. *Cold Spring Harbor Symposia on Quantitative Biology*, *18*, 123–131.
1083. Wayment-Steele, H., Kladwang, W., Strom, A., Lee, J., Treuille, A., Becka, A., Participants, E., & Das, R. (2022). RNA secondary structure packages evaluated and improved by high-throughput experiments. *Nature Methods*, *19*, 1234–1242.
1084. Wei, L., Xing, P., Shi, G., Ji, Z., & Zou, Q. (2017). Fast prediction of protein methylation sites using a sequence-based feature selection technique. *IEEE/ACM Transactions on Computational Biology and Bioinformatics*, *16*, 1264–1273.
1085. Wei, L., Tang, J., & Zou, Q. (2017). SkipCPP-Pred: An improved and promising sequence-based predictor for predicting cell-penetrating peptides. *BMC Genomics*, *18*, 1–11.
1086. Wei, L., Zhou, C., Chen, H., Song, J., & Su, R. (2018). ACPred-FL: A sequence-based predictor using effective feature representation to improve the prediction of anti-cancer peptides. *Bioinformatics*, *34*, 4007–4016.
1087. Wei, M., Yu, C., Li, L., You, Z., Ren, Z., Guan, Y., Wang, X., & Li, Y. (2023). LPIH2V: LncRNA-protein interactions prediction using HIN2Vec based on heterogeneous networks model. *Frontiers in Genetics*, *14*, 1122909.
1088. Weiss, S., Xu, Z., Peddada, S., Amir, A., Bittinger, K., Gonzalez, A., Lozupone, C., Zaneveld, J., Vázquez-Baeza, Y., Birmingham, A., Others. (2017). Normalization and microbial differential abundance strategies depend upon data characteristics. *Microbiome*, *5*, 27. PUBMED.
1089. Werner, J., Koren, O., Hugenholtz, P., DeSantis, T., Walters, W., Caporaso, J., Angenent, L., Knight, R., & Ley, R. (2011). Impact of training sets on classification of high-throughput bacterial 16s rrna gene surveys. *The Isme Journal*, *6*, 94–103.
1090. Wert, S., Whitsett, J., & Nogee, L. (2009). Genetic disorders of surfactant dysfunction. *Pediatric and Developmental Pathology*, *12*, 253–274.
1091. Westhof, E., & Auffinger, P. (2000). RNA tertiary structure. *Encyclopedia of Analytical Chemistry* (pp. 5222–5232). John Wiley Sons Ltd.
1092. Whissell, J., & Clarke, C. (2011). Improving document clustering using Okapi BM25 feature weighting. *Information Retrieval*, *14*, 466–487.
1093. White, G., & Seffens, W. (1998). Using a neural network to backtranslate amino acid sequences. *Electronic Journal of Biotechnology*, *1*, 17–18.
1094. Whitford, D. (2013). *Proteins: Structure and function*. John Wiley & Sons.
1095. Wilhelm, A., Zabalawi, M., Owen, J., Shah, D., Grayson, J., Major, A., Bhat, S., Gibbs, D., Thomas, M., & Sorci-Thomas, M. (2010). Apolipoprotein AI modulates regulatory T cells in autoimmune LDLr-/-,ApoA-I-/- mice. *Journal of Biological Chemistry*, *285*, 36158–36169.
1096. Williamson, J. (1994). G-quartet structures in telomeric DNA. *Annual Review of Biophysics and Biomolecular Structure*, *23*, 703–730.
1097. Wishart, D., Knox, C., Guo, A., Cheng, D., Shrivastava, S., Tzur, D., Gautam, B., & Hassanali, M. (2008). DrugBank: A knowledgebase for drugs, drug actions and drug targets. *Nucleic Acids Research*, *36*, D901–D906.
1098. Wold, S., Jonsson, J., Sjöström, M., Sandberg, M., & Rännar, S. (1993). DNA and peptide sequences and chemical processes multivariately modelled by principal component analysis and partial least-squares projections to latent structures. *Analytica Chimica Acta*, *277*, 239–253.
1099. Wolfram Saenger, W., & Verlag, S. (1984). *Principles of Nucleic Acid Structure. Chapter*, *7*, 159.
1100. Wong, K., Zhang, J., Yan, S., Li, X., Lin, Q., Kwong, S., & Liang, C. (2019). DNA sequencing technologies: Sequencing data protocols and bioinformatics tools. *ACM Computing Surveys (CSUR)*, *52*, 1–30.
1101. Wu, K., Zhou, D., Slonim, D., Hu, X., & Cowen, L. (2023). Melissa: Semi-supervised embedding for protein function prediction across multiple networks. *BioRxiv*.

1102. Wu, J., Vallenius, T., Ovaska, K., Westermarck, J., Mäkelä, T., & Hautaniemi, S. (2009). Integrated network analysis platform for protein-protein interactions. *Nature Methods, 6*, 75–77.
1103. Wu, S., Li, Q., Yin, C., Xue, W., & Song, C. (2020). Advances in CRISPR/Cas-based gene therapy in human genetic diseases. *Theranostics, 10*, 4374.
1104. Wu, H., Liu, M., Zhang, P., & Zhang, H. (2023). iEnhancer-SKNN: A stacking ensemble learning-based method for enhancer identification and classification using sequence information. *Briefings in Functional Genomics, 22*, 302–311.
1105. Wuerer, J., Gran, M., & Held, T. (1994). Analytical Design Package (ADP2): A computer aided engineering tool for aircraft transparency design. *NASA, Washington, Technology 2003: The fourth national technology transfer conference and exposition* (Vol. 2).
1106. Wyatt, J., Vickers, T., Roberson, J., Buckheit, Jr., R., Klimkait, T., DeBaets, E., Davis, P., Rayner, B., Imbach, J., & Ecker, D. (1994). Combinatorially selected guanosine-quartet structure is a potent inhibitor of human immunodeficiency virus envelope-mediated cell fusion. *Proceedings of the National Academy of Sciences, 91*, 1356–1360.
1107. Xenarios, I., Salwinski, L., Duan, X., Higney, P., Kim, S., & Eisenberg, D. (2002). DIP, the Database of Interacting Proteins: A research tool for studying cellular networks of protein interactions. *Nucleic Acids Research, 30*, 303–305.
1108. Xia, S., Feng, J., Chen, K., Ma, Y., Gong, J., Cai, F., Jin, Y., Gao, Y., Xia, L., Chang, H., Others. (2018). CSCD: A database for cancer-specific circular RNAs. *Nucleic Acids Research, 46*, D925–D929.
1109. Xia, J., Han, K., & Huang, D. (2010). Sequence-based prediction of protein-protein interactions by means of rotation forest and autocorrelation descriptor. *Protein and Peptide Letters, 17*, 137–145.
1110. Xia, S., Xia, Y., Xiang, C., Wang, H., Wang, C., He, J., Shi, G., & Gu, L. (2022). A virus-target host proteins recognition method based on integrated complexes data and seed extension. *BMC Bioinformatics, 23*, 256.
1111. Xia, L., Xu, L., Pan, S., Niu, D., Zhang, B., & Li, Z. (2023). Drug-target binding affinity prediction using message passing neural network and self supervised learning. *BMC Genomics, 24*, 557.
1112. Xiang, X., Corsi, G., Anthon, C., Qu, K., Pan, X., Liang, X., Han, P., Dong, Z., Liu, L., Zhong, J., Others. (2021). Enhancing CRISPR-Cas9 gRNA efficiency prediction by data integration and deep learning. *Nature Communications, 12*, 3238.
1113. Xiao, X., Shao, Y., Cheng, X., & Stamatovic, B. (2021). iAMP-CA2L: A new CNN-BiLSTM-SVM classifier based on cellular automata image for identifying antimicrobial peptides and their functional types. *Briefings in Bioinformatics, 22*, bbab209.
1114. Xiao, L., Wan, Y., & Jiang, Z. (2021). AttCRISPR: A spacetime interpretable model for prediction of sgRNA on-target activity. *BMC Bioinformatics, 22*, 589.
1115. Xiao, W., Zhang, X., & Xiao, W. (2020). A deep learning framework for predicting human essential genes by integrating sequence and functional data. *BioRxiv*.
1116. Xiao, X., Shao, S., Huang, Z., & Chou, K. (2006). Using pseudo amino acid composition to predict protein structural classes: approached with complexity measure factor. *Journal of Computational Chemistry, 27*, 478–482.
1117. Xie, B., Ding, Q., Han, H., & Wu, D. (2013). miRCancer: A microRNA-cancer association database constructed by text mining on literature. *Bioinformatics, 29*, 638–644.
1118. Xie, F., Yang, Z., Song, J., Dai, Q., & Duan, X. (2021). DHNLDA: A novel deep hierarchical network based method for predicting lncRNA-disease associations. *IEEE/ACM Transactions on Computational Biology and Bioinformatics, 19*, 3395–3403.
1119. Xie, X., Yu, T., Li, X., Zhang, N., Foster, L., Peng, C., Huang, W., & He, G. (2023). Recent advances in targeting the "undruggable" proteins: from drug discovery to clinical trials. *Signal Transduction and Targeted Therapy, 8*, 335.
1120. Xie, P., Zhuang, J., Tian, G., & Yang, J. (2023). Emvirus: An embedding-based neural framework for human-virus protein-protein interactions prediction. *Biosafety and Health, 5*, 152–158.

1121. Xie, H., Ding, Y., Qian, Y., Tiwari, P., & Guo, F. (2024). Structured sparse regularization based random vector functional link networks for DNA N4-methylcytosine sites prediction. *Expert Systems with Applications*, *235*, Article 121157.
1122. Xu, Y., Xiaohan, Z., Shuai, L., & Wen, Z. (2020). Predicting long non-coding RNAs through feature ensemble learning. *BMC Genomics 21*(13).
1123. Xu, C., Chambers, J., Nicolai, H., Brown, M., Hujeirat, Y., Mohammed, S., Hodgson, S., Kelsell, D., Spurr, N., Bishop, D., Others. (1997). Mutations and alternative splicing of the BRCA1 gene in UK breast/ovarian cancer families. *Genes, Chromosomes And Cancer*, *18*, 102–110.
1124. Xu, H., Jia, P., & Zhao, Z. (2021). Deep4mC: Systematic assessment and computational prediction for DNA N4-methylcytosine sites by deep learning. *Briefings in Bioinformatics*, *22*, bbaa099.
1125. Xu, J., Li, F., Li, C., Guo, X., Landersdorfer, C., Shen, H., Peleg, A., Li, J., Imoto, S., Yao, J., Others. (2023). iAMPCN: A deep-learning approach for identifying antimicrobial peptides and their functional activities. *Briefings in Bioinformatics*, *24*, bbad240.
1126. Xu, K., Li, C., Tian, Y., Sonobe, T., Kawarabayashi, K., & Jegelka, S. (2018). Representation learning on graphs with jumping knowledge networks. *International conference on machine learning* (pp. 5453–5462).
1127. Xu, B., Wang, N., Chen, T., & Li, M. (2015). Empirical evaluation of rectified activations in convolutional network. Preprint. ArXiv:1505.00853.
1128. Xu, Y., Wang, C., Xu, K., Ding, Y., Lyu, A., & Zhang, L. (2023). TRAFICA: Improving transcription factor binding affinity prediction using deep language model on ATAC-seq data.
1129. Xu, H., Xiao, T., Chen, C., Li, W., Meyer, C., Wu, Q., Wu, D., Cong, L., Zhang, F., Liu, J., Others. (2015). Sequence determinants of improved CRISPR sgRNA design. *Genome Research*, *25*, 1147–1157.
1130. Xu, J., Xu, N., Xie, W., Zhao, C., Yu, L., & Feng, W. (2024). BERT-siRNA: siRNA target prediction based on BERT pre-trained interpretable model. *Gene*, 148330.
1131. Xu, Z., Zhong, H., He, B., Wang, X., & Lu, T. (2024). PTransIPs: Identification of phosphorylation sites enhanced by protein PLM embeddings. *IEEE Journal of Biomedical and Health Informatics, 28,* 3762–3771.
1132. Xu, J. (2024). Scmformer integrates large-scale single-cell proteomics and transcriptomics data by multi-task transformer. *Advanced Science*, *11*.
1133. Xu, S., & Onoda, A. (2023). Accurate and fast prediction of intrinsically disordered protein by multiple protein language models and ensemble learning. *Journal of Chemical Information and Modeling*, *64*, 2901–2911.
1134. Xu, Y., Wen, X., Wen, L., Wu, L., Deng, N., & Chou, K. (2014). iNitro-Tyr: Prediction of nitrotyrosine sites in proteins with general pseudo amino acid composition. *PloS One*, *9*, Article e105018.
1135. Xu, W., Dong, Y., Guan, J., & Zhou, S. (2022). Identifying essential proteins from protein-protein interaction networks based on influence maximization. *BMC Bioinformatics*, *23*, 339.
1136. Xu, Z., Wang, X., Meng, J., Zhang, L., & Song, B. (2023). m5U-GEPred: prediction of RNA 5-methyluridine sites based on sequence-derived and graph embedding features. *Frontiers in Microbiology*, *14*, 1277099.
1137. Xu, J., Sun, W., Li, K., Zhang, W., Zhang, W., Zeng, Y., Wong, L., & Zhang, P. (2024). MNESEDA: A prior-guided subgraph representation learning framework for predicting disease-related enhancers. *Knowledge-Based Systems*, *294*, Article 111734.
1138. Xuan, J., Sun, W., Lin, P., Zhou, K., Liu, S., Zheng, L., Qu, L., & Yang, J. RMBase v2. 0: Deciphering the map of RNA modifications from epitranscriptome sequencing data. *Nucleic Acids Research*, *46*, D327-D334 (2018)
1139. Xuan, P., Zhang, X., Zhang, Y., Hu, K., Nakaguchi, T., & Zhang, T. (2022). Multi-type neighbors enhanced global topology and pairwise attribute learning for drug–protein interaction prediction. *Briefings In Bioinformatics*, *23*, bbac120.

1140. Xue, L., Tang, B., Chen, W., & Luo, J. (2018). Prediction of CRISPR sgRNA activity using a deep convolutional neural network. *Journal of Chemical Information and Modeling, 59*, 615–624.
1141. Yadav, A., & Shakya, D. (2016). *Digital filtering approach for prediction of hotspot locations using nucleotide sequences of protein*. IET.
1142. Yaish, O., Asif, M., & Orenstein, Y. (2022). A systematic evaluation of data processing and problem formulation of CRISPR off-target site prediction. *Briefings in Bioinformatics, 23*, bbac157.
1143. Yamada, Y., Kakikawa, M., Kubo, M., Purnama, B., Delimayanti, M., Nguyen, N., Mahmudah, K., Shimaguchi, Y., & Satou, K. (2019). Prediction of subnuclear location for nuclear protein. *BIOINFORMATICS* (pp. 276–280).
1144. Yamada, H. (2021). *Geary'sc and spectral graph theory. Mathematics, 9*, 2465.
1145. Yan, J., Zhang, B., Zhou, M., Kwok, H., & Siu, S. (2022). Multi-Branch-CNN: Classification of ion channel interacting peptides using multi-branch convolutional neural network. *Computers in Biology and Medicine, 147*, Article 105717.
1146. Yan, Y., Li, W., Wang, S., & Huang, T. (2024). Seq-RBPPred: Predicting RNA-binding proteins from sequence. *ACS Omega, 9*, 12734–12742.
1147. Yang, M., Huang, H., Huang, L., Zhang, N., Wu, J., Yang, H., & Mu, F. (2021). LOGO, a contextualized pre-trained language model of human genome flexibly adapts to various downstream tasks by fine-tuning.
1148. Yang, B., Khatri, M., Zheng, J., Deogun, J., & Yin, Y. (2023). Genome mining for anti-CRISPR operons using machine learning. *Bioinformatics (Oxford, England), 39*.
1149. Yang, G., Li, J., Hu, J., & Shi, J. (2024). Recognition of cyanobacteria promoters via Siamese network-based contrastive learning under novel non-promoter generation. *Briefings in Bioinformatics, 25*, bbae193.
1150. Yang, Y., Li, J., Zou, Q., Ruan, Y., & Feng, H. (2023). Prediction of CRISPR-Cas9 off-target activities with mismatches and indels based on hybrid neural network. *Computational and Structural Biotechnology Journal, 21*, 5039–5048.
1151. Yang, Z., Ren, F., Liu, C., He, S., Sun, G., Gao, Q., Yao, L., Zhang, Y., Miao, R., Cao, Y., Others. (2010). dbDEMC: A database of differentially expressed miRNAs in human cancers. *BMC Genomics, 11*, 1–8 .
1152. Yang, Q., Wu, L., Meng, J., Ma, L., Zuo, E., & Sun, Y. (2022). EpiCas-DL: Predicting sgRNA activity for CRISPR-mediated epigenome editing by deep learning. *Computational and Structural Biotechnology Journal, 21*, 202–211.
1153. Yang, X., Wuchty, S., Liang, Z., Ji, L., Wang, B., Zhu, J., Zhang, Z., & Dong, Y. (2024). Multi-modal features-based human-herpesvirus protein–protein interaction prediction by using LightGBM. *Briefings in Bioinformatics, 25*, bbae005.
1154. Yang, J., Roy, A., & Zhang, Y. (2012). BioLiP: A semi-manually curated database for biologically relevant ligand-protein interactions. *Nucleic Acids Research, 41*, D1096–D1103.
1155. Yang, A., Zhang, W., Wang, J., Yang, K., Han, Y., & Zhang, L. (2020). Review on the application of machine learning algorithms in the sequence data mining of DNA. *Frontiers in Bioengineering and Biotechnology, 8*, 1032.
1156. Yang, R., Wu, F., Zhang, C., & Zhang, L. (2021). iEnhancer-GAN: A deep learning framework in combination with word embedding and sequence generative adversarial net to identify enhancers and their strength. *International Journal of Molecular Sciences, 22*, 3589.
1157. Yang, S., Huang, J., & He, B. (2021). CASPredict: A web service for identifying Cas proteins. *PeerJ, 9*, Article e11887.
1158. Yang, B., Zheng, J., & Yin, Y. (2022). AcaFinder: Genome mining for anti-CRISPR-associated genes. *Msystems, 7*, e00817-22.
1159. Yang, M., Huang, L., Huang, H., Tang, H., Zhang, N., Yang, H., Wu, J., & Mu, F. (2022). Integrating convolution and self-attention improves language model of human genome for interpreting non-coding regions at base-resolution. *Nucleic Acids Research, 50*, e81–e81.

1160. Yang, S., Yang, Z., & Yang, J. (2023). 4mCBERT: A computing tool for the identification of DNA N4-methylcytosine sites by sequence-and chemical-derived information based on ensemble learning strategies. *International Journal of Biological Macromolecules*, *231*, Article 123180.
1161. Yates, A., Achuthan, P., Akanni, W., Allen, J., Allen, J., Alvarez-Jarreta, J., Amode, M., Armean, I., Azov, A., Bennett, R., Others. (2020). Ensembl 2020. *Nucleic Acids Research*, *48*, D682–D688.
1162. Ye, G., Wu, H., Huang, J., Wang, W., Ge, K., Li, G., Zhong, J., & Huang, Q. (2020). LAMP2: A major update of the database linking antimicrobial peptides. *Database*, *2020*, baaa061.
1163. Ye, C., Wu, Q., Chen, S., Zhang, X., Xu, W., Wu, Y., Zhang, Y., & Yue, Y. (2024). ECDEP: Identifying essential proteins based on evolutionary community discovery and subcellular localization. *BMC Genomics*, *25*, 117.
1164. Yi, H., Huang, L., Yang, B., Gomez, J., Zhang, H., & Yin, Y. (2020). AcrFinder: Genome mining anti-CRISPR operons in prokaryotes and their viruses. *Nucleic Acids Research*, *48*, W358–W365.
1165. Yılmaz, A. (2020). Assessment of mutation susceptibility in DNA sequences with word vectors. *Journal of Intelligent Systems: Theory and Applications*, *3*, 1–6.
1166. Yin, Z., Lai, F., & Gao, F. (2024). Unveiling human origins of replication using deep learning: Accurate prediction and comprehensive analysis. *Briefings in Bioinformatics*, *25*, bbad432.
1167. Yoneda, Y. (1997). How proteins are transported from cytoplasm to the nucleus. *The Journal of Biochemistry*, *121*, 811–817.
1168. Yoo, P., Zhou, B., & Zomaya, A. (2008). Machine learning techniques for protein secondary structure prediction: An overview and evaluation. *Current Bioinformatics*, *3*, 74–86.
1169. Youmans, M., Spainhour, J., & Qiu, P. (2019). Classification of antibacterial peptides using long short-term memory recurrent neural networks. *IEEE/ACM Transactions on Computational Biology and Bioinformatics*, *17*, 1134–1140.
1170. Yu, L., & Marchisio, M. (2020). Types I and V anti-CRISPR proteins: From phage defense to eukaryotic synthetic gene circuits. *Frontiers in Bioengineering and Biotechnology*, *8*, 575393.
1171. Yu, L., & Marchisio, M. (2020). Types I and V anti-CRISPR proteins: From phage defense to eukaryotic synthetic gene circuits. *Frontiers in Bioengineering and Biotechnology*, *8*, Article 575393.
1172. Yu, B., Yu, Z., Chen, C., Ma, A., Liu, B., Tian, B., & Ma, Q. (2020). DNNAce: Prediction of prokaryote lysine acetylation sites through deep neural networks with multi-information fusion. *Chemometrics and Intelligent Laboratory Systems*, *200*, Article 103999.
1173. Yu, T., Chen, M., & Wang, C. (2020). An improved method for identification of pre-mirna in Drosophila. *IEEE Access*, *8*, 52173–52180.
1174. Yu, Y., He, W., Jin, J., Xiao, G., Cui, L., Zeng, R., & Wei, L. (2021). iDNA-ABT: Advanced deep learning model for detecting DNA methylation with adaptive features and transductive information maximization. *Bioinformatics*, *37*, 4603–4610.
1175. Yu, D., Yu, Z., Han, G., Li, J., & Anh, V. (2021). Heterogeneous types of miRNA-disease associations stratified by multi-layer network embedding and prediction. *Biomedicines*, *9*, 1152.
1176. Yu, B., Zhang, Y., Wang, X., Gao, H., Sun, J., & Gao, X. (2022). Identification of DNA modification sites based on elastic net and bidirectional gated recurrent unit with convolutional neural network. *Biomedical Signal Processing and Control*, *75*, Article 103566.
1177. Yu, D., Liu, H., & Yao, S. (2024). Drug-target interaction prediction based on improved heterogeneous graph representation learning and feature projection classification. *Expert Systems with Applications*, *252*, Article 124289.
1178. Yuan, M., Chen, L., & Deng, M. (2022). Clustering single-cell multi-omics data with moclust. *Bioinformatics*, *39*.

1179. Yuan, Q., Tian, C., Song, Y., Ou, P., Zhu, M., Zhao, H., & Yang, Y. (2024). GPSFun: Geometry-aware protein sequence function predictions with language models. *Nucleic Acids Research*, gkae381.
1180. Yuan, G., Wang, Y., Wang, G., & Yang, L. (2023). RNAlight: A machine learning model to identify nucleotide features determining RNA subcellular localization. *Briefings in Bioinformatics*, *24*, bbac509.
1181. Yusuf, S., Zhang, F., Zeng, M., & Li, M. (2021). DeepPPF: A deep learning framework for predicting protein family. *Neurocomputing*, *428*, 19–29.
1182. Zambryski, P. (2004). Cell-to-cell transport of proteins and fluorescent tracers via plasmodesmata during plant development. *The Journal of Cell Biology*, *164*, 165.
1183. Zarate, O., Yang, Y., Wang, X., & Wang, J. (2022). BoostMEC: Predicting CRISPR-Cas9 cleavage efficiency through boosting models. *BMC Bioinformatics*, *23*, 446.
1184. Zdrazil, B., Felix, E., Hunter, F., Manners, E., Blackshaw, J., Corbett, S., Veij, M., Ioannidis, H., Lopez, D., Mosquera, J., Others. (2024). The ChEMBL database in 2023: A drug discovery platform spanning multiple bioactivity data types and time periods. *Nucleic Acids Research*, *52*, D1180–D1192.
1185. Zeller, M., Gauger, P., Arendsee, Z., Souza, C., Vincent, A., & Anderson, T. (2021). Machine learning prediction and experimental validation of antigenic drift in H3 influenza A viruses in swine. *MSphere*, *6*, e00920-20.
1186. Zeng, W., Gautam, A., & Huson, D. (2023). MuLan-Methyl—multiple transformer-based language models for accurate DNA methylation prediction. *GigaScience*, *12*, giad054.
1187. Zeng, M., Wu, Y., Li, Y., Yin, R., Lu, C., Duan, J., & Li, M. (2023). LncLocFormer: A transformer-based deep learning model for multi-label lncRNA subcellular localization prediction by using localization-specific attention mechanism. *Bioinformatics*, *39*, btad752.
1188. Zeng, R., & Liao, M. (2020). Developing a multi-layer deep learning based predictive model to identify DNA N4-methylcytosine modifications. *Frontiers in Bioengineering and Biotechnology*, *8*, 274.
1189. Zeng, W., Rao, N., Zheng, J., Wan, Y., Wang, G., Li, Z., & Li, S. (2017). Identification of candidate targeted genes in molecular subtypes of gastric cancer. *Journal of Biomedical Science and Engineering*, *10*, 45–53.
1190. Zhang, F., Chang, S., Wang, B., & Zhang, X. (2024). DSSGNN-PPI: A protein-protein interactions prediction model based on double structure and sequence graph neural networks. *Computers in Biology and Medicine*, 108669.
1191. Zhang, W., Cheng, B., & Xu, B. (2017). Application of next-generation sequencing technology in forensic science. *Chinese Journal of Forensic Medicine*, 40–43.
1192. Zhang, G., Dai, Z., & Dai, X. (2020). A novel hybrid CNN-SVR for CRISPR/Cas9 guide RNA activity prediction. *Frontiers in Genetics*, *10*, 1303.
1193. Zhang, X., Guo, H., Zhang, F., Wang, X., Wu, K., Qiu, S., Liu, B., Wang, Y., Hu, Y., & Li, J. (2023). HNetGO: Protein function prediction via heterogeneous network transformer. *Briefings in Bioinformatics*, *24*, bbab556.
1194. Zhang, T., Jia, Y., Li, H., Xu, D., Zhou, J., & Wang, G. (2022). CRISPRCasStack: A stacking strategy-based ensemble learning framework for accurate identification of Cas proteins. *Briefings in Bioinformatics*, *23*.
1195. Zhang, Z., Lamson, A., Shelley, M., & Troyanskaya, O. (2023). Interpretable neural architecture search and transfer learning for understanding CRISPR/Cas9 off-target enzymatic reactions. https://lens.org/136-526-001-641-407.
1196. Zhang, T., Li, L., Sun, H., & Wang, G. (2023). DeepITEH: A deep learning framework for identifying tissue-specific eRNAs from the human genome. *Bioinformatics*, *39*, btad375.
1197. Zhang, G., Liu, Z., Dai, J., Yu, Z., Liu, S., & Zhang, W. (2019). ItLnc-BXE: A Bagging-XGBoost-ensemble method with multiple features for identification of plant lncRNAs. Preprint. ArXiv:1911.00185.
1198. Zhang, G., Luo, Y., Dai, X., & Dai, Z. (2023). Benchmarking deep learning methods for predicting CRISPR/Cas9 sgRNA on-and off-target activities. *Briefings in Bioinformatics*, *24*, bbad333.

1199. Zhang, X., Wang, S., Xie, L., & Zhu, Y. (2023). PseU-ST: A new stacked ensemble-learning method for identifying RNA pseudouridine sites. *Frontiers In Genetics*, *14*.
1200. Zhang, X., Wei, L., Ye, X., Zhang, K., Teng, S., Li, Z., Jin, J., Kim, M., Sakurai, T., Cui, L., Others. (2023). SiameseCPP: A sequence-based Siamese network to predict cell-penetrating peptides by contrastive learning. *Briefings in Bioinformatics*, *24*, bbac545.
1201. Zhang, X., Wu, F., Yang, N., Zhan, X., Liao, J., Mai, S., & Huang, Z. (2022). In silico methods for identification of potential therapeutic targets. *Interdisciplinary Sciences: Computational Life Sciences* (pp. 1–26).
1202. Zhang, H., Yan, J., Lu, Z., Zhou, Y., Zhang, Q., Cui, T., Li, Y., Chen, H., & Ma, L. (2023). Deep sampling of gRNA in the human genome and deep-learning-informed prediction of gRNA activities. *Cell Discovery*, *9*, 48.
1203. Zhang, Y., Yu, S., Xie, R., Li, J., Leier, A., Marquez-Lago, T., Akutsu, T., Smith, A., Ge, Z., Wang, J., Others. (2020). PeNGaRoo, a combined gradient boosting and ensemble learning framework for predicting non-classical secreted proteins. *Bioinformatics*, *36*, 704–712.
1204. Zhang, G., Zeng, T., Dai, Z., & Dai, X. (2021). Prediction of CRISPR/Cas9 single guide RNA cleavage efficiency and specificity by attention-based convolutional neural networks. *Computational and Structural Biotechnology Journal*, *19*, 1445–1457.
1205. Zhang, R., & Zhang, C. (2014). A brief review: The z-curve theory and its application in genome analysis. *Current Genomics*, *15*, 78–94.
1206. Zhang, R., Ou, H., & Zhang, C. (2004). DEG: A database of essential genes. *Nucleic Acids Research*, *32*, D271–D272.
1207. Zhang, S., Hu, H., Jiang, T., Zhang, L., & Zeng, J. (2017). TITER: Predicting translation initiation sites by deep learning. *Bioinformatics*, *33*, i234–i242.
1208. Zhang, L., Yu, G., Xia, D., & Wang, J. (2019). Protein-protein interactions prediction based on ensemble deep neural networks. *Neurocomputing*, *324*, 10–19.
1209. Zhang, F., Song, G., & Tian, Y. (2019). Anti-CRISPRs: The natural inhibitors for CRISPR-Cas systems. *Animal Models and Experimental Medicine*, *2*, 69–75.
1210. Zhang, S., Wang, Y., Zhang, X., & Wang, J. (2019). Prediction of the RBP binding sites on lncRNAs using the high-order nucleotide encoding convolutional neural network. *Analytical Biochemistry*, *583*, Article 113364.
1211. Zhang, X., Xiao, W., & Xiao, W. (2020). DeepHE: Accurately predicting human essential genes based on deep learning. *PLOS Computational Biology*, *16*, Article e1008229.
1212. Zhang, M., Hu, Y., & Zhu, M. (2021). Epishilbert: Prediction of enhancer-promoter interactions via hilbert curve encoding and transfer learning. *Genes*, *12*, 1385.
1213. Zhang, W., Yin, J., Zhang-Ding, Z., Xin, C., Liu, M., Wang, Y., Ai, C., & Hu, J. (2021). In-depth assessment of the PAM compatibility and editing activities of Cas9 variants. *Nucleic Acids Research*, *49*, 8785–8795.
1214. Zhang, Y., Chu, X., Jiang, Y., Wu, H., & Quan, L. (2022). SemanticCAP: Chromatin accessibility prediction enhanced by features learning from a language model. *Genes*, *13*, 568.
1215. Zhang, P., Zhang, H., & Wu, H. (2022). iPro-WAEL: A comprehensive and robust framework for identifying promoters in multiple species. *Nucleic Acids Research*, *50*, 10278–10289.
1216. Zhang, Z., Li, F., Zhao, J., & Zheng, C. (2023). CapsNetYY1: Identifying YY1-mediated chromatin loops based on a capsule network architecture. *BMC Genomics*, *24*, 448.
1217. Zhang, T., Jia, J., Chen, C., Zhang, Y., & Yu, B. (2023). BiGRUD-SA: Protein S-sulfenylation sites prediction based on BiGRU and self-attention. *Computers in Biology and Medicine*, *163*, Article 107145.
1218. Zhang, C., Zhang, X., Freddolino, P., & Zhang, Y. (2024). BioLiP2: An updated structure database for biologically relevant ligand-protein interactions. *Nucleic Acids Research*, *52*, D404–D412.
1219. Zhang, Z., Zhang, Z., Ye, X., Sakurai, T., & Lin, H. (2024). A BERT-based model for the prediction of lncRNA subcellular localization in Homo sapiens. *International Journal of Biological Macromolecules*, *265*, Article 130659.

1220. Zhang, J., Zhu, H., Liu, Y., & Li, X. (2024). miTDS: Uncovering miRNA-mRNA interactions with deep learning for functional target prediction. *Methods, 223*, 65–74.
1221. Zhao, S., Pan, Q., Zou, Q., Ju, Y., Shi, L., Su, X., Others. (2022). Identifying and classifying enhancers by dinucleotide-based auto-cross covariance and attention-based Bi-LSTM. *Computational and Mathematical Methods in Medicine, 2022*.
1222. Zhao, W., Liang, G., Chen, Y., & Yang, L. (2011). A new quantitative structure-retention relationship model for predicting chromatographic retention time of oligonucleotides. *Science China Chemistry, 54*, 1064–1071.
1223. Zhao, X., Zhang, W., Xu, X., Ma, Z., & Yin, M. (2012). *Prediction of protein phosphorylation sites by using the composition of k-spaced amino acid pairs*. Public Library of Science San Francisco.
1224. Zhao, X., Wu, H., Lu, H., Li, G., & Huang, Q. (2013). LAMP: A database linking antimicrobial peptides. *PloS One, 8*, Article e66557.
1225. Zhao, J., Jiang, H., Zou, G., Lin, Q., Wang, Q., Liu, J., & Ma, L. (2022). CNNArginineMe: A CNN structure for training models for predicting arginine methylation sites based on the One-Hot encoding of peptide sequence. *Frontiers in Genetics, 13*, 1036862.
1226. Zhao, Z., Lin, J., Wang, Z., Guo, J., Zhan, X., Huang, Y., Shi, C., & Huang, W. (2023). SEBGLMA: Semantic embedded bipartite graph network for predicting lncRNA-miRNA associations. *International Journal of Intelligent Systems, 2023*, 2785436.
1227. Zheng, J., Yang, X., Huang, Y., Yang, S., Wuchty, S., & Zhang, Z. (2023). Deep learning-assisted prediction of protein-protein interactions in Arabidopsis thaliana. *The Plant Journal, 114*, 984–994.
1228. Zhou, B., Ji, B., Liu, K., Hu, G., Wang, F., Chen, Q., Yu, R., Huang, P., Ren, J., Guo, C., Others. (2021). EVLncRNAs 2.0: An updated database of manually curated functional long non-coding RNAs validated by low-throughput experiments. *Nucleic Acids Research, 49*, D86–D91.
1229. Zhou, Z., Liao, Q., Wei, J., Zhuo, L., Wu, X., Fu, X., & Zou, Q. (2024). Revisiting drug–protein interaction prediction: A novel global–local perspective. *Bioinformatics, 40*, btae271.
1230. Zhou, C., Peng, D., Liao, B., Jia, R., & Wu, F. (2022). ACP_MS: Prediction of anticancer peptides based on feature extraction. *Briefings in Bioinformatics, 23*, bbac462.
1231. Zhou, G., Soufan, O., Ewald, J., Hancock, R., Basu, N., & Xia, J. (2019). Networkanalyst 3.0: A visual analytics platform for comprehensive gene expression profiling and meta-analysis. *Nucleic Acids Research, 47*, W234–W241.
1232. Zhou, C., Wang, C., Liu, H., Zhou, Q., Liu, Q., Guo, Y., Peng, T., Song, J., Zhang, J., Chen, L., Others. (2018). Identification and analysis of adenine N6-methylation sites in the rice genome. *Nature Plants, 4*, 554–563.
1233. Zhou, Z., Xiao, C., Yin, J., She, J., Duan, H., Liu, C., Fu, X., Cui, F., Qi, Q., & Zhang, Z. (2024). PSAC-6mA: 6mA site identifier using self-attention capsule network based on sequence-positioning. *Computers in Biology and Medicine* (p. 108129). Elsevier.
1234. Zhou, J., You, Z., Cheng, L., & Ji, B. (2021). Prediction of lncRNA-disease associations via an embedding learning HOPE in heterogeneous information networks. *Molecular Therapy-Nucleic Acids, 23*, 277–285.
1235. Zhou, L., Wang, Z., Tian, X., & Peng, L. (2021). LPI-deepGBDT: A multiple-layer deep framework based on gradient boosting decision trees for lncRNA-protein interaction identification. *BMC Bioinformatics, 22*, 1–24.
1236. Zhou, F., Gan, R., Zhang, F., Ren, C., Yu, L., Si, Y., & Huang, Z. (2022). PHISDetector: A tool to detect diverse in silico phage-host interaction signals for virome studies. *Genomics, Proteomics and Bioinformatics, 20*, 508–523.
1237. Zhou, L., Peng, X., Zeng, L., & Peng, L. (2024). Finding potential lncRNA-disease associations using a boosting-based ensemble learning model. *Frontiers in Genetics, 15*, 1356205.
1238. Zhu, L., Wang, X., Li, F., & Song, J. (2022). PreAcrs: A machine learning framework for identifying anti-CRISPR proteins. *BMC Bioinformatics, 23*, 444.

1239. Zhu, J. Movicshiny: an interactive website for multi-omics integration and visualisation in cancer subtyping. *Clinical and Translational Medicine*, *14* (2024)
1240. Zhu, J., Zheng, Z., Yang, M., Fung, G., & Huang, C. (2019). Protein complexes detection based on semi-supervised network embedding model. *IEEE/ACM Transactions on Computational Biology and Bioinformatics*, *18*, 797–803.
1241. Zhu, W., Guo, Y., & Zou, Q. (2021). Prediction of presynaptic and postsynaptic neurotoxins based on feature extraction. *Mathematical Biosciences and Engineering*, *18*, 5943–5958.
1242. Zhu, L., Wang, X., Li, F., & Song, J. (2022). PreAcrs: A machine learning framework for identifying anti-CRISPR proteins. *BMC Bioinformatics*, *23*, 444.
1243. Zhu, Y., Liu, Y., Chen, Y., & Li, L. (2022). ResSUMO: A deep learning architecture based on residual structure for prediction of lysine SUMOylation sites. *Cells*, *11*, 2646.
1244. Zhu, H., Ao, C., Ding, Y., Hao, H., & Yu, L. (2022). Identification of D modification sites using a random forest model based on nucleotide chemical properties. *International Journal of Molecular Sciences*, *23*, 3044.
1245. Zhu, W., Xie, H., Chen, Y., & Zhang, G. (2024). CrnnCrispr: An interpretable deep learning method for CRISPR/Cas9 sgRNA on-target activity prediction. *International Journal of Molecular Sciences*, *25*, 4429.
1246. Zhuang, Z., Shen, X., & Pan, W. (2019). A simple convolutional neural network for prediction of enhancer-promoter interactions with DNA sequence data. *Bioinformatics*, *35*, 2899–2906.
1247. Zhuo, L., Wang, R., Fu, X., & Yao, X. (2023). StableDNAm: Towards a stable and efficient model for predicting DNA methylation based on adaptive feature correction learning. *BMC Genomics*, *24*, 742.
1248. Zimmerman, S., Cohen, G., & Davies, D. (1975). X-ray fiber diffraction and model-building study of polyguanylic acid and polyinosinic acid. *Journal of Molecular Biology*, *92*, 181–192.
1249. Zitnik, M., Sosic, R., & Leskovec, J. (2018). BioSNAP datasets: Stanford biomedical network dataset collection (Vol. 5). http://snap.stanford.edu/biodata.
1250. Zou, H., Ji, B., Zhang, M., Liu, F., Xie, X., & Peng, S. (2024). MHGTMDA: Molecular heterogeneous graph transformer based on biological entity graph for miRNA-disease associations prediction. *Molecular Therapy-Nucleic Acids, 35,* 1–9.
1251. Zou, H. (2022). Identifying blood-brain barrier peptides by using amino acids physicochemical properties and features fusion method. *Peptide Science*, *114*, Article e24247.
1252. Zou, L., Wang, Z., & Wang, Y. (2008). Prediction of outer membrane proteins using support vector machine with combined features. *Chinese Journal of Biotechnology*, *24*, 651–658.
1253. Zulfiqar, H., Ahmed, Z., Ma, C., Khan, R., Grace-Mercure, B., Yu, X., & Zhang, Z. (2022). Comprehensive prediction of lipocalin proteins using artificial intelligence strategy. *Frontiers in Bioscience-Landmark*, *27*, 84.
1254. Zulfiqar, H., Sun, Z., Huang, Q., Yuan, S., Lv, H., Dao, F., Lin, H., & Li, Y. (2022). Deep-4mCW2V: A sequence-based predictor to identify N4-methylcytosine sites in Escherichia coli. *Methods*, *203*, 558–563.
1255. Zuo, Y., Li, Y., Chen, Y., Li, G., Yan, Z., & Yang, L. (2017). PseKRAAC: A flexible web server for generating pseudo K-tuple reduced amino acids composition. *Bioinformatics*, *33*, 122–124.

www.ingramcontent.com/pod-product-compliance
Ingram Content Group UK Ltd.
Pitfield, Milton Keynes, MK11 3LW, UK
UKHW022203230426
470311UK00001BA/4

9783031904493